Multidisciplinary Design Optimization

Multidisciplinary Design Optimization

Editor: Tommy Haynes

NY RESEARCH PRESS
New York

Published by NY Research Press
118-35 Queens Blvd., Suite 400,
Forest Hills, NY 11375, USA
www.nyresearchpress.com

Multidisciplinary Design Optimization
Edited by Tommy Haynes

International Standard Book Number: 978-1-63238-528-4 (Hardback)

Cataloging-in-publication Data

Multidisciplinary design optimization / edited by Tommy Haynes.
p. cm.
Includes bibliographical references and index.
ISBN 978-1-63238-528-4
1. Multidisciplinary design optimization. 2. Engineering design. I. Haynes, Tommy.
TA658.8 .M85 2017
624.17--dc23

Printed in the United States of America.

Contents

Preface

Multidisciplinary Design Optimization is a rapidly growing field of study. It falls under the umbrella of engineering and focuses on solving design related problems with the help of optimization methods. The techniques are also helpful and useful in other fields like automobile design, electronics, computers, etc. This book aims to elaborately discuss the various problem solving techniques under the broader category of design optimization like gradient based methods, population based methods and gradient free methods, etc. As this field is emerging at a rapid pace, the contents of this book will help the readers understand the modern concepts and applications of the subject. From theories to research to practical applications, case studies related to all contemporary topics of relevance to the field of multidisciplinary design optimization have been included in this book.

This book has been the outcome of endless efforts put in by authors and researchers on various issues and topics within the field. The book is a comprehensive collection of significant researches that are addressed in a variety of chapters. It will surely enhance the knowledge of the field among readers across the globe.

It gives us an immense pleasure to thank our researchers and authors for their efforts to submit their piece of writing before the deadlines. Finally in the end, I would like to thank my family and colleagues who have been a great source of inspiration and support.

Editor

1

3D BEM-based cooling-channel shape optimization for injection molding processes

N. Pirc[1,2,a], F. Schmidt[1], M. Mongeau[2] and F. Bugarin[2]

[1] CROMeP – École des Mines d'Albi-Carmaux, Campus Jarlard, 81013 Albi, Cedex 9, France
[2] Université de Toulouse, LAAS-CNRS, and Institut de Mathématiques, UPS 31062 Toulouse Cedex 9, France

Abstract – Today, around 30% of manufactured plastic goods rely on injection molding. The cooling time can represents more than 70% of the injection cycle. Moreover, in order to avoid defects in the manufactured plastic parts, the temperature in the mold must be homogeneous. We propose in this paper a practical methodology to optimize both the position *and* the shape of the cooling channels in 3D injection molding processes. For the evaluation of the temperature required both by the objective and the constraint functions, we must solve 3D heat-transfer problems via numerical simulation. We solve the heat-transfer problem using Boundary Element Method (BEM). This yields a reduction of the dimension of the computational space from 3D to 2D, avoiding full 3D remeshing: only the surface of the cooling channels needs to be remeshed at each evaluation required by the optimization algorithm. We propose a general optimization model that attempts at minimizing the desired overall low temperature of the plastic-part surface subject to constraints imposing homogeneity of the temperature. Encouraging preliminary results on two semi-industrial plastic parts show that our optimization methodology is viable.

Key words: BEM; injection molding; SQP; cooling channel optimization

1 Introduction

Today, around 30% of manufactured plastic goods rely on injection molding, which is based on the injection of a fluid plastic material into a closed mold (Fig. 1) (Madehow [1]) displays an illustrative injection molding process). The cooling time can represent more than 70% of the injection cycle. Moreover, in order to avoid defects in the manufactured plastic parts, the temperature in the mold must be homogeneous. Thus, the design and the position of the cooling channels are crucial elements in the design of the mold. In order to decide the position and the shape of the cooling channels in the mold, designers commonly rely on experience and intuition within a costly trial-and-error design process. This manual design process becomes inadequate and unpractical for complex problems. This is particularly true nowadays with rapid prototyping processes such as layered design or selective laser sintering that enable manufacturers to build almost any desired shape of cooling channel geometry in the mold. As a consequence, designers need a more powerful tool integrating the cooling analysis, its numerical simulation, and even optimization algorithms into the design process.

We propose in this paper a practical methodology to optimize both the position and the shape of the cooling channels in 3D injection molding processes. For the evaluation of the temperature, required both by the objective and the constraint functions, we must solve 3D heat-transfer problems via numerical simulation. Severe numerical methods such as Finite Element Method (FEM) [2] or Boundary Element Method (BEM) [3] can be used for solving the heat-transfer problem. Silva [4] used 3D FEM software to model the injection molding cycle for complex geometries of molds. However, the computational burden of 3D FEM makes the integration of an optimization algorithm unpractical. Indeed, a FEM approach to the 3D heat-transfer problem imposes severe storage and CPU requirements, even if moderately complex industrial parts were to be targeted. An optimization algorithm requires numerous (and here expensive in CPU time) objective-function evaluations. Moreover, unless one knows a very good initial design, from one optimization iteration to the next the design is generally significally different. As a consequence, the capability of FEM to handle (small) shape perturbations for 3D mesh cannot be advantageously exploited here. BEM is a method that was popularized by Brebbia [5]. It is used in many applications, such as gas-assisted injection molding processes [6], or groundwater flow and mass transport problems [7]. BEM transforms domain integrals into boundary element integrals, involving therefore a discretization restricted only to the external and internal

[a] Corresponding author: `pircpirc@enstimac.fr`

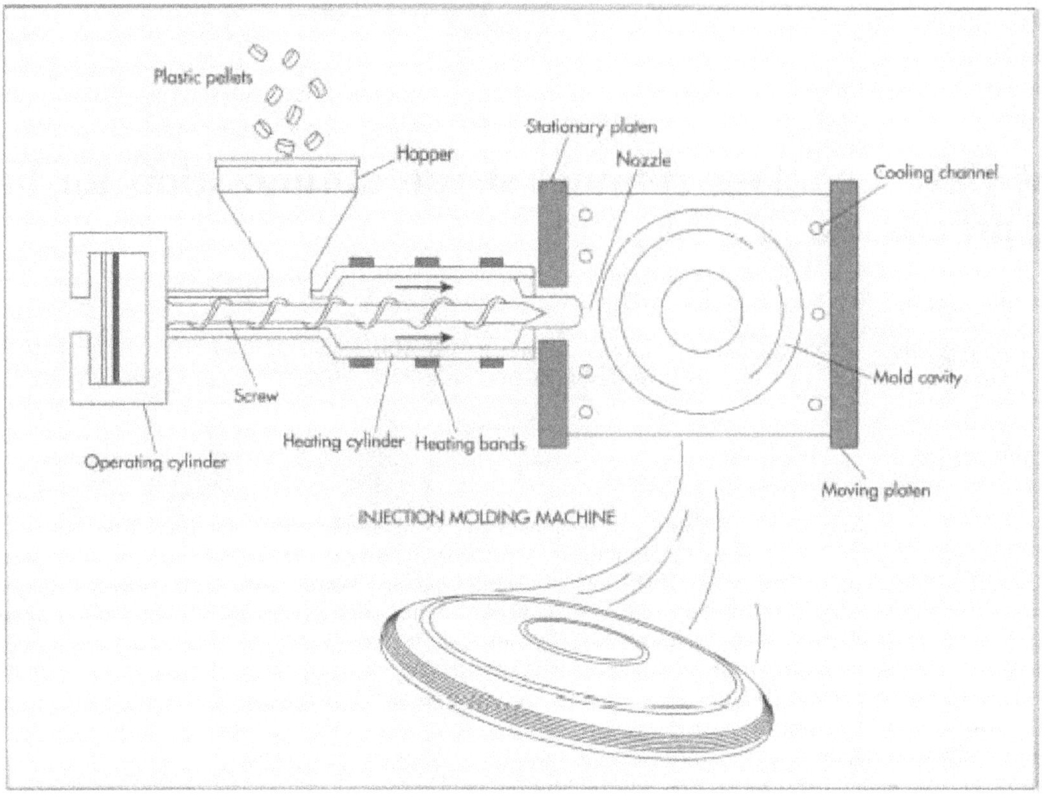

Fig. 1. Injection unit.

field boundary. An optimization method can therefore be envisaged to modify the position and also the shape parameters of the cooling channels in order to improve the cooling performance of the mold.

Park [8] proposed such a BEM-based procedure for the position of the cooling channels relying on augmented-Lagrangian optimization method. However, his study is restricted to molds that are generalized cylinders, amenable to 2D molds. His optimization model minimizes the variations of the temperature distribution on the cavity surface with respect to the average temperature. As a consequence, his optimal configuration provides a uniform but high average temperature, leading a very long cooling time which is not desirable in the context of large-scale manufacturing. Mathey [9] also used BEM to solve the heat-transfer problems with a Sequential Quadratic Programming (SQP) [10] algorithm to improve mold injection cooling. She minimizes an objective function which is the weighted sum of two criteria. Her first criterion is the average temperature at the plastic-part surface. Her second criterion is the sum of the temperature variations with respect to the average temperature. Moreover, Mathey optimizes both the position and the shape of the cooling channels. However, her approach is also restricted to 2D molds (as for Parks [8]), BEM then reduces the dimension of the computation space from 2D to 1D).

Our contribution is threefold. First, we address 3D mold geometries with a BEM approach reducing the dimension of the computation space from 3D to 2D, avoiding full 3D remeshing: only the surface of the cooling channels needs to be re-meshed at each evaluation required by the optimization algorithm. Secondly, we propose a general optimization models that attempts at minimizing the desired overall low temperature of the plastic-part surface subject to constraints imposing homogeneity of the temperature. Thirdly, we demonstrate that our optimization methodology is viable with encouraging preliminary results on two semi-industrial plastic parts. The paper is organized as follows. We detail in the next section the 3D heat-transfer problem that we must solve for every optimization evaluation of the temperature required by the optimization algorithm. Section 3 describes our overall optimization methodology. In order to validate our approach, we report encouraging preliminary results in Section 4. We conclude in Section 5.

2 The heat transfer problem

This section describes the heat-transfer problem that must be solved at every temperature evaluation required by the optimization algorithm:

As shown in Figure 2, after a few cycles the variation of the temperature in the injection mold in production can be considered to be quasi-stationary [11]. Once the average temperature of the mold is stabilized, the cycle-averaged approach can predict well the overall performance of the cooling system. Thus, we can consider a stationary regime neglecting the transitory oscillations of the temperature. For a metallic mold, when between 0 °C

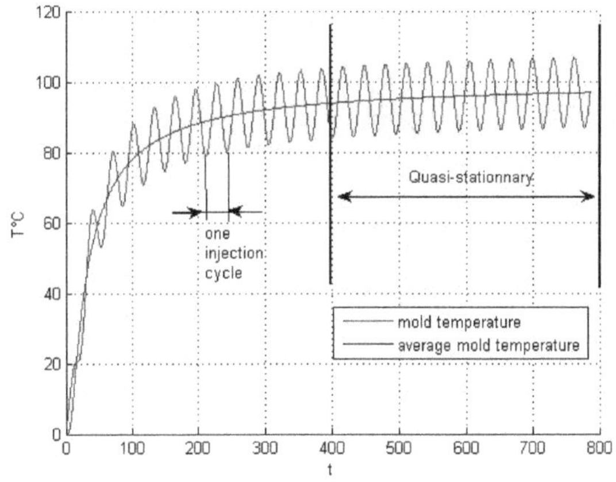

Fig. 2. Temperature at the surface of the mold versus time (in seconds).

to 200 °C, the thermal conductivity can be considered as a constant. Thus, the stationary heat conduction problem reduces to the following Laplace equation:

$$\Delta T = 0,$$

where Δ is the Laplacian operator, T is a temperature vector. Multiplying this equation by a weighting function T^*, and using Green's theorem, we obtain the well-known Somigliana's equation [5]. The following Green integral representation formula gives the value of T in terms of integral equations involving the fundamental solution T^* of the Laplace equation:

$$C.T + \int_\Gamma T.(\nabla T^*.n).d\Gamma = \int_\Gamma (\nabla T.n).T^*.d\Gamma. \quad (1)$$

Here, n is the unit normal at one element, Γ is the boundary of the domain, C is equal to 1 inside the domain Ω and to 0.5 on its boundary Γ. The weighting function T^* and the flux q^* are the fundamental solutions of the heat-problem, the so-called Green's functions [5]:

$$T^* = \frac{-1}{2.\pi.r} \text{ and } q^* = \frac{-r.n}{4.\pi.r^2}, \quad (2)$$

where r is the distance from the point of application of the concentrated unit source to any other point under consideration. Note that in equation (1) all the integrals are taken over the boundary of the domain. For the computation of the heat-transfer problem, the contour Γ is discretized and the integrals in the above equation are defined in terms of nodal values by means of interpolation functions. After reorganization of terms in equation (2) using the boundary conditions, we obtain a non-symmetrical linear system. We shall need to remesh the cooling channel surfaces using 2D elements, and to solve this system of equations at every optimization evaluation.

Figure 3 shows the boundary conditions on the mold. We detail the boundary conditions of the heat transfer problem in Figure 3. The following equation relates the

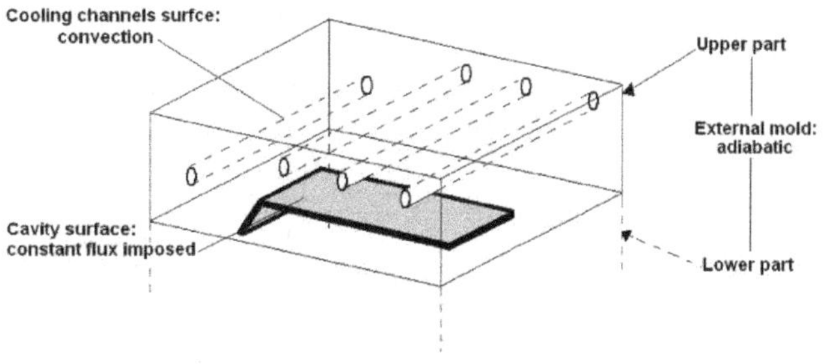

Fig. 3. Boundary conditions.

Table 1. Parameter values for heat-transfer problem (from [9])

	Units	Polymer (PP)	Mold (steel 40cmd8s)
λ	W m^{-1} K^{-1}	0.63	34
ρ	kg m^{-1}	891	7800
C_p	J kg^{-1} K^{-1}	2740	460

temperature of the coolant T_c, and the heat-transfer coefficient h, with the coolant flow rate (via Colburn correlation coefficient) [12]:

$$\lambda.q = h(T_a - T_c),$$

where λ is the conductivity, q is the heat flux, and T_a is the ambient temperature. The following polymer properties are referenced in Table 1: ρ is the density and C_p is the heat capacity. The flux density at the cavity surface, ϕ_{cavity}, is computed from the cooling cycle time, $t_{cooling}$, and the polymer properties (heat due to polymer crystallization is neglected) [12]:

$$\phi_{cavity} = \frac{Q}{t_{cooling}.\Gamma_{cavity}},$$

where Q is the heat evacuated by the plastic part during one injection cycle and Γ_{cavity} is the area of the cavity.

Finally, remark that for our problem, we only need to mesh *once* the cavity surface and the external surface of the mold. Then, each time the optimization algorithm requires to evaluate the objective and the constraint functions, we must mesh the surface of the cooling channels. The output of the heat-transfer problem resolution is a set of temperature measurements, one for each surface element: cavity, external mold surface, and cooling channels. However, we only need the vector T of temperature measurements, $\{T_i(x)\}_{i \in S}$, at the *cavity* surface elements (S denotes the index set of the cavity surface elements).

3 Overall optimization methodology

We first present in this section how we formulate our problem under a mathematical programming form. Then, we detail our overall computational methodology.

In the sequel, x will denote the vector of optimization variables (position and shape parameters for the cooling

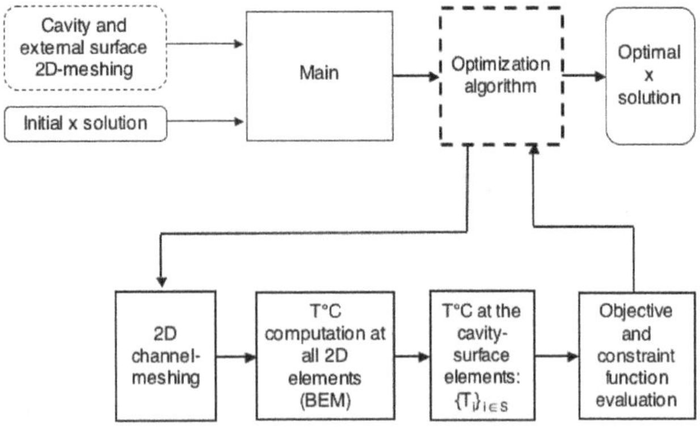

Fig. 4. The heat-transfer simulation coupled with the optimization algorithm.

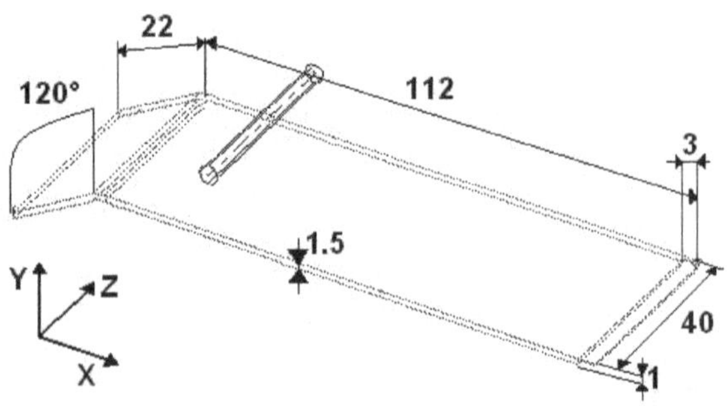

Fig. 5. Plastic part dimensions and one cooling channel.

channels). Since the output of the heat-transfer problem is function of x, we shall make explicit the dependence of the temperature measurements upon the position and shape parameters $x : \{T_i(x)\}_{i\in S}$.

Most practical optimization problems involve several (often contradictory) objective functions. It is the case here as one aims at minimizing the temperature of the plastic-part surface while minimizing the variation of the temperature along this surface. The simplest way to proceed in such a multicriterion context is to consider as objective function a weighted sum of the various criteria, which involves choosing appropriate weighting parameter values (more on multicriterion optimization can be found in [13]). An obvious alternative is to use one criterion as objective function while requiring in constraints maximal threshold levels for the remaining criteria. We choose here the latter approach because in our application context, it is more desirable to consider the temperature requirement as a constraint. Indeed, we do know a threshold level value for the maximal temperature variation under which any variation is equally acceptable. More precisely, we formulate our problem under the form:

$$\min_{x} \|T(x)\|$$
$$\text{subject to } f(T(x)) \leq 0 \quad (3)$$
$$g(x) \leq 0, \quad (4)$$

where f is a real-valued function used to stipulate the uniformity-temperature constraint, and $g(x)$ is a general vector-valued non-linear function. Remark that $\|\cdot\|$ stands for any norm such as the l_1 norm (sum of absolute values), the standard Euclidean (l_2) norm, or the max (l_∞) norm. The general constraints (4) represent any geometry-related or other industrial constraints, such as:

- upper/lower-bound constraints on the x_i's;
- keeping the cooling channels within the mold;
- technically-forbidden zones where we cannot position the cooling channels (for instance due to the presence of ejectors);
- constraints stipulating a minimal distance between every pair of cooling channels to avoid inter-channels collision.

Figure 4 shows the coupling between the thermal solver and the optimization algorithm.

4 Applications

In this section, we present computational experiments on two semi-industrial plastic parts.

For both applications, we specify the following implementation strategies. We use commercial software for the tasks represented by dashed-line boxes in Figure 4:

- IDEAS for the initial 2D meshing of the cavity and the external surface (note that this meshing remains constant throughout the optimization process);
- Matlab's fmincon Sequential Quadratic Programming subroutine [10] for the optimization algorithm (with finite-difference computation of the gradients for these preliminary experiments – in future work, we intend to compute exact gradients).

We programmed every other task (represented by full-line boxes in Fig. 4) in Matlab. The system of linear equations involved in the heat-transfer problem is solved by the LApack subroutine included in Matlab 7.0 [14]. We use here the l_∞ (max) norm for the objective function:

$$\|T(x)\|_\infty := \max_{i\in S} T_i(x). \quad (5)$$

All numerical results reported in the sequel are obtained on a Macintosh 1.83 GHz Intel core 2 duo.

4.1 EuroTooling mold with straight cooling channels

For our first application, we consider a semi-industrial injection mold design for the European project: EuroTooling 21. The plastic part produced from this mold is a bended plate whose dimensions are shown in Figure 5 (in 10^{-3} m). We want to optimize the position of the cooling channels. The channels are simple straight horizontal cylinders of constant length as illustrated in Figure 6. In order to simplify the presentation, in this application we only optimize the position of the cylinders (we could also consider optimizing the radius of the cross-section

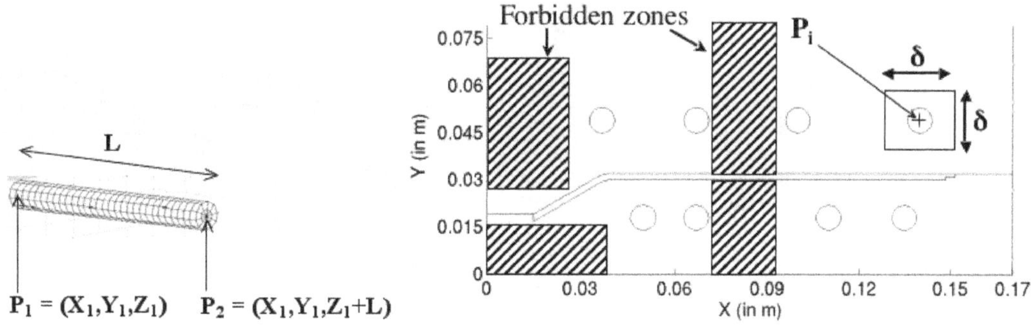

(a) (b)

Fig. 6. Cooling channel parameters.

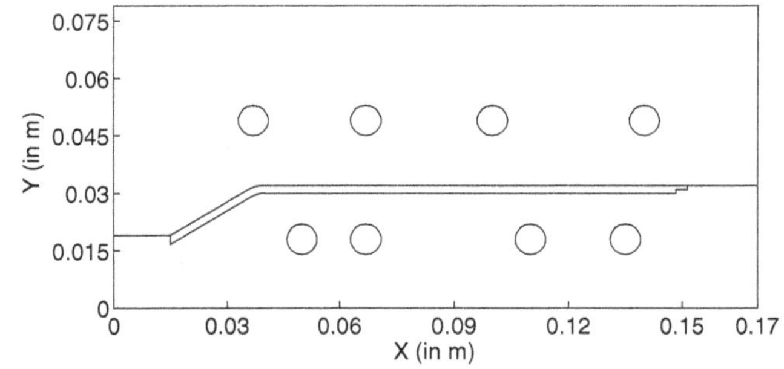

Fig. 7. Initial and optimized (dashed lined) positions of the cooling channels.

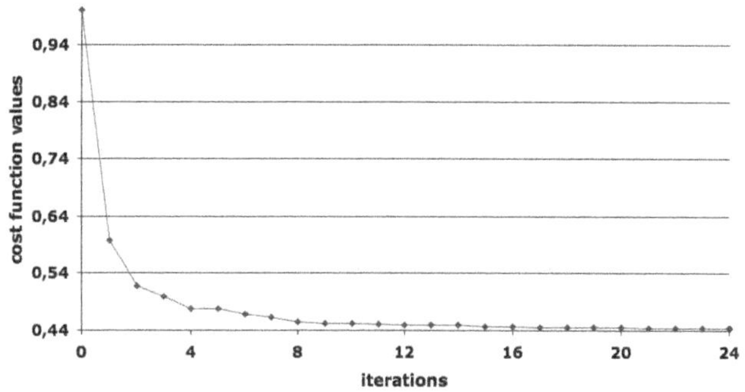

Fig. 8. Objective function versus iterations.

of the channels and the number of channels, etc.). As a consequence, three parameters suffice to describe the position of each cooling channel. For illustration purposes we consider here 8 cooling channels. The components of the optimization vector x are here the coordinates of each end point (X_i, Y_i), $i = 1 \ldots 8$. (in our application Z_i is fixed). Thus, our problem involves 16 optimization variables. For this application, we use the constraint function $F(T(x)) = \|T(x) - \overline{T}(x)\|_\infty - \sigma$, where the difference $T(x) - \overline{T}(x)$ stands here for the vector whose ith component is $T_i(x) - \overline{T}(x)$, $\overline{T}(x) := \frac{1}{|S|} \sum_{i \in S} T_i(x)$ is the average of the T_i's, and σ is a user-defined temperature uniformity tolerance. We choose here $\sigma = 4$ °C. In other words, we do not accept variations of the temperature that are above 4 °C with respect to the average temperature and this, everywhere on the plastic-part surface. The remaining constraints here simply stipulate lower/upper bounds and also forbid the grey zones displayed in Figure 6.

The 2D meshing of the mold surfaces is displayed in Figure 10. The surface of each channel is discretized into 320 quadrangles. We use as starting point, a heuristic solution provided by an experienced engineer. On average, one objective-function evaluation (i.e. the computation of the temperatures) requires 86 s of CPU time. Since for these preliminary experiments we are content with computing gradients using finite-difference approximations, one optimization iteration involves 26 min of CPU time on average. Figure 7 displays initial (full lines) and optimal (dashed lines) positions of the cooling channels. Figure 8 displays the decrease of the objective-function value in terms of the number of optimization iterations. We observe in Figure 8 that the objective function is reduced by 90% of its initial value within the first four optimization iterations.

Figure 9 shows the temperature distribution along the surface of the cavity mold before and after optimization (points A, B and B' refer to positions in Fig. 7). We observe that both the temperature variance and the average temperature decreased significantly. The temperature distribution after optimization can be seen in Figure 10

4.2 Plastic-cup mold with a helix cooling channel

We now report computational results on a 3D plastic part: a plastic cup. The height of the mold and that of the plastic cup are respectively 0.12 and 0.08 m. The upper radius of the cup is 0.06 m, its lower radius is 0.04 m, and its thickness is 3×10^{-4} m. In this preliminary study, we are content with considering a helix-shape cooling channel with circular cross section, with only three degrees of freedom: the section diameter of the cooling channel, the frequency of the turn and its overall diameter. The cooling channel is therefore defined in $\Re^3$ using the single

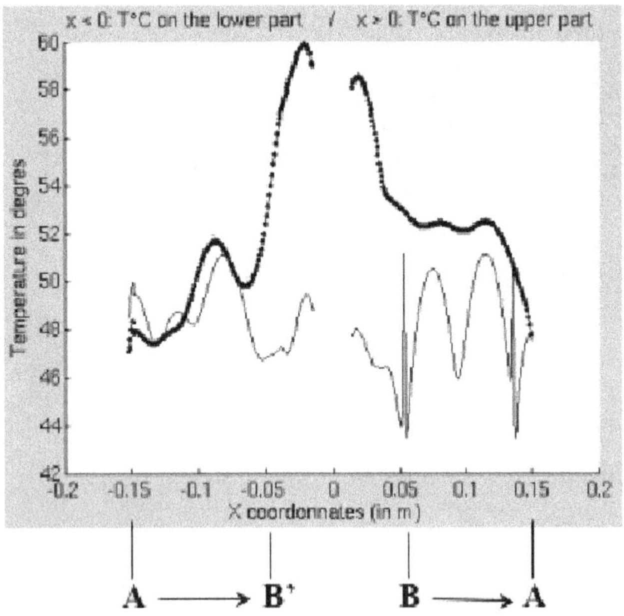

Fig. 9. Temperature measurements along the surface of the mold cavity before (black) and after (grey) optimization.

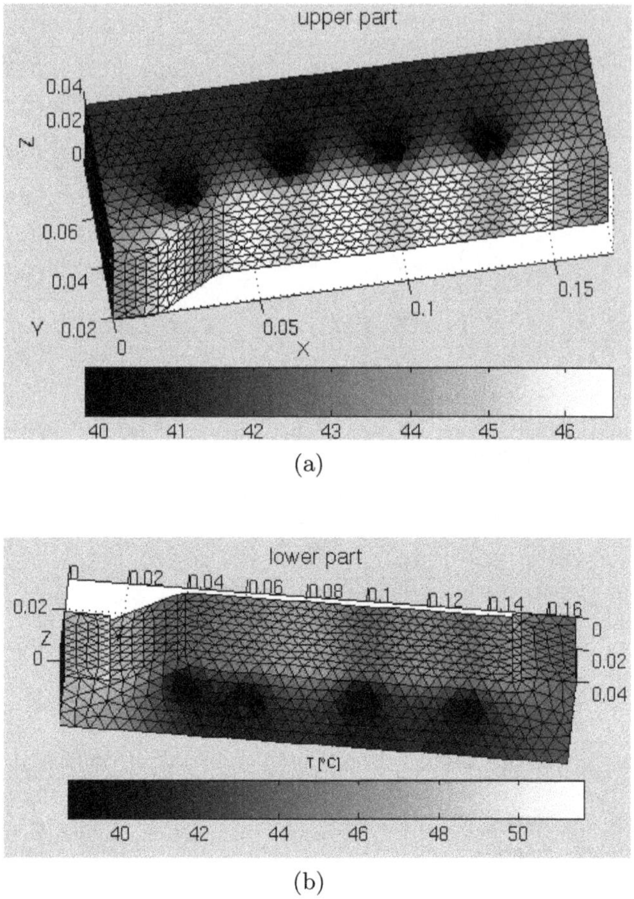

Fig. 10. Temperature at the surface of the 3D mold after optimization.

parameter algebraic equation:

$$\begin{aligned} X(t) &= r_s \cos t \\ Y(t) &= r_s \sin t \\ Z(t) &= \quad t\,, \end{aligned} \tag{6}$$

where r_s is the radius of the spiral. We discretize this curve into 63 straight-cylinder channels (Fig. 12) with radius r_c. Each of these cylinders is further discretized into 20 quadrangle elements as shown in Figure 13. The 2D meshing of the mold surfaces is displayed in Figure 11. The optimization variables here are:

- the geometrical parameters that control the shape of the helix, that is to say: the radius, r_s of the spiral and the number, n, of helix turns;
- the radius r_c of the circular cross section of the cylinder channels.

Our temperature homogeneity constraint here is:

$$\max_{i \in S} T_i(x) - \min_{i \in S} T_i(x) \leq \sigma,$$

with $\sigma = 4$ °C. The meshing of the mold surfaces is shown in Figure 12. On average, one objective-function evaluation (temperature computation) requires 82 s of CPU time. Again, because of the fact that we compute gradients using finite-difference approximation, one optimization iteration involves, on average, 7 min of CPU time. Figure 14 shows the helix curve before and after optimization. Figure 15 displays the decrease of the objective-function value in terms of the number of optimization iterations. We observe a rapid convergence during the first five iterations. Table 2 displays the initial and optimal solutions, and Figure 16 shows the temperature distribution at the surface of the mold after optimization.

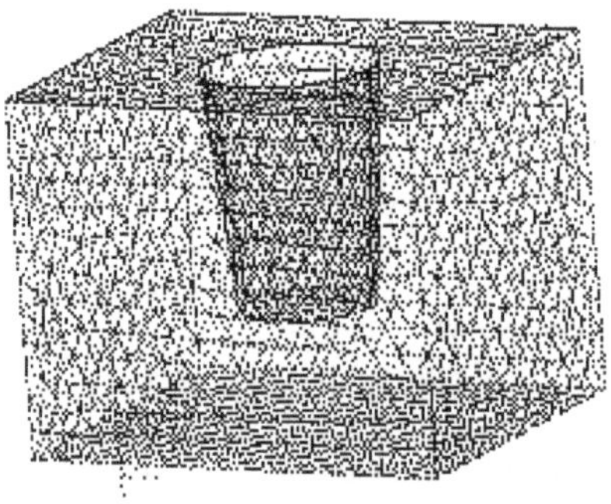

Fig. 11. Meshing of the surfaces for the plastic-cup mold.

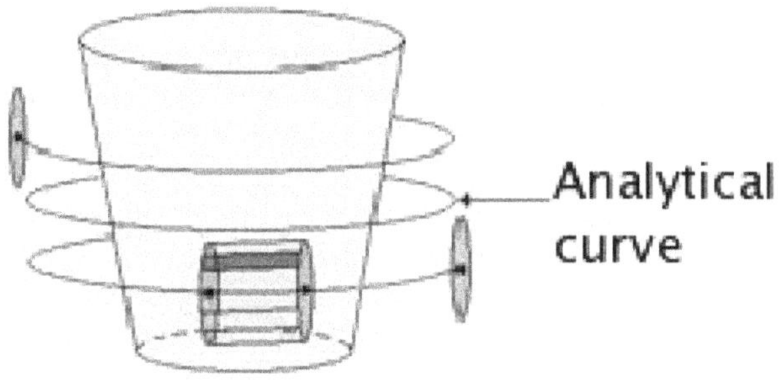

Fig. 12. Discretization into straight cylinder channels.

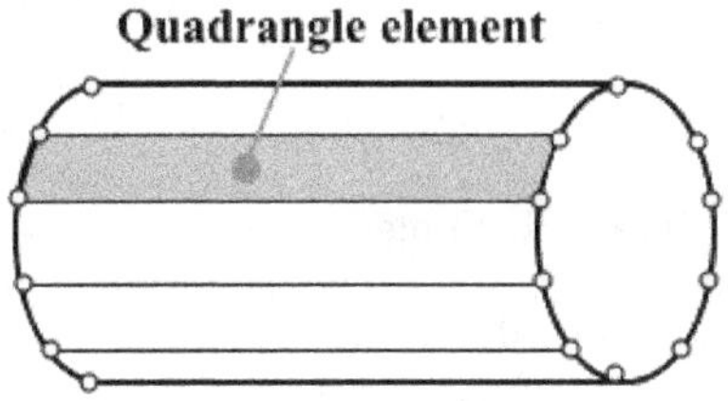

Fig. 13. Cooling-channel surface meshing.

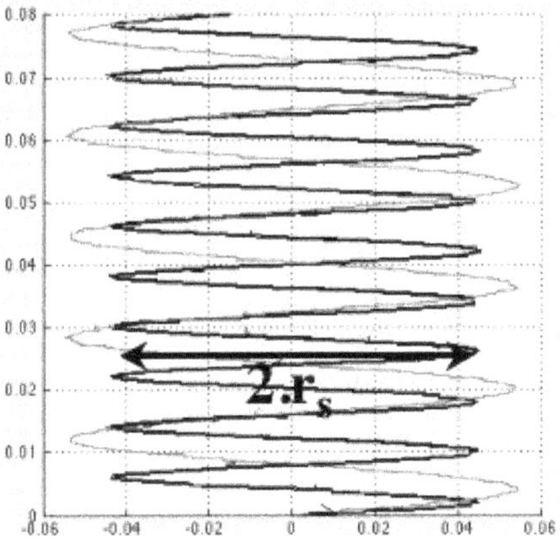

Fig. 14. Helix curve before (black) and after (grey) optimization.

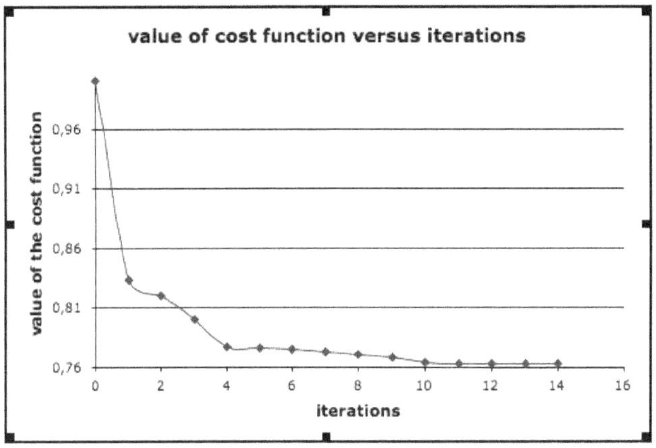

Fig. 15. Objective-function value versus iterations.

Table 2. initial and optimal solutions.

	Before optimization	After optimization
r_s (m)	0.045	0.055
r_c (m)	0.003	0.001
n	10	5
objective-function value	35 °C	28 °C
$\max T_i - \min T_i$	5 °C	2 °C

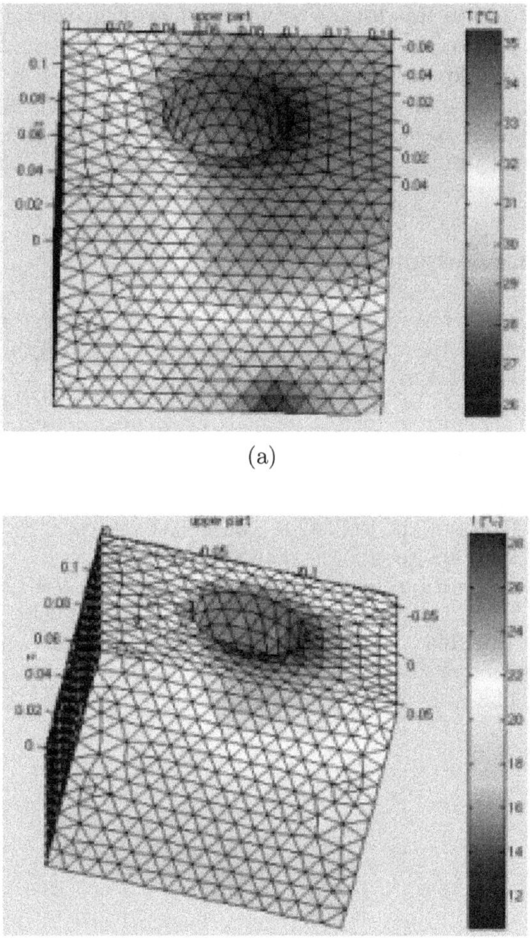

(a)

(b)

Fig. 16. Temperature at the surface of the 3D mold after optimization.

5 Conclusion

We proposed in this paper a practical methodology to optimize both the position and the shape of the cooling channels in industrial 3D molds. This was made possible through the use of the boundary elements method that avoids full 3D remeshing. Indeed, only the *surface* of the cooling channels needs to be remeshed to solve the heat-transfer problem involved each time the optimization algorithm needs to evaluate the temperature at the plastic part surface. Our optimization model can account for various ways to address the overall desired low-temperature criterion and the temperature-homogeneity constraint. Encouraging computational experiments on two semi-industrial plastic-parts, demonstrated the viability of our approach that is intended to be used as a decision-analysis tool for designing new, original mold geometries.

We are currently implementing exact gradients to replace finite-difference approximations in order to improve the efficiency of the optimization. Future work could allow more complex shape variations. Recall that we considered in this preliminary study only three degrees of freedom in order to test the viability of our optimization methodology. Other degrees of freedom for our plastic-cup application could for instance include the flux in the cooling channel and a vertical angle so that the helix runs along a cone (closer to the actual cup shape). Finally, further work could formulate our design problem as a topology optimization problem in order to offer even larger design freedom. The wide range of geometries we can tackle for the cooling channels will bring us to address more complex industrial molds. This will undoubtly yield challenging global optimization problems, as this can be anticipated based on the numerical experiments we reported [15,16].

Acknowledgements

This study was conducted within the framework of the European project EuroTooling 21 (IP 505901-5), www.eurotooling21.com.

References

1. www.madehow.com/Volume-5/Frisbee.html
2. S. Kolossov, E. Boillat, R. Glardon, P. Fischer, Internat. J. Machine Tools **44**, 117 (2004)
3. A. Polynkin, Intern. Polymer Process. **XIX**, 108 (2004)
4. L. Silva, Intern. Polymer Process. **XX**, 265 (2005)
5. C. Brebbia, J. Domiguez, *Boundary elements: An introductory course*, WIT Press/Computational Mechanics Publication, 1992
6. K.E. Khayat, J. Non-Newtonian Fluid Mechanics **57**, 253 (1994)
7. K.L. Katsifarakis, Engineering Anal. Boundary elements **23**, 555 (1999)
8. S.J. Park, Intern. J. Numer. Methods Eng. **43**, 1109 (1998)
9. E. Mathey, Automatic optimization of the cooling of injection mold based on the boundary element method. *NumiForm 2004 Proceedings* (2004) pp. 222–227
10. J. Nocedal, S.J. Wright, *Numerical optimization.* Springer, 2nd ed., 2006
11. I. Catic, A. Abadzic, M. Rujnic-Sokele, *J. Injection Molding Technology* **3**, 194 (1999)
12. L. Jui-Ming, *J. Injection Molding Technology* **6**, (2002)
13. M. Ehrgott, *Multicriteria Optimization* Springer 2nd ed., 2005
14. LApack, www.netlib.org/lapack/
15. A. Carlos, D. Alan, *Engineering Optim.*, **31**, 337 (1999)
16. H.J. Lee, Y.D. Kim, *Engineering Optimization* **40**, 17 (2008)
17. The MathWorks, User's Guide. *Optimization Toolbox For Use With Matlab*, Version 2
18. Y. Wen-Hsien, Three-dimensional simulation of injection-compression molding of a compact disc. *ANTEC 2001 Conference Proceedings* (2001)

2

A new smooth contact element: 3D diffuse contact element

D. Chamoret[1,a], A. Rassineux[2] and J.M. Bergheau[3]

[1] Laboratoire M3M, UTBM, Site de Sévenans, 90010 Belfort Cedex, France
[2] Laboratoire Roberval, FRE 2833 CNRS/UTC, BP 20.529, 60205 Compiègne Cedex, France
[3] LTDS, UMR5513, CNRS/ECL/ENISE, 58 rue Jean Parot, 42023 Saint-Étienne, France

Abstract – Many difficulties due to geometrical and material non-linearities arise when dealing with numerical simulation of contact problems. Within a finite element context, the contact interface is usually represented by a piecewise differentiable surface. Numerical problems due to the non-smoothness of the contact surface may occur especially when large slips are considered. Major changes of normal and tangential vectors may impede both convergence and precision. In order to smooth the contact interface and to release constraints due to the mesh, we propose a technique in which diffuse approximation is combined with a determination of neighboring nodes by a convex hull strategy. The formulation is developed for three-dimensional applications with frictionless contact and a 3D diffuse contact element has been developed. The efficiency of the approach has been validated with industrial frictionless contact problems.

Key words: Contact modelling; finite element method; implicit method; smooth contact surface; diffuse approximation; contact element; neighbourhood criterion; convex hulls

1 Introduction

The simulation of contact remains an important challenge for engineers. A variety of problems are concerned with contact phenomenom for instance metal forming processes (stamping, metal bottle cap as shown in Fig. 1), rubber sealings, crash analysis of cars, rolling contact between car tyres and the road. In this context, numerical instabilties may occur and impede the convergence process. Moreover contact problems can be associated either with large elastic or inelastic deformations. Several approaches have been proposed. Small deformation contact problems can be suitably modelled by node-to-node contact element. When dealing with large slips, node-on-segment contact algorithms can be used [4]. In this context, powerful contact research algorithms have been presented [3, 11, 13, 17, 21, 24, 34, 35]. Most of these algorithms have a complex implementation and any lack of robusteness may impede the convergence of the process especially when the equilibrium equation are handled by implicit schemes. Numerical problems due to the non-smoothness of the contact surface may occur especially when large slips are considered. Major changes of normal and tangential vectors may impede both convergence and precision. Moreover, these effects may alter the residual vector and impede convergence. A number of authors have proposed techniques to smooth or average the normal vector [14, 26, 30]. The determination of a pseudo normal vector varying continuously in the neighbourhood of contact node have been proposed in [22]. Interpolation techniques based on curved patches can be used [28]. Rigid contact surfaces are smoothed by techniques used

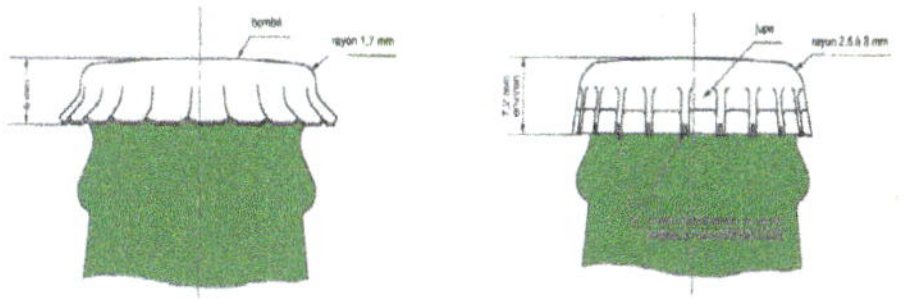

(a) Geometrical model

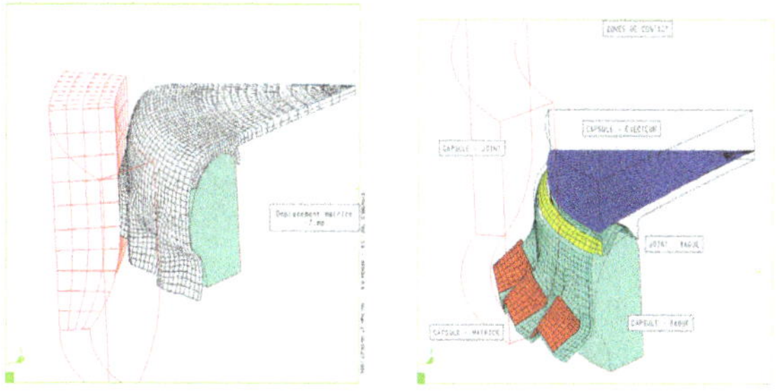

(b) Finite element model

Fig. 1. Metal bottle cap.

[a] Corresponding author: dominique.chamoret@utbm.fr

in CAD surface modeling [15]. Bézier or B-Splines can be used with efficiency to represent deformable contact surfaces [12,18,25,27,32].

The approach presented here consists in creating a smooth representation of the contact surface with the use of a meshfree technique denoted as diffuse approximation [23]. The interpolation of a geometrical model using a second order equation has been introduced by Rassineux in a remeshing context of discrete data [29]. We have extended this technique to the smoothing of the contact geometry [7]. Belytschko has proposed a similar approach [2]. The discretization of the problem leads to the creation of a 3D diffuse contact element. When dealing with meshfree techniques, nodal shape functions are not given by element connectivity but by a set of neighboring nodes. When dealing with the Finite Element Method, the influence of the orientation of the mesh on the solution is a well-known academic problem. In order to overcome this problem in a meshfree context, we propose a searching technique based on convex hulls determination [9]. The determination of all the contact elementary quantities is detailed and the efficiency of this original approach illustrated by examples.

The following outline is used in the presentation. Basic concepts of contact modelling are reviewed in part 2. The contact surface smoothing procedure is presented in part 3. New contact geometry is illustrated in part 4. The finite element method and our approach are combined to construct a new contact element proposed in part 5. Examples which demonstrate the efficiency of the proposed algorithms are discussed in part 6.

2 Contact modelling

We provide a brief presentation of techniques frequently used to solve contact problems in a FEM context [1,10,16,19,20,31].

2.1 Contact geometry and gap function

Let us consider two bodies in potential contact. In a master-slave approach, the slave body is denoted as Ω^s and the master body is denoted as Ω^m. The same terminology is used for surfaces. The potential contact surfaces are noted γ_c^i with $i = s, m$ in the current configuration. The master body is assumed to be the reference body which means that the relative motion of the slave body is described with respect to the master body. The condition to ensure is that the slave body must not penetrate the master body and all displacements are computed with respect to this assumption. This situation introduces dissymmetry in the description of the problem. However this drawback can be reduced numerically by alternating the role of the bodies with the help of specific algorithms [4].

The aim is to measure the penetration by computing the minimal distance between the two surfaces into contact. At each slave point $\boldsymbol{x}^c$ of the surface γ_c^s is associated

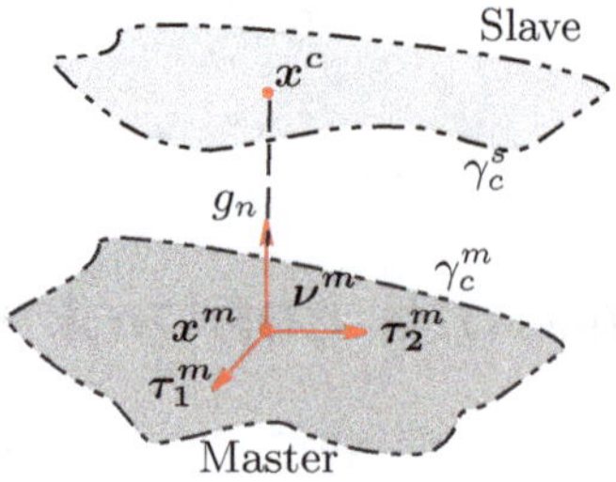

Fig. 2. Representation of the normal gap.

a point $\boldsymbol{x}^m$ belonging to the master surface γ_c^m given by the following minimization problem:

$$\|\boldsymbol{x}^c - \boldsymbol{x}^m\| = \min_{x \in \gamma_c^m} \|\boldsymbol{x}^c - \boldsymbol{x}\|. \tag{1}$$

This point is the closest projection point of $\boldsymbol{x}^c$ onto the master surface. Consequently, vector $\boldsymbol{x}^c - \boldsymbol{x}^m$ and normal vector $\boldsymbol{\nu}^m$, have the same direction and we can establish the relation:

$$\boldsymbol{x}^c - \boldsymbol{x}^m = g_n \boldsymbol{\nu}^m \tag{2}$$

$b^m = (\boldsymbol{\tau}_1^m, \boldsymbol{\tau}_2^m, \boldsymbol{\nu}^m)$ is the natural basis associated with the master surface in the current configuration at point $\boldsymbol{x}^m$ (see Fig. 2). The normal gap function, denoted as g_n is given by:

$$g_n = (\boldsymbol{x}^c - \boldsymbol{x}^m) \cdot \boldsymbol{\nu}^m. \tag{3}$$

The normal gap has a major importance in the description of the penetration between the two bodies. The sign of the gap provides the geometrical status of a contact point. Three situations may occur:

$$\begin{cases} \text{No contact} & g_n > 0 \\ \text{Perfect contact} & g_n = 0 \\ \text{Interpenetration} & g_n < 0. \end{cases}$$

2.2 Numerical schemes to solve contact problems

The numerical treatment of the contact constraints is essentialy based on two main strategies: penalty and Lagragian multiplier based methods. Both approaches have their advantages and their drawbacks. A penalty method can be easily implemented in an existing finite element code. This is the reason why it is used so frequently. The penalty method estimates the normal contact traction as:

$$T_n = \begin{cases} -\varepsilon_n g_n & \text{if} \quad g_n \leq 0 \\ 0 & \text{otherwise} \end{cases} \tag{4}$$

where ε_n is the normal penalty parameter and g_n the normal gap.

The main disadvantage of this technique is the adequate choice of the penalty parameter and the convergence of the technique is highly dependent on this choice. An unwise choice may lead to ill-conditioned stiffness matrix if the penalty parameter is too important or to unacceptable penetration if it is too small. Solutions to adjust the penalty parameter have been proposed [8].

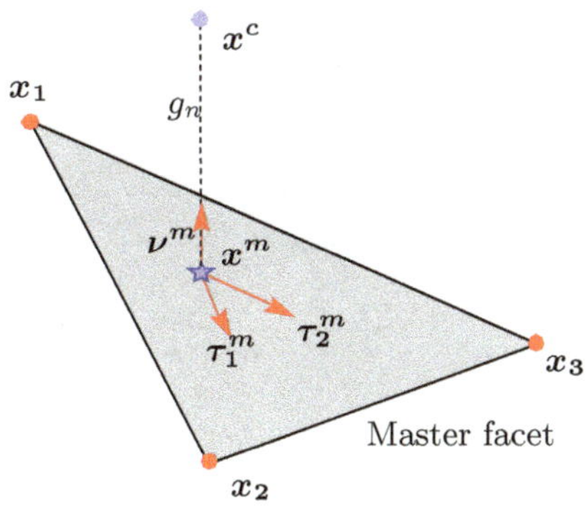

Fig. 3. Node-to-facet contact element.

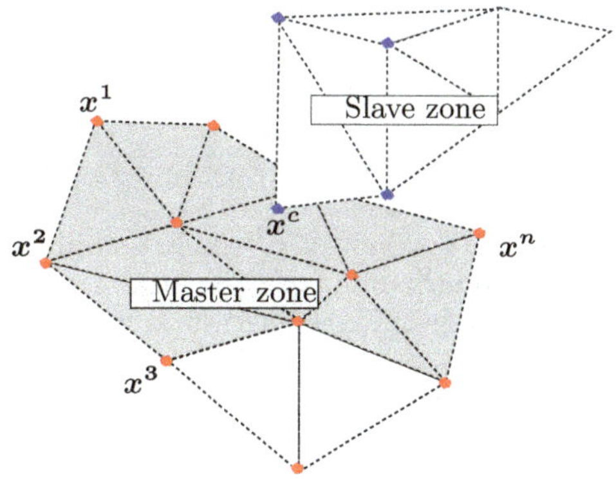

Fig. 4. Contact zones.

2.3 Weak form

The variational problem is classicaly written as

$$G(\boldsymbol{u},\ \boldsymbol{\delta u}) + G_c(\boldsymbol{u},\ \boldsymbol{\delta u}) = 0 \tag{5}$$

where $\boldsymbol{u}$ is the solution displacement field. Function $\boldsymbol{\delta u}$ can be seen as virtual displacements. First term G is the virtual work associated with finite deformation of solids mechanics problems.

Second functional G_c contains the contact contributions. These contributions [19, 31] can be expressed as follows:

$$G_c(\boldsymbol{u},\ \boldsymbol{\delta u}) = \int_{\gamma_c^s} \varepsilon_n\, g_n\, \delta g_n\, d\gamma \tag{6}$$

where δg_n is the first derivative of g_n.

The problem introduced by relation (5) is nonlinear and can be solved by a Newton Raphson procedure which requires the computation of derivatives. Using a continuous description of the contact, the linearization of contact contributions can be written as:

$$\Delta G_c(\boldsymbol{u},\ \boldsymbol{\delta u}) = \int_{\gamma_c^s} (\varepsilon_n \Delta g_n\, \delta g_n + \varepsilon_n g_n\, \Delta\delta g_n\,)\, d\gamma \tag{7}$$

where $\Delta\delta g_n$ is the second derivative of g_n. We remark the crucial role of the gap and its variations in the contact contributions.

2.4 Discretization: contact element

When the finite element method is used, contact surfaces are discrete surfaces on which relation (6) must be discretized. Only one slave node is supposed to be in contact with the discretized master surface. Due to the variety of contact situations, specific contact elements have been developed such as node-to-node contact element for small deformations and node-to-face contact element for large deformations [33]. This element is composed of a slave node and nodes representing the master contact face (Fig. 3). The difficulty consists in expressing g_n and its variations.

3 Diffuse approximation and contact

In this part, the building of the smooth contact surface is developed. Let us consider two potential contact zones denoted as master and slave zone. The master zone is made of a set of faces (such as triangles or quadrangles in a 3D context) from the finite element mesh of the master solid. Let $\boldsymbol{x^i}$ denotes the position of a node i from this set. The slave zone is made of a set of nodes likely to come into contact with the master zone (Fig. 4). The aim is to determine an approximation S_g^d of the contact surface with the only data of nodes $\boldsymbol{x^i}$.

A moving least square approach denoted as diffuse approximation technique has been used. The diffuse surface is built from a succession of local approximations. However, the coefficients of the surface depend on the evaluation point. The method guarantees that the coefficients vary in a continuous way with respect to the location of the point. The diffuse interpolation is given by the minimization of a criterion based on the interpolation of the set of neighboring nodes and can be written S_g^d. A local approximation of the contact surface S^d and therefore, a set of neighboring nodes is associated to each slave node $\boldsymbol{x^c}$.

3.1 Local approximation

We suppose that we have determined a set S_i of nodes $\boldsymbol{x^i}$ belonging to the master surface in the neighborhood of $\boldsymbol{x^c}$, the evaluation point. We remark that the master surface is used to create the diffuse interpolated surface. The computation of an interpolated surface requires a set a 2D parametric coordinate system of the surface.

We suppose that at each point of the discrete surface, we can find a neighborhood and therefore a local coordinate system on which the surface can be locally defined by a Monge patch

$$\mathrm{f}(X_1, X_2) = X_3 \tag{8}$$

where f is a C^2 function defined on a planar domain.

Therefore, the first step consists in the determination of a moving least square plane (average plane) with the use of the interpolation set S_i. The surface equation is evaluated through a second order equation and can be expressed in the local coordinate system as

$$f(X_1, X_2) = \langle 1,\ X_1,\ X_2,\ X_1^2,\ X_1X_2,\ X_2^2 \rangle\ \boldsymbol{\alpha}\ =\ \mathbf{p}^t\ \boldsymbol{\alpha}. \tag{9}$$

In a local set of coordinates R^d linked to all nodes $\boldsymbol{x^i}$ and in the neighbourhood of $\boldsymbol{x^c}$, the local approximation associated with node $\boldsymbol{x^c}$ is defined by a Monge patch:

$$f_{\boldsymbol{x}^c}(X_1, X_2) = X_3 \tag{10}$$

where $f_{\boldsymbol{x}^c}$ is a C^2 function defined on a planar domain and (X_1, X_2, X_3) are the coordinates of a point in R^d. $f_{\boldsymbol{x}^c}$ can be expressed in the local coordinate system as

$$f_{\boldsymbol{x}^c}(X_1, X_2) = \mathbf{p}^t(\boldsymbol{x} - \boldsymbol{x}^c)\boldsymbol{\alpha} \tag{11}$$

where (X_1^c, X_2^c, X_3^c) are the coordinates of $\boldsymbol{x^c}$ in R^d.

For each node $\boldsymbol{x^c}$, the neighbourhood is composed of all nodes $\boldsymbol{x^i}$ with $i = 1, \ldots, n$ belonging to the opposite master zone. Coordinates of a master node $\boldsymbol{x^i}$ in this reference frame are written in block letters X_k^i with $k = 1, \ldots, 3$. In order to define the approximation, the number of nodes $\boldsymbol{x^i}$ must be at least equal to the size of the polynomial basis, i.e. 6. The search of neighboring nodes is extended until the number of nodes is sufficient. Whenever the master contact zone does not contain enough nodes, a linear polynomial basis is used.

3.2 Determination of the coefficient of the diffuse surface

All quantities thereafter are computed in the local coordinates system R^d. The 6 coefficients are calculated by a moving least squares method, based on the minimisation of the difference between the altitude X_3^i of a master node $\boldsymbol{x^i}$ and function $f_{\boldsymbol{x}^c}$ evaluated at this node, which leads to the following criterion:

$$J_{\boldsymbol{x}^c}(\boldsymbol{\alpha}) = \sum_{i=1}^{i=n} w(\boldsymbol{x^i},\ \boldsymbol{x^c})\left(\mathbf{p}^t(\boldsymbol{x^i} - \boldsymbol{x^c})\,\boldsymbol{\alpha} - X_3^i\right)^2 \tag{12}$$

where $w(\boldsymbol{x^i}, \boldsymbol{x^c})$ for $i = 1, \ldots, n$ are the weighting function associated with node $\boldsymbol{x^i}$.

Vector $\boldsymbol{\alpha}$ is given by the minimization of criterion $J_{\boldsymbol{x}^c}$ and must be the solution of the system defined by:

$$P^t\ W\ P\ \boldsymbol{\alpha} = P^t\ W\ \mathbf{Z} \tag{13}$$

$\mathbf{Z}$ is the vector of the altitudes of all nodes $\boldsymbol{x^i}$ and P is composed of polynomial basis $\mathbf{p}$ evaluated at each node $\boldsymbol{x^i}$. W is the diagonal matrix of the weights.

The resolution of 6×6 system (13) leads to the determination of $\boldsymbol{\alpha}$ and thus to the knowledge of the local interpolation of the contact surface associated with node $\boldsymbol{x^c}$.

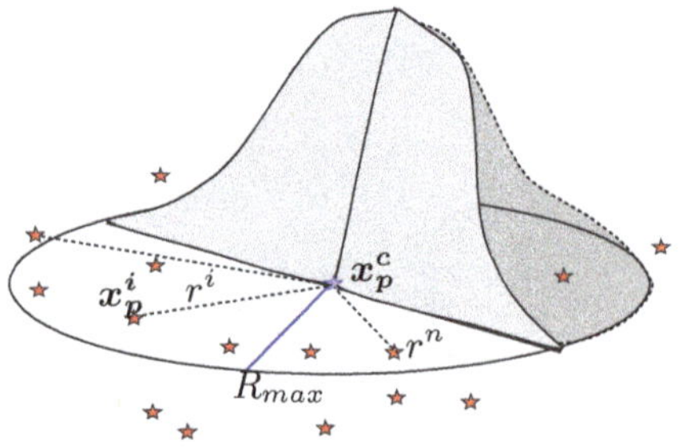

Fig. 5. Domain of influence.

3.3 Weighting functions

The computation of the weighting function has a major influence on the determination of the diffuse surface.

3.3.1 Domain of influence

For each point $\boldsymbol{x^c}$, only the nearest nodes $\boldsymbol{x^i}$ are taken into account. The contribution of each nodal value to the approximation is influenced by a weighting function w^i such that $w^i > 0$ inside the domain of influence of node i and $w^i = 0$ otherwise, providing a local character to the approximation. Fundamental properties related to MLS approximation, such as locality and continuity mainly depend on an appropriate choice of the weighting functions. In order to limit the number of nodes used for the local evaluation, the support of the approximation must be bounded. The issues relative to the construction and to the choice of different weighting functions are detailed in references [5, 6]. In our context, the existence of the approximation requires a number of nodes at least equal to 6 at each evaluation point. Weighting function are radial function, the influence of which decrease with the distance to the node as shown in Figure 5. The domain of influence is centered at $\boldsymbol{x_p^c}$ which is the projection of $\boldsymbol{x^c}$ on the diffuse plane, and we assume that in the projection plane, the set of nodes is fully contained in a circle of radius R_{max}. r^i denotes the distance between $\boldsymbol{x_p^c}$ and $\boldsymbol{x_p^i}$, the projection of node $\boldsymbol{x^i}$ on the diffuse plane. This distance is calculated in the projection plane.

3.3.2 Value of the weight

In a practical way, we have chosen a cubic polynomial function which is set to zero outside the domain of influence. The value of the weight is usually given by

$$w^i = \begin{cases} (1-r)^2(1+2r) & \text{if } 0 \le r < 1 \\ 0 & \text{if } r \ge 1 \end{cases} \quad \text{where } r = \frac{r^i}{R_{max}}. \tag{14}$$

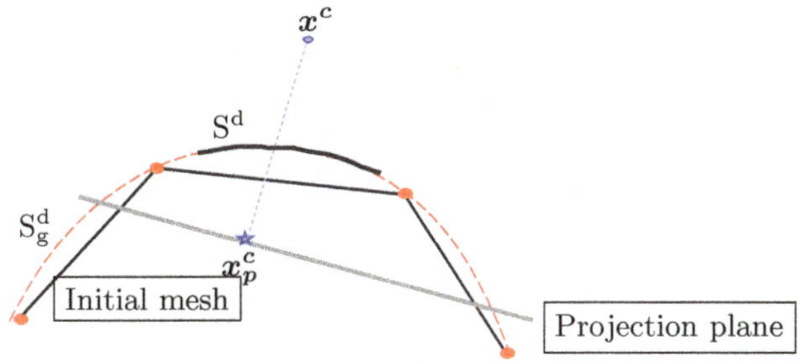

S^d_g: diffuse surface

S^d: local approximation at note x^c

Fig. 6. Local or global approximation?

3.3.3 Diffuse approximation or diffuse interpolation?

Diffuse approximation in its standard form does not interpolate data. The interpolation property is commonly obtained with weighting functions which has infinite value at the node:

$$x^c \to x^i \Rightarrow w^i \to \infty. \tag{15}$$

These functions are defined using the following substitution (Shepard):

$$w^i \to \frac{w^i}{1 - w^i}. \tag{16}$$

4 New contact geometry

A smooth contact surface has been built. We argued that the gap controls the whole contact process. We propose therefore to compute the gap using the intrinsic properties of the diffuse surface.

4.1 Notations

f_{x^c} is dependent on local coordinates X_1^c, X_2^c and X_3^c, and on the coordinates of set of nodes x^i, therefore, we use the following notations:

$$\alpha = \alpha(X_1^c,\ X_2^c,\ X_3^c,\ X_1^1,\ X_2^1,\ \ldots,\ X_2^n,\ X_3^n) \tag{17}$$

$$f_{x^c}(X_1,\ X_2) = f(X_1,\ X_2,\ X_1^c,\ X_2^c,\ \ldots,\ X_2^n,\ X_3^n). \tag{18}$$

4.2 Notion of diffuse gap

The gap is defined as the minimal distance between the slave node x^c and the master surface. It is necessary to associate any slave node with a point belonging to the diffuse surface. This point is the projection which minimize the distance between x^c and the diffuse surface S^d_g. The equation of the diffuse surface is not explicit and a succession of local approximations is necessary to define it (Fig. 6). The idea is to use the local approximation at node x^c as the approximation of the surface.

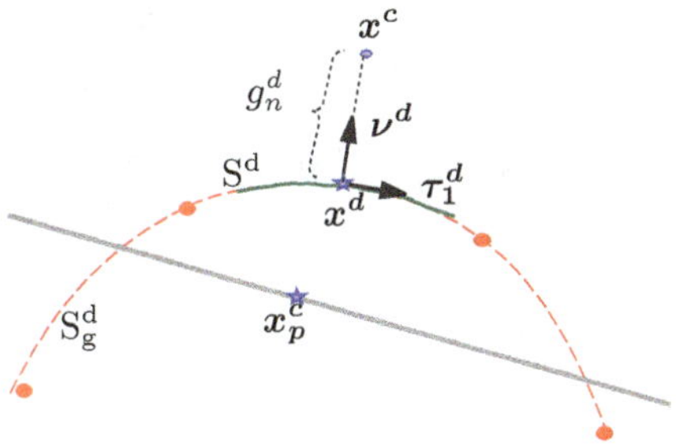

x^d: projection of x^c on S^d

g_n^d: diffuse gap

ν^d: normal vector at the point x^d

τ_1^d: tangent vector at the point x^d

Fig. 7. Diffuse gap (2D representation).

A new point x^d defined as the projection of x^c on S^d is introduced. A statement is made that the vector between x^c and x^d and the normal vector ν^d (associated with the surface S^d at the point x^d) are collinear what leads to:

$$x^c - x^d = g_n^d\, \nu^d. \tag{19}$$

Scalar g_n^d (diffuse gap) describes the contact state on the regularized surface (Fig. 7).

4.3 Local basis in x^d

The Newton method can be used to determine x^d. Its coordinates are noted $(\overline{X}_1, \overline{X}_2, \overline{f})$ where

$$\overline{f} = f(\mathbb{X}) \tag{20}$$

$$\text{with}\quad \mathbb{X} = (\overline{X}_1,\ \overline{X}_2,\ X_1^c,\ \ldots,\ X_i^j,\ \ldots,\ X_3^n).$$

The tangent vectors at the point x^d are given by:

$$\tau_1^d = \begin{bmatrix} 1 \\ 0 \\ \dfrac{\partial f}{\partial X_1}(\mathbb{X}) \end{bmatrix} \quad \text{and} \quad \tau_2^d = \begin{bmatrix} 0 \\ 1 \\ \dfrac{\partial f}{\partial X_2}(\mathbb{X}) \end{bmatrix}. \tag{21}$$

The normal vector ν^d is the inner product of these two vectors:

$$\nu^d = \frac{\tau_1^d \wedge \tau_2^d}{\|\tau_1^d \wedge \tau_2^d\|} = \frac{1}{\|\tau_1^d \wedge \tau_2^d\|} \begin{bmatrix} -\dfrac{\partial f}{\partial X_1}(\mathbb{X}) \\ -\dfrac{\partial f}{\partial X_2}(\mathbb{X}) \\ 1 \end{bmatrix}. \tag{22}$$

5 3D diffuse contact element

We propose to combined the regularization of the contact surface with a specific search of neighboring nodes based

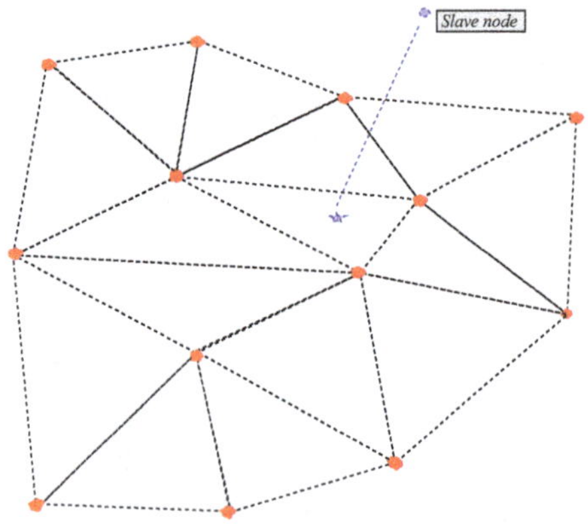

Fig. 8. $3D$ diffuse contact element.

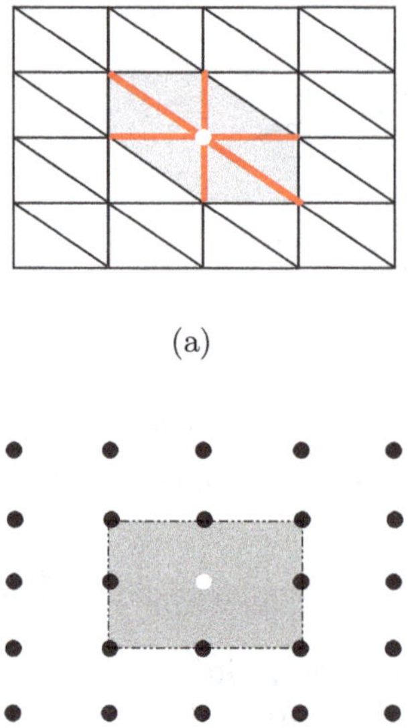

(a)

(b)

Fig. 9. Anisotropic neighborhood determination.

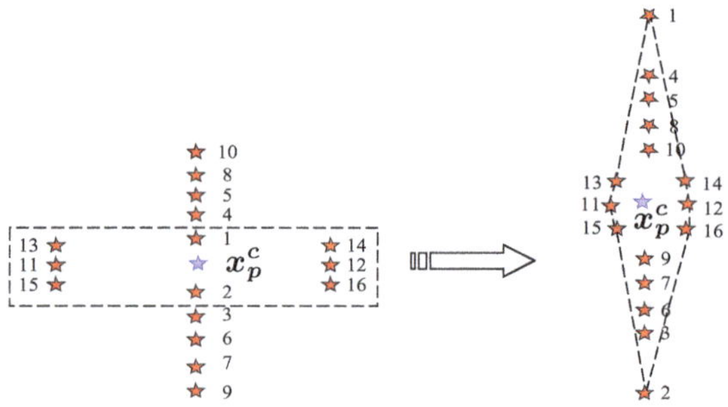

(a) (b)

Fig. 10. Neighborhood determination by convex hull.

on convex hull strategy in order to create a 3D diffuse contact element.

The main difference between a traditional node-facet approach and our technique is that the contact is driven by potential contact nodes and not only by the nodes of the potential contact facet. Indeed, the slave and master surfaces can be seen as a set of nodes: a slave node $\boldsymbol{x}^c$ is supposed in contact with a master zone composed of n nodes $\boldsymbol{x}^i$. The contact element we want to build is composed of a slave node $\boldsymbol{x}^c$ and n neighbor nodes $\boldsymbol{x}^i$ (Fig. 8). A neighbourhood criterion based on convex hulls determination is used to select nodes $\boldsymbol{x}^i$ used to build the local approximation.

5.1 Neighbourhood criterion: convex hulls determination

Contact areas are unknown a priori and may change considerably with the load step especially when large deformations occurs. The use of convex hulls in order to improve the quality of the approximation must be justified. The ideas is to capture the same number of nodes in all directions what cannot be obtained with a search based on a Euclidian metric. The computation of the distances is performed on the plane on which the interpolation surface has been determined. In order to optimize this step, we propose a searching technique based on convex hulls determination. In a first step, we solve the difficulties due to the anisotropy of the data points. When dealing with the Finite Element Method, the influence of the orientation of the mesh on the solution is a well-known academic problem. If a mesh-based approach is used to determine the neighborhood, the procedure can be carried out with the help of edges sharing the same node. However, we can remark that even in this context, the neighborhood depends on the swapping of the edges. Figure 9(a) shows that a search based on the nearest nodes of the mesh leads to the determination of 6 neighbors, when 8 (Fig. 9(b)) could be expected. In order to give to our work much emphasis, we decided to extend our techniques to set of points. Figure 10 illustrates the main steps of the technique. In a first step, m points are selected with respect to their Euclidian distance at the evaluation point. Points are sorted and numbered with respect to their distance to the evaluation point. Each point is repositioned by inverting its distance to the origin. Then, the convex hull (using Graham scan) is determined in the inverted space as shown in Figure 10(a). The result of this procedure, a set of points in side a rectangular area, is displayed in Figure 10(b).

5.2 Notations

If $\boldsymbol{x}$ denotes a vector composed of the position of all the nodes, the first and the second variation of this vector are written:

$$\boldsymbol{x}=\begin{bmatrix}\boldsymbol{x}^{c}\\ \boldsymbol{x}^{1}\\ \boldsymbol{x}^{2}\\ \cdots\\ \cdots\\ \boldsymbol{x}^{n}\end{bmatrix}\quad \delta\boldsymbol{x}=\begin{bmatrix}\delta\boldsymbol{x}^{c}\\ \delta\boldsymbol{x}^{1}\\ \delta\boldsymbol{x}^{2}\\ \cdots\\ \cdots\\ \delta\boldsymbol{x}^{n}\end{bmatrix}\quad \text{and}\quad \Delta\boldsymbol{x}=\begin{bmatrix}\Delta\boldsymbol{x}^{c}\\ \Delta\boldsymbol{x}^{1}\\ \Delta\boldsymbol{x}^{2}\\ \cdots\\ \cdots\\ \Delta\boldsymbol{x}^{n}\end{bmatrix}. \tag{23}$$

In order to describe a 3D contact element, elementary residual vector and the tangent matrix must be determined using the features of the diffuse surface.

5.3 Contact elementary residual vector

In Section 2.3, we have shown that the first variation of g_n^d must be computed to determine the contact residual vector.

5.3.1 First variation of g_n^d

Using relation (19), we can write

$$\begin{aligned}\delta g_n^d &= \delta\left\{(\boldsymbol{x^c}-\boldsymbol{x^d})\cdot\boldsymbol{\nu^d}\right\}\\ &= \delta\left(\boldsymbol{x^c}-\boldsymbol{x^d}\right)\cdot\boldsymbol{\nu^d}+g_n^d\,\boldsymbol{\nu^d}\cdot\delta\boldsymbol{\nu^d}.\end{aligned} \tag{24}$$

However $\|\boldsymbol{\nu^d}\|=1$ which leads to:

$$\boldsymbol{\nu^d}\cdot\delta\boldsymbol{\nu^d}=0. \tag{25}$$

Finaly, δg_n^d can ne expressed as

$$\delta g_n^d=\delta\boldsymbol{x^c}\cdot\boldsymbol{\nu^d}-\delta\boldsymbol{x^d}\cdot\boldsymbol{\nu^d}. \tag{26}$$

In order to calculate $\delta\boldsymbol{x^d}$, the variation of $\bar{\mathrm{f}}$ must be determined. Using (17), it can be proved that

$$\delta\bar{\mathrm{f}}=\frac{\partial\mathrm{f}}{\partial X_1}(\mathbb{X})\,\delta\overline{X}_1+\frac{\partial\mathrm{f}}{\partial X_2}(\mathbb{X})\,\delta\overline{X}_2+\delta\boldsymbol{x}^t\,\underset{\boldsymbol{x}}{\boldsymbol{\nabla}}\bar{\mathrm{f}} \tag{27}$$

where

$$\underset{\boldsymbol{x}}{\boldsymbol{\nabla}}\bar{\mathrm{f}}=\begin{bmatrix}\dfrac{\partial\mathrm{f}}{\partial X_1^c}(\mathbb{X})\\ \dfrac{\partial\mathrm{f}}{\partial X_2^c}(\mathbb{X})\\ \dfrac{\partial\mathrm{f}}{\partial X_3^c}(\mathbb{X})\\ \dfrac{\partial\mathrm{f}}{\partial X_1^1}(\mathbb{X})\\ \dfrac{\partial\mathrm{f}}{\partial X_2^1}(\mathbb{X})\\ \vdots\\ \dfrac{\partial\mathrm{f}}{\partial X_3^n}(\mathbb{X})\end{bmatrix}. \tag{28}$$

Using the definition of tangent vector and relation (27), we have

$$\delta\boldsymbol{x^d}=\delta\overline{X}_1\,\boldsymbol{\tau_1^d}+\delta\overline{X}_2\,\boldsymbol{\tau_2^d}+\mathrm{d}\bar{\mathbf{f}} \tag{29}$$

$$\text{with}\quad \mathrm{d}\bar{\mathbf{f}}=\begin{bmatrix}0\\0\\ \delta\boldsymbol{x}^t\,\underset{\boldsymbol{x}}{\boldsymbol{\nabla}}\bar{\mathrm{f}}\end{bmatrix}.$$

The expressions $\dfrac{\partial\mathrm{f}}{\partial X_k^c}(\mathbb{X})$ and $\dfrac{\partial\mathrm{f}}{\partial X_k^i}(\mathbb{X})$ for $k=1,2,3$ and $i=1,\dots,n$ are obtained with the help of relation (9):

$$\begin{aligned}&\frac{\partial\mathrm{f}}{\partial X_k^c}(\mathbb{X})=\frac{\partial\overline{\mathbf{p}}}{\partial X_k^c}^t\,\boldsymbol{\alpha}+\overline{\mathbf{p}}^t\,\frac{\partial\boldsymbol{\alpha}}{\partial X_k^c}\ \text{ for } k=1,2\\ &\frac{\partial\mathrm{f}}{\partial X_3^c}(\mathbb{X})=0\\ &\frac{\partial\mathrm{f}}{\partial X_k^i}(\mathbb{X})=\overline{\mathbf{p}}^t\,\frac{\partial\boldsymbol{\alpha}}{\partial X_k^i}\end{aligned} \tag{30}$$

where $\overline{\mathbf{p}}$ is the basis $\mathbf{p}$ evaluated at $\overline{X}_1$ and $\overline{X}_2$. The above expression show that $\dfrac{\partial\boldsymbol{\alpha}}{\partial X_k^c}$ and $\dfrac{\partial\boldsymbol{\alpha}}{\partial X_k^i}$ must be calculated.

5.3.2 Discrete expression

Introducing the value of $\delta\boldsymbol{x^d}$ in the definition of g_n^d, we obtain a discretized expression of the first variation of g_n^d:

$$\delta g_n^d=\delta\boldsymbol{x}^t\,\mathbf{N_c^d} \tag{31}$$

where $\mathbf{N_c^d}$ is a $3(n+1)$ component vector,

$$\mathbf{N_c^d}=\frac{-1}{\|\boldsymbol{\tau_1^d}\times\boldsymbol{\tau_2^d}\|}\begin{bmatrix}\dfrac{\partial\mathrm{f}}{\partial X_1^c}(\mathbb{X})+\dfrac{\partial\mathrm{f}}{\partial X_1}(\mathbb{X})\\ \dfrac{\partial\mathrm{f}}{\partial X_2^c}(\mathbb{X})+\dfrac{\partial\mathrm{f}}{\partial X_2}(\mathbb{X})\\ -1\\ \dfrac{\partial\mathrm{f}}{\partial X_1^1}(\mathbb{X})\\ \dfrac{\partial\mathrm{f}}{\partial X_2^1}(\mathbb{X})\\ \vdots\\ \dfrac{\partial\mathrm{f}}{\partial X_3^n}(\mathbb{X})\end{bmatrix}. \tag{32}$$

We remark that the first three components of this vector take into account only the contributions of the slave node and the others components take into account the contributions of master nodes.

When a penalty method is used, the elementary residual contact vector is thus a $3(n+1)$ component vector given by:

$$\mathbf{R_c^d}=-\varepsilon_n\,g_n^d\,\mathbf{N_c^d} \tag{33}$$

where ε_n is the penalty parameter.

5.4 Contact elementary tangent matrix

The evaluation of the contact elementary tangent matrix requires the determination of the second variation of g_n^d.

5.4.1 Second variation of g_n^d

The derivation of relation (26) leads to

$$\Delta\delta g_n^d = \Delta\left\{\left(\delta\boldsymbol{x}^{\mathbf{c}} - \delta\boldsymbol{x}^{\mathbf{d}}\right)\cdot\boldsymbol{\nu}^{\mathbf{d}}\right\} \tag{34}$$

which is equivalent to:

$$\Delta\delta g_n^d = g_n^d\,\delta\boldsymbol{\nu}^{\mathbf{d}}\cdot\,\Delta\boldsymbol{\nu}^{\mathbf{d}} - \Delta\delta\boldsymbol{x}^{\mathbf{d}}\cdot\boldsymbol{\nu}^{\mathbf{d}}. \tag{35}$$

If we introduce (29), $\Delta\delta\boldsymbol{x}^{\mathbf{d}}$ can be evaluated:

$$\begin{aligned}\Delta\delta\boldsymbol{x}^{\mathbf{d}} &= \Delta\left\{\delta\overline{X}_1\,\boldsymbol{\tau}_{\mathbf{1}}^{\mathbf{d}} + \delta\overline{X}_2\,\boldsymbol{\tau}_{\mathbf{2}}^{\mathbf{d}} + \mathrm{d}\bar{\mathrm{f}}\right\}\\ &= \delta\overline{X}_1\,\Delta\boldsymbol{\tau}_{\mathbf{1}}^{\mathbf{d}} + \delta\overline{X}_2\,\Delta\boldsymbol{\tau}_{\mathbf{2}}^{\mathbf{d}}\\ &\quad + \Delta\delta\overline{X}_1\,\boldsymbol{\tau}_{\mathbf{1}}^{\mathbf{d}} + \Delta\delta\overline{X}_2\,\boldsymbol{\tau}_{\mathbf{2}}^{\mathbf{d}} + \Delta\mathrm{d}\bar{\mathrm{f}}.\end{aligned} \tag{36}$$

The problem is to determine $\Delta\mathrm{d}\bar{\mathrm{f}}$ and particularly the third component. The variation of the gradient can be calculated as

$$\Delta\mathrm{d}\bar{\mathrm{f}} = \begin{bmatrix} 0 \\ 0 \\ \delta\boldsymbol{x}^t\left\{\mathrm{H}_{\bar{\mathrm{f}}}\,\Delta\boldsymbol{x} + \mathbf{T_1}\,\Delta\overline{X}_1 + \mathbf{T_2}\,\Delta\overline{X}_2\right\}\end{bmatrix} \tag{37}$$

where

$$\mathbf{T}_l = \begin{bmatrix} \dfrac{\partial^2 \mathrm{f}}{\partial X_l \partial X_1^c}(\mathbb{X}) \\ \dfrac{\partial^2 \mathrm{f}}{\partial X_l \partial X_2^c}(\mathbb{X}) \\ 0 \\ \dfrac{\partial^2 \mathrm{f}}{\partial X_l \partial X_1^1}(\mathbb{X}) \\ \dfrac{\partial^2 \mathrm{f}}{\partial X_l \partial X_2^1}(\mathbb{X}) \\ \vdots \\ \dfrac{\partial^2 \mathrm{f}}{\partial X_l \partial X_3^n}(\mathbb{X}) \end{bmatrix} \quad \text{with } l = 1, 2. \tag{38}$$

$\mathrm{H}_{\bar{\mathrm{f}}}$ is the Hessian $(n+1), 3(n+1)$ matrix of f.

$$\mathrm{H}_{\bar{\mathrm{f}}} = \begin{bmatrix} \dfrac{\partial^2 \mathrm{f}}{\partial X_1^c \partial X_1^c}(\mathbb{X})\cdots & \dfrac{\partial^2 \mathrm{f}}{\partial X_k^i \partial X_1^c}(\mathbb{X})\cdots & \dfrac{\partial^2 \mathrm{f}}{\partial X_3^n \partial X_1^c}(\mathbb{X}) \\ \vdots & \vdots & \vdots \\ \dfrac{\partial^2 \mathrm{f}}{\partial X_1^c \partial X_3^n}(\mathbb{X})\cdots & \dfrac{\partial^2 \mathrm{f}}{\partial X_k^i \partial X_3^n}(\mathbb{X})\cdots & \dfrac{\partial^2 \mathrm{f}}{\partial X_3^n \partial X_3^n}(\mathbb{X}) \end{bmatrix}.$$

5.4.2 Discrete expression and contact tangent matrix

The variations of $\overline{X}_1$, $\overline{X}_2$ and vector $\boldsymbol{\nu}^{\mathbf{d}}$ can be expressed as follows:

$$\begin{aligned}&\delta\overline{X}_l = \delta\boldsymbol{x}^t\,\mathbf{S}_l \quad et \quad \Delta\overline{X}_l = \mathbf{S}_l{}^t\,\Delta\boldsymbol{x} \quad l = 1, 2\\ &\delta\boldsymbol{\nu}^{\mathbf{d}} = \mathrm{U}\,\delta\boldsymbol{x} \ \text{ and } \Delta\boldsymbol{\nu}^{\mathbf{d}} = \mathrm{U}\,\Delta\boldsymbol{x}\end{aligned} \tag{39}$$

where $\mathbf{S}_l$ $(l = 1, 2)$ is a $3(n+1)$ component vector and U a matrix belonging to $\mathbb{M}_{3,3(n+1)}$.

Using equation (21), it is possible to obtain the variation of the tangent vectors:

$$\Delta\boldsymbol{\tau}_l^{\mathbf{d}} = \begin{bmatrix} 0 \\ 0 \\ \mathbf{D}_{\boldsymbol{\tau}_l}{}^t\,\Delta\boldsymbol{x} \end{bmatrix} \quad l = 1, 2. \tag{40}$$

In this relation, $\mathbf{D}_{\boldsymbol{\tau}_l}$ is a $3(n+1)$ component vector depending $\mathbf{S}_l$ and $3(n+1)$ and is given by:

$$\begin{aligned}\mathbf{D}_{\boldsymbol{\tau}_1} &= \frac{\partial^2 \mathrm{f}}{\partial X_1 \partial X_1}(\mathbb{X})\,\mathbf{S_1} + \frac{\partial^2 \mathrm{f}}{\partial X_1 \partial X_2}(\mathbb{X})\,\mathbf{S_2} + \mathbf{T_1}\\ \mathbf{D}_{\boldsymbol{\tau}_2} &= \frac{\partial^2 \mathrm{f}}{\partial X_2 \partial X_1}(\mathbb{X})\,\mathbf{S_1} + \frac{\partial^2 \mathrm{f}}{\partial X_2 \partial X_2}(\mathbb{X})\,\mathbf{S_2} + \mathbf{T_2}.\end{aligned} \tag{41}$$

The relations introduced before leads to the following discrete expression of $\Delta\delta g_n^d$

$$\Delta\delta g_n^d = \delta\boldsymbol{x}^t\,\mathrm{M}_c^d\,\Delta\boldsymbol{x} \tag{42}$$

where M_c^d is a matrix $\mathbb{M}_{3(n+1),3(n+1)}$.

$$\begin{aligned}\mathrm{M}_c^d\ &\frac{-1}{\|\boldsymbol{\tau}_{\mathbf{1}}^{\mathbf{d}}\wedge\boldsymbol{\tau}_{\mathbf{2}}^{\mathbf{d}}\|}\left(\mathrm{H}_{\bar{\mathrm{f}}} + \mathbf{T_1}\,\mathbf{S_1}{}^t\right.\\ &\left.+\ \mathbf{T_2}\,\mathbf{S_2}{}^t + \mathbf{S_1}\,\mathbf{D}_{\boldsymbol{\tau}_1}{}^t + \mathbf{S_2}\,\mathbf{D}_{\boldsymbol{\tau}_2}{}^t\right) + g_n^d\,\mathrm{U}^t\,\mathrm{U}.\end{aligned}$$

Finally, when a penalty method is used, K_c^d, the elementary diffuse contact tangent matrix belonging to $\mathbb{M}_{3(n+1),3(n+1)}$ is:

$$\mathrm{K}_c^d = \varepsilon_n\,g_n^d\,\mathrm{M}_c^d + \varepsilon_n\,\mathbf{N}_{\mathbf{c}}^{\mathbf{d}}\,\mathbf{N}_{\mathbf{c}}^{\mathbf{d}t}. \tag{43}$$

6 Numerical examples

The approach presented here has been implemented in the finite element code SYSTUS developped by the company ESI Group and tested on both academic and industrial examples. We assume the contact to be frictionless.

6.1 Contact between a plate and a rigid cylinder

We consider the contact between a plate and a rigid cylinder. The rigid cylinder is choosen as the master body. The plate is submitted to an imposed pressure on a part, S_p, of its boundary and is clamped on another part (See Fig. 11). The pressure is applied gradually until $P = 200$ MPa by steps of 20 MPa. The mechanical properties of the plate are: $E = 200\,000$ MPa and $\nu = 0.3$.

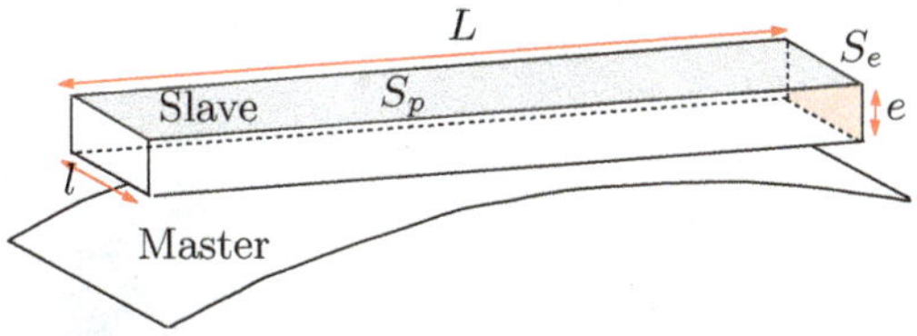

$L = 70mm$
$e = 4mm$
$l = 20mm$

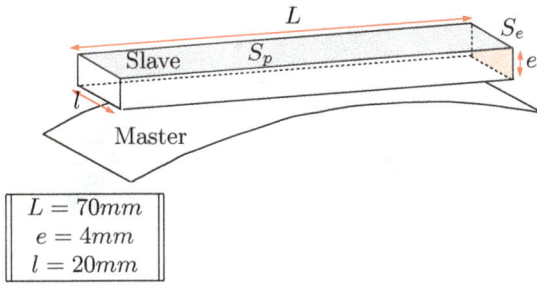

Fig. 11. Contact between a plate and a rigid cylinder: geometry.

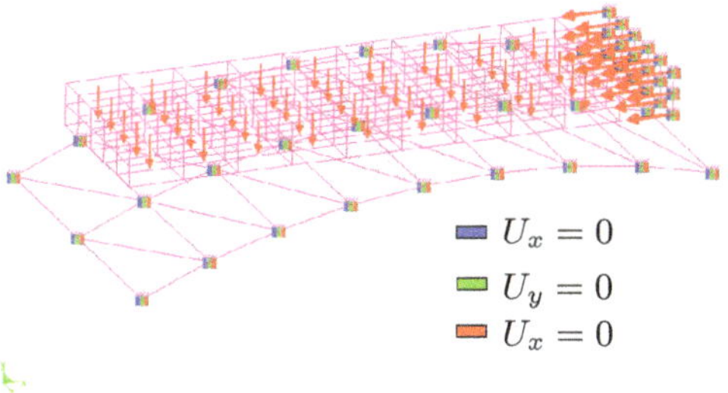

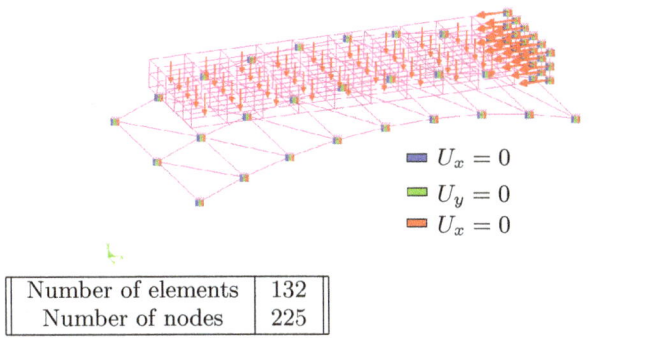

Number of elements	132
Number of nodes	225

Fig. 12. Contact between a plate and a rigid cylinder: finite element model.

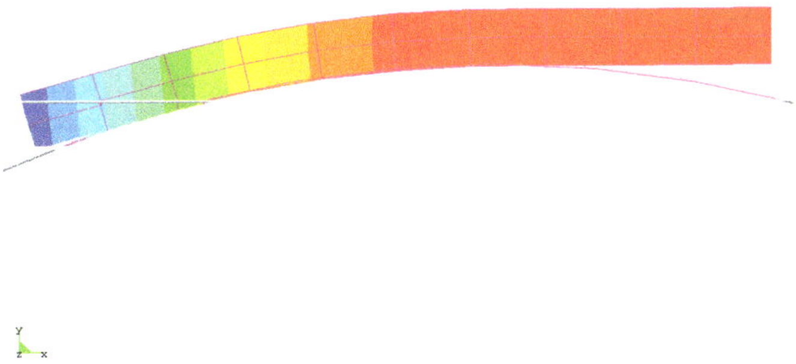

Fig. 13. Contact between a plate and a rigid cylinder: deformed configuration.

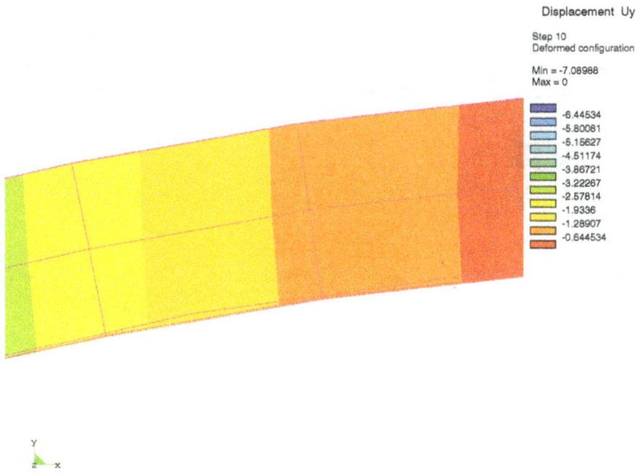

(a) Classic contact element

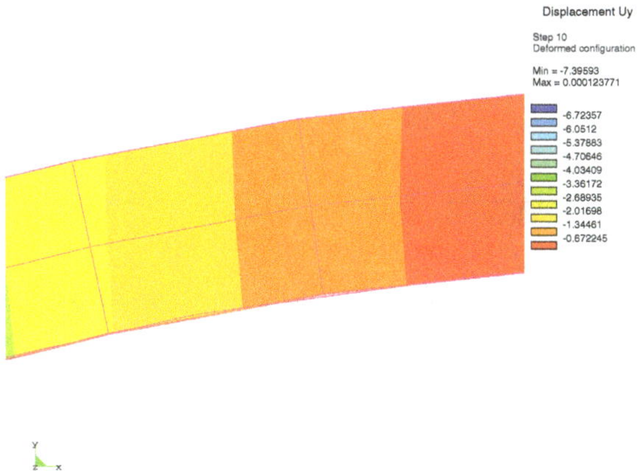

(b) 3D diffuse contact element

Fig. 14. Zoom in deformed configuration.

The finite element model can be seen in Figure 12: the master surface is composed of 32 triangular facets and the slave body of 100 quadrangle facets. There are 225 nodes.

The shape of the master surface can generate difficulties. To point out these problems, only a part of the master surface is studied (Fig. 13). A zoom on the deformed configuration obtained with a classic contact algorithm is represented in Figure 14(a). It appears clearly that the non smoothness of the cylinder induces an important interpenetration. This problem is strongly attenuated when the new contact element is used. We can see in Figure 14(b) that the contact does not occur on a flat facet, but on a surface which fits the shape of the cylinder. This problem is strongly attenuated by using the new contact element. It can be seen especially in figure. The contact does not take place on a fat facet, but on a surface which completely follows the shape of the cylinder

6.2 Pinch of pipes

This example deals with contact problem between two deformable bodies in 3D. A pipe is pinched between two identical parallel plates (Fig. 15). Only one quarter of the model is used because of symmetry reason (Fig. 16). The plates and the pipe have the same elastoplastic properties:

– Young Modulus: $E = 200\,000$ MPa.

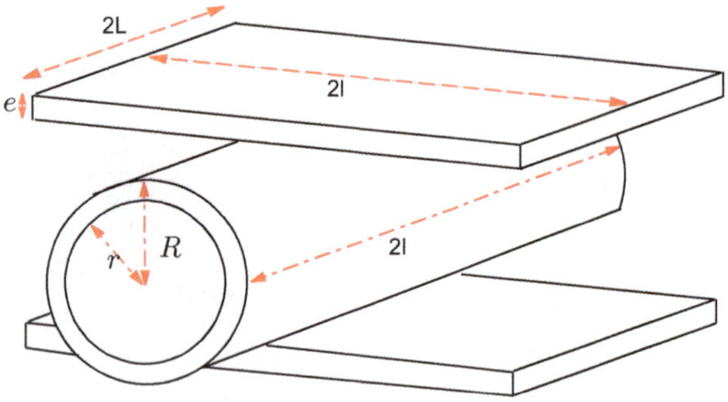

Plates	Pipe
$l = 200$ mm	$l = 200$ m
$L = 75$ mm	$r = 90$ mm
$e = 20$ mm	$R = 100$ mm

Fig. 15. Pinch of pipes: geometrical model.

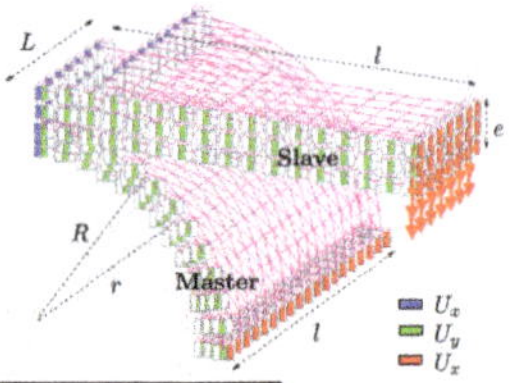

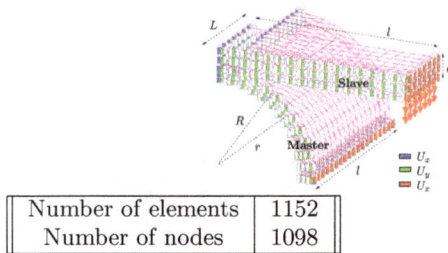

Number of elements	1152
Number of nodes	1098

Fig. 16. Pinch of pipes: finite element model.

- Poisson ration: $\nu = 0.30$.
- Yield stress: $\sigma_y = 400$ MPa.
- Slope hardening: $H = 3450$ MPa.

The load consists in a vertical imposed displacement of 100 mm at the end of the plate (Fig. 16). In the elastic study, the load step is equal to 2.5 mm and reduced to 1 mm when dealing with elastoplasticity. This test has been performed with a traditional contact algorithm and with our approach. The computations have been performed for both cases.

We observe that the number of iterations is reduced of 14% when the surface smoothing procedure is used. The CPU time is not increased. The deformed configuration can be seen in Figure 17.

7 Conclusion

In this work, a 3D smoothing contact surface procedure based on diffused approximation associated with an original contact search algorithm by convex hulls has been presented. Our approach proved to overcome the numerical instabilities and lacks of precision encountered when dealing with contact problems. We also experienced that the robustness and the efficiency of the contact search are considerably improved. Problems due to strong variations of vector fields are reduced. Nodes may slide smoothly on the diffuse surface without impeding the convergence of the global process.

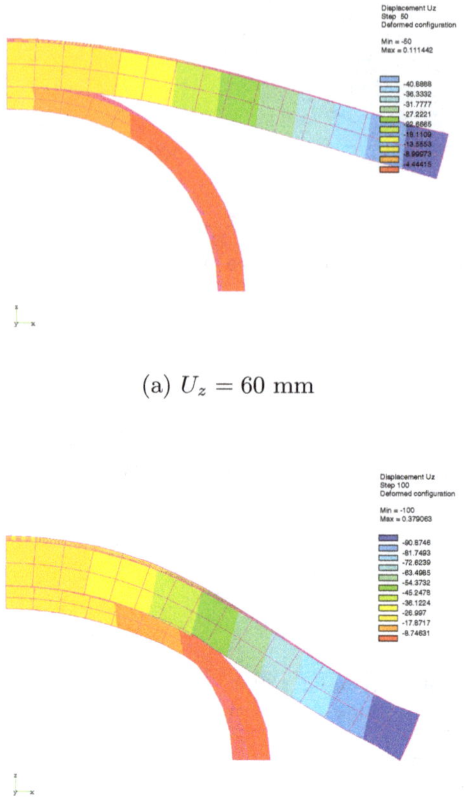

(a) $U_z = 60$ mm

(b) $U_z = 100$ mm

Fig. 17. Pinch of pipes: deformed configurations.

References

1. P. Alart, A. Curnier, A mixed formulation for frictional contact problems prone to Newton like solution methods. Comput. Meth. Appl. Mech. Eng. **92**, 353–375 (1991).
2. T. Belytschko, W.J.T. Daniel, G. Ventura, A monolithic smoothing-gap algorithm for contact-impact based on the signed distance function. Int. J. Numer. Meth. Eng. **55**, 101–125 (2002).
3. T. Belytschko, M.O. Neal, Contact-impact by the piball algorithm with penalty and lagrangian methods. Int. J. Numer. Meth. Eng. **31**, 547–572 (1991).
4. D.J. Benson, J.O. Hallquist, A single surface contact algorithm for the post-buckling analysis of shell structures. Comput. Meth. Appl. Mech. Eng. **78**, 141–163 (1990).
5. P. Breitkopf, A. Rassineux, J.M. Savignat, P. Villon, Integration and convergence constraints in Diffuse Element Method. Comput. Meth. Appl. Mech. Eng. **193**, 1203–1220 (2004).
6. P. Breitkopf, A. Huerta, *Meshfree and Particle Based Approaches in Computational Mechanics*. Kogan Page Science, ISBN 1903996457 (2004).
7. D. Chamoret, A. Rassineux, J.M. Bergheau, P. Villon, Modelling of contact surface by local hermite diffuse interpolation. *Proceedings of ESAFORM 2001, The*

4th *International ESAFORM Conference on Material Forming*, University of Liège, Belgium **1**, 179–182 (2001).
8. D. Chamoret, P. Saillard, A. Rassineux, J.M. Bergheau, New smoothing procedures in contact mechanics. J. Comput. Appl. Math. **168**, 107–116 (2004).
9. C. Chappuis, A. Rassineux, P. Breitkopf, P. Villon, Improving surface remeshing by feature recognition. Eng. Comput. **20**, 202–209 (2004).
10. A. Curnier, Q.C. He, A. Klarbring, Continuum mechanics modelling of large deformation contact with friction, in *Contact mechanics*, M. Raous, M. Jean and J.J Moreau, Eds. Plenum Press (1995), pp. 145–158.
11. R. Diekmann, J. Hungershö, M. Lux, L. Taenzer, J.M. Wierum, Efficient contact searching for finite element Analysis. *European Congress on Computational Methods in Applied Sciences and Engineering*, Barcelona, Spain (2000).
12. N. El-Abbasi, S.A. Meguid, A. Czekanski, On the modelling of smooth contact surfaces using cubic splines. Int. J. Numer. Meth. Eng. **50**, 953–967 (2001).
13. E.G. Nezami, Y.M.A. Hashash, D. Zhao, J. Ghaboussi, A fast contact detection algorithm for 3-D discrete element method. Comput. Geotech. **31**, 575–587 (2004).
14. L. Fourment, J.L. Chenot, K. Mocellin, Numerical formulations and algorithms for solving contact problems in metal forming simulation. Int. J. Numer. Meth. Eng. **46**, 1435–1462 (1999).
15. A. Heege, P. Alart, A Frictional contact element for strongly curved contact problems. Int. J. Numer. Meth. Eng. **39**, 165–184 (1996).
16. A. Klarbring, Large displacement frictional contact: a continuum framework for finite element discretization. Eur. J. Mech. A **14**, 237–253 (1995).
17. G. Kloosterman, A.H. Van Den Boogaard, J. Huetink, An efficient contact search algorithm. *The* 5th *international ESAFORM conference on MATERIAL FORMING.* M. Pietrzyk, Z. Mitura, J. Kaczmar, Cracow, Poland (2002), pp. 99–102.
18. L. Krstulovic-Opara, P. Wriggers, J. Korelc, A C^1-continuous formulation for $3D$ finite deformation frictional contact. Comput. Mech. **29**, 27–42 (2002).
19. T.A. Laursen, *Computational Contact and Impact Mechanics.* Springer Verlag (2002).
20. T.A. Laursen, J.C. Simo, A Continuum-based finite element formulation for the implicit solution of multibody, large deformation frictional contact problems. Int. J. Numer. Meth. Eng. **36**, 3451–3485 (1993).
21. Li S, Dong Qian, Wing Kam Liu, Belytschko T. A meshfree contact-detection algorithm. Comput. Meth. Appl. Mech. Eng. **190**, 3271-3292.
22. W.N. Liu, G. Metschke, H.A. Mang, A note on the algorithmic stabilization of 2D contact analyses, in *Computational Methods in Contact Mechanics*, L. Gaul, C.A. Brebbia, Wessex Institute of Technology, WITpresss (1999), pp. 231–240.
23. B. Nayrolles, G. Touzot, P. Villon, Generalizing the finite element method: diffuse approximation and diffuse elements. J. Comput. Mech. **10**, 307–138 (1992).
24. M. Oldenburg, L. Nilsson, The position code algorithm for contact searching. Int. J. Numer. Meth. Eng. **37**, 359–386 (1994).
25. V. Padmanabhan, T.A. Laursen, A framework for development of surface smoothing procedures in large deformation frictional contact analysis. Finite Elem. Anal. Design **37**, 173–198 (2001).
26. P. Papadopoulos, R.L. Taylor, A mixed formulation for the finite element solution of contact problems. Comput. Meth. Appl. Mech. Eng. **94**, 373–389 (1992).
27. G. Pietrzak, A. Curnier, Large deformation frictional contact mechanics: continuum formulation and augmented Lagrangian treatment. Comput. Meth. Appl. Mech. Eng. **177**, 351–381 (1999).
28. M.A. Puso, T.A. Laursen, A 3D contact smoothing method using Gregory patches. Int. J. Numer. Meth. Eng. **54**, 1161–1194 (2002).
29. A. Rassineux, P. Villon, J.M. Savignat, O. Stab, Surface remeshing by local hermite diffuse interpolation. Int. J. Numer. Meth. Eng. **49**, 31–49 (2000).
30. Sheng Ping Wang, Eiiji Nakamachi. The inside-outside contact search algorithm for finite element analysis. Int. J. Numer. Meth. Eng. **40**, 3665–3685 (1997).
31. P. Wriggers, *Computat. Contact Mechanics.* Wiley (2002).
32. P. Wriggers, L. Krstulovic-Opara, J. Korelc, Smooth C^1-interpolations for two-dimensional frictional contact problems. Int. J. Numer. Meth. Eng. **51**, 1469–1495 (2001).
33. P. Wriggers, T. Vu Van, E. Stein, Finite element formulation of large deformation impact-contact problems with friction. Comput. Struct. **37**, 319–331 (1990).
34. Z.H. Zhong, L. Nilsson, A contact searching algorithm for general 3-D contact-impact problems. Comput. Struct. **34**, 327–335 (1990).
35. Z.H. Zhong, L. Nilsson, Automatic contact searching algorithm for dynamic finite element analysis. Comput. Struct. **52**, 187–197 (1994).

3

Finite element modeling of abradable materials – Identification of plastic parameters and issues on minimum hardness against coating's thickness

F. Peyraut[1,a], J.-L. Seichepine[2], C. Coddet[2] and M. Hertter[3]

[1] M3M, Belfort-Montbéliard University of Technology, Belfort, France
[2] LERMPS, Belfort-Montbéliard University of Technology, Belfort, France
[3] MTU Aero Engines GmbH, Dachauer Straße 665, 80995 München, Germany

Abstract – Abradable materials are used to decrease the gas consumption of aircraft engines by minimizing the gap between the blade tips and the stator. The key idea consists in using the blades themselves to machine the gap on the abradable coating. The best compromise between soft and hard coating properties has to be reached to avoid blades wear and prevent coating erosion by gas flux and particles. The plastic parameters of abradable coating were identified by using an optimization process directly connected to FEA. The first order optimization method (conjugate gradient strategy + golden section algorithm) was applied to achieve the optimal solution. A good agreement was found between experimental and numerical results. The plastic parameters were used to study the hardness variability of abradable materials with the coating thickness. Surprisingly, a minimum hardness value was found while it was expected that hardness should be always decreasing with thickness. It has been demonstrated that this minimum is produced by the boundary conditions influence on hardness measurement. This research work was completed within the Seal-Coat project funded by the European Commission under the FP5 Growth Program.

Key words: Finite element method; optimization; HR15Y indentation test; abradable materials

1 Introduction

Abradable materials are used to decrease the gas consumption of aircraft engines by minimizing the gap between the blade tips and the stator [1]. These materials are located on the static parts of gas turbines, in front of blades, in order to control the overtip leakage. The key idea consists in machining the gap on an abradable coating with the blades themselves. The best compromise between soft and hard coating properties must be reached to avoid blades wear on the one hand and prevent coating erosion by gas flux and particles on the other.

In the last decades, many attempts have been made to model indentation problems with the finite element method. A detailed bibliographical review of papers published in 1997-2000 and dealing with FEM and BEM simulations of indentation problems can be found in [2]. Most of these works use general purpose FE commercial software like ABAQUS [3], ANSYS [4], MARC [5], CASTEM 2000 [6], NASTRAN [7] or ADINA [8]. To identify plastic parameters by indentation test, material properties are fitted in such a way that measured and computed hardness match. However, this fitting can be performed by various means. Cao et al. [3] have proposed to solve the inverse problem of identifying the plastic parameters by using an analytical dimensionless function relating the indentation response to the material properties. This approach requires preliminary computational work to fit the coefficients of the dimensionless function. Moreover, the results are only available for a restricted class of materials. To cover a wider range of alloys and pure metals, Zhao et al. [9] have extended the work of Cao and Lu to materials involved in engineering design, such as copper, aluminium, tin and tin alloys. Lee et al. [10] have proposed numerical formulas based on the incremental theory of plasticity for material property evaluation. The material properties used for the FE analysis cover the property range of general metals with a Young's modulus varying from 70 to 400 GPa. It does not include abradable materials for which it was found that calculated Young's moduli are closer to about 10 rather than 70 GPa. Tang and Arnell [11] have proposed a new mathematical model for the effective Young's modulus of the combined substrate and coating and the mean pressure as a measure of indentation hardness. But this model, which provides coating modulus and hardness independent on substrate effects, is only applicable for elastic-perfectly plastic materials. In [12], FEM is employed to obtain indentation parameters like maximum and residual penetration depths.

[a] Corresponding author: francois.peyraut@utbm.fr

These parameters are used in the indentation force equation which is solved by the linear regression method. However, this method only concerns materials exhibiting exponential hardening and can not be extended to others without loss of accuracy. More recently, Beghini et al. [13] have proposed an inverse procedure, starting from experimentally measured pairs of load and depth of penetration, for deducing the material parameters by means of an optimization algorithm. This procedure requires preliminary FEA to build up a database of numerical curves relating the load to the depth of penetration. Moreover, it can only be applied to plastic behaviour involving three parameters such as the Hollomon like power law or a linear strain hardening.

The limited range of applicability is generally considered as the main drawback of methods using response surface even if most classic alloys and metals are accounted. To provide a general approach applicable to any kind of material, including the abradables, an alternative is to directly connect optimization algorithms to FE analysis. Bouzakis et al. [14] have proposed this kind of approach by using FEM to extract material parameters for a multilinear plastic law. The optimization algorithm was developed by using minimization technique supported by the FEM package ANSYS. The material parameters were updated and FE computation repeated until a criterion between measured and calculated penetration forces will be fulfilled. We have selected a similar approach by using the FE code ANSYS [15]. The difference between measured and computed data was minimized to find the appropriate material properties satisfying the similarity of computational and experimental hardness. The first order optimization method (conjugate gradient strategy + golden section algorithm) was applied to achieve the optimal solution. This method is time consuming but converges accurately towards the global minimum of the objective function. Two indentation depths corresponding to two different coating thicknesses have been considered instead of the complete load-depth indentation curve related to a single coating thickness. Since two different coating thicknesses are considered for the reverse analysis, the proposed approach likely works as the dual method [16] used for sharp indentation to provide a single combination of appropriate material parameters. This ensures that the relation between material parameters and hardness is one-to-one. The proposed method can be directly applied to any kind of bilinear plastic material and does not need any preliminary computational work to fit coefficients and material constants as required by other methods.

Thanks to the identification of the plastic parameters, an extensive study of hardness against thickness was performed by using an axisymmetrical FE model of the HR15Y indentation test. Several hardness tests are available to perform hardness measurement, each one providing different hardness value (Vickers, Rockwell and Rockwell superficial, Brinell ...). Among these various hardness tests, the HR15Y indentation test [17] is the most widely used by industry manufacturers to control

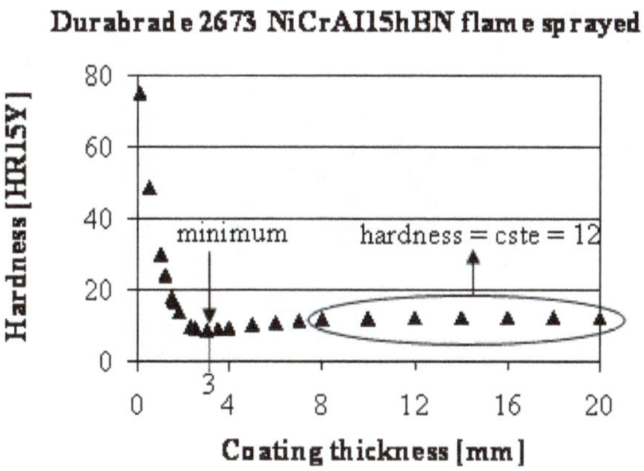

Fig. 1. Micrograph of a typical abradable coating structure; left: surface view – right: cut view.

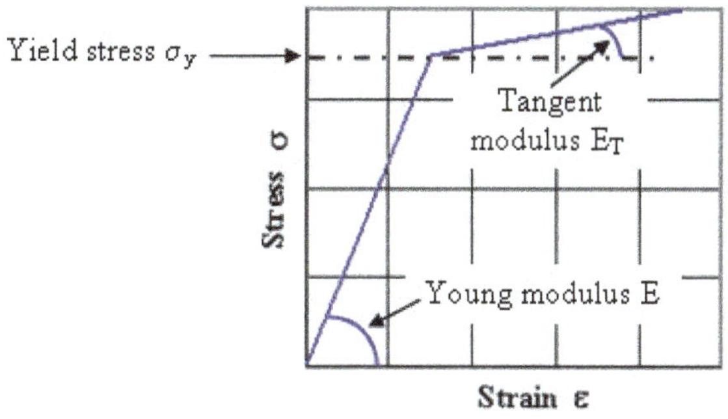

Fig. 2. Elastic-plastic behaviour law.

the quality of thermally sprayed abradable coating. The general principle is the measurement of the permanent increase in depth of penetration with a 12.70 mm diameter steel ball indenter, a preliminary test force of 29 N and a total test force of 147 N. The FE results show a fairly good agreement with measured hardness. However, a minimum hardness value was found while it was expected that hardness should be always decreasing with thickness. By focussing the study on the central part of the coating, which is the most significant region to account for plastic effects, we have demonstrated that this minimum hardness is produced by the boundary conditions influence on hardness measurement.

2 Problem presentation

Abradable coatings are made of complex 3-D heterogeneous composite materials randomly distributed (Fig. 1). A bilinear plastic law was used to model the mechanical behaviour of these coatings (Fig. 2). To take into account the scale of the heterogeneities (micron) and the scale of the coating (millimetre), the elastic properties were computed thanks to a multi-scaled approach [18]. The Young's modulus and the Poisson ratio calculated with this approach for the abradable coating considered in this work are respectively 15 GPa and 0.2.

The plastic parameters were identified by minimizing the difference between computed and measured

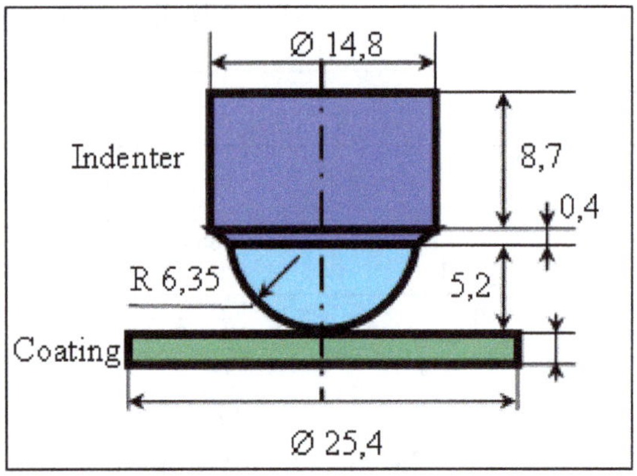

Fig. 3. Schematic of HR15Y test rig.

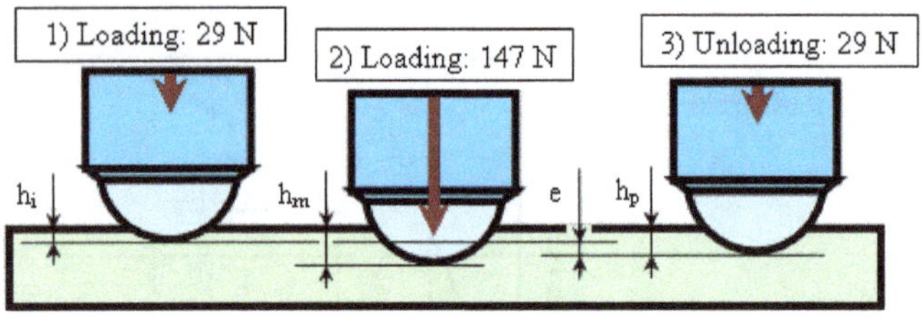

Fig. 4. Multi-step loading.

hardness's. The general principle is the measurement of the permanent increase in depth of penetration under a multi-step load. Various geometrical indenters can be used as spheroconical tip or ball with various diameters. In [19], the effects of different ball materials on HRB hardness measurement values are discussed. In our work, the standardized HR15Y test (Fig. 3) is selected because it is the most widely used by manufacturers of abradable coatings for aero-engines application. This test uses a 12.70 mm diameter steel ball indenter, a preliminary test force of 29 N and a total test force of 147 N (Fig. 4). The preliminary test force is applied to cause an initial penetration h_i of the specimen in accordance with recommendations provided by hardness measurement standards [17]. The total force is next applied to provide the maximum depth of penetration h_m. The permanent depth of penetration h_p is obtained by unloading the system from 147 N to 29 N. By introducing the difference e between the permanent h_p and the initial h_i depths of penetration with a μm scale, the material hardness H is defined as $H = 100$-e.

3 Numerical procedures

3.1 Finite element model

The FE code ANSYS was used to build a bi-dimensional model of the coating and the indenter (Fig. 5). Both of them are meshed in the (x, y) plane by four nodes PLANE42 elements used with axisymmetrical option. The conventional notations used by ANSYS were adopted to represent the radius and the rotation axis (respectively x and y directions). The FE model is made of 2176 nodes and 2163 elements for the lowest thickness used in the present work (0.1 mm) and 14893 nodes and 14723 elements for the highest (20 mm). This model includes local non linearity to model frictionless contact between the indenter and the coating and material non linearity to represent the plastic behaviour of abradable materials. Frictionless contact is assumed because it has been observed that friction produces negligible effects on the curve relating the load to the depth of penetration [10]. TARGE169 and CONTA171 elements were used to model contact between the coating and the indenter and the mesh density is refined in the vicinity of the contact area to account for high stress gradient growing in this region. TARGE169 element represents the "target" surface and CONTA171 is a "surface-to-surface" contact element used to represent contact and sliding between 2-D "target" surfaces and a deformable surface defined by this element. The bottom of the coating was constrained against displacement in the y-direction and the vertical symmetry axis, defined by $x = 0$, was constrained against displacement in the x-direction to model axisymetrical condition. To model the plastic behaviour of the coating, a bilinear isotropic law was used (Fig. 2). The material parameters are the yield stress σ_y, the tangent modulus E_T and the Young modulus E. A preliminary FEA is performed on a 2D model of the material microstructure in order to compute E [18]. σ_y and E_T are next identified by using a reverse procedure described in Section 3.2.

3.2 Optimization

The optimization process used to identify the yield stress σ_y and the tangent modulus E_T values is presented in this subsection. The idea is to determine σ_y and E_T in order to minimize the difference D between measured and computed hardness

$$\underset{\sigma_y,\, E_T}{\mathrm{Min}}\ [D]. \tag{1}$$

The yield stress σ_y and the tangent modulus E_T represent the design variables and the objective function D is defined by the distance between experimental and numerical hardnesses. This distance is expressed in terms of percentage by using the usual Euclidian norm

$$D = 100 \times \frac{\|\mathbf{H}^c - \mathbf{H}^m\|}{\|\mathbf{H}^m\|}. \tag{2}$$

Measured and computed hardnesses are stored with two vectors $\mathbf{H}^m$ and $\mathbf{H}^c$

$$\mathbf{H}^m = \begin{Bmatrix} 24.1 \\ 10 \end{Bmatrix};\ \mathbf{H}^c = \begin{Bmatrix} H^c_{1.2} \\ H^c_{2.3} \end{Bmatrix}. \tag{3}$$

Each component corresponds to a different thickness. Two components are needed since two plastic parameters have to be found. In order to extract the most relevant information from experimental data's, the lowest and highest coating thicknesses were logically considered (1.2 and 2.3 mm).

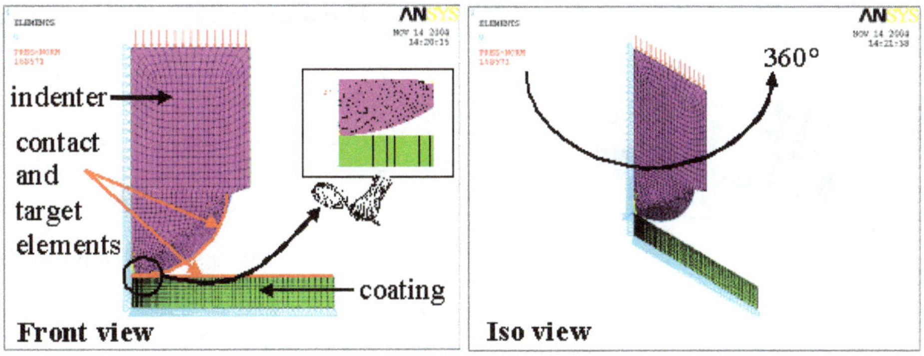

Fig. 5. Axisymmetrical finite element model.

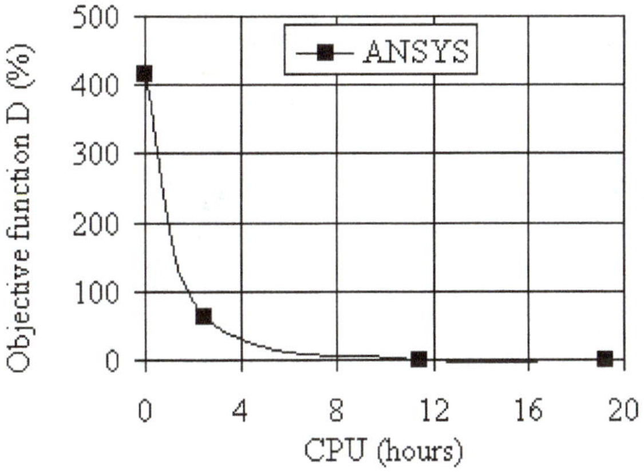

Fig. 6. Computation time.

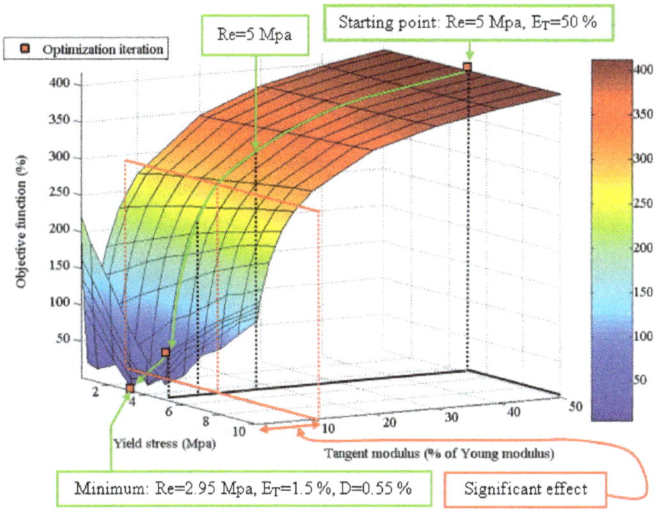

Fig. 7. Objective function versus design variables.

The first order optimization method was applied to achieve the optimal solution. This method combines a conjugate gradient strategy to determine the descent direction and a golden section algorithm to perform a line search along the descent direction. It is noted in Figure 6 that nearly twenty hours are required to achieve the optimal solution on a PC (Pentium 4/2.8 GHz). The optimization procedure is thus computationally demanding. However, it is possible to reduce the computation time by about a half (11 h instead of 19 h) without losing accuracy ($D = 0.86$ instead of 0.55%). Moreover, the method is efficient and provides accurate results as it will be demonstrated in the next section.

4 Computational results

4.1 Plastic parameters

The objective function surface is depicted in Figures 7 and 8. This surface was built by computing hardness against design variables (σ_y, E_T) satisfying 0.5 Mpa $\leqslant \sigma_y \leqslant$ 10 Mpa and 1% $\leqslant E_T/E \leqslant$ 50%. The iterative solutions provided by the optimization procedure are represented by squares. It is observed that the procedure converges perfectly well towards a minimum defined by: $\sigma_y = 3$ MPa and $E_T = 1.5\%\ E$. It is also noticed that the first iterations move from the starting point along a straight line defined by $\sigma_y = 5$ MPa. The objective function is thus more sensitive to the tangent modulus than the yield stress. The most significant effect is found for E_T changing from 0 to 10% E corresponding to the stiffest slope of the objective function. This observation is in agreement with the results obtained by Gan et al. in [20]. It is also noticed that there is only one global minimum in the considered range of parameters variation. Non-uniqueness, i.e. identical hardness's obtained from different combination of plastic parameters [21–24], is not observed. Since two thicknesses are used to perform the reverse analysis, the proposed approach likely works as the dual method [16] used for sharp indentation to provide a single combination of appropriate material parameters. It is actually noticed that two different thicknesses provide two different load-indentation depth curves (Fig. 9). This ensures that the relation between material parameters and hardness is one-to-one.

4.2 Hardness against thickness

The plastic parameters identified in Section 4.1 and the FE model described in Section 3.1 were used to compute hardness against coating thickness. Comparison between

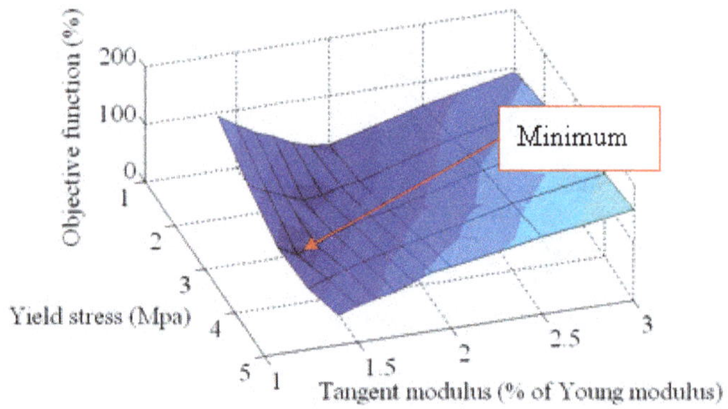

Fig. 8. Zoom on the minimum of the objective function.

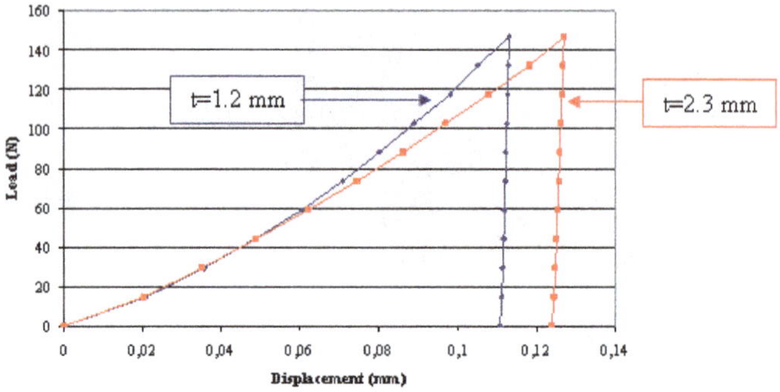

Fig. 9. Numerical load-indentation depth curves.

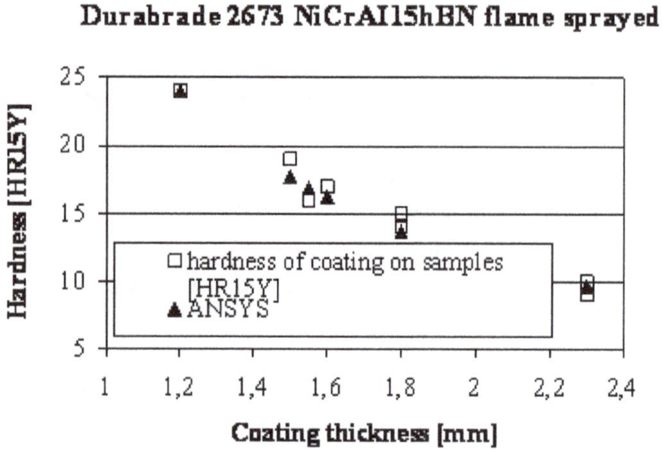

Fig. 10. Comparison between measured and computed hardness.

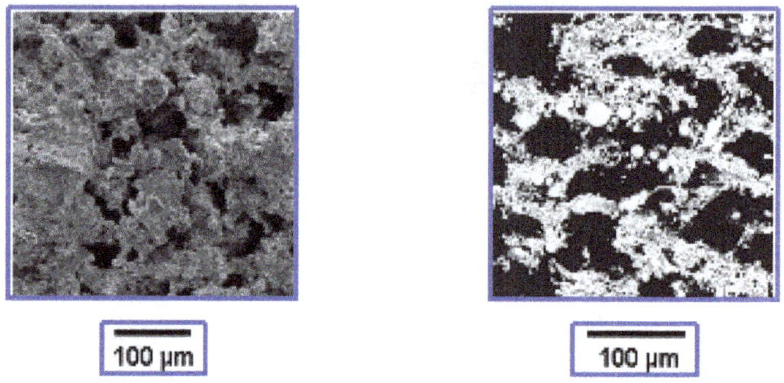

Fig. 11. Computed hardness.

measured and computed hardness is presented in Figure 10. A perfect agreement is noticed if the coating thickness is equal to 1.2 or 2.3 mm. This result was expected because these thicknesses were used to equate to zero the difference between tested and computed hardness (see Sect. 3.2). A very good agreement between numerical and experimental results is also observed for other thicknesses since the difference never exceeds 6.7% (Tab. 1). If the thickness is less than 3 mm, it is noted that hardness decreases sharply with the thickness up to a minimum value (Fig. 11). On the contrary, if the thickness is more than 3 mm, the hardness firstly increases and then reaches a thickness independent value. The variability of hardness against thickness has been already observed for real material and is not induced by a numerical artefact [25].

4.3 Relationship between hardness measurements and residual stresses in the coating

It was observed in Figure 11 that the hardness reaches a minimum for a thickness value equal to 3 mm. By investigating the relationship between hardness and stress field, we will demonstrate in this section that the minimum hardness is produced by the boundary conditions influence on hardness measurement.

It is firstly noticed that the central part of the coating is the most significant region to account for plastic effects. The Von Mises plastic strain is actually concentrated within the range $x \in [0\text{–}2.54 \text{ mm}]$ (Fig. 12). In this region, the plastic strain extends radially from the contact zone to the coating bottom. This general trend of radial expanding has been already observed in [5,19,26]. By focusing on the central part of the coating, it is noted that the coating thickness giving the maximum reaction force is equal to 3 mm (Fig. 13). This corresponds to the thickness value providing the minimum hardness (Fig. 11). Since the reaction force results from the constrained displacement applied to the coating bottom, this demonstrates that hardness variability, and particularly occurrence of minimum hardness, are ascribed to the boundary condition influence on hardness measurement. The distribution of the vertical stress component σ_{yy} on the coating bottom was also studied to understand why the maximum force is related to a 3 mm coating thickness. Three stress curves are plotted against x coordinate in Figure 14. These three stress curves are related to a 3 mm thickness and to two thicknesses close to 3 mm (2.5 and 3.5 mm). These curves intersect each other nearly x equal to 1.6 mm. The lowest thickness provides the lowest stress value up to the intersection while the opposite trend is observed beyond the intersection. A similar conclusion was found for the reaction forces acting within the range $x \in [0,1.6 \text{ mm}]$ and $x \in [1.6, 2.54 \text{ mm}]$ (Fig. 15). The maximum force then results from the competition between two opposite trends and depends on the stress distribution on the coating bottom. The minimal hardness observed in this work

Table 1. Comparison between measured and computed hardness.

Coating thickness [mm]	Measured hardness	Computed hardness	Difference (%)
1,2	24,1	24	0,4
1,5	19	17,72	6,7
1,55	16	16,85	−5,3
1,6	17	16,2	4,7
1,8	14,5	13,62	6,1
2,3	9,5	9,6	−1,1

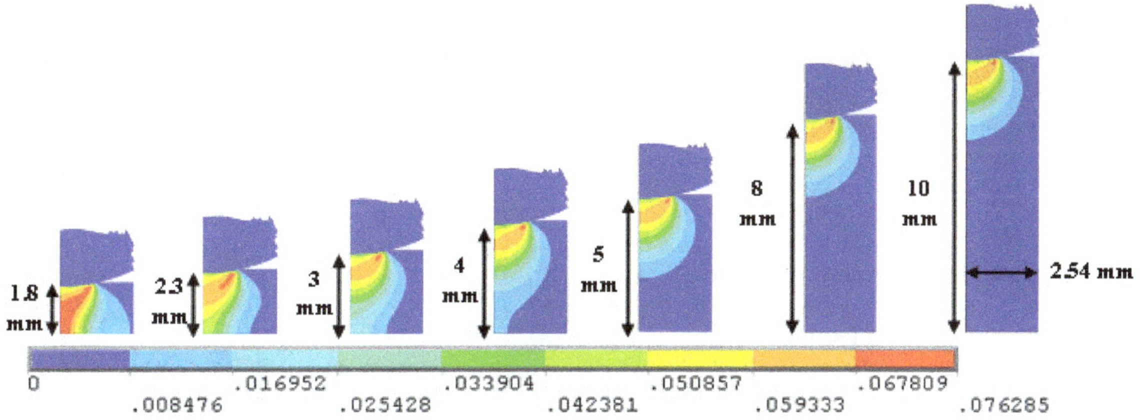

Fig. 12. Von Mises plastic strain component for various thicknesses.

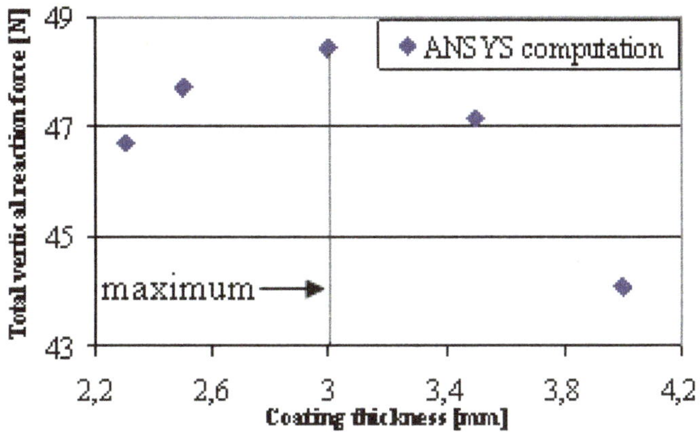

Fig. 13. Reaction force applied to the coating bottom within the range $x \in [0{,}2.54]$.

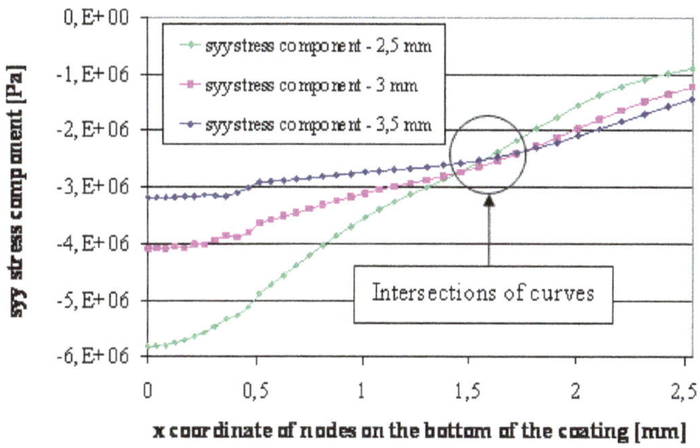

Fig. 14. σ_{yy} stress component distributed on the central part of the bottom coating.

is thus produced by the boundary conditions influence on hardness measurement.

5 Conclusion

A finite element model of the HR15Y hardness test was developed to identify the plastic parameters of abradable materials. To provide a general approach applicable to any kind of material, including the abradables, the parameters identification was performed by connecting optimization algorithms to FE analysis. The first order optimization method was used in this work. This method is time consuming but has been found accurate and efficient to provide the global minimum of the objective function. In the considered range of parameters variation, non-uniqueness has not been observed (i.e. identical hardness's obtained from different combination of plastic parameters). Since two different load-indentation depth curves corresponding to two different thicknesses are considered for the reverse analysis, the proposed approach likely works as the dual method used for sharp indentation to provide a single combination of appropriate material parameters.

The plastic parameters were used to study the variation of hardness with the coating thickness. The results show a fairly good agreement between computed and measured hardness. However, a minimum hardness value was

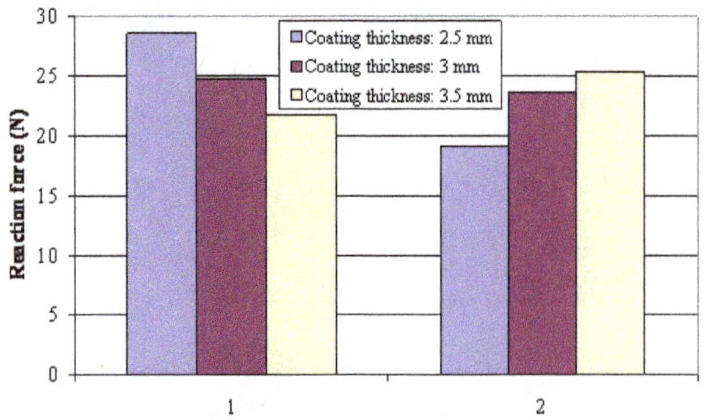

Fig. 15. Reaction force applied to the coating bottom within the range: (1) $x \in [0, 1.6]$ – (2) $x \in [1.6, 2.54]$.

found while it was expected that hardness should be always decreasing with thickness. By focussing the study on the central part of the coating, which is the most significant region to account for plastic effects, we have demonstrated that this minimum hardness is produced by the boundary conditions influence on hardness measurement.

The good agreement found between numerical and experimental results demonstrates that a bilinear hardening model is sufficient to simulate the loading cycle of the HR15Y hardness test applied to the abradable layers studied in this work. This simple model may be proposed as a starting assumption for the simulation of other experiments, to be validated afterwards. However, in order to investigate more accurately a widest range of materials and loading conditions, the proposed method could be easily extended to multilinear plastic materials. To include such materials in the identification process, the number of design variables used for the optimization stage should be increased and a multilinear plastic law used rather than a bilinear one for the finite element analysis.

Acknowledgements

This research work was completed from May 2002 to May 2006 within the European Seal-Coat project funded by the European Commission under the FP5 Growth Program. The Seal-Coat project has involved UTBM (Peter Chandler, in charge of the Project Management, and the authors of this paper) and the following partners:

ESIL, Dublin/IE (Barry O'Reilly)
Rolls-Royce plc, Derby/GB (Noel Hopkins and Chris Sellars)
MTU Aero Engines GmbH, München/DE (Manuel Hertter and Ulrike Hain)
Euromat GmbH, Heinsberg/DE (Ulrich Morkramer)
RWTH Aachen University, Aachen/DE (Jochen Zwick)
Neomet Limited, Manchester/GB (Dieter Sporer)
Institute of Plasma Physics, Praha/CZ (Jiri Matejicek)
RALSA, Langreo/ES (Ramiro Rilla)

References

1. R.K. Schmid, F. Gharispoor, M. Dorfman, X. Wei, An Overview of Compressor Abradables, *International Thermal Spray Conference 2000* (Montréal, 2000) pp. 1087–1093
2. J. Mackerle, Finite Elements in Analysis and Design **37**, 811 (2001)
3. Y.P. Cao, J. Lu, Acta Materialia **52**, 4023 (2004)
4. J.L. He, S. Veprek, Surface and Coatings Technology **163–164**, 374 (2003)
5. X. Cai, H. Bangert, Thin Solid Films **264**, 59 (1995)
6. A. Nayebi, O. Bartier, G. Mauvoisin, R. El Abdi, Internat. J. Mech. Sci. **43**, 2679 (2001)
7. I.H. Choi, C.H. Lim, Composite Structures **66**, 125 (2004)
8. J.M. Olaf, E. Sommer, Finite-element analysis of indentation experiments in surface-coated materials, *Experimental mechanics* (1993), pp. 201–204
9. M. Zhao, N. Ogasawara, N. Chiba, X. Chen, Acta Materialia **54**, Issue 1, 23 (2006)
10. H. Lee, J.H. Lee, G.M. Pharr, J. Mech. Phys. Solids **53**, 2037 (2005)
11. K.C. Tang, R.D. Arnell, Thin Solid Films **355–356**, 263 (1999)
12. S. Kucharski, Z. Mroz, Materials Sci. Eng. **A318**, 65 (2001)
13. M. Beghini, L. Bertini, V. Fontanari, Internat. J. Solids Struct. **43**, 2441 (2006)
14. K.D. Bouzakis, N. Michailidis, G. Erkens, Surface and Coatings Technology **142** 102 (2001)
15. ANSYS HTML Online Documentation, Release 8.1, 2004
16. S. Swaddiwudhipong, K.K. Tho, Z.S. Liu, K. Zeng, Internat. J. Solids Struct. **42**, 69 (2005)
17. ASTM International, Standard Test Methods for Rockwell Hardness and Rockwell Superficial Hardness of Metallic Materials, E18-05e1, 2005
18. H.I. Faraoun, J.L. Seichepine, C. Coddet, H. Aourag, J. Zwick, N. Hopkins, D. Sporer, M. Hertter, 2006, Surface and Coatings Technology **200**, Issues 22–23, 6578 (2006)
19. L. Ma, S.R. Low, J. Song, Journal of Testing and Evaluation **31**, 514 (2003)
20. L. Gan, B. Ben-Nissan, Computational Materials Science **8**, 273 (1997)
21. M. Zhao, N. Ogasawara, N. Chiba, X. Chen, Acta Materialia **54**, Issue 1, 23 (2006)
22. N. Ogasawara, N. Chiba, X. Chen, Scripta Materialia **54**, 65 (2006)
23. K.K. Tho, S. Swaddiwudhipong, Z.S. Liu, K. Zeng, Materials Sci. Eng. A **390**, 202 (2005)
24. Y.-T. Cheng, C.-M. Cheng, Mat. Sci. Eng.: R: Reports **44**, 91 (2004)
25. D.-C. Lim, G.-C. Chen, S.-B. Lee, J.-H. Boo, Surface and Coating Technology **163–164**, 318 (2003)
26. N. Ye, K. Komvopoulos, Journal of Tribology, **125**, 685 (2003)

4

A new algorithm for the problem of robust single objective optimization

A. Noriega[a], R. Vijande, E. Rodríguez, J.L. Cortizo and J.M. Sierra

Department of Mechanical Engineering, University of Oviedo, Gijón, Spain

Abstract – This paper propounds a new algorithm, the Sub-Space Random Search (SSRS) for the problem of single-objective optimization, with the aim of improving the robustness and the precision of classical methods of global optimization. The new algorithm is compared with a genetic algorithm (GA), on a set of four scaleable test functions and with the number of variables changing from 1 to 5. A new test function called Deceptive-bimodal (DB) is proposed. Results indicate that, with the same total number of function evaluations, SSRS is about 50% faster than GA. Moreover, SSRS shows a greater precision and similar ability to find the global optimum than GA with 1, 2 and sometimes 3 variables. But this advantage diminishes when the number of variables increases on multimodal and narrow-flat valley functions. Finally, SSRS is successfully applied to a problem of dynamical synthesis of a mechanism.

Key words: Meta-heuristic; unconstrained optimization; stratified random search; synthesis of mechanisms.

1 Introduction

Stratified Sampling Methods are widely used today in the design of experiments and allow the amount of important information which can be extracted from the output data of a process to be maximized, see [1]. The application of these methods with refinement on global optimization problems allows regions with a high probability of containing the global optimum to be identified. Moreover, the finer the stratification is, the minimum is delimited with more precision.

The paper propounds a meta-heuristic optimization algorithm based on search space stratification similar to the methods cited before. But the sampling and the choice of the zone where the resolution must be increased are completely different to those methods.

The paper is organized as follows: in Section 2, a detailed description of algorithm implementation and its basic concepts is given. In Section 3, the experimental validation of this algorithm is described. Test conditions and functions are indicated in it, including a new test function proposed and an application to a problem of dynamical synthesis of a mechanism. In Section 4, the results are shown and discussed, extracting some conclusions about the new algorithm and possible future work-lines in Section 5.

[a] Corresponding author: noriegaalvaro@uniovi.es

2 Algorithm description

For this algorithm, it is supposed that the optimization problem is formulated as follows:

$$\min f(\boldsymbol{x}) \tag{1}$$

with $\boldsymbol{x} \in \Re^n$ and $l_i \leqslant x_i \leqslant u_i$ being $i = 1, ..., n$.

The algorithm Sub-Space Random Search (SSRS) is based on the hypothesis that there is a search space stratification in sub-spaces of equal size and whose union makes the initial space of every iteration. In every sub-space, a sample of individuals is generated and evaluated to get a measure of the behaviour of the objective function in this sub-space. Selecting the most promising sub-space and re-initiating the process on it is proposed to increase the resolution in the optimum search. Figure 1 shows a graphic representation of this idea with an example of 2 variables.

However, to do this refinement process, it is necessary to determine a reference value which estimates the probability that a sub-space contains the global minimum. This reference value allows it to be compared with other sub-spaces and thus being able to select the most suitable to continue the process. To calculate this reference value, a random sample of individuals is generated and evaluated in every sub-space because a random sample is the easiest way of generating individuals with the sample size being a parameter defined by the user. Furthermore, this random sample has the same average probability of approaching the unknown minimum of the sub-space as any other method of generating individuals as is shown in [2]. Finally, this randomized approach to estimate the

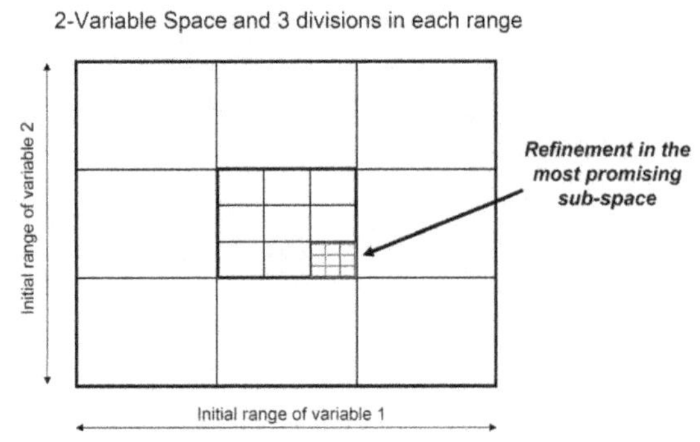

Fig. 1. Reiterated stratification and selection of the most promising sub-space.

reference value, is similar to Monte-Carlo techniques used in robust optimization [3].

Then, the minimum value of the individuals' evaluations can be taken as a reference value for the comparison of sub-spaces. Or the average value of the evaluations of all the individuals of the sample can be calculated. A priori, it is not known which one is the best option to calculate the reference value because it depends on the landscape of the objective function in this sub-space. For a continuous and monotonous function, a sample with few individuals generates a probably correct estimation of the function behaviour in the sub-space. But the same thing does not happen with a multimodal function (with several minima) since there exists the probability of not generating individuals in the influence area of the global minimum of sub-space when it represents a small percentage regarding the whole space. For the influence area of a minimum, we understand the minimum's neighbourhood where the function is unimodal.

The problem is that, for having a precise description of the behaviour of the function in the sub-space and, therefore, the high security that the election of the most suitable region is correct, it is necessary to increase the size of the random sample a great deal which will penalize the general efficiency of the search. However, this action improves the robustness of the search to Type II variations just as they are referred to in [4].

The number of selected individuals to calculate the reference value (parameter p) can be modified to compensate that difficulty without increasing the size of the sample in every sub-space. If the sample is poor, a small number of individuals will be selected. In the case of having a big sample size, the number of selected individuals must increase to improve the security of the estimation.

But the landscape of the objective function in the sub-space also has an influence on choosing the number of individuals selected to calculate the reference value. If there are sharp minima, it is necessary to generate a sample with a big number of individuals and select a small number of them to increase the probability of obtaining almost one individual closer to these minima.

These values directly influence the efficiency of the algorithm and its robustness and globality in an opposed way. However, it would also allow the search towards a type of certain minimum (more or less sharp) to be guided.

The stratification of the search space is an important factor since the more divisions in the range of every variable, the finer the global stratification is. Therefore, the search is more global because the complexity of the function in every sub-space diminishes and, then, it is easier to have a correct evaluation of them. But it is a very delicate parameter since it has a negative influence on the efficiency of the algorithm, because it increases the number of sub-spaces exponentially with the number of variables. For instance, if there are three divisions in every variable's range of a two-dimensional space, the result is $3^2 = 9$ sub-spaces to study, but if the space is three-dimensional, the result is $3^3 = 27$ sub-spaces. Therefore, this is the most critical parameter in the algorithm and hence it would limit its application to optimization in spaces with many variables.

Thus, the sample's individuals are evaluated to calculate three values in every sub-space: the smallest value obtained in the objective function (bvs), the individual of the sample corresponding to this minimum (bis) and the comparison value for the sub-space (cvs). This last value is calculated taking the mean of the p % of smallest values of the objective function, with p being a parameter defined by the user.

Once these three values of every sub-space have been obtained, the sub-spaces are compared among themselves to obtain the final values of this iteration: the best individual obtained (bii), its corresponding value of the objective function (bvi) and the most promising sub-space to look for the global minimum of the function (psi). It can be seen in Figure 2 with an example.

The sub-space with smaller cvs is considered as the most promising and the process explained above is repeated again on it. The number of times that this refinement process is repeated is externally set by the user, so that in every step, the precision of the obtained solution increases. The flowchart of the algorithm is shown in Figure 3.

Then, the proposed algorithm has 4 parameters:

Number of divisions per variable (ndv): this parameter is related with the complexity of the objective function and it is good that it is a small integer number because the number of sub-spaces depends on it by means of the following equation:

$$t = ndv^n \tag{2}$$

where n is the number of variables.

Population size (ps): this parameter indicates the number of individuals randomly generated in every sub-space. It measures the quality of the function description in every sub-space and it is also related with the complexity of the objective function.

% pop for reference value (p): this parameter indicates the % of the sample smallest values taken to calculate the comparison value of every sub-space (cvs). This value

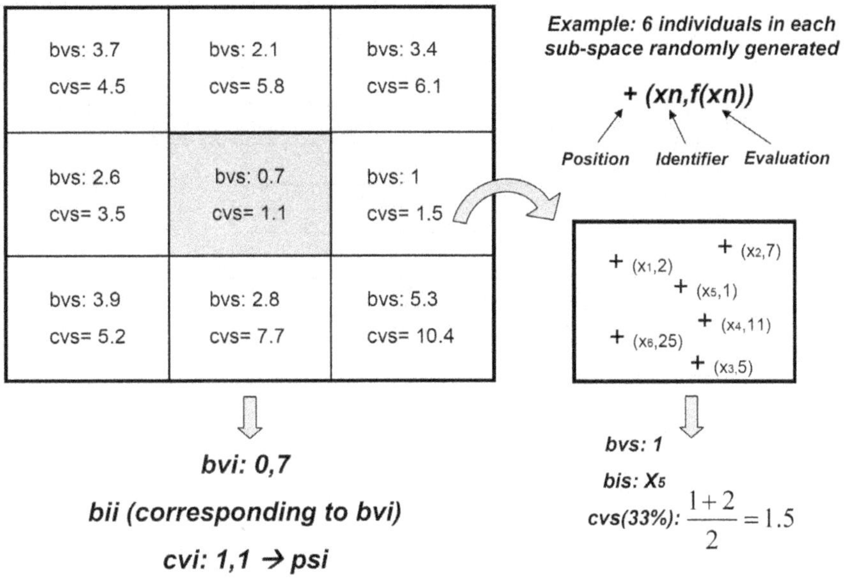

Fig. 2. Sub-space evaluation and selection of the most promising sub-space.

is calculated by means of the average of selected values.

Number of iterations (*itermax*): this parameter indicates the number of iterations that will be carried out; it must not be not too big an integer because this algorithm converges very fast to the minimum.

The best individual obtained (*bii*) will usually be in the most promising sub-space (*psi*) but it may not be so. In this case, it is recommended to increase the population size because the function can have a very complicated landscape in the search space.

The whole explained process has the advantage of allowing a very simple implementation. Furthermore, it also reduces the number of the individuals' comparisons to small groups corresponding to the sub-spaces populations and later among sub-spaces. The result is that the general speed of this algorithm is increased.

3 Experimental validation

The validation of SSRS will be done by means of the benchmarking of this algorithm with another similar algorithm. In the field of robust optimization, there are two main philosophies [5]. The first one is called *simplification strategy* whose aim is the transformation of the problem to another one that can be solved using standard techniques of mathematical programming. Its drawback is that these techniques are usually of local search and they need additional information like derivatives of objective function. On the other hand, there exist the *simulation optimization techniques* which are less efficient than the first ones but they do not need additional information and, moreover, they can use randomized approaches, like Genetic Algorithms, which can do a global search.

Because of the populational nature of SSRS and its general guidance, it is considered to compare it with an algorithm with similar features, selecting a Genetic Algorithm (GA) as a competitor. In this case, Genetic Algorithm and Direct Search Toolbox of MATLAB ® is used.

3.1 Test functions

To test the SSRS, two types of test are proposed. First, standard test functions are used. To select them, the basic features of a function to optimize are identified. Then, four benchmark functions which represent some of these features have been selected [6]. These functions will be scaleable, i.e., the number of variables of the function (n) can be changed but the features are maintained. These features are:

Number of minima: the number of local and global minima and their relative values determine the complexity of the search and they allow the robustness of the optimization techniques to be verified. A distinction can made between Unimodal (a single minimum) and Multimodal (two or more minima). For these types, De Jong's function F1 is selected as unimodal function (UM) and Rastrigin's function as multimodal function (MM).

Narrow and flat valley: the minimum is located in a very narrow and sharp valley with an almost flat bottom in which is very difficult to determine whether you have arrived or not at the proximity of the minimum. For this type, Rosenbrock's function (NFV) is selected.

Flat surfaces: flat surfaces are obstacles for optimization algorithms based on derivatives, because they do not give any information of which direction is favourable.

Deception: this feature arises when in a multimodal function the global minimum influence area represents a small

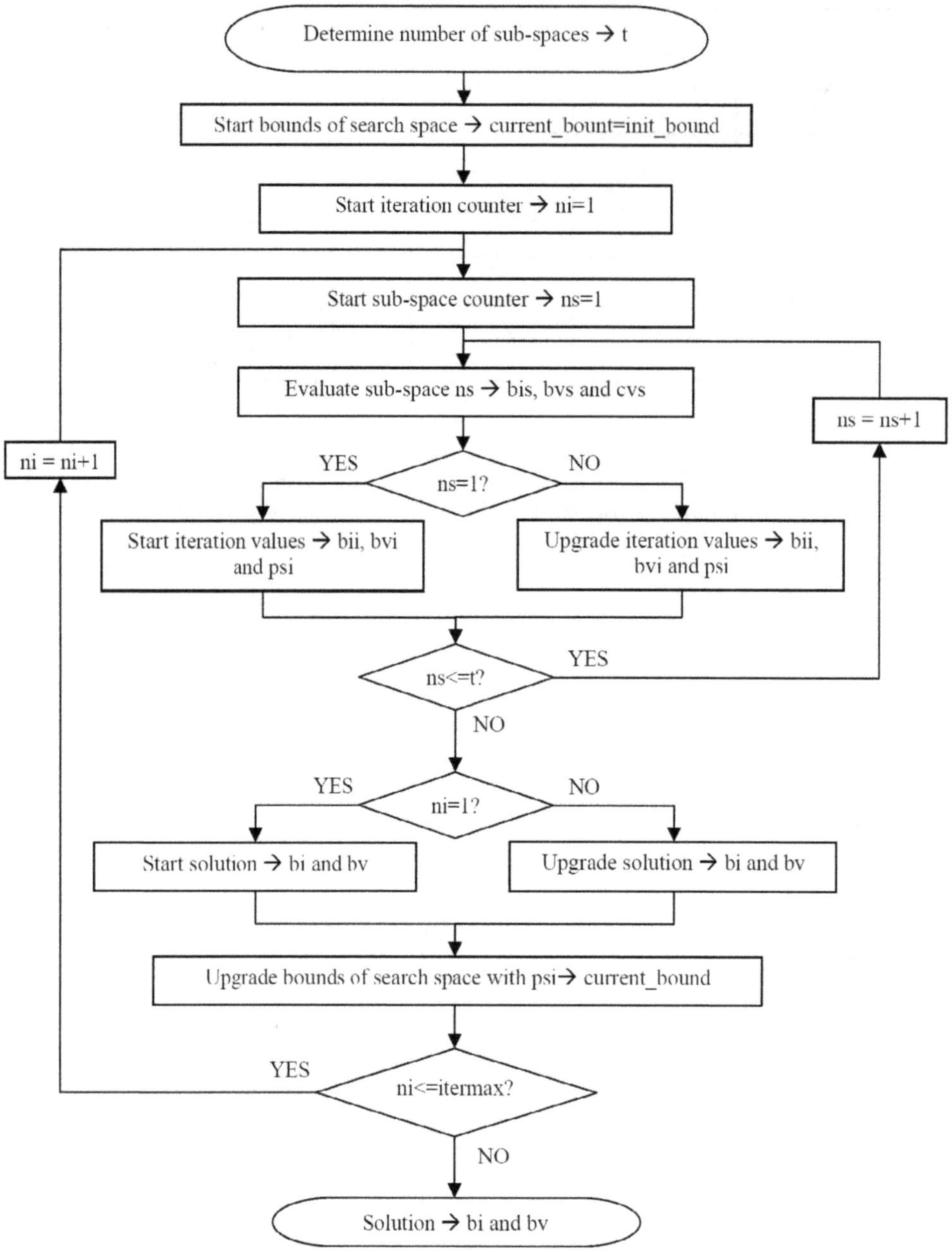

Fig. 3. Flowchart of SSRS.

percentage of all the search space which is dominated by local minima influence areas. This feature complicates the search a great deal and it allows the robustness of an optimization technique to be verified. For this type, the Deceptive-bimodal function (DB) is proposed, inspired by one in Deb [7] but with different parameters and modified to make it scaleable. It is a bimodal function with a local minimum in $x_i = 7$ with a big influence area and a global minimum in $x_i = 1$ with a small influence area and where n is the number of variables. A graphic representation in 2 variables is shown in Figure 4.

$$f(\boldsymbol{x}) = 1.2 - e^{-10\cdot\sum_{i=1}^{n}(x_i-1)^2} - 0.7\cdot e^{-0{,}1\cdot\sum_{i=1}^{n}(x_i-7)^2} \qquad (3)$$

$$0 \leqslant x_i \leqslant 10.$$

The function value in the global minimum is $0.2 - 0.7 \cdot e^{-3.6\cdot n}$.

In second place, a real optimization problem is proposed. In this case an application to a dynamical synthesis of a mechanism is selected. The mechanism proposed is a very simple system composed by a mass, a spring and a damper, see Figure 5. The mass can move along the Y axis by means of a vertical guide. The problem consists in adjusting the values of the mass (M), the spring stiffness (K) and the damper coefficient (C) so that the acceleration in Y of the mass has a certain response regarding time. To define this response and to ensure that the problem has a well-known solution, a previous simulation of the system is made with the following variables: $M = 30$ kg, $K = 3$ N mm^{-1} and $C = 0.1$ N s mm^{-1}. A

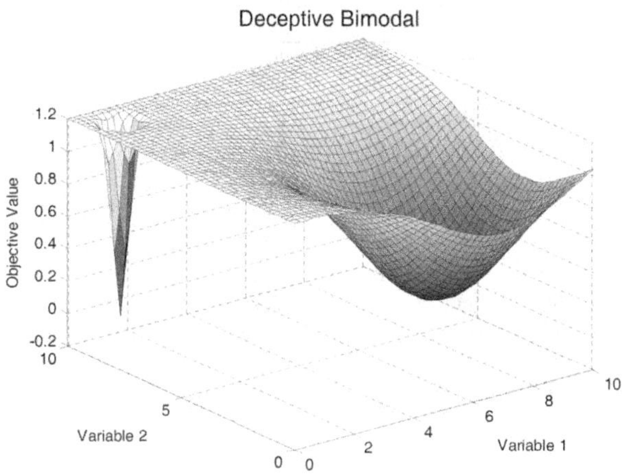

Fig. 4. Deceptive-bimodal function in 2 variables.

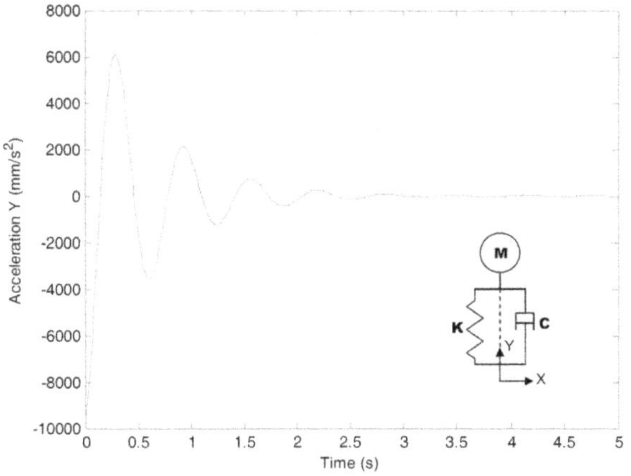

Fig. 5. Dynamical system and desired response.

graphical view of this response can be seen in Figure 5. The objective is to minimize the average of absolute errors in every simulation frame between the obtained and the desired responses. The bounds of the variables are the followings: $M \in [10\ 50]$, $K \in [1\ 10]$ and $C \in [0.01\ 1]$.

3.2 Test conditions

A benchmark comparison between the SSRS and a GA is made on the group of test functions cited above. With the standard test functions, the number of variables changes from 1 to 5 in a similar way as can be seen in Elbeltagi [8] except for Rosenbrock's function which varies from 2 to 5 variables because it is not defined for 1 variable. Experiments are repeated 100 times. Both the average error and its standard deviation and the average run time are calculated as in Deb [9].

Parameters which control the population size and the generations in both algorithms are tuned so that, keeping logical values for every test problem, the total number of function evaluations is equal in both algorithms. These parameters are shown in Table 1.

In the case of dynamical synthesis, there is a simulation of the behaviour of the system done with ADAMS ® in the objective function. Since the time cost of the objective function evaluation is around 4 seconds, only one run with both algorithms will be done. The problem is solved with SSRS with the following parameters: $ndv = 3$, $ps = 20$, $p = 10\%$ and $itermax = 4$, using 2160 function evaluations while GA uses a population size of 48 individuals and 45 generations with the same total function evaluations.

The implementation of SSRS and benchmark test has been carried out in MATLAB ® 7.0. The computer used for the tests has a processor Intel ® Celeron ® 2.0 GHz and 512 Mb of RAM.

4 Results and discussions

The outcomes of the benchmarking with the four standard test functions are the mean and the standard deviation of the absolute error and the mean of runtime. The error mean gives an idea of the value of the minimum reached. It can be related with the globality and the precision of the algorithm while the standard deviation gives an idea of the repeatability of the algorithm. In the case of runtime value, comparison among the averages obtained by the algorithms gives an idea of the cost of their internal operations since the number of function evaluations are the same in both cases.

Thus, the most important result which can be seen in Table 2 is that SSRS is faster than GA for all the cases. SSRS run time is about 40–60% less than GA run time with an equal number of function evaluations over 3, 4 and 5 variables. This ratio is even better for 1 and 2 variables.

Watching mean and standard deviation of error, a different behaviour is observed in different test functions. On the UM function, algorithm SSRS shows a greater precision than GA. In NFV function, SSRS presents a better behaviour than GA for 1, 2 and 3 variables but for 4 and 5 this tendency is inverted and the advantage of GA over SSRS increases as the number of variables increases.

In the MM function something similar happens. With 1 and 2 variables, SSRS does a more global search and obtains better precision than GA but this behaviour changes when the number of variables increases. In this case, GA works better with more variables while SSRS shows the opposite behaviour.

In the DB function, both algorithms present a similar behaviour, converging to the local minimum almost every time but SSRS has better repeatability than GA.

To have an idea about the robustness it would be possible to see the evolution of the *cvs* in the most promising sub-space and to relate this value with the size of the sub-space in every iteration.

For the problem of dynamical synthesis, the SSRS obtains the best solution $M = 31.269$, $K = 3.132$ and $C = 0.103$ with an error of 7.066. The GA obtains the best solution $M = 49.260$, $K = 4.912$ and $C = 0.167$ with an error of 13.351. Then, the SSRS is clearly better than the GA obtaining the correct solution with more precision than the GA and in a more efficient way.

Table 1. Algorithm parameters.

Function	Var.	SSRS				GA			Function Eval.
		ndv	ps	p	itermax	Pop	Gen	Elite	
UM	1	2	10	25	5	10	10	1	100
	2	2	10	25	5	20	10	1	200
	3	2	20	25	5	50	16	1	800
	4	2	50	25	5	100	40	1	4000
	5	2	100	25	5	200	80	1	16000
NFV	2	3	20	10	2	20	18	1	360
	3	3	50	10	2	50	54	1	2700
	4	3	100	10	2	100	162	1	16200
	5	3	200	10	2	200	486	1	97200
MM	1	3	20	10	2	10	12	1	120
	2	3	20	10	2	20	18	1	360
	3	3	50	10	2	50	54	1	2700
	4	3	100	10	2	100	162	1	16200
	5	3	200	10	2	200	486	1	97200
DB	1	5	20	1	2	20	10	2	200
	2	5	20	1	2	50	20	2	1000
	3	5	40	1	2	100	100	2	10000
	4	5	60	1	2	500	150	2	75000
	5	5	80	1	2	1000	500	2	500000

Table 2. Results.

Function	Var.	SSRS			GA		
		Error Mean	Error Std	Time Mean	Error Mean	Error Std	Time Mean
UM	1	0.0023	0.004735	0.0175	0.0697	0.156271	0.1267
	2	0.0122	0.011414	0.0371	0.2137	0.298253	0.1356
	3	0.0172	0.011193	0.1223	0.2611	0.238042	0.2717
	4	0.0233	0.012047	0.5502	0.2076	0.154413	1.0549
	5	0.0277	0.012261	2.0028	0.2083	0.141052	3.8506
NFV	2	0.2501	0.351123	0.0498	0.2701	0.292721	0.1679
	3	0.8793	0.274383	0.3312	1.1223	0.875923	0.7321
	4	1.8731	0.398454	1.7983	1.3866	1.081146	3.8459
	5	3.0801	0.410591	10.8642	1.7645	0.969627	20.0491
MM	1	0.3578	1.133234	0.0183	1.5934	1.618239	0.1411
	2	2.1151	1.719442	0.0619	2.7411	2.079256	0.1869
	3	2.3418	1.001727	0.4152	1.7676	0.917453	0.8573
	4	3.3164	1.117348	2.5969	0.9946	0.664095	4.5358
	5	4.1791	1.206706	14.2511	0.3617	0.284541	25.9321
DB	1	0.1216	0.112855	0.0339	0.1381	0.099891	0.1133
	2	0.2011	0.000224	0.1725	0.1931	0.031975	0.3522
	3	0.2005	0.000288	1.6128	0.1985	0.017446	2.6995
	4	0.2009	0.000449	11.6342	0.2039	0.004432	19.1681
	5	0.2013	0.000572	75.7334	0.2052	0.014621	124.8826

5 Conclusions and future work-lines

A new meta-heuristic algorithm, the Sub-Space Random Search, is proposed for the global optimization problem and a benchmarking with a GA is made to estimate its behaviour. This benchmark is made on a basic casuistry of scaleable optimization problems with the number of variables changing from 1 to 5.

Results indicate that SSRS shows a greater precision than GA on unimodal functions. In multimodal and narrow and flat valley functions its precision is better on low dimensions but it get worse as the number of dimensions increases. In deceptive function the behaviour of both algorithms is similar. An application of SSRS to a problem of dynamical synthesis of mechanism also shows the better performance of this algorithm regarding a GA in problems with few variables. Furthermore, SSRS is faster than GA in all the cases studied and it is worth noting the great simplicity of the code needed to implement this algorithm.

Then, the SSRS is showed as a good algorithm for relatively easy problems (unimodal and low multimodality functions) with few variables since it can get the global optimum and it also allows the robustness during the process to be studied.

At present, the authors are working on improving this behaviour when the number of variables increases and on implementing a parallel search in several equally promising sub-spaces.

Acknowledgements

The first author is supported by the Ministry of Education of Spain (FPU grant AP-2004-6492)

References

1. C. Tong, Reliab. Eng. Syst. Saf. **91**, 1257 (2006)
2. D.H. Wolpert, W.G. MacReady, IEEE Trans. Evol. Comput. **1**, 67 (1997)
3. S. Andradóttir, *A review of simulation optimization techniques*, Proceedings of the 1998 Winter Simulation Conference (Piscataway, NJ, 1998) pp. 151–158.
4. W. Chen, J. Allen, K.-L. Tsui, F. Mistree, J. Mech. Design **118**, 478 (1996)
5. H-G. Beyer, B. Sendhoff, Comput. Method. Appl. M. **196**, 3190 (2007)
6. D. Goldberg, *Genetic algorithms in search, optimization, and machine learning* (Addison-Wesley, New York, 1989)
7. K. Deb, Evol. Comput. **7**, 205 (1999)
8. E. Elbeltagi, T. Hegazy, D. Grierson, *Adv. Eng. Inform.* **19**, 43 (2005)
9. K. Deb, A. Pratap, S. Agarwal, T. Meyarivan, A fast and elitism multi-objective genetic algorithm: NSGA-II, Technical Report N° 2000001, Kanpur Genetic Algorithm Laboratory, 2000

5

Reliability approach for fibre-reinforced composites design

H. Dehmous[1,2], Hélène Welemane[1,a], Moussa Karama[1] and Kamel Aît Tahar[2]

[1] Laboratory of Engineering Production, National Engineers School of Tarbes, 47 Avenue d'Azereix, BP 1629, 65016 Tarbes, France
[2] Laboratory of Mechanics and Structures, University of Tizi-Ouzou, Algeria

Abstract – This study aims at investigating the mechanical reliability of unidirectional fibre-reinforced composites in view of their design optimization. Owing to the manufacturing process, many uncertainties affect inherently the properties of these materials. Besides, the brittle character of the fibre failure leads to important safety factors that limit their development for engineering applications. The objective is to get a realistic evaluation of the mechanical reliability of such structures which accounts for the various uncertainties involved. The effects of random design parameters on the composite failure are then investigated for different kinds of reinforcement (nature of fibre, mono-material or hybrid structures) with finite-element probabilistic solver PERMAS.

Key words: Composite materials; reliability engineering; mechanical failure; unidirectional fibre reinforcement; failure probability; sensitivity analysis

1 Introduction

Fibre-reinforced composites are gradually becoming more popular for several engineering applications and especially for mechanical structures by providing enhanced performances in comparison to classical materials and consequently new design perspectives [1]. Indeed, composite materials allow significant progress in weight-strength ratio, durability under thermal and fatigue solicitations and corrosion resistance which are of crucial interest in the fields of aeronautics (plane aileron, helicopter blade), transport (line shaft, suspension) or civil engineering (concrete pre-stressed cables, bridge suspension stays).

The counterpart of composite materials is related to the important scatter of their mechanical properties due to the manufacturing process. Besides, fibre materials (mostly used are carbon, Kevlar and glass) may exhibit a brittle behaviour for which failure occurs suddenly without critical signs. According to these uncertainties and risks, the traditional design of such structures takes into account important safety factors on fixed parameters (deterministic approach): at least a reduction of 50% of the characteristic strength for composite for only 30% in the case of steel. This severe safety margin generates a major increase of the structures dimensions and leads consequently to limiting costs for these design solutions.

In order to optimize this approach and then develop the use of composite materials for mechanical structures, it is essential to get a realistic evaluation of their reliability which especially accounts for the various uncertainties that affect their structural behaviour (material properties, loads and physical mechanisms of damage) [2–8]. In this way, reliability analysis offers a very useful tool in the conception phase as well as for the maintenance program:

- for structural design as it provides the range of use to achieve a specified reliability level, which consequently helps in the design optimization;
- for risk control on existing structures by evaluating the failure probability (equally the security level) and, if necessary, by defining the crucial parameters which mainly influence this phenomenon.

Such probabilistic analyses build a stochastic modelling that requires three important steps:

- first, the choice of input random variables that describe the various sources of uncertainties involved;
- then, the choice of a performance function (or failure criterion) which mathematically defines the structure failure;
- finally, the calculation of indicators (failure probability, failure index) which provide a quantitative and qualitative evaluation of the structure reliability.

In particular, the failure probability P_f of a structure according to a vector X of random variables representing the uncertainties in the model is defined by the following integral:

$$P_f = \int_{G(X)\leq 0} f_X(X)\; dX \tag{1}$$

where f_X denotes the associated distribution function and G is the performance function: $G(X) = 0$ is the limit

[a] Corresponding author: Helene.Welemane@enit.fr

P_f	failure probability
β	reliability index
X	random variables vector
f_X	associated distribution function
G	performance function
N	number of evaluations of function G
L	length of composite structure
ϕ (ϕ_A)	diameter (diameter of composite A in hybrid composite)
F	axial tension force in the longitudinal direction
q	uniform pressure in the transverse direction
$J_2(\overline{\overline{\sigma}})$	second invariant of the stress tensor $\overline{\overline{\sigma}}$
F_e	initial plastic yield (elastic limit)
E_i	Young modulus in direction i
ν_{ij}	Poisson ratio in plane ij
G_{ij}	Shear modulus in plane ij
σ_1, σ_2, τ_{12}	stresses components in the principal material directions
T1	tension ultimate stresses in the fibre direction
$T2$	tension ultimate stresses in the transverse direction
$C1$	compression ultimate stresses in the fibre direction
$C2$	compression ultimate stresses in the transverse direction
S	shear ultimate stresses

List of the symbols.

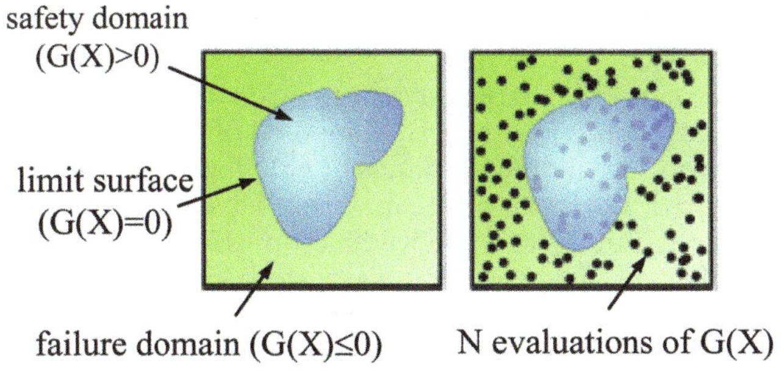

Fig. 1. Monte Carlo simulation method.

surface between the failure domain ($G(X) \leq 0$) and the safety domain ($G(X) > 0$).

The mechanical systems considered usually exhibit a degree of complexity (many random variables whose joint density of probability f_X is not available, integration domain with complex shape) that prevents the direct calculation of integral (1). Accordingly, various methods have been developed [9]:

- simulation methods (Monte Carlo for example) in which N evaluations of function G for various vectors realizations X are performed and treated with statistical methods (P_f is defined as the ratio of failing cases to N, Fig. 1). Such method gives in all cases a good estimation of P_f if the number of simulations N is significant, which remains usually quite time expensive;

- approximation methods which represent the limit surface with linear (FORM) or second-order (SORM) polynomials to define the reliability index β, that is the shorter distance between the origin and the failure domain in the normalized space obtained through an isoprobabilistic transformation (Fig. 2). Such approaches provide also the dependence of β according to the different random variables, which is very interesting for the designer and for risk control.

This paper aims at associating a mechanical modelling with a reliability approach on the case of unidirectional fibre-reinforced composites. The objective is to investigate the nature of the composite material used and for each case, the impact of the variable design parameters on the structure failure.

2 Structure description

For this study, we consider a cylindrical composite structure (length $L = 152$ mm, diameter $\phi = L/10$) composed of an epoxy matrix reinforced by unidirectional fibres in the longitudinal direction (Fig. 3). The cylinder is embedded on one side and subjected to two loads (Fig. 4):

- an axial tension force F in the longitudinal direction which represents the main solicitation of the structure;
- a uniform pressure q in the transverse direction that we consider as a secondary load.

In a civil engineering context, such a structure is for example representative of a bridge stay submitted to axial tension through the bridge deck and to bending induced by lateral wind.

In order to investigate the reinforcement nature, the reliability analysis is performed for three different constitutive materials:

- steel as a reference;
- composite materials with a single kind of fibre reinforcement;

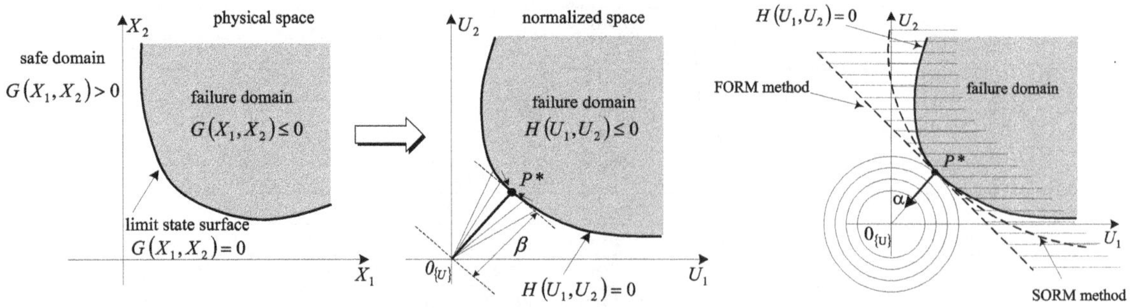

Fig. 2. Definition of the reliability index β and approximation methods FORM/SORM [9].

Table 1. Elastic properties of fibre-reinforced composites (epoxy matrix).

Nature of fibre	carbon HRT800	glass R	Kevlar 49
Elastic modulus E_1 in fibre direction (GPa)	135	52	72
Elastic modulus $E_2 = E_3$ in transverse direction (GPa)	10	13.8	5.5
Poisson ratio $\nu_{12} = \nu_{13}$	0.34	0.25	0.34
Poisson ratio ν_{23}	0.4	0.4	0.4
Shear modulus $G_{12} = G_{13}$ (GPa)	5	4.5	2.1

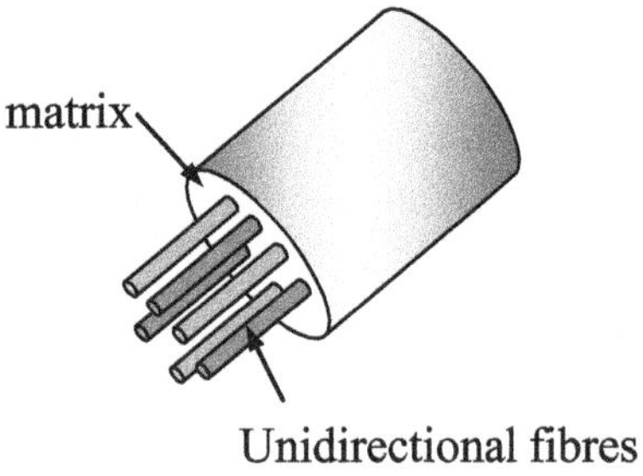

Fig. 3. Composite structure.

- hybrid material composed of two single fibre-reinforced composites.

We consider here an homogeneous isotropic elasto-plastic steel with density of 7.85. According to the low variability observed during experimental results, mechanical properties take the following fixed values:

- Young modulus: $E = 200$ GPa;
- Poisson ratio: $\nu = 0.3$;
- initial plastic yield (elastic limit) observed in an uniaxial tension test: $F_e = 1600$ MPa.

All composite materials are composed of 60% by volume of different unidirectional fibres in an epoxy resin matrix:

- carbon (density 1.6);
- glass (density 2);
- Kevlar (density 1.38).

Within the framework of this study, homogenized properties of these composite materials are considered (macroscopic description). As a matter of fact, they exhibit isotropic transverse anisotropy with axis of symmetry the fibre direction 1, then direction 2 = direction 3 (Fig. 5). Assuming stable values for elastic properties, data related to the three composite materials are specified in Table 1 [10].

The case of an hybrid structure made up of two coaxial cylinders (denoted A and B) composed of fibre-reinforced composites with different kinds of fibre for each is also investigated. Both composite materials have the same fibre direction, namely the longitudinal direction of the cylinders (Fig. 6). The composite A with diameter ϕ_A is embedded in the composite B with external diameter ϕ, and adhesion between A and B is supposed to be perfect.

3 Reliability model

As already mentioned, the reliability analysis requires first the choice of the random variables. Here the geometry, boundary conditions (connections), steel properties and composites elastic properties are considered as fixed parameters. Within random variables, we include the loads (F and q) and composites failure parameters that enter the failure criteria expressions (see Eq. (3)), as we consider that they play a crucial role in the structure reliability. The probabilistic model retained for each random variable should derive in practice from statistical studies carried out on sufficiently representative data. In this paper, the probabilistic data are inspired by the bibliography devoted to the composites [10]. Accordingly, all the random parameters used for simulations are considered as lognormal distributed random variables, characterized by a mean value and a standard deviation defined in Table 2.

The second stage consists in choosing a mathematical representation of the materials failure. In the case of steel, the failure domain is defined as the initial plastic domain

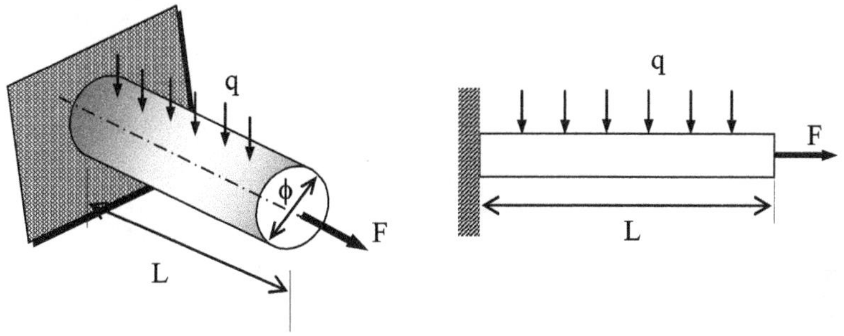

Fig. 4. Structure and loads.

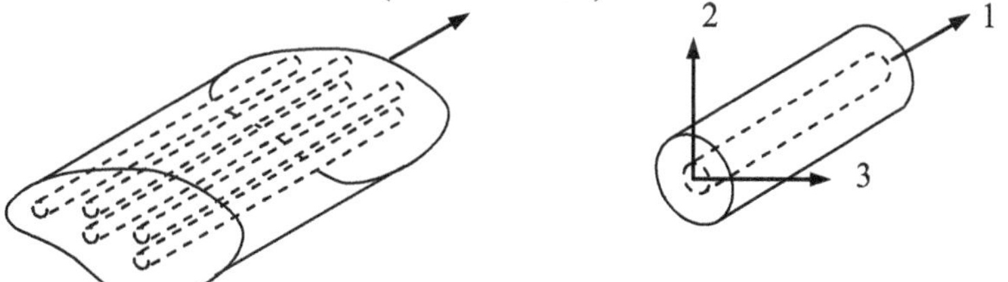

Fig. 5. Principal material coordinate system for unidirectional composite.

Table 2. Random variables.

	Random variables	Mean value		
Loads	F q	Various mean values investigated		
Composites		carbon-epoxy	glass-epoxy	Kevlar-epoxy
	$T1$ (MPa)	2550	1900	1380
	$C1$ (MPa)	1470	970	280
	$T2$ (MPa)	60	41	41
	$C2$ (MPa)	270	138	138
	S (MPa)	100	70	60

The standard deviation is equal to 10% for all random variables.

and described through the Von Mises criterion:

$$G(\overline{\overline{\sigma}}) = 1 - \frac{J_2(\overline{\overline{\sigma}})}{F_e} \tag{2}$$

where $J_2(\overline{\overline{\sigma}})$ denotes the second invariant of the stress tensor $\overline{\overline{\sigma}}$. For composite materials, many criteria for unidirectional brittle elastic composite materials have been proposed in the literature [11,12]. In this study, reliability analysis is performed with the following Tsai-Wu criterion [7,8] which allows to account for different behaviour in tension and compression:

$$\mathrm{G}(\overline{\overline{\sigma}}) = 1 - \frac{\sigma_1^2}{T1 \cdot C1} + \frac{\sigma_2^2}{T2 \cdot C2} + \frac{\tau_{12}^2}{S^2} - \frac{\sigma_1 \sigma_2}{\sqrt{T1 \cdot C1 \cdot T2 \cdot C2}} + \left(\frac{1}{T1} - \frac{1}{C1}\right)\sigma_1 + \left(\frac{1}{T2} - \frac{1}{C2}\right)\sigma_2 \tag{3}$$

where σ_1, σ_2 and τ_{12} are the stresses components in the principal material directions, T and C denote respectively the tension and compression ultimate stresses (index 1 corresponds to the fibre direction, index 2 to the transverse direction), and S represents the shear ultimate stress related to directions 1-(2=3).

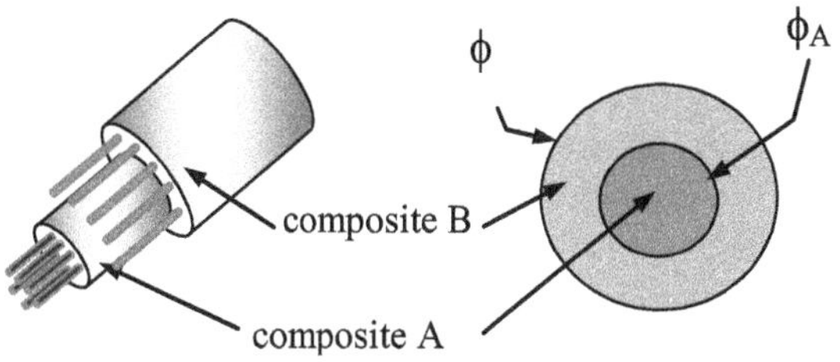

Fig. 6. Hybrid structure.

4 Simulation procedure

Like mostly mechanical studies, the formulation of the limit state function G is here implicit as the stress state $\overline{\overline{\sigma}}$ depends on the vector X of random variables. Only particular evaluations of this function can be obtained by

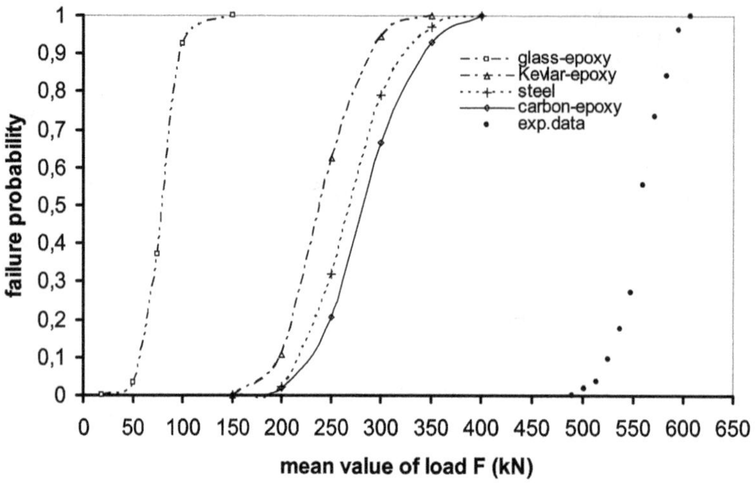

Fig. 7. Failure probability according to axial load F ($q = 0$ MPa).

numerical means, generally through finite element simulations. In order to perform the reliability analysis, it is then necessary to couple two numerical solvers each dedicated to a particular role:

- a probabilistic solver that generates the random variables vector realizations, carries out the isoprobabilistic transformation and finally validates the calculation steps and convergence to obtain the failure probability, reliability index β and sensitivity analysis;
- a finite element solver that provides on the other hand the evaluations of the limit state function G and its gradients.

In the present case, we use the PERMAS coupled code (see [13] for a review) which associates in a single numerical tool these two aspects. Such solver uses the Rosenblatt isoprobabilistic transformation and the classical Hasofer-Lind-Rackwitz-Fiessler convergence algorithm to determinate the reliability index. In view of the quadratic form of the failure criteria retained, the SORM approximation method is used here to perform the reliability calculations. Note finally that the expression of the limit state function G is introduced through a subroutine written in FORTRAN.

5 Results and discussion

The objective is to quantify here the effects of the random variables retained (loads, materials failure parameters) on the mechanical response of the structure and then to provide design recommendations to enhance its reliability.

5.1 Mono-material structure

Let examine first the case of the single material structure, either made of steel or single fibre-reinforced composite. Figure 7 shows the failure probability for the four materials studied (steel, carbon-epoxy, glass-epoxy and Kevlar-epoxy) when the structure is subjected only to the axial force F (pressure $q = 0$ MPa). For example for a load $F = 200$ kN, the probability that the carbon-epoxy, steel and Kevlar-epoxy structures fail are respectively 2%, 2.25% and 10.5%. On the other hand, the glass-epoxy composite has 100% chances of failing. If experimental data related to the composites investigated in this paper are not available, tests results obtained for another carbon epoxy composite (fibre volume ratio of 67%, mean elastic modulus $E_1 = 155$ GPa, mean tension ultimate stress in fibre direction $T1 = 3075$ MPa [14]) are however represented in Figure 7 for comparison. We can note that the model seems to predict correctly the failure probability trend as the carbon composite studied here exhibits lower reinforcement ratio and mechanical performances (see Sect. 2).

For a given reliability level, one can compares the maximum force admissible for each material: taking steel as a reference, carbon-epoxy can support 5% more, Kevlar-epoxy 12% less and glass-epoxy 70% less for a failure probability $P_f = 0.5$. The failure mode of the carbon-epoxy and Kevlar-epoxy composites under longitudinal tension is induced by the fibre failure. Considering the properties of carbon and Kevlar fibers, the carbon-epoxy composite presents logically better performances. For glass-epoxy composite, failure mode is controlled by the transverse failure of matrix, accordingly this composite presents weak properties compared to others.

Figure 8 represents the evolution of the reliability index β according to F. Specific to approximation methods, such index constitutes another reliability indicator that can be linked to the failure probability [9]: precisely, more β is high for a specified load, more the structure is reliable. For example for a load $F = 200$ kN, reliability index of carbon-epoxy, steel and Kevlar-epoxy structures are respectively 3.19, 2.53 and 1.25. Carbon-epoxy composite presents then the highest reliability level, which confirms the tendency observed in Figure 7.

It is also interesting to investigate the effect of lateral pressure q on the structure reliability. Figure 9 illustrates

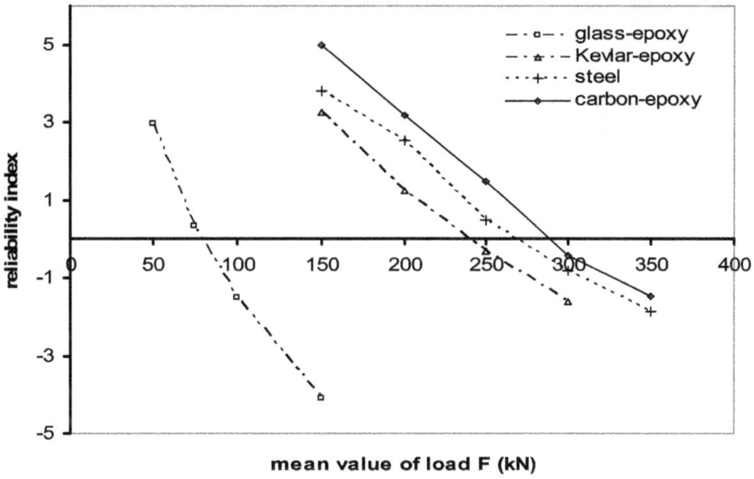

Fig. 8. Reliability index β according to axial load F ($q = 0$ MPa).

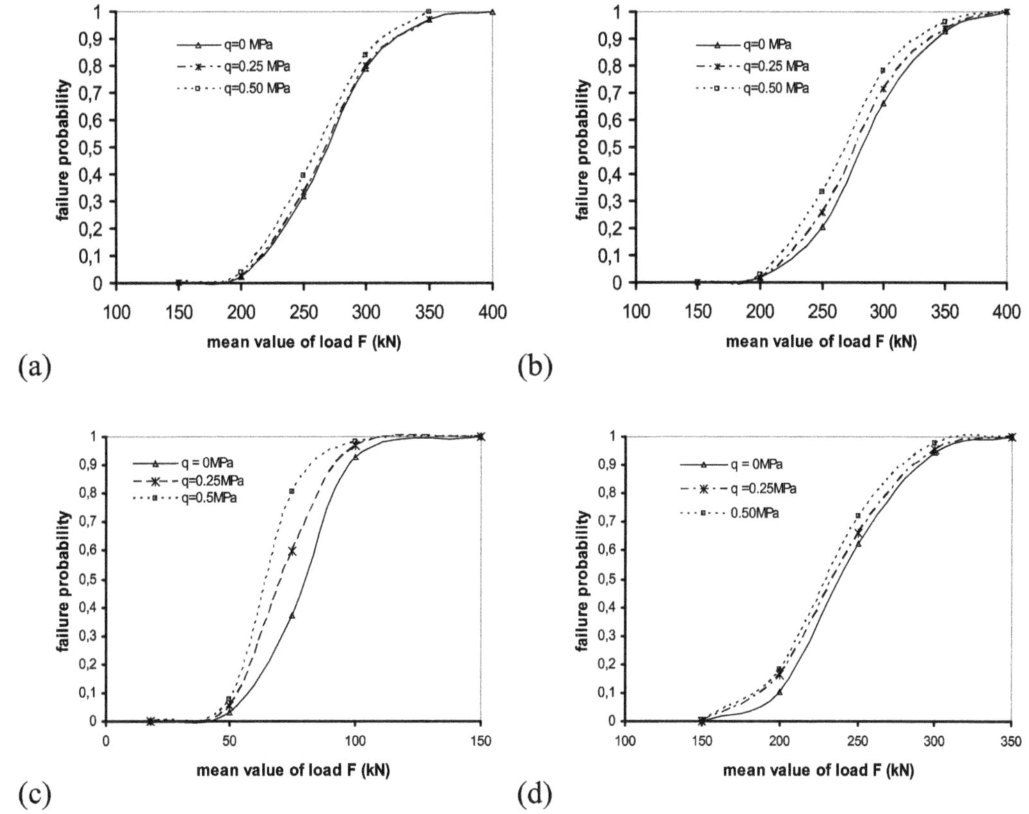

Fig. 9. Failure probability according to loads F and q: (a) steel, (b) carbon-epoxy, (c) glass-epoxy, (d) Kevlar-epoxy.

such impact for the different materials with failure probabilities according to loads F and q. Note that values of q are taken such that the axial stress induced by q in the cylinder remains inferior to the one generated by load F ($\leq$400 kN). We observe in all cases that an increase of q makes the structure less reliable. For example, the load F associated to a probability of failure $P_f = 0.5$ when $q = 0$ MPa leads respectively to $P_f = 0.58$ for steel, $P_f = 0.65$ for carbon-epoxy, $P_f = 0.88$ for glass-epoxy and $P_f = 0.6$ for Kevlar-epoxy when $q = 0.5$ MPa. Kevlar-epoxy appears then as the most stable composite according to the lateral load and should be used when many uncertainties may occur on this point.

In order to highlight the influence of composite failure parameters, a sensitivity study of index β according to the mean value (Fig. 10) and to the standard deviation (Fig. 11) of these random variables is carried out for $P_f \approx 10^{-2}$: the impact of the average values provides an indication for the more safety range of use, and the effect of the standard deviation underlines the role of the quality control. In particular, when the sensitivity factor of a parameter is positive (respectively negative), it

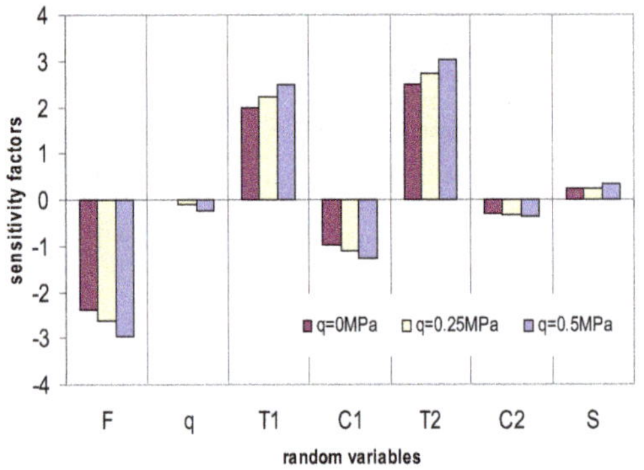

(a)

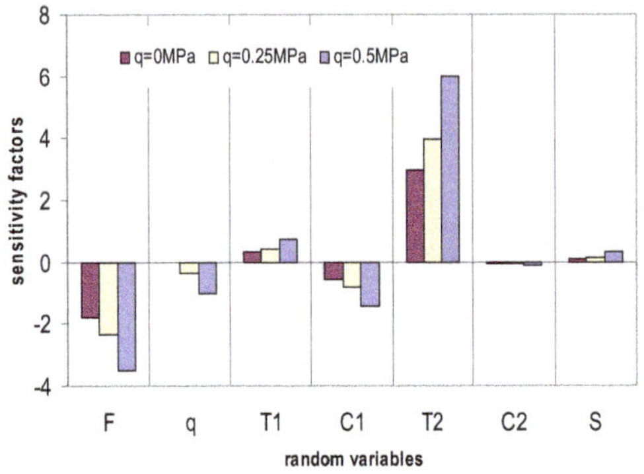

(b)

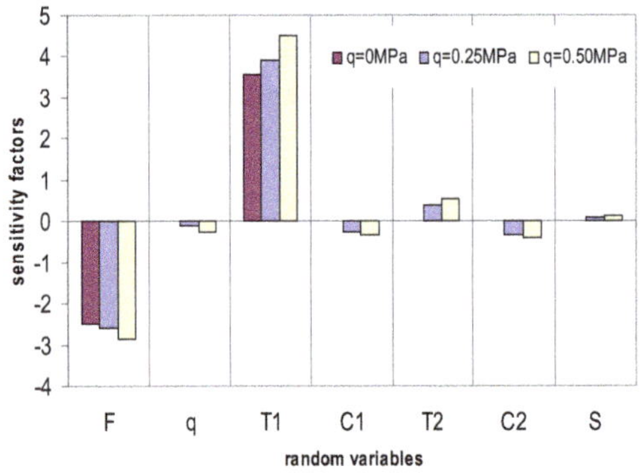

(c)

Fig. 10. Sensitivity analysis with respect to mean value of random variables ($P_f \approx 10^{-2}$): (a) carbon-epoxy, (b) glass-epoxy, (c) Kevlar-epoxy.

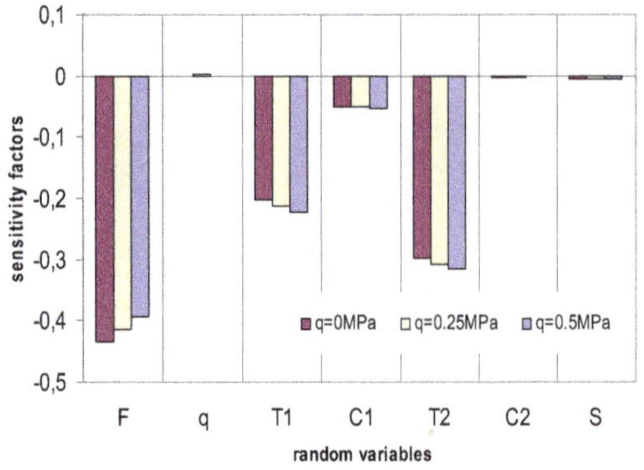

(a)

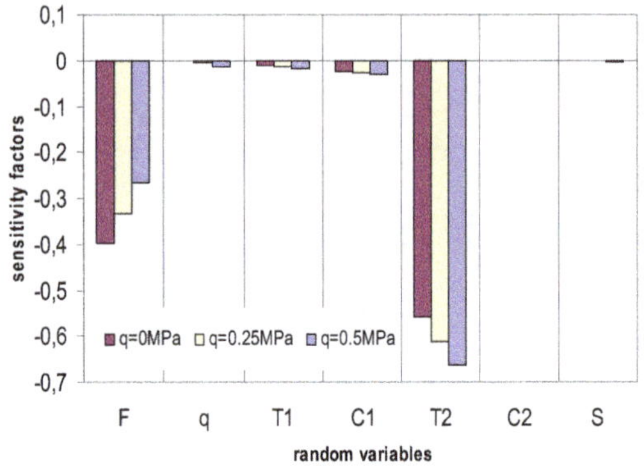

(b)

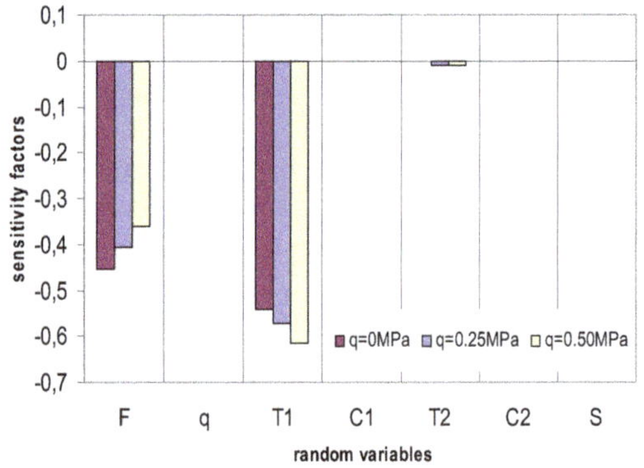

(c)

Fig. 11. Sensitivity analysis with respect to standard deviation of random variables ($P_f \approx 10^{-2}$): (a) carbon-epoxy, (b) glass-epoxy, (c) Kevlar-epoxy.

means that an increase of this parameter increases (resp. decreases) reliability. Note finally that PERMAS solver performs each sensitivity analysis on the most probable failure point of the structure, which is generally close to the structure embedded side.

In Figure 10, we observe again that in all cases an increase of both loads tends to decrease the reliability, especially axial load F. The failure parameters influence is quite different for the three composites and an increase of the lateral load q just accentuates their trends. In particular, the variables favourable to reliability are the tension ultimate stresses in fibre direction ($T1$) and transverse direction ($T2$): $T1$ and $T2$ for carbon-epoxy, $T2$ for glass-epoxy and $T1$ for Kevlar-epoxy. To understand these results, stress components (σ_{11}, σ_{22} and τ_{12} in the principal material coordinate system) induced in the carbon-epoxy structure are evaluated through deterministic calculations corresponding to $P_f \approx 10^{-2}$ (Tab. 3). For all values of pressure q, the stress state belongs to the tension domain ($\sigma_1 \geq 0$, $\sigma_2 \geq 0$). Furthermore, the previously mentioned favourable parameters correspond for each material to the respective ultimate stresses (mean value defined at Tab. 2) that stresses σ_1, σ_2 or τ_{12} are closer to. Besides, the Tsai-Wu failure criterion applied to carbon-epoxy composite is drawn in the stress space (σ_1, σ_2) for different values of limit stresses $T1$ (mean value given at Tab. 2 ± 10%, Fig. 12). In the tension domain, we see that an increase of $T1$ tends actually to

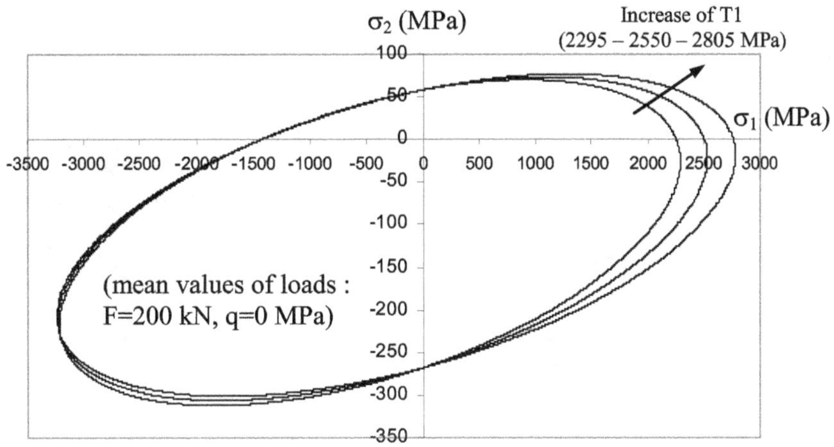

Fig. 12. Failure criteria according to the limit stress $T1$ (carbon-epoxy).

Table 3. Deterministic calculations : stress ratios for carbon-epoxy ($F = 200$ kN).

	$\sigma_1/T1$ (%)	$\sigma_2/T2$(%)	τ_{12}/S (%)
$q = 0$ MPa	43.6	77.7	20.1
$q = 0.25$ MPa	45.2	80.7	20.3
$q = 0.5$ MPa	46.8	83.5	22.2

extent the safety domain, and then to reduce the failure probability.

Sensitivity factors related to the standard deviation of all random variables are negative which confirms the idea that the variability of a parameter reduces the structure reliability. The minimization of uncertainties by a good systematic quality control is then essential for the structure assessment. Figure 11 shows the respective effect for each random parameter and follows the trends observed for the sensitivity analysis on mean values. Reliability is then essentially improved when the axial load F and tension ultimate stresses are estimated with the best precision ($T1$ and $T2$ for carbon-epoxy, $T2$ for glass-epoxy and $T1$ for Kevlar-epoxy).

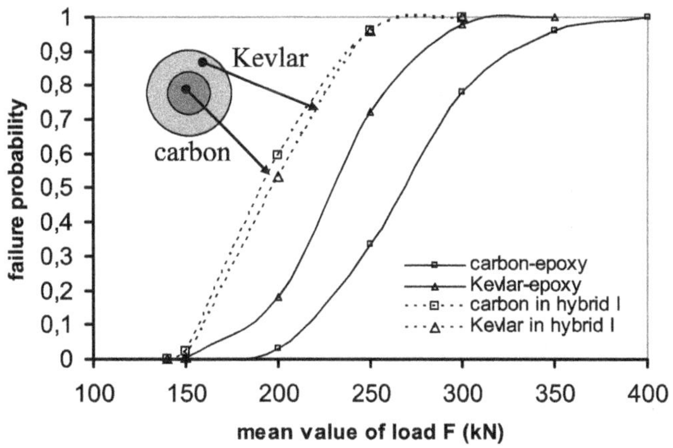

Fig. 13. Failure probability of hybrid I according to axial load F ($q = 0.5$ MPa).

5.2 Hybrid structure

In this part, we focus our attention on composite materials that lead previously to the best performances, namely the carbon-epoxy and Kevlar-epoxy. The internal diameter is fixed to $\phi_A = \phi/2$ and two configurations are considered:

- hybrid I: carbon-epoxy inside (A), Kevlar-epoxy outside (B);
- hybrid II: Kevlar-epoxy inside (A), carbon-epoxy outside (B).

To study the reliability of such hybrid structures, we adopt the same approach as [8] by evaluating the failure probability of each single fibre-reinforced composite (A and B) within the structure. This is an independent analysis of both parts of the structure which do not take into account any load transfer when one part fails. Such study provides then the weakest composite material (between A and B) within the hybrid structure which exhibits the biggest failure probability for a given load. Accordingly, if we consider that the failure of the structure occurs when one layer fails, the structure reliability is then directly given by the curve relative to the weakest material. The failure probabilities of hybrid I (Fig. 13) and hybrid II (Fig. 14) according to axial load F are presented with the previous results on mono-material cylinders (with diameter ϕ) for each composite as a reference.

We can point out that configuration I is much more critical than composite material alone as P_f is much more important in both Kevlar-epoxy and carbon-epoxy parts. Especially, this geometry leads to extra stresses causing failure of carbon-epoxy sooner. Looking at the other configuration, we note that hybrid II allows an improvement as the failure probability of carbon-epoxy in the structure decreases slightly. It may then be interesting to investigate such design option to optimise the structure strength.

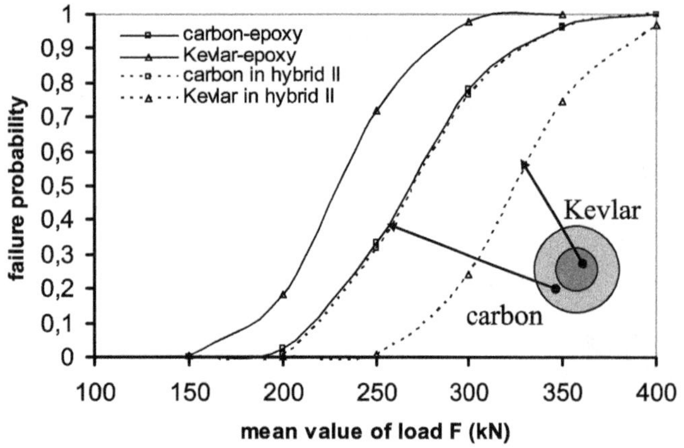

Fig. 14. Failure probability of hybrid II according to axial load F ($q = 0.5$ MPa).

6 Conclusion and perspectives

The objective of this study is to highlight the interest in introducing composite materials into the field of engineering mechanical design. In this way, probabilistic simulations with finite element solver PERMAS are performed to analyse the mechanical reliability of unidirectional fibre-reinforced composite structures. The failure probability of different kinds of single fibre reinforcement is then evaluated: carbon-epoxy exhibits the best performance related to axial load (better than steel), the behaviour of Kevlar-epoxy is particularly stable according to lateral pressure, whereas glass-epoxy is definitively weak on all these points. The sensitivity analysis provides also interesting information for the design optimisation, namely that ultimate tension stresses take a significant part and should be as big as possible to improve the structure reliability.

The results obtained on the hybrid composite seem attractive at first hand, especially as Kevlar fibres very much less expensive than carbon. However we suppose here a perfect adherence between the two layers of the composite. In the real case, it is thus possible that interface between these two layers fails. This point needs to be checked through experimental tests before further modelling investigations.

In order to complete this work, it may be interesting to look at the influence of the kind of statistical models introduced in the probabilistic solver and also to investigate others hybrid solutions such as mono-material composite made of epoxy resin reinforced by different kinds of fibres.

References

1. U. Meier, Carbon fibre reinforced polymers, Modern materials in bridge engineering. Structural Engineering International **1**, 1–12 (1992).
2. C. Boyer, A. Béakou, M. Lemaire, Design of a composite structure to achieve a specified reliability level. Reliability Engineering and System Safety **56**, 273–283 (1997).
3. S. Guedes, Reliability of components in composite materials. Reliability Engineering and System Safety **55**, 171–177 (1997).
4. M. Walker, R. Smith, A methodology to design fibre reinforced laminated composite structures for maximum strength. Composites Part B: Engineering **34**, 200–214 (2003).
5. J.L. Pelletier, S.S. Vel, Multi-objective optimisation of fiber reinforced composite laminates for strength, stiffness and minimal mass. Computer and Structures **84**, 2065–2080 (2006).
6. J. Renard, A. Thionnet, Damage in composites: From physical mechanisms to modelling. Composites Science and Technology **66**, 642–646 (2006).
7. T.P. Philippidis, D.J. Lekou, Probabilistic failure prediction for FRP composites. Composites Science and Technology **58**, 1973–1982 (1998).
8. D.M. Frangopol, S. Recek, Reliability of fiber reinforced composite laminate plates. Probabilistic Engineering Mechanics **18**, 119–137 (2003).
9. O. Ditlevsen, H. Madsen, *Structural reliability analyses*. J. Wiley & Sons (1996).
10. F.L. Matthews, R.D. Rawlings, *Composite Materials: Engineering and Science*. Woodhead Publishing Ltd and CRC Press LLC (2003).
11. A.S. Kaddour, M.J. Hinton, P.D. Soden, A comparison of the predictive capabilities of current failure theories for composite laminates: additional contributions. Composites Science and Technology **64**, 449-476 (2004).
12. M.J. Hinton, A.S. Kaddour, P.D. Soden, Evaluation of failure prediction in composite laminates: background to 'part C' of the exercise. Composites Science and Technology **64**, 321–327 (2004).
13. M.F. Pellissetti, G.I. Schuëller, On general purpose software in structural reliability. Structural Safety **28**, 3–16 (2006).
14. H. Dehmous, H. Welemane, I. Zivanovic, M. Karama, *Reliability analysis of composite structures: application to the Laroin footbridge*, in Proceedings of ICSAM Conference. Patras, September (2007).

6

Computational system for multi disciplinary optimization at conceptual design stage

P. Fantini[1], L.K. Balachandran[2] and M.D. Guenov[2,a]

[1] Computational Aerodynamics, Aircraft Research Association Ltd., Bedford, UK
[2] Aerospace Engineering, Cranfield University, Bedford, UK

Abstract – Presented is a computational system for multidisciplinary design optimization (MDO) at the conceptual design stage. During this phase, hundreds of low-fidelity models such as equations and compiled code, and thousands of variables are used to describe a complex product such as aircraft. In this context the paper presents a novel computational approach associated with the complete MDO process. The first aspect of the proposed approach is the dynamic derivation of the optimal computational plan for each design study, given the designer's choice of independent variables. The second aspect is the effectiveness with which the trade-off landscape is obtained. This is crucial from an engineering point of view, since such information will be used for selecting a baseline design. The approach is demonstrated with an aircraft design test case consisting of 96 models and 120 variables.

Key words: Multidisciplinary Design and Optimization (MDO); multi-objective optimization, computational process planning; design structure matrix (DSM), model based design, aircraft conceptual design.

1 Introduction

The aim of the conceptual phase of the design of complex products such as aircraft is to deliver a sound baseline which is then 'frozen' and passed onto the subsequent more detailed design stages. This is a crucial phase since the decisions taken at this stage commit the majority of the product lifecycle costs. It is therefore essential that the underpinning computational processes offer maximum flexibility in creating appropriate performance studies. This includes the dynamic assembly of a computational plan according to the designer's choice of input (independent) variables and also performing effective trade-off analysis for selecting the baseline design.

In this context the objective of this paper is to present a novel computational approach associated with this process. The first aspect of the proposed approach is the dynamic derivation of the optimal computational plan for each design study, given the designer's choice of independent variables. This is not a trivial task since during the conceptual phase, hundreds of low-fidelity models such as equations and compiled code, and thousands of variables are used to describe the complex product. The second aspect is the effectiveness with which the trade-off landscape is obtained. This is crucial from an engineering point of view since the decision maker (DM) needs this information for selecting the baseline design. It also implies that the DM should be able to consider all solutions of potential interest, even local ones, since these could become very attractive when taking into account decision factors not initially taken into consideration, for example, design robustness.

The following section briefly describes the developed procedure for producing an optimal computational process plan. Section three presents a novel method for generating well distributed points describing the entire Pareto frontier, as well as the local Pareto frontiers. Section four presents test results from our ongoing work, aiming at integrating the computational process with multi-objective optimization and uncertainty analysis. The example uses an aircraft design test case consisting of 96 models and 120 variables. Finally conclusions are drawn and future work outlined.

2 Computational process plan

Computational process modelling is the process of organising a complex system of models, in order to compute the output variables, according to specific input variables. An optimised computational plan for executing the models has to be attained each time the designer wishes to perform a computation associated with a different choice of input variables. To obtain such an optimal computational plan a novel procedure has been developed which is briefly explained as follows (Fig. 1):

1. Once the designer has chosen the independent variables, variable flow modelling is performed using the

[a] Corresponding author: m.d.guenov@cranfield.ac.uk

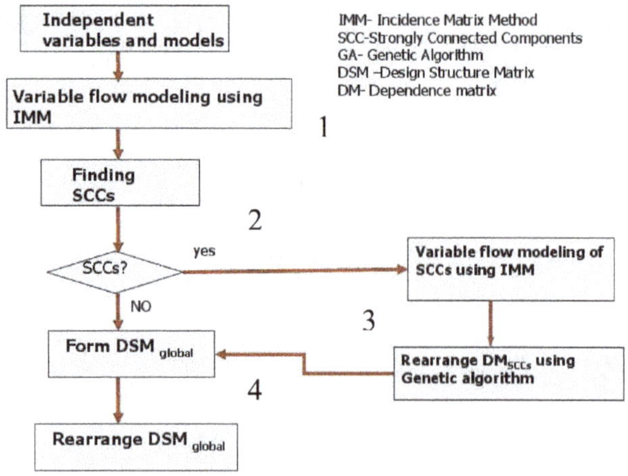

Fig. 1. Flow chart for process management.

incidence matrix method (IMM) to determine the information (data) flow between the models.

2. The next step is to establish the existence of any strongly connected components (SCCs). A SCC is a set of models which are strongly coupled through shared variables.
3. If SCCs exist then variable flow modelling of the models belonging to each SCC is performed. The different variable flow models obtained for each SCC are then populated in corresponding dependence matrices also known as design structure matrices or DSM [1]. Each dependence matrix is rearranged using a genetic algorithm, with feedback length as the objective function to be minimized. The variable flow model with the minimum feedback length is selected for further solving.
4. The SCCs (if any) and the remaining models are now populated in a global DSM. This matrix is further rearranged into upper triangular form, which determines the optimal sequence for the execution of the models.

The following sub-sections explain the procedure in more detail.

2.1 Variable flow modelling using Incidence matrix method

Variable flow modelling is the process of identifying the information flow between the models in the system, according to the input (independent) variables selected by the designer. The information flow between the models is required in order to determine the output variables to be computed, according to the input variables provided by the designer. This subsection explains a novel incidence matrix method (IMM) which allows to obtain the information flow dynamically.

The incidence matrix is used to represent the relationship between the models and their associated variables. The rows of the matrix represent the models, while the columns denote the variables. The association of a variable in a column with a model in a row is denoted by '*' in the corresponding entry (cell). For solving the incidence matrix, the *'s in each cell have to be substituted either with an 'i' (input) or with an 'o' (output), depending on whether the variable in the column should be an input to or an output from the model in the row. The incidence matrix is solved or populated according to the rules stated below.

1. An independent variable should be always an input to a model. (This implies that all the '*'s in the column of the independent variable should be replaced by 'i's.)
2. If a variable is associated only with one model and if it is not an independent variable, it should be an output from that model. In the same way if a model is associated with only one variable, that variable should be the output from that model (This implies that if a column/row has only one '*' marked in it and if the corresponding data variable in the column is not an independent variable, then the '*' should be replaced with an 'o'.)
3. Each variable should be output from only one model. (This implies that except for the columns of the independent variables, all the other columns should have an 'o' in only one cell.)
4. The number of outputs identified for a particular model as a result of variable flow modelling should correspond to the number of outputs from the original model. (This implies that every row should have the same number of 'o's as the number of outputs from the associated model)
5. After the four rules above have been applied to a SCC it is possible that there will be some models remaining, for which not all '*' have been replaced with either an 'i' or an 'o'. Of these, the models with inputs differing from the original ones are selected for further population. If no such a model exists, the incidence matrix is populated with the original inputs and outputs from the model.

A basic example demonstrating the above rules is shown in Figure 2 where a simple set of models for balancing the weight and lift of an aircraft is converted into its corresponding initial incidence matrix. Data variables entering the models are the inputs and data variables leaving the models are the outputs. In this example Ws and V are considered the independent variables selected by the designer.

The final populated incidence matrix is shown in Figure 3, after applying the rules stated earlier. The numbers in the curly brackets represent the sequence in which the matrix was populated: {1}-Rule1, {2}-Rule2, {3}-Rule-3, {4}-Rule4, {5}-Rule3 and {6}-Rule4. Thus the final outcome of incidence matrix method is: model1 has Ws and q as input and C_L as output, model2 has ρ and V as input and q as output, and finally, model3 produces ρ as output.

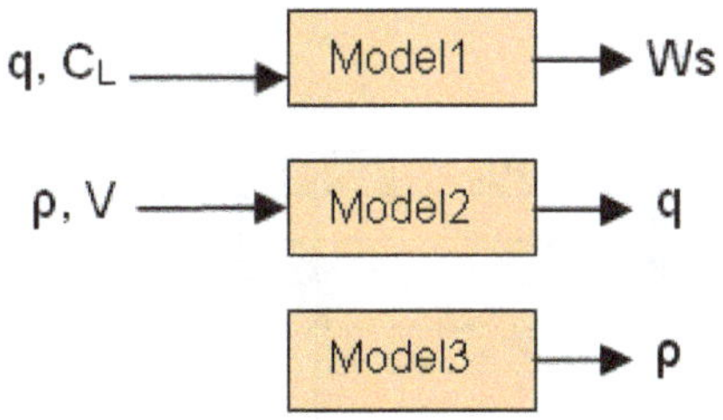

(a)

	Ws	C_L	q	ρ	V
Model1	*	*	*		
Model2			*	*	*
Model3				*	

(b)

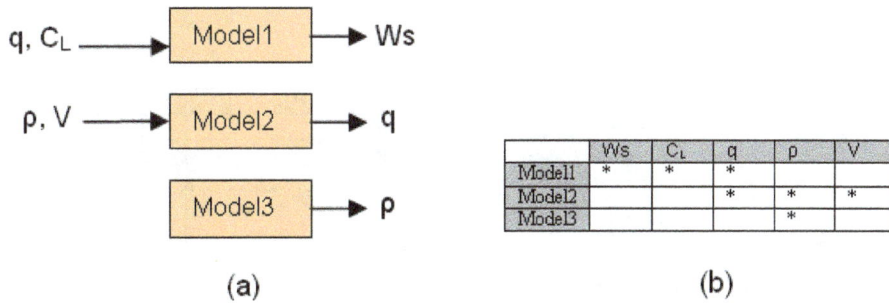

Fig. 2. (a) Models balancing the weight of aircraft with its lift, (b) Initial incidence matrix.

	Ws	C_L	q	ρ	V
Model1	i {1}	o {6}	i {5}		
Model2			o {4}	i {3}	i {1}
Model3				o {2}	

Fig. 3. Final populated incidence matrix.

2.2 Variable flow modelling in strongly connected components

A strongly connected component (SCC) is a set of models which are strongly coupled through shared variables. The term SCC is derived from graph theory and more formally, an SCC is a subset of nodes in a directed graph such that there is a path from every node to every other node [2]. In the context of a system of models, a SCC is a cluster of models in which each model requires input from one or more models from the same cluster. Presence of SCCs in a system adds complexity to the computational process modelling and also to the execution and solving of the models belonging to the same SCC.

In the above simple example only the first four rules were applied to populate the matrix. Figure 4 shows a system of models with its corresponding populated incidence matrix with X_3 chosen as the independent variable. It can be noted that there are some cells in which the '*' is not yet replaced with either an 'i' or an 'o'. This situation arises as a result of the presence of a SCC. In the figure, the models which belong to the SCC are those which have at least one variable undefined (i.e. still labelled as '*'), after applying the first four rules. Thus in this example, models 1, 2, 3, 5 and 6 are strongly connected and are coupled through variables X_1, X_4, X_5, X_6, and X_7. To generalise, the models of a SCC have their own unique set of data variables through which they are coupled. A design problem can have more than one mutually exclusive SCCs.

Returning to the example (Fig. 4), the solution of the incidence matrix of the SCC is obtained by choosing the outputs from the constituent models according to rule 5. According to this rule model6 is chosen for further solving, since X_3 is an output of the original model (Fig. 4a), but after applying the four rules X_3 has become an input (Fig. 4b). Given the choice of model6, then variables X_1, X_5 or X_6 can now be selected as the preferred output. The set of variable flow models obtained for each case is shown in Figure 5.

2.3 Rearrangement of SCCs using genetic algorithm

The next step following the variable flow modelling is to schedule the model execution sequence while accounting for any feedback loops. Feedback loops are generated when models require inputs from other models which are run further down in the execution sequence. Thus the aim of the rearrangement is to reduce the length and number of feedback loops or to eliminate them altogether if possible. In complex design systems a complete elimination of iterative feedback loops is not always possible, especially in the SCCs. Here we propose an optimisation based approach for rearranging the SCCs which utilises a genetic algorithm. The objective function which the genetic algorithm minimises is shown in equation (1), where $DX(i,j)$ denotes the value of the (i,j)th element of the dependence matrix DX, J is the feedback length and n is the number of models.

$$J = \sum_{i=2}^{n} \sum_{j=1}^{i=1} DX(i,j) \cdot (i-j) + \sum_{k=1}^{n} DX(k,k). \qquad (1)$$

The feedback length of the dependence matrix represents an approximate estimate of the time required for solving the SCCs if a fixed point iterative scheme is applied. Figure 6 shows the dependence matrix for the variable flow model in Figure 5a.

In the matrix, '1' marked in a cell above the main diagonal symbolises a feed forward loop and a '1' below the main diagonal represents a feedback loop. In addition, '0.5' marked in a cell on the main diagonal itself denotes that the model in the corresponding row has its inputs and outputs interchanged as a result of variable flow modelling. In such cases numerical methods have to be applied to solve those particular models and the value '0.5' signifies the additional time required for solving these. Each of the dependence matrices for the variable flow models in Figure 5 is rearranged using genetic algorithm with equation (1) as the fitness function to be minimized. The value of the objective function is considered as the criteria for choosing the variable flow model which leads to faster convergence of the SCC. After rearranging the dependence matrix of each of the four variable flow models

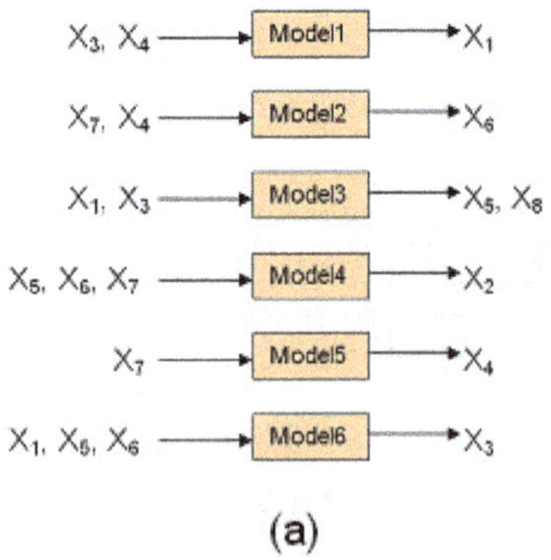

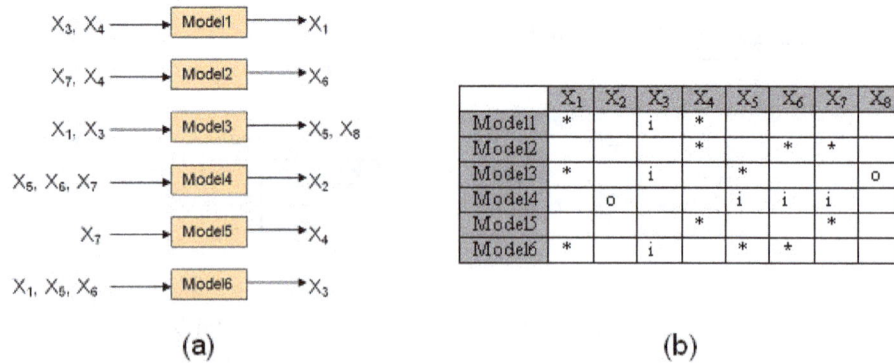

	X_1	X_2	X_3	X_4	X_5	X_6	X_7	X_8
Model1	*		i	*				
Model2				*		*	*	
Model3	*		i		*			o
Model4		o			i	i	i	
Model5				*			*	
Model6	*		i		*	*		

(b)

Fig. 4. (a) System of models (b) Corresponding populated incidence matrix by applying first four rules.

	X_1	X_2	X_3	X_4	X_5	X_6	X_7	X_8
Model1	i		i	o				
Model2				i		o	i	
Model3	i		i		o			o
Model5				i			o	
Model6	o		i		i	i		

(a)

	X_1	X_2	X_3	X_4	X_5	X_6	X_7	X_8
Model1	i		i	o				
Model2				i		o	i	
Model3	o		i		i			o
Model5				i			o	
Model6	i		i		o	i		

(b)

	X_1	X_2	X_3	X_4	X_5	X_6	X_7	X_8
Model1	o		i	i				
Model2				o		i	i	
Model3	i		i		o			o
Model5				i			o	
Model6	i		i		i	o		

(c)

	X_1	X_2	X_3	X_4	X_5	X_6	X_7	X_8
Model1	o		i	i				
Model2				i		i	o	
Model3	i		i		o			o
Model5				o			i	
Model6	i		i		i	o		

(d)

Fig. 5. Incidence matrices with the following variables chosen as output from model6: (a) X_1; (b) X_5; (c) X_6 then X_4; (d) X_6 then X_7.

from the example, the corresponding minimised feedback lengths are: 5.5, 6, 5.5 and 5. This shows that the fourth variable flow model (Fig. 5d) will converge faster than the rest and therefore will be chosen as the solution. The final rearranged dependence matrix of the fourth variable flow model is shown in Figure 7.

	Model1	Model2	Model3	Model5	Model6
Model1	0.5	1	0	1	0
Model2	0	0	0	0	1
Model3	0	0	0	0	1
Model5	0	1	0	0.5	0
Model6	1	0	1	0	0.5

Fig. 6. Dependence matrix for the incidence matrix from Figure 5a.

2.4 Global rearrangement

After the rearrangement of a SCC is completed, it is considered as single sub-system and is reintroduced into the global Design Structure Matrix (DSM) along with the remaining models which were not part of any SCC. By applying the graph theoretical algorithm of Tang et al. [3], the global DSM is rearranged into an upper-triangular form, thus ensuring all loops are feed forward. Hence, the final model computation sequence obtained for the above example will be: SCC (1→3→5→2→6) → 4. A more detailed description of the procedure with examples can be found in [4].

3 Double hyper cone boundary intersection method

When performing a multi-objective optimization the first task of the DM is that of setting up the optimization problem. This involves defining design variables, constraints and objectives. An optimization algorithm will then be applied in order to obtain a set of Pareto solutions. Finally the DM will have to evaluate the results, eventually making additional considerations not originally taken in account when setting up the optimization problem. When the DM is unsure of how to define the optimization problem it can be useful to identify all the solutions of potential interest, rather than only those that are optimal with respect to a particular setup of the optimizer. The local Pareto frontier concept will be used for this purpose.

3.1 The algorithm

The Double Hyper-cone Boundary Intersection (DHCBI) method has been developed for obtaining the global as well as the local Pareto frontiers [5]. The DHCBI method shares conceptual similarities with the NBI and NC methods [6–10] for which an optimization is performed for each

	Model1	Model3	Model5	Model2	Model6
Model1	0	1	0	0	1
Model3	0	0	0	0	1
Model5	1	0	0	1	0
Model2	0	0	1	0.5	0
Model6	0	0	0	1	0.5

Fig. 7. Rearranged dependence matrix of variable flow model of Figure 5d.

of the Pareto points sought. However, as it will be described below, the DHCBI is designed to improve the effectiveness of the search.

For the optimization problem:

$$\min_{\mathbf{x}} \ \mathbf{F}(\mathbf{x})$$
$$\text{subject to } K \text{ inequality constraints: } g_k(\mathbf{x}) \leqslant 0, \quad k = 1, 2, ..., K$$
$$\text{and } P \text{ equality constraints: } h_p(\mathbf{x}) = 0, p = 1, 2, ..., P \tag{2}$$

where $F_i : R^N \to R^M$, the NBI and NC algorithms can be summarised as follows:

1. A single-objective minimization is performed for each of the M objectives yielding, as termed by Mattison et al. [11], *anchor points*, $\boldsymbol{\mu}_i \in R^M$, $i \in \{1, ..., M\}$. Das and Dennis [6] refer to the polygon of vertices $\boldsymbol{\mu}_i$, $i \in \{1, ..., M\}$ as *Convex Hull of Individual Minima* (CHIM). Messac et al. [8] terms *utopia plane*, the hyper-plane to which the M anchor points belong.
2. In the NBI method, K evenly distributed *utopia plane points* $\mathbf{p}_k$, belonging to the polygon of vertices $\boldsymbol{\mu}_i$, are defined as:

$$\mathbf{p}_k = \sum_{i=1}^{M} \alpha_{ki} \boldsymbol{\mu}_i, \quad k = \{1, 2, ..., K\}. \tag{3}$$

   In the NBI method α_{ki} is such that $0 \leqslant \alpha_{ki} \leqslant 1 \quad \forall k, i$ and $\sum_{i=1}^{M} \alpha_{ki} = 1$. In contrast, the NC method takes in consideration the fact that for $M > 2$ the projection of portions of the Pareto front on the utopia plane is external to the polygon of vertices $\boldsymbol{\mu}_i$ [6, 9]. This region will be referred to as *peripheral region*. In order to deal with this problem Messac et al. solve a set of computationally benign optimizations, which allow them to obtain the lower and upper limits of α_{ki} while still imposing $\sum_{i=1}^{M} \alpha_{ki} = 1$.
3. For utopia plane point $\mathbf{p}_k$, the optimization problem is reformulated and solved yielding a Pareto point.

In the NBI method the kth optimization problem is reformulated as follows:

$$\min_{\mathbf{x}} \ t(\mathbf{x})$$
$$\text{subject to } K \text{ inequality constraints: } g_k(\mathbf{x}) \leqslant 0, \quad k = 1, 2, ..., K,$$
$$\text{subject to } P \text{ equality constraints: } h_p(\mathbf{x}) = 0, \quad p = 1, 2, ..., P$$
$$\text{and subject to the additional equality constraint: } \mathbf{p}_k + t\mathbf{n} = \mathbf{F}(\mathbf{x}) \tag{4}$$

where $\mathbf{n}$ is the normal to the utopia plane and t is the projection of $\mathbf{F}(\mathbf{x})$ on $\mathbf{n}$. The kth optimization problem for the NC method is instead the following:

$$\min_{\mathbf{x}} \ f_M(\mathbf{x})$$
$$\text{subject to } K \text{ inequality constraints: } g_k(\mathbf{x}) \leqslant 0, \quad k = 1, 2, ..., K,$$
$$\text{subject to } P \text{ equality constraints: } h_p(\mathbf{x}) = 0, \quad p = 1, 2, ..., P$$
$$\text{with the M} - 1 \text{ additional equality constraint: } \mathbf{q}_j \cdot (\mathbf{p}_k - \mathbf{F}(\mathbf{x})) \leqslant 0, \forall j : \ j \in \{1, 2, ..., M-1\} \tag{5}$$

where $\mathbf{q}_j = \frac{\boldsymbol{\mu}_M^* - \boldsymbol{\mu}_j^*}{\|\boldsymbol{\mu}_M^* - \boldsymbol{\mu}_j^*\|}$ are $M - 1$ unit vectors.

Even though the DHCBI approach is similar to the NBI and NC methods, it differs in two main aspects, one relative to the reformulation of each optimization problem and one relative to the *anchor points*.

3.2 DHCBI reformulation for solving sub-problem k

We propose the following alternative reformulation for sub-problem $k$with respect to utopia plane point $\mathbf{p}_k$ (Fig. 8):

$$\min_{\mathbf{x}} \left[t(\bar{\mathbf{F}}(\mathbf{x}), \mathbf{p}_k) + q(\bar{\mathbf{F}}(\mathbf{x}), \mathbf{p}_k)\right]$$
$$\text{subject to } K \text{ inequality constraints: } g_k(\mathbf{x}) \leqslant 0, k = 1, 2, ..., K, \tag{6}$$
$$\text{subject to } P \text{ equality constraints: } h_p(\mathbf{x}) = 0, p = 1, 2, ..., P$$
$$\text{and subject to the additional constraint: } c \leqslant 0$$

where $\bar{f}_i(\mathbf{x}) = (f_i(\mathbf{x}) - \mu_{ii}^*)/(\max\{\mu_{1i}^*, \ldots, \mu_{Mi}^*\} - \mu_{ii}^*)$,

$$\gamma_{min} = \gamma_{min}\left(\bar{\mathbf{F}}(\mathbf{x}), \mathbf{p}_k\right) = \min\{\gamma_1, \gamma_2\} \tag{7}$$

with

$$\gamma_i = \arccos\left(\frac{\bar{\mathbf{F}}(\mathbf{x}) - \mathbf{p}_k}{\left\|\bar{\mathbf{F}}(\mathbf{x}) - \mathbf{p}_k\right\|} \cdot \mathbf{l}_i\right),$$
$$t(\bar{\mathbf{F}}(\mathbf{x}), \mathbf{p}_k) = \left\|\bar{\mathbf{F}}(\mathbf{x}) - \mathbf{p}_k\right\| \cos(\gamma_{min}),$$
$$n(\bar{\mathbf{F}}(\mathbf{x}), \mathbf{p}_k) = \left\|\bar{\mathbf{F}}(\mathbf{x}) - \mathbf{p}_k\right\| \sin(\gamma_{min}),$$
$$q(\bar{\mathbf{F}}(\mathbf{x}), \mathbf{p}_k) = q\left(n\right) \tag{8}$$

and

$$c = \begin{cases} t_c \dfrac{e^{\ln(\tan_{\gamma_c}+1)\frac{t}{t_c}} - 1}{\ln(\tan_{\gamma_c}+1)} - t + n_c & t \leqslant t_c \\ (t - t_c)\tan\gamma_c + t_c\left(\frac{\tan\gamma_c}{\ln(\tan\gamma_c+1)} - 1\right) + n_c & t > t_c \end{cases} \tag{9}$$

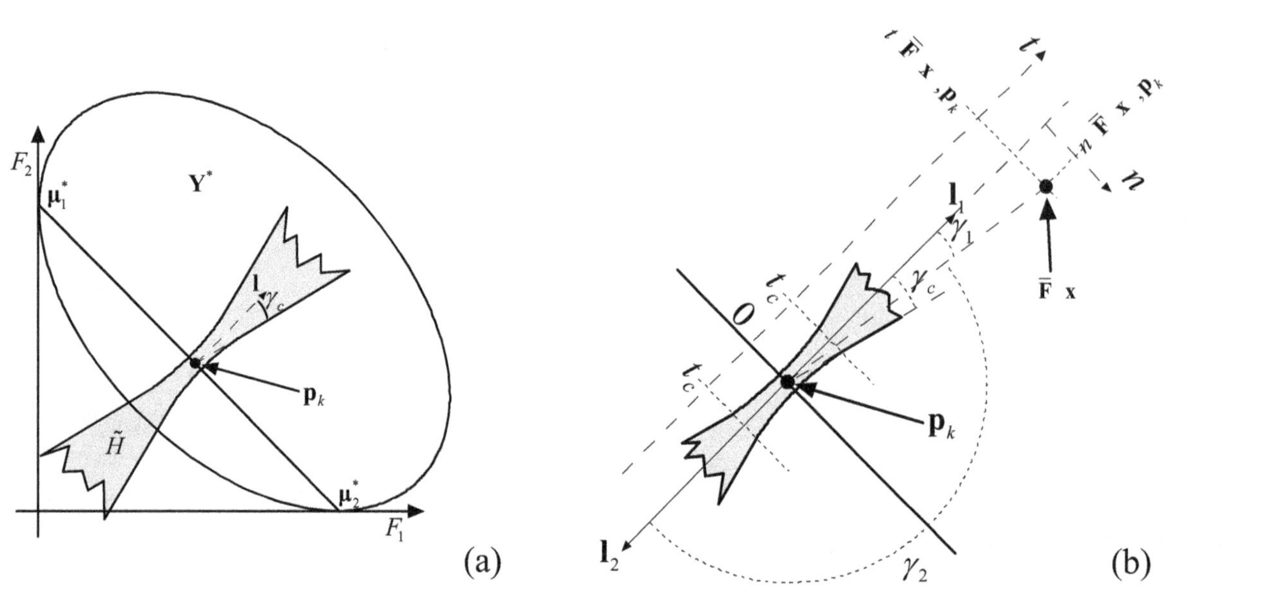

Fig. 8. DHCBI: a) problem formulation; b) detail of the DHCBI formulation.

where n_c is a fraction of the Euclidean distance between two contiguous utopia plane points, γ_c is the cone angle and $t_c = \frac{n_c}{\tan \gamma_c}$ is the distance from $\mathbf{p}_k$ along $\pm\mathbf{l}$ at which the constraint becomes a hyper-cone. The formulation of the additional constraint, c, allows maintaining the continuity of the first derivative $t = 0$ and $t = t_c$:

$$c'(t) = \begin{cases} e^{\ln(\tan\gamma_c+1)\frac{t}{t_c}} - 1 & t \leqslant t_c \\ \tan\gamma_c & t > t_c. \end{cases} \tag{10}$$

In its formulation the DHCBI method (6) differs from the NBI method (4), in that the additional equality constraint of the NBI method is substituted with the inequality constraint $c(9)$.This results in an increased size of the feasible region with respect to the NBI approach. Furthermore, while the NC method 3.2 requires $M-1$ additional constraints, the DHCBI requires just one.

Finally function $q(\bar{\mathbf{F}}(\mathbf{x}), \mathbf{p}_k)$ in (6) and (8) is chosen as:

$$q(\bar{\mathbf{F}}(\mathbf{x}), \mathbf{p}_k) = q(n) = \begin{cases} \frac{n^{c+1}}{n_c^c},\ n \leqslant n_c \\ (1-c)n + \frac{cn^2}{n_c},\ n > n_c \end{cases} \tag{11}$$

to direct the solver towards the axis of the hyper-cone.

For obtaining a complete representation of the Pareto frontier it is necessary to obtain also those Pareto points that belong to the peripheral region. Here we propose an alternative approach for obtaining such points with respect to the one proposed by Messac and Mattson [9]. Since the points belonging to the peripheral region are external to the polygon of vertices $\boldsymbol{\mu}_i^*$, a possible solution is that of generating the peripheral region utopia plane points $\mathbf{p}^+$ from the utopia plane points belonging to the edges of the polygon.

Let us consider an utopia plane point $\mathbf{p}_k^*$ belonging to the kth edge of the polygon identified by the two anchor points μ_{j+1} and μ_j. Then $\mathbf{p}_k^* = \sum_{i=1}^{M} \alpha_{ki}\mu_i^*$ with $\alpha_{ki} = 0$ for $i \neq j, j+1$ and $0 < \alpha_{ki} < 1$ for $i = j, j+1$. Similarly to [10], a vector $\mathbf{s}$ lying in the utopia plane, orthogonal to the kth edge and pointing outwards with respect to the polygon, can be defined as:

$$\mathbf{s} = \frac{\boldsymbol{\nu}_{j-1} + \beta_j \boldsymbol{\nu}_j}{|\boldsymbol{\nu}_{j-1} + \beta_j \boldsymbol{\nu}_j|}, \quad \text{with} \quad \beta_j = -\frac{(\boldsymbol{\nu}_{j-1}, \boldsymbol{\nu}_j)}{(\boldsymbol{\nu}_j, \boldsymbol{\nu}_j)} \tag{12}$$

where $\boldsymbol{\nu}_j = \boldsymbol{\mu}_{j+1}^* - \boldsymbol{\mu}_j^*$.

Then for each of the utopia plane points $\mathbf{p}_k^*$ belonging to an edge, the peripheral region utopia plane points can be obtained as:

$$\mathbf{p}^+ = \mathbf{p}_k^* + kn_d\mathbf{s}, \quad k = \{1, 2, ...\} \tag{13}$$

where n_d corresponds to the distance between to adjacent anchor points $\mathbf{p}^+$ and k is increased until the optimizer fails to obtain a solution. By following this approach, each new peripheral region utopia plane point $\mathbf{p}^+$ is generated from an initial utopia plane point $\mathbf{p}_k^*$, moving orthogonally to the edge of the polygon to which $\mathbf{p}_k^*$ belongs and parallel to the utopia plane.

3.3 Local Pareto frontiers

With regard to the local and global minima of a single-objective optimization problem, we can define as local Pareto frontier, a Pareto frontier within a particular neighbourhood of the design space. The procedure presented bellow shows how such frontiers can be obtained. The procedure has two main advantages, the first is that a more complete investigation of the design space is performed improving the chances of obtaining the global Pareto frontier. Secondly, it allows the DM to obtain and analyse a set of solutions that can be of interest from an engineering point of view, for example, when accounting for additional considerations such as design robustness.

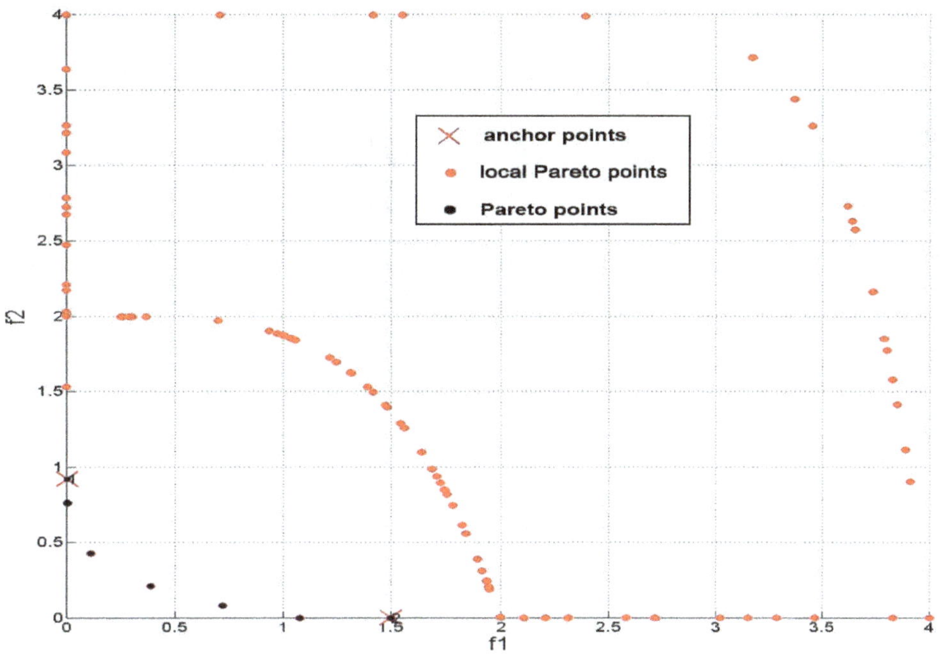

Fig. 9. Criterion space.

The procedure for obtaining the local Pareto frontiers is the following:

1. A Latin-hypercube sampling procedure is used for obtaining L well distributed points in the design space.
2. L single-objective minimizations are performed for each of the M objectives, yielding the *local anchor points* $\boldsymbol{\mu}_{ij} \in R^M$, $i \in \{1, ..., M\}$, $j \in \{1, ..., L_i\}$, where L_i is the number of local minima found for objective i.
3. Among the $\sum_{i=1}^{M} L_i$ local minima, the M *global anchor points* $\boldsymbol{\mu}_i \in R^M$, $i \in \{1, ..., M\}$ are identified.
4. For objective i and local minima j, *local anchor points* $\boldsymbol{\mu}_1, ..., \boldsymbol{\mu}_{ij}, ..., \boldsymbol{\mu}_M$ are used for defining the K utopia plane points:

$$\mathbf{p}_{kij} = \sum_{l=1, l \neq i}^{M} \alpha_{kl}\boldsymbol{\mu}_l + \alpha_{ki}\boldsymbol{\mu}_{ij}, \quad k = \{1, 2, ..., K\}. \tag{14}$$

5. For utopia plane point $\mathbf{p}_{kij}$ sub-problem k is reformulated and solved.
6. Steps 4 through 6 are repeated L_i times for each of the M objectives.

Similarly to the NBI and NC methods, the order in which the sub-problems are solved in the DHCBI is chosen to minimize the distance between each pair of subsequent utopia plane points considered. This allows for using the solution of the previous utopia plane point as a starting point for the subsequent sub-problem, thus minimizing the number of iterations required for convergence and producing a solution in its neighbourhood.

Starting the sequence of optimizations from the utopia plane point closest to the local anchor point $\boldsymbol{\mu}_{ij}$ and moving towards the other $M-1$ global anchor points allows to obtain the local Pareto in the neighbourhood of $\boldsymbol{\mu}_{ij}$.

3.4 Example of solution obtained with the DHCBI

The following test case is a multimodal multi-objective optimization problem developed by Deb [12] which has been widely used for evaluating evolutionary algorithms.

$$\min[F_1(x_1, x_2), F_2(x_1, x_2)] \quad \text{for} \quad 0 < x_i < 1 \tag{15}$$

where

$$F_1 = 4x_1, \quad F_1 = g_1 g_2,$$

$$g_1 = \begin{cases} 4 - 3e^{-\left(\frac{x_2 - 0.2}{0.02}\right)^2} & (\mathrm{x}_2 < 0.4) \\ 4 - 3e^{-\left(\frac{x_2 - 0.7}{0.02}\right)^2} & (\mathrm{x}_2 \geqslant 0.4) \end{cases} \quad \text{and}$$

$$g_2 = \begin{cases} 1 - \left(\frac{F_1}{g_1}\right)^{0.25 + 3.75(g_1 - 1)} & F_1 \leqslant g_1 \\ 0 \; F_1 > g_1. \end{cases} \tag{16}$$

Figure 9 shows the criterion space, including the various local Pareto frontiers obtained with the DHCBI method. The global Pareto frontier is identified by the black dots (lower left corner). It must be noted that weak Pareto points have not been removed. In this test case it can be demonstrated that the global Pareto frontier is extremely sensitive to variations of variable x_2. In such a case, if the DM wished to consider robustness, then solutions from the second best local Pareto frontier could be considered as a better choice.

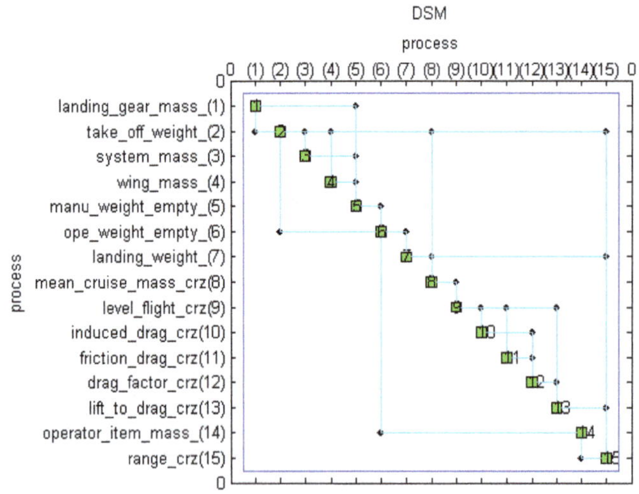

Fig. 10. Results from the computational flow modelling process. Due to the size of the problem, shown is only the Design Structure Matrix of the Strongly Connected Component with the shortest feedback length.

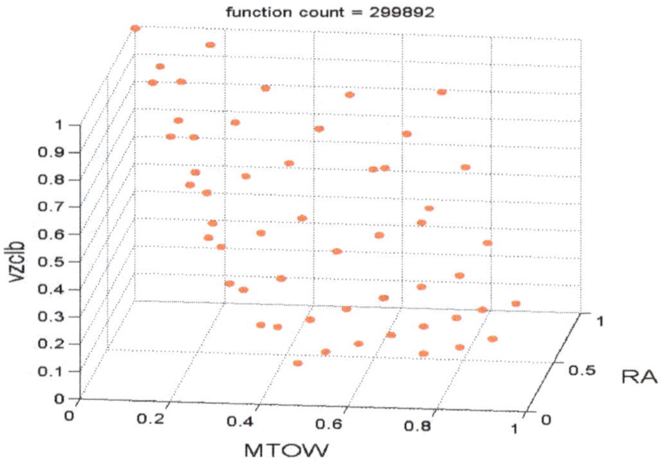

Fig. 11. Normalised Pareto frontier.

4 Test results

A simplified set of models which represent a real aircraft is considered for testing the computational process concept developed in this research. The test case contains 96 models and 126 variables. The multi-objective optimization problem is set as follows:

- Objectives: range (RA) to be maximized, takeoff weight (MTOW) to be minimized and climb rate (vzclb) to be maximized.
- Constraints: takeoff field length $\leqslant$ 2000 m, approach speed $\leqslant$ 120 kts, climb rate $\geqslant$ 500 ft/min, cruise thrust coefficient $\leqslant$ 1 and wing fuselage fuel ratio $\geqslant$ 0.75.
- Independent variables: engine thrust, wing span, sweep angle, thickness to chord, fuel quantity and wing span.
- Constant variables: number of first class passengers, number of standard class passengers, number of aisles, engine bypass ratio, number of engines, cruise altitude, cruise mach, takeoff altitude and approach altitude.

Following the procedure described in Figure 1, an optimal computational plan (process) for the test case was obtained. It was found that, out of the 96 models, 15 were strongly connected thus forming a SCC with 15 models. For doing this several candidate variable flow models were rearranged using a genetic algorithm to minimize the feedback length. Figure 10 shows the SCC with minimum feedback. Figures 11 and 12 show two views of the 30 Pareto points (each one representing an aircraft configuration) generated by the application of DHCBI on the optimal computational process.

While the emphasis of the current work on the DHCBI has been on the effectiveness (quality) of the solution, the focus of the work on process modeller has been on flexibility and efficiency. The extensive testing of the process modeller demonstrated that the selections which it made were always amongst the best in terms of convergence.

5 Conclusions

A novel approach for obtaining optimal computational plans for conceptual design studies is proposed in this paper. It incorporates a novel variable flow method based on the incidence matrix concept. Unlike other existing methods the variable flow method can handle multiple outputs of a particular constituent model as well as the identification of strongly connected components (SCCs) in the entire multidisciplinary set of models and equations describing the aircraft. Traditionally, variable flow modelling and rearrangement of SCCs are performed separately and the only link is the transfer of variable flow model results to the rearrangement process. Our approach is capable of exploring a number of feasible variable flow models according to the objectives of the particular design study. The computational process is being integrated with the proposed novel Double Hyper Cone Intersection Method (DHCBI) for obtaining the Pareto frontier. The advantage of this method is that the number of analyses to be performed increases automatically with the number of local minima of the objective functions. Thus the DHCBI algorithm adapts to the complexity of the problem to be solved. Since a local (gradient based) optimizer such as the sequential quadratic programming (SQP) algorithm needs to be used for obtaining local Pareto solutions, the determination of the gradients would increase the number of analysis to be performed. This problem could be alleviated by the application of automatic differentiation (AD) for obtaining partial derivatives.

Future work will advance further the research into convergence issues associated with the SCCs and the application of AD for robust optimization. Also a significant effort is underway to create an object-oriented conceptual design framework incorporating workflow management, multi-objective optimisation and uncertainty analysis.

Acknowledgements

This research was conducted as part of VIVACE (Value Improvement through a Virtual Aeronautical Collaborative

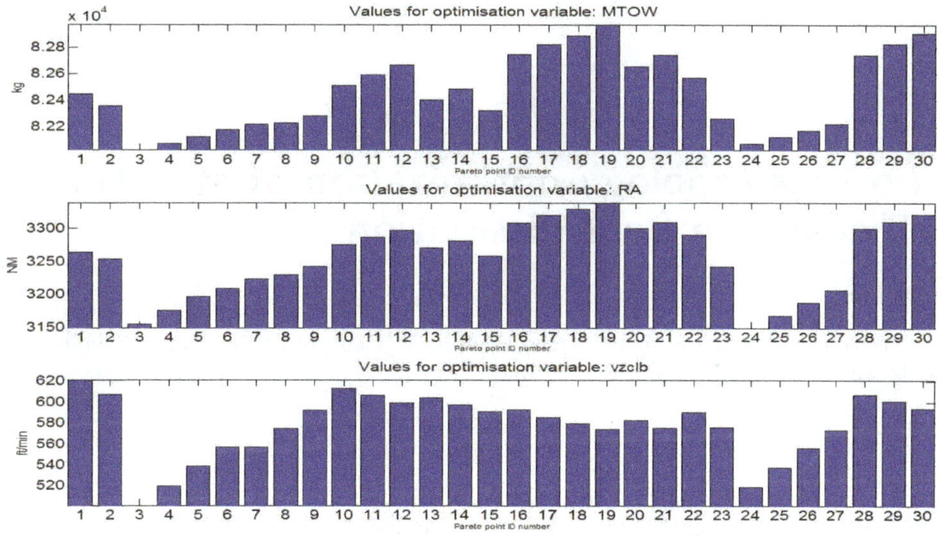

Fig. 12. Pareto frontier – absolute values by point.

Enterprise)- integrated project AIP3 CT-2003-502917, partly sponsored by the Sixth Framework Programme of the European Community under priority 4, "Aeronautics and Space".

References

1. D.V. Steward, *Systems Analysis and Management: Structure, Strategy and Design* (Petrocelli Books, Inc., New York, 1981)
2. M.J. Buckley, K.W. Fertig, D.E. Smith, Design sheet: an environment facilitating flexible trade studies during conceptual design, in *1992 Aerospace Design Conference*, 3–6 February, 1992 (Irvine, Ca), AIAA 92-1191
3. D. Tang, L. Zheng, Z. Li, Computers and Industrial Engineering **38**, 479 (2000)
4. M.D. Guenov, T.D. Libish, H. Lockett, Computational design process modelling, in *25th International Congress of the Aeronautical Sciences conferences (ICAS)*, 3–8 September, 2006 (Hamburg, Germany), ISBN 0-9533991-7-6
5. P. Fantini, L.K. Balachandran (Libish), M.D. Guenov, Computational Intelligence in Multi Disciplinary Optimization at Feasibility Design Stage, in *First International Conference on Multidisciplinary Design Optimization and Applications*, 17–20 April, 2007, Besancon, France (EDP Sciences), ISBN 978-2-7598-0023-0
6. I. Das, J.E. Dennis, SIAM J. Optimiz. **8**, 631 (1998)
7. I. Das, An Improved Technique for Choosing Parameters for Pareto Surface Generation Using Normal-Boundary Intersection, in *Third World Congress of Structural and Multidisciplinary Optimization (WCSMO-3),* edited by C.L. Bloebaum, et al., 17–21 May, 1999 (Buffalo, NY, University of Buffalo), Vol. 2, pp. 411–413
8. A. Messac, A. Ismail-Yahaya, C.A. Mattson, Structural and Multidisciplinary Optimization, Journal of the International Society of Structural and Multidisciplinary Optimization (ISSMO) **25**, 86 (2003)
9. A. Messac, C. Mattson, AIAA J. **42**, 2101 (2004)
10. S.V. Utyuzhnikov, P. Fantini, M.D. Guenov, Numerical Method for Generating the Entire Pareto Frontier in Multiobjective Optimization, in *EUROGEN 2005*, 12–14 September, 2005 (Munich, Germany) ISBN: 3-00-017534-2
11. C.A. Mattison, A.A. Mullur, A. Messac, Minimal Representation of Multiobjective Design Space Using Smart Pareto Filter, in *9th AIAA/ISSMO Symposium on Multidisciplinary Analysis and Optimization*, 4–6 September, 2002. Atlanta, GA, AIAA-2002-5458
12. K. Deb, *Multi-objective genetic algorithms: problem difficulties and construction of test problems*, Technical Report CI-49/98, Dortmund: Department of Computer Science/LS11, University of Dortmund, Germany, 1998

BESO method for topology optimization of structures with high efficiency of heat dissipation

Dan He[a] and Shutian Liu

State Key Laboratory of Structutal Analysis for Industrial Equipment, Dept. of Engineering Mechanics, Dalian Univerisity of Technology, 116023 Dalian, P.R. China

Abstract – The purpose of this paper is to present a Bi-Directional Evolutionary Structural Optimization (BESO) method for topology optimization of heat conduction structures. In BESO method the elements are allowed to be added as well as removed. Focused on the heat performance of structure, the additive criterion and rejection criterion were proposed respectively. With the limit volume of the high conductive material, the optimal layout of structure with high efficiency of heat dissipation and uniform temperature distribution can be obtained efficiently by BESO procedure.

1 Introduction

The design process is generally divided into two stages: conceptual design and detailed design. The structural layout can be obtained in the conceptual design stage. Then the shape and the size of the structure are acquired in the detailed design stage. It has been recognized that by using the topology optimization techniques in the conceptual design stage, we can yield substantial improvements in the performance of structures.

Relative to the shape and topology optimization of structures with mechanical properties (Bendsoe [1], Li et al. [2], Rozvany et al. [3]), the thermal conducting solid issue has received relatively less attention in spite of its significance. Cooling fins, thermal diffusers and moulding dies are examples of shape optimization in this category (Lee [4]). With the rapid development of microelectronic and micro processing technology, how to remove the heat as rapidly as possible from the package becomes further important. One solution is to insert a finite amount of high-conductivity material in the structure to increase the conductivity of the structure. Optimizing the topology (the distribution of the high conductivity material) can not only reduce the cost of high-conductivity material and manufacture, but also suit for further miniaturization (Shutian and Yongcun [5]). Bejan [6] puts forward the "tree-like network" construct method based on the construction theory which derived the optimal high effective conduction channel distribution with uniform heat resource and high conductivity ratio of high conductive material to conductivity of substrate material. Cheng et al. [7] constructed the high effective conduction channel in different conditions by bionic optimization based on the biological evolution principle. The bionic optimization has no limitations that the heat generation is uniform and the ratio of two kind's materials is high, however, the construct obtained by bionic optimization is similar to that obtained by Bejan's "tree-like network" construct method both in structure features and heat transfer performance. In Wu et al. [8], the "tree-like network" construct method was improved and the high effective conduction channel distribution was optimized again without the premise that the new order assembly construct must be assembled by the optimized last order construct. A better construct was obtained and the limit of the minimum heat resistance is derived when the thermal conductivity and the proportion of the two heat conduction materials are constant. Li et al. [9] developed an efficient finite element based computational procedure for the topology design of heat conducting fields. By removing or degenerating the conductive material of the elements with the most negative sensitivity the temperature objective at the control point can be most efficiently reduced.

The evolutionary structural optimization (ESO) method has been studied in recent years (Li et al. [2]). Originally the method was conceived from the engineering perspective that the topology and shape of structures were naturally conservative for safety reasons and therefore contained an excess of material. To move from the conservative design to a more optimum design would therefore involve the removal of material. On the contrary, by adding material onto the most efficient regions begins from a minimum structure to obtain optimum is the additive ESO (AESO) method (Querin et al. [10]).

The validity of the ESO method has been examined critically and several arguments have been made against it (Zhou et al. [17], Rozvany et al. [3]), since it may lead to highly non-optimal solutions in same circumstances.

[a] Corresponding author:
stliu@dlut.edu.cn, hedan@student.dlut.edu.cn

However, this method could be justified if a user wishes to achieve rapid design improvement at a low computational cost. Recently, it has been stated that ESO has an actual theoretical basis (Tanskanen [18]).

A new development in ESO is the recent introduction of bidirectional evolutionary structural optimization (BESO), which is the combination of ESO and AESO. In BESO, elements are allowed to be added as well as removed (Querin et al. [11], Huang and Xie [12], Zhu et al. [13]). In recent years the ESO method has been developed into an effective engineering design tool and successfully extended allowing different structural constraints to be incorporated into the optimization process, such as stiffness (Chu et al. [14]), frequency (Li et al. [2]) or buckling (Manickarajah et al. [15]). Li et al. [16] extend the algorithm of Evolutionary Structural Optimization to shape and topology design problems subjected to steady heat conduction. Li et al. [9] developed an efficient finite element based computational procedure for the topology design of heat conducting fields. Based on the authors' knowledge, the Bi-Directional Evolutionary Structural Optimization (BESO) method has not been used in topology design for heat dissipation.

The purpose of this paper is to present a Bi-Directional Evolutionary Structural Optimization (BESO) method for topology optimization of structures with high efficiency of heat dissipation. The BESO method is the combination of ESO and AESO, which makes elements allowed to be added and rejected. So it has two evolution procedures: the AESO procedure and the ESO procedure. Considering the design object of high efficiency of heat dissipation and uniform distribution of temperature, sensitivity of heat resistance was taken as the element rejection criterion in the ESO procedure, which makes the efficiency of heat dissipation high; and the elemental temperature gradient level was taken as the element additive criterion which makes the temperature distribution of structure uniform.

The design domain of the structure is depicted by the relative density of high-conductivity material at any point and the relative densities assigned to the finite elements are chosen as design variables. Being similar with the compliance in structural problems we introduce a factor of heat resistance which is defined as the integration of the product of heat flux and temperature over the domain. Heat resistance generally described the efficiency of heat dissipation of structure. The element sensitivity of heat resistance was taken as element rejection criterion in the ESO procedure, which means the elements with lowly sensitivity of heat resistance are regarded as inefficient region and should be removed from structure. The elemental temperature gradient level was taken as the element additive criterion in the AESO procedure, which means elements with highly temperature gradient are considered as efficient region and should be added material on that. The checkerboard patterns (Bourdin [19], Zhou et al. [17]) for topology optimization of mechanical problems, can also be observed in heat conduction problems based on the ESO procedure. To avoid the checkerboard design, an intuitive smoothing technique is introduced. Numerical examples are also presented to demonstrate the capabilities of the method proposed in this paper.

2 Heat resistance and sensitivity analysis

The governing equation and boundary condition for general steady state heat conduction problems are well known as

$$\frac{\partial}{\partial \mathbf{x}_i}\left(\mathbf{k}_{ij}\frac{\partial \mathbf{T}}{\partial \mathbf{x}_j}\right)+\mathbf{Q}=0, \quad \text{on } \Omega \tag{1}$$

$$\mathbf{T}\,|_{t=0}=\mathbf{T_0}\,(x,y,z)\,, \quad \text{on } \Gamma_1 \tag{2}$$

$$\left(\alpha_i\frac{\partial \mathbf{T}}{\partial \mathbf{x}_i}n_i\right)_s=q_s, \quad \text{on } \Gamma_2 \tag{3}$$

$$\alpha_i\frac{\partial \mathbf{T}}{\partial \mathbf{x}_i}n_i=h(\mathbf{T}_f-\mathbf{T}), \quad \text{on } \Gamma_3. \tag{4}$$

Where $\mathbf{k}_{ij}$ denotes heat conductivities, $\mathbf{Q}$ the heat energy generated per unit volume, $\mathbf{T}$ the temperature field. Equation (2) described the temperature condition on boundary Γ_1, and $\mathbf{T}_0\,(x,y,z)$ denotes the given temperature. Equation (3) described the heat flux condition on boundary Γ_2, and q_s the external heat flux. Equation (4) described the convectional condition on boundary Γ_3, and h the convection heat transfer coefficient, and T_f the temperature of coolant.

Equation (1) can be approximated by means of a finite element formulation as

$$\mathbf{KT}-\mathbf{Q}=0, \tag{5}$$

where $\mathbf{K}$ represents the global conductivity matrix, $\mathbf{T}$ the global nodal temperature vector and $\mathbf{Q}$ the applied heat load vector.

To represent the efficiency of heat dissipation, a factor of heat resistance is introduced here, which is defined as the integration of the product of generating heat rate in unit volume $Q(x)$ and temperature $\bar{T}(x)$ over the domain, and it is chosen as the design objective function of the ESO procedure.

$$D=\int_{\Omega}\bar{\mathbf{T}}\mathbf{Q}d\boldsymbol{\Omega}+\int_{\Gamma_2+\Gamma_3}q_n\bar{\mathbf{T}}ds. \tag{6}$$

When Q is specific, reducing the temperature T will lead to the reduction of the factor of heat resistance D. Generally, to minimize the temperature can be implemented by minimizing this factor D. In the other hand, the factor of heat resistance described the heat energy stored in the structures when the heat exchange process reaches the steady state. Thus, D can be chosen as the represented index of the heat dissipation property of the structures. When the heat conduction problem is solved with finite element method, and only the temperature boundary condition is considered, the factor of heat resistance can be expressed in teams of the nodal temperature vector as

$$D = \mathbf{T}^{\mathrm{T}}\mathbf{K}\mathbf{T}. \qquad (7)$$

Based on the basic idea of structural topology optimization, the topology of the structure is depicted by the relative density of materials at any point and the relative density assigned to each element is chosen as design variable (Bendsoe and Sigmund [20]).

$$\rho_i = \begin{cases} 1 & \text{if} \quad i\text{th element} \in \Omega_s \\ 0 & \text{if} \quad i\text{th element} \in \Omega/\Omega_s \end{cases} \quad i = 1, 2, n_e, \qquad (8)$$

where ρ_i denotes the density of the materials at ith element, and the element which has $\rho_i = 1$ means there high conductive material exists; the element which has $\rho_i = 0$ means there a void exists. Ω denotes the design domain, Ω_s denotes the area occupied by solid material, and n_e denotes the total number of elements in the design domain. The heat conductive matrix of each element $\mathbf{k}_i$ is linearly depended on the density of the ith element, which is shown as:

$$\mathbf{k}_i = \mathbf{k}_{i0} \times \rho_i, \qquad (9)$$

where $\mathbf{k}_{i0}$ is the conductive matrix of the ith element when it was an entity element.

The sensitivity of the factor of heat resistance of heat conduction problem can be obtained by differentiating equation (7).

$$\frac{\partial D}{\partial \rho_i} = \frac{\partial \mathbf{T}^T}{\partial \rho_i}\mathbf{K}\mathbf{T} + \mathbf{T}^T\frac{\partial \mathbf{K}}{\partial \rho_i}\mathbf{T} + \mathbf{T}^T\mathbf{K}\frac{\partial \mathbf{T}}{\partial \rho_i}. \qquad (10)$$

Differentiating equation (5) yields

$$\frac{\partial \mathbf{K}}{\partial \rho_i}\mathbf{T} + \mathbf{K}\frac{\partial \mathbf{T}}{\partial \rho_i} = 0. \qquad (11)$$

Introducing above equation into equation (9), the sensitivity of resistance factor can be expressed as

$$\frac{\partial D}{\partial \rho_i} = -\mathbf{T}^T\frac{\partial \mathbf{K}}{\partial \rho_i}\mathbf{T} = -\mathbf{T}^T\mathbf{k}_{i0}\mathbf{T}. \qquad (12)$$

Thus, the sensitivity analysis of the objective function can be obtained mathematically and in an efficient manner in the iteration process.

3 BESO method for topology optimization of heat conduction dissipation

Here we present a BESO method for topology optimization of heat conduction dissipation. The BESO method is the combination of ESO and AESO, which makes elements allowable to be added and rejected. So it has two evolution procedures: the AESO procedure and the ESO procedure. Considering the design purpose being to find an appropriate distribution of high conductive material to improve the efficiency of heat dissipation and the fact that minimum difference of the maximum and minimum temperature in the domain is required, sensitivity of heat resistance was taken as the element rejection criterion in the ESO procedure, which makes the efficiency of heat dissipation high; and the elemental temperature gradient level was taken as the element additive criterion which makes the temperature distribution uniform.

4 Element rejection criterion and the procedure of ESO

It is clear that removing those conductive elements which have the lowest value of sensitivity of the factor of heat resistance will result in the least contribution to the addition of the energy of system. According the brief idea of ESO, by gradually rejecting the inefficient element, the optimal structure will be obtained. The evolutionary criterion for such a purpose is determined by comparing the heat resistance sensitivity of each element. So the element should be removed from domain field if its sensitivity of heat resistance is the lowest. The ESO procedure in each iteration was shown as follows:

(1) FEA analysis and sensitivity analysis;
(2) remove those entity elements which have the lowest value of sensitivity of heat resistance;
(3) calculate the numbers of element rejected in this iteration α;
(4) if $\alpha < \alpha_{ESO}$, then repeat (1)–(4);
(5) if $\alpha \geqslant \alpha_{ESO}$, then finish this iteration of ESO.

Here the α_{ESO} was the element rejection ratio in each iteration, which is given to control the process of optimization. And by numerical experience, the ratio of element rejection was given as $\alpha_{ESO} = 0.5\%$, that's to say, there are 0.5% of all the elements in the domain field will be rejected during each iteration. Here the rejection of element doesn't means the density value was given as 0, but the density of rejected element was given a very low value (as 1.0E-5) for the numerical continuity.

5 Element additive criterion and the procedure of AESO

Structure with best heat performance can be immediately described as: minimize the maximum temperature. Based on the intuitive thinking, the process of reducing the temperature as low as possible by adding material can be implemented by making the temperature distribution uniform. However, in the process of adding elements, it is best favoured for averaging the system temperature to add material onto the void element which has the largest temperature gradient to the next entity element.

The temperature of each element is described as the average value of the temperature of its nodes.

$$T_i = \sum_{j=1}^{m} t_j, \qquad (13)$$

where T_i denotes the temperature of the ith element, t_j denotes the temperature of the jth node of the element, and m denotes the node number of the ith element.

In the process of AESO, we compute the temperature gradient of every entity element and the next void element, choose the void element with largest temperature gradient to develop. In the optimize process, the temperature of entity element may be lower than the next void element. In that case, we compute the void element with highest temperature and make it as entity element, and then continue. The AESO procedure was shown as follows:

(1) FEA analysis;
(2) calculate the temperature of each element;
(3) calculate the temperature grad between entity element and its neighbour hole element;
(4) add material onto the hole which has the largest temperature gradient to the next entity element;
(5) calculate the number of element added in this iteration β;
(6) if $\beta < \beta_{AESO}$, then repeat (1)–(6);
(7) if $\beta \geqslant \beta_{AESO}$, then finish this iteration of AESO.

Here the β_{AESO} was the element added ratio in each iteration, which is given to control the process of optimization. By numerical experience, the ratio of element added was given as $\beta_{AESO} = 1.0\%$, that's to say, there are 1.0% of all the elements in the domain field will be added during each iteration. Here the addition of element means the density of the hole element was changed to 1.0.

6 Evolutionary procedure

The BESO method is the combination of ESO and AESO, so in the BESO procedure, elements of the structure can either be added or removed. The evolutionary procedure was shown as follows:

(1) the maximum allowable domain the structure can occupy must be specified;
(2) the physical domain is subdivided with a densely populated regular finite element mesh, and all the elements are hole elements in initial;
(3) define all kinematic boundary constraints, loads, material properties and volume constraint;
(4) AESO procedure;
(5) ESO procedure;
(6) examine the volume constraint;
(7) repeat (4)–(6) until the volume constraint was reached.

What should be noticed is that, at the beginning of the optimization, all the elements are holes, we calculate the elements with highest temperature and develop them to entity elements in the first AESO process.

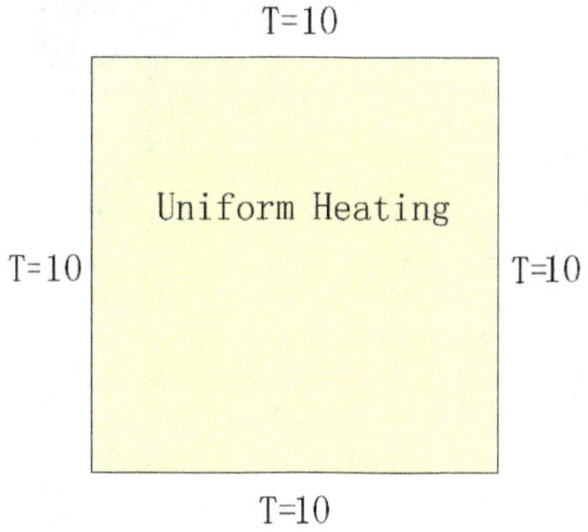

Fig. 1. A square plate with uniformly distributed heat resource and boundaries with prescribed temperatures.

7 Smoothing technique

The checkerboard patterns for topology optimization of mechanical problems, can also be observed in heat conduction problems based on the BESO procedure. To avoid the checkerboard design, an intuitive smoothing technique is introduced, in brief whose idea demonstrate as follow. The volume fraction of each element is affected by that of correlation elements, which reads as

$$\overline{f_s^m} = \sum_{j=1}^{N} W_{mj} f_s^m;$$
$$\overline{f_1^m} = \sum_{j=1}^{N} W_{mj} f_1^m W_{mj} = (r_0 - r_{mj}) \Big/ \sum_{k=1}^{N} (r_0 - r_{mk}), \tag{14}$$

where N denotes total number of correlation elements to element m, which is determined by the factor r_0. W_{mj} is the weight factor of element j in element m, which is determined by the distance between centroids of two elements. The weight factor is shown as follows:

$$W_{mj} = (r_0 - r_{mj}) \Big/ \sum_{k=1}^{N} (r_0 - r_{mk}), \tag{15}$$

where r_{mj} is the distance between the centroids of element m and j.

8 Illustrative examples

Here we consider a square of plate (length of sides is 0.1 m) with uniform heating resource (3600 kw/m^2), and temperature of 4 sides were predicated (10 °C). The conductivity of base material is $k_{low} = 2.1$ w/(m.°C), and the design target is to design the layout of high conductive material ($k_{high} = 210$ w/(m.°C)) on the plate which can reduce the temperature of structure and improve the efficiency of dissipation of structure.

The geometry model was shown in Figure 1, the volume constraint of the high conductivity material was 50%.

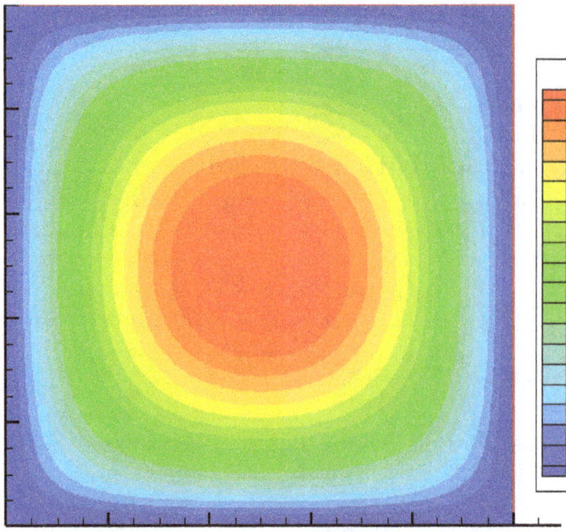

Fig. 2. Temperature distribution when high conductive material is uniformally distributed.

Fig. 3. Optimal distribution of high conductive material based on BESO method.

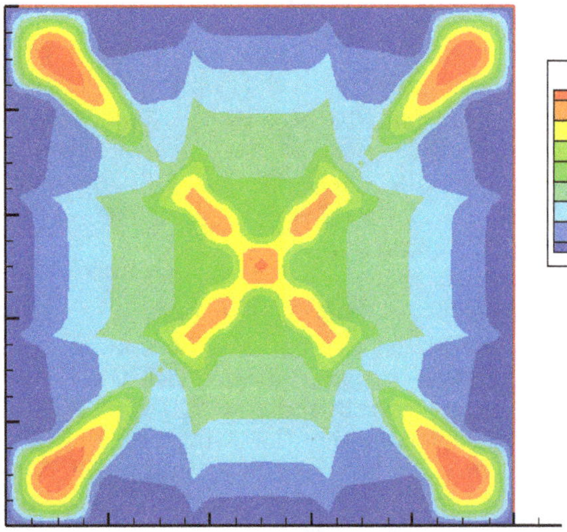

Fig. 4. Temperature distribution of optimal design based on BESO method.

Fig. 5. Optimal distribution of high conductive material based on SIMP method.

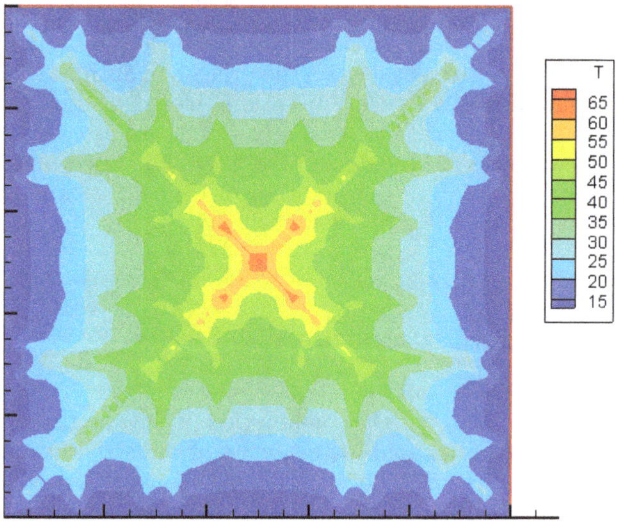

Fig. 6. Temperature distribution of optimal design based on SIMP method.

The design domain is divided into $60 \times 60 = 3600$ elements, and the layout of high conductive material was calculated by the proposed BESO method. The optimal result is shown in the Figure 3. The optimal results give out the path that collect and transfer the energy. From the temperature contour map shown in Figure 4, it can be seen that the temperature distribution is more uniformly distributed, and the maximum temperature is reduced about 50%. And compare with the result based on SIMP method, the topology based on BESO method is similar with the optimal result based on SIMP method, which can demonstrate that the proposed method is valid. And the BESO result seemed more easily to manufacture.

9 Concluding remarks

It can be concluded from this work that the topology optimization in the presence of heat conduction can be easily solved by the proposed BESO method. It can offer the designer with optimal solutions in the design stage which can make significant improvement in the thermal performance. The numerical experiments demonstrate that the proposed method is valid and effective. Compare with the result based on SIMP method, the topology based on BESO method is more easy to manufacture.

References

1. M.P. Bendsoe, *Optimization of Structure Topology, Shape and Material*, Springer-Verlag, Berlin (1995)
2. Q. Li, G.P. Steven, Y.M. Xie, A simple checkerboard suppression algorithm for evolutionary structural optimization. Structural and Multidisciplinary Optimization **22**, 230–239 (2001)
3. G.I.N. Rozvany, M.P. Bendsoe, U. Kirsch, Layout optimization of structures. Applied Mechanics Review **48**, 41–118 (1995)
4. B.Y. Lee, Shape sensitivity formulation for an axisymmetric thermal conducting solids. J. Mech. Eng. Sci. **207**, 209–216 (1993)
5. L. Shutian, Z. Yongcun, *Design of high-conduction paths based on pology optimization*, in *Proc The Fourth China-Japan-Korea Joint Symposium on Optimization of Structural and Mechanics System Kunming*, China (2006) pp. 449–454
6. A. Bejan, *Shape and Structure, from Engineering to Nature*, Cambrige University Press (2000)
7. X.G. Cheng, Z.X. Li, Z.Y. Guo, Constructs of highly effective heat transport paths by bionic optimization. Science in China Series E-Technological Sciences **46**, 296–302 (2003)
8. W.J. Wu, L.G. Chen, F.R. Sun, Improvement of tree-like network constructal method for heat conduction optimization. Science in China Series E-Technological Sciences **49**, 332–341 (2006)
9. Q. Li, G.P. Steven, Y.M. Xie, O.M. Querin, Evolutionary topology optimization for temperature reduction of heat conducting fields. International Journal of Heat and Mass Transfer (2004) **47**, 5071–5083
10. O.M. Querin, G.P. Steven, Y.M. Xie, Evolutionary structural optimisation using an additive algorithm. Finite Elements in Analysis and Design **34**, 291–308 (2000)
11. O.M. Querin, V. Young, G.P. Steven, Y.M. Xie, Computational efficiency and validation of bi-directional evolutionary structural optimisation. Computer Methods in Applied Mechanics and Engineering **189**, 559–573 (2000)
12. X. Huang, Y.M. Xie, Bidirectional evolutionary topology optimization for structures with geometrical and material nonlinearities. AIAA Journal **45**, 308–313 (2007)
13. J.H. Zhu, W.H. Zhang, K.P. Qiu, Bi-directional evolutionary topology optimization using element replaceable method. Computational Mechanics **40**, 97–109 (2007)
14. D.N. Chu, Y.M. Xie, A. Hira, G.P. Steven, Evolutionary structural optimization for problems with stiffness constraints. Finite Elements in Analysis and Design **21**, 239–251 (1996)
15. D. Manickarajah, Y.M. Xie, G.P. Steven, An evolutionary method for optimization of plate buckling resistance. Finite Elements in Analysis and Design **29**, 205–230 (1998)
16. Q. Li, G.P. Steven, O.M. Querin, Y.M. Xie, Shape and topology design for heat conduction by evolutionary structural optimization. International Journal of Heat and Mass Transfer **42**, 3361–3371 (1999)
17. M. Zhou, Y.K. Shyy, H.L. Thomas, Checkerboard and minimum member size control in topology optimization. Structural and Multidisciplinary Optimization **21** 152–158 (2001)
18. P. Tanskanen, The evolutionary structural optimization method: theoretical aspects. Computer Methods in Applied Mechanics and Engineering **191**, 5485–5498 (2002)
19. B. Bourdin, Filters in topology optimization. International Journal for Numerical Methods in Engineering **50**, 2143–2158 (2001)
20. M.P. Bendsoe, O. Sigmund, Material interpolation schemes in topology optimization. Archive of Applied Mechanics **69**, 635–654 (1999)

8

Development of a collaborative optimization tool for the sizing design of aerospace structures

L. Guadagni[1,a]

CNES, Rond-point de l'Espace, 91023 Evry, France

Abstract – The work introduces a computational method for the optimization of aerospace preliminary designs. It was developed considering a collaborative approach to solve the sizing optimization of a structure model defined by finite element method and by a large number of subsystems. A brief introduction explains how the collaborative optimization method can be used for general structural problem and how this approach can offer many advantages to the computation analyses. Central discussion is the description of an automatic procedure defined by three deterministic optimization codes and their interface systems with the MSC.Nastran environment. They decompose the structure complexity executing an interface algorithm for FE models searching an optimized solution for the structural subsystems respecting different kinds of constraints. Furthermore, in order to show the uses of the procedure in an industrial context, two realistic applications for the minimum design are also discussed. The results show a good comparison with the commercial solvers, taking out better results in less time processing and finding new alternative design configurations.

Key words: Collaborative optimization method; FE analysis; aerospace application

1 Introduction

Aerospace engineering is a complex system of different sciences working towards the definition of an aerospace design. Since the advent of aviation the structural problem was one of the more critical tasks for the engineers. In general a good structural solution was a good starting point from which all the other disciplines could adapt. The earliest studies, acting on the interaction between all the aerospace branches, show that a different approach is necessary. As the aerospace design is a multidisciplinary problem, the single disciplines are inserted in an iterative design process: any decision or choice made on a discipline has an effect on the other ones. Nowadays, this fact is well known in the engineering area and with the increase of the aerospace activities, more complex computational methods are been developed. Some good samples and solutions founded to solve complex phisical systems in the aerospace demain are available in different works [1–3].

In a structural approach this means to take into account more detailed model in order to consider the phenomena inherent to all structural aspects; consequently the structural complexity of the design become an important aspect to face because of its influence on the solution method. In the aerospace industry, for example, looking for a structural solution often means considering a large number of design requirements that can heavily modify the solution during the computation. In a mathematical context the design process turns out to be an optimization task definition; defining the structural design with a mathematical model, the target becomes to reach an optimal condition (i.e. a lighter structural configuration in a structural task) where the performances and the requirements have to be feasible and satisfied. There are a lot of programs available today and most of them are general purpose software based on finite element method (FEM) working with non-specific models. Sometimes the response of the optimization is too generic and a more specific solution is required.

In this article an optimization procedure for preliminary aerospace design is presented. The activity is focused on the sizing optimization of complex system defined by a large number of structural subsystems that are improved with dedicated external solvers using the finite element analysis (FEA) only for the static equilibrium determination. The process is iterative and in order to guarantee the effects of each subsystem analysis, the overall model is updated with the previous local results for each iteration. The procedure is a tool of external software working under the control of Multi Level Architectures (MLA) called Collaborative Optimization (CO). The last section of this work is dedicated to the analyses of complex aeronautical and space structures and to the results discussion.

[a] Corresponding author: luca.guadagni@gmail.com

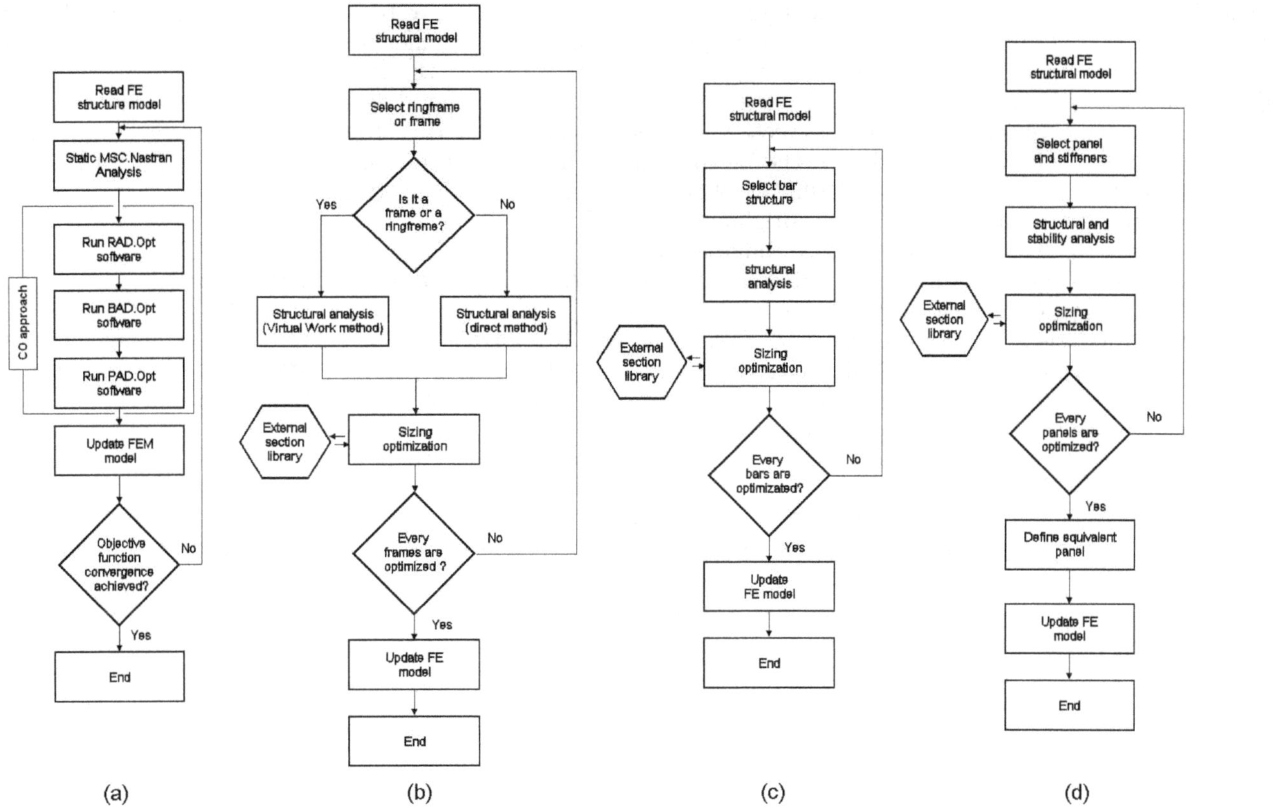

Fig. 1. The flow chart for the global procedure OPT.TOOL (a) and for the local optimization software: RAD.Opt (b), BAD.Opt (c) and PAD.Opt (d).

2 The OPT.TOOL procedure

In this article a Multidisciplinary Design Optimization (MDO) procedure with a CO architecture is used offering many advantages. The main idea is to disassemble the system to different structural subsystems and to consider these as single disciplines. In a CO context the theory imposes that the discipline solvers work as stand-alone, changing the whole process as well as a system of independent subsystem software [4]. The procedure developed here is an OPTimization TOOL (OPT.TOOL) dedicated to the aerospace applications. The software is not a general purpose optimizer, but a manager for local optimization solvers. The procedure was created to deal with sizing optimization tasks where: (1) the global structure is defined by a FEM with high level of complexity; (2) local solvers for optimal sizing are developed for specific subsystems; (3) the local optimal solutions are derived without using the FEA; (4) the optimization goal is the global weight minimization; (5) a large number of design variables and constraints are taken into account; (6) the FE model is updated with all optimal solutions from each external solvers. At the first step of its flow chart (Fig. 1a) a FEM properties reading is provided and in order to evaluate the stress conditions, a static equilibrium of the global model is performed using MSC.Nastran. Furthermore the execution of some external solvers in a CO environment are activated and the process ends with the optimal results updating in a FEM file. Three optimization analyses for aerospace structures are available at the local level: (1) wing and fuselage frame (RAD.Opt); (2) beam and stringer elements (BAD.Opt); (3) stiffened and curved panels (PAD.Opt). All the external codes, the software manager and the algorithms were written by the author in C language using the *Numerical Recipes Library in C*[1].

2.1 The external optimization software

In this paragraph a description of the main common and specific properties for the external optimizators is reported. Each optimization program operates on the suitable elements only: after the property reading from the FE input file and the recovery of the stress conditions, a optimization solution is performed. At the end of the process the optimal results are updated into a new FEM. The flow charts are an assembly of modules defined by several complex algorithms interfacing with the FEM. The codes can distinguish between different type of FE elements (with one and two dimensions only) and automatically translates the mechanical properties for the solver. As the optimization considers the forces acting on the elements, the rest of the data are taken from the static FEA performed at the beginning. Afterwards the codes search

[1] *Numerical Recipes Library in C (2nd ed.)* is a registered trade-mark by Cambridge University Press.

an optimal configuration using the feasible direction algorithm (based on the gradient method) [5] respecting specific requirements. As a commercial software gives a single solution of the optimization problem, the codes are provided with an external library of cross section shapes. When needed it can be used to change the mechanical characteristics (i.e. for a stringer or stiffener element), besides having an improvement on the original model a set of alternative configurations is also available. The process ends with the updating of the optimal data into the global FEM properties. The first local software is a Ring frame Analysis and Design Optimizer (RAD.Opt), its target is to determie an optimal sizing of a frame with a specific shape without using the FEA computation, but resolving inherent closed formulation. As the structural analysis is tightly linked to the frame shape, and to give to the procedure more generality, two closed shapes are taken into account (Fig. 1b): (1) circular shape (*ringframe*) and (2) non-circular shape (*frame*). For both cases a specific structural analysis for the stress distribution determination is available and discussed in a previous work [6]. The second local solvers is the Beam Analysis and Design Optimizer (BAD.Opt) for the computation of structural optimization of longitudinal beam elements, like wing stringers or stiffened panel stiffeners (Fig. 1c). The last code is a Panel Analysis and Design Optimizer (PAD.Opt). The program's goal is the sizing optimization of stiffener panels in the case of a preliminary design (Fig. 1d). This means the software hasn't a complete theory for panel analysis, but it is only a way to predict an optimal configuration considering: (1) several geometrical (flat and curved panels, stiffened panels with a orthogonal grid) and load configurations; (2) different material characteristics (isotropic and composite material defined with the smeared theory); (3) failure modes for local skin, stiffener and global buckling (performed solving handbook formulas). Three groups of variables designs are considered for the optimum panel weight: (1) the skin thickness; (2) the stiffener cross sections; (3) the stiffener distance (between two subsequent stiffeners). For each program an optimization task on the specific structural components is performed. The algorithm can be defined by the following common system of equations:

$$\begin{cases} W(\rho, d_i, l_e) \rightarrow W_{min} & \text{objective function} \\ d_{i,min} \leq d_i \leq d_{i,max} & \text{geometric constraints} \\ \sigma_e(s) \leq \sigma_{amm} & \text{stress constraints} \\ \sigma_e(s) \leq \sigma_{l,cr} & \text{local stability constraints} \\ \sigma_e(s) \leq \sigma_{g,cr} & \text{global stability constraints} \end{cases} \quad (1)$$

where ρ is the material density, e the number of subsystem finite elements (located by the geometrical variable s) and l_e the FE element lengths. The target of the optimization is to minimize the subsystem weight W with respect to the design variables d_i of the cross section and to geometrical ($d_{i,min}$ and $d_{i,max}$ is the dimensional range by user's defined) and structural constraints, comparing these last ones with the ultimate yield stress of the materials (σ_{amm}). There is also a third group of constraints relative to the instability conditions. Each solver has some specific features to evaluate the local critical load $\sigma_{l,cr}$ on particular elements (i.e. the local instability of a panel stiffener) or the global one $\sigma_{g,cr}$ on the overall subsystems (i.e. the global instability of a stiffened panel). At the present time the software capabilities are limited to the structural analysis of metal alloy with isotropic or anisotropic properties.

2.2 The collaborative approach

OPT.TOOL is a multilevel software manager for the execution of the local optimizers working on specific substructures. The use of this method is based on the intuition to consider a problem classified as disciplinary with a multidisciplinary approach. This idea should appear in contradiction and counterproductive, but the CO can bring a lot of advantage to the structural optimization for complex FE systems. The three programs work on the same disciplines, but on different components. They appear to be independent, but as the components are together assembled and the global structure must be statically determined, a set of shared parameters are identifiable (i.e. all the forces and flows at the interface FE grids). These parameters influence the behaviour of the local optimizers acting on the local stress and instability constraints and realizing a little interdisciplinary data exchange. In relation to two different definitions of the interdisciplinary consistency constraints, two different CO methods (CO_1 and CO_2) are developed by Alexandrov and Lewis [7, 8]. In this work a CO_1 approach is used (Fig. 2) bringing several benefits: (1) the local optimization always obtains a feasible design; (2) the interdisciplinary parameters are seen at the local level only (independent disciplines); (3) the interdisciplinary consistency constraints are defined with a linear function allowing the use a gradient based algorithm.

3 OPT.TOOL analyses

In this paragraph two OPT.TOOL analyses are discussed. As the software tool is developed to deal with realistic structural designs, the results are based on complex FE model and show the main OPT.TOOL features: (1) providing a multiconfiguration structural solution; (2) satisfying the requirements initially defined. The first problem refers to an aeronautical application, the second one to a space module and, in order to evaluate the procedure, a comparison analysis with a commercial optimizer is also presented on the last problem.

3.1 Aeronautical wing structure

The example presents a wing structure of a utility aircraft (Fig. 3a), loaded with a set of realistic loads: aerodynamic pressure, gravitational forces and inside system weights. The loads are performed defining the load factor n in different flight condition as: normal ($n = +1$), upside down

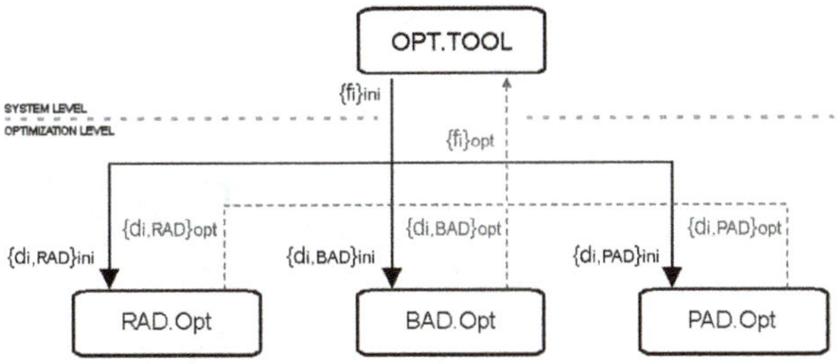

Fig. 2. The OPT.TOOL collaborative architecture. it reads the local design varables $\{f\}_{ini}$ and takes out the optimal values $\{f\}_{opt}$.

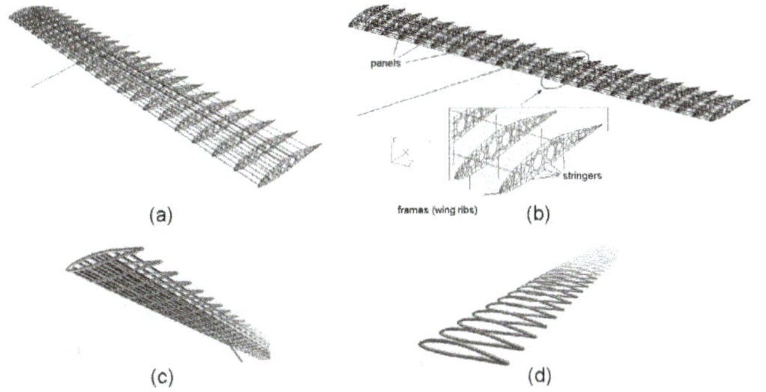

Fig. 3. Outline of an aircraft wing (a) with the details of three structural groups: the panels (b), the stringers (c) and the wing ribs (d).

($n = -1$), hard landing ($n = +2.5$) and over the maximum load factor ($n = +3.9$). The following substructures are present in the model: 48 panels (Fig. 3b), 17 longitudinal stringers (Fig. 3c), 23 wing frames (Fig. 3d). For this last group only the cross section of the closed-shaped frame around the wing web is taken into account for the analysis because of the small value of the wing web thickness. No ring frames or stiffened panels are in the model. The material is an aluminium alloy for all groups. In relation to a simuilar application developed by Piperni [9] and Craig [10], the aim of these examples is to show the feature of the OPT.TOOL procedure to determine a optimal configuration in relation to the following requirements: (1) all the load cases are acting (improbable situation); (2) specific range of values defined on the cross section and panel thickness (respectively $9\ \text{mm}^2 \leq A_i\,(d_i) \leq 900\ \text{mm}^2$ and $1\ \text{mm} \leq t_p \leq 100\ \text{mm}$); (3) ultimate yield stress imposed to 270 MPa; (4) different safety factors for the structural (s_1) and stability constraints (s_2) (respectively $|\sigma(d_i)| \leq \frac{\sigma_{mat}}{s_1}$ for the stress constraint and $\sigma(d_i) \leq \frac{\sigma_{cr}}{s_2}$ for stability one). A total number of 288 design variables are available for the optimization. Three analysis cases are performed in relation to different security factors: $s_1 = s_2 = 1.0$ (case 1), $s_1 = s_2 = 1.2$ (case 2) and $s_1 = s_2 = 1.3$ (case 3), where the graphs in Figure 4 are drawing respect to the computational iterations nedeed for the objective function convergence. Because of the acting loads, the global weight increase for the three cases (Tab. 1), while a light increase is available for the "set 1", a reference analysis with single load $n = +1$ and security factors $s_1 = s_2 = 1.2$. The result shows that, in

Table 1. OPT.TOOL wing analysis: results for multi-load and single (set 1) load cases.

	case 1	case 2	case 3	set 1
Start [kg]	81.21	81.21	81.21	81.21
OPT.TOOL [kg]	128.03	128.06	128.52	105.79
time	1^h02^m	1^h05^m	1^h03^m	0^h25^m

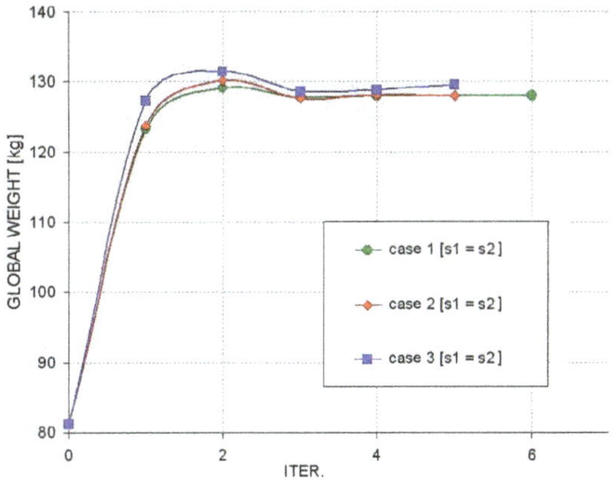

Fig. 4. OPT.TOOL analyses for a wing aircraft. Global weight graphs for different cases.

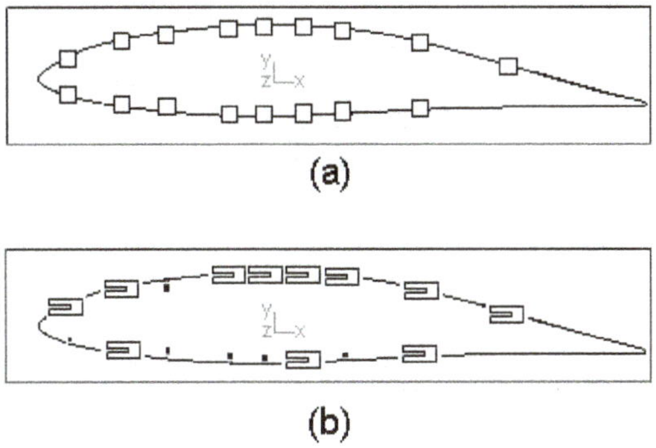

Fig. 5. OPT.TOOL wing analysis. Cross section (case 2) for initial (a) and optimal configuration (b).

order to consider a safe configuration and respecting all loads taken into account, a 20 kg of structure mass must be added. Another OPT.TOOL feature is the performing of a multiconfiguration solution. In this example the OPT .TOOL can generate different cross section typologies for the design variables of the wing section. From a homogeneous condition (Fig. 5a, rectangular cross sections) the solver goes towards a heterogeneous one (Fig. 5b rectangular and "C" cross section).

3.2 Space module

The example shows the OPT.TOOL computation performances for space applications. The structure is a space module of the International Space Station (ISS) designed by ALCATEL ALENIA Space. An example of the use of a CO approach to the aerospace application is available in Brown [11] (where a more generic optimization on

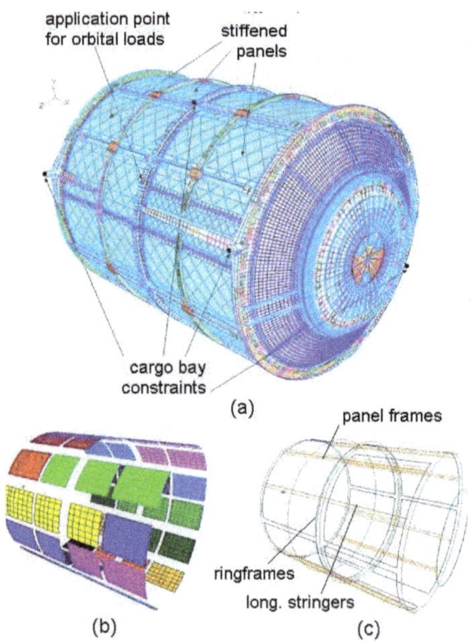

Fig. 6. Space structure model (a) with the panels (b) and the bar elements (c) details.

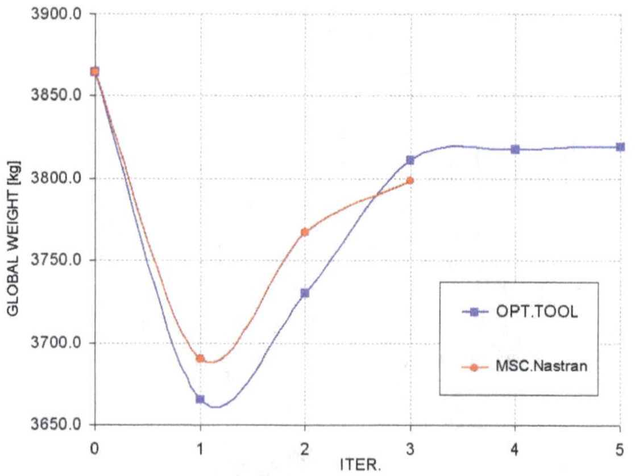

Fig. 7. Result comparison of the global weight on a space structure.

the global design of a space vehicle, with a comparison of three different methods, is reported) and in Wallace [12], where the philosophy of the CO design is applied as well as it was done for this work. The FE model[2] has an high definition degree and was created taking into account (1) the primary structure (no inside structures are presented) and (2) a large set of loads (Fig. 6a). These are taken from a realistic situation of typical space mission: lift- off and cargo bay loads, internal pressure, berthing conditions and static connection forces to the Space Station. Two results are performed: (a) a sample computation of the optimal configuration and (b) a comparison with a commercial optimizer. The first analysis is based on the sizing optimization of the following three design structural groups: 44 biaxial blade stiffened panels (Fig. 6b), 32 longitudinal stringers, 13 ring frames and general shape frames (Fig. 6c). All the materials are aerospace isotropic alloy and composites. As reported before, for the wing example, also for this problem different requirements are imposed: (1) all the load cases are activated; (2) specific range defined on the design variables of the cross section of stringers, frames, panel stiffeners and panel thickness (respectively 8 mm $\leq d_{st}, d_{fr}, d_{ps}, t_p \leq$ 50 mm); (3) ultimate yield stress imposed to 270 MPa; (4) structural and stability security factors fixed respectively to $s_1 = s_2 = 1.0$. Overall the procedure has to solve an optimization problem with 666 variables designs. In Figure 8a two lines are reported to indicate the initial FE model weight (dotted line) and the actual one considering the mass of the all component features that, requiring a high FEM definition, they are described into the OPT.TOOL input file (i.e. the stiffened panels etc.). The results show a small increase of the global model weight. The single code behaviours show (Fig. 8b) an increase of the panel group weight (dotted line), on the contrary a decrease for the stringer and frames groups appears (Tab. 2a). Some considerations about this analysis follow. The optimal result is achieved with not violated constraints, so the new configuration is feasible in relation to the load cases and the requirement. The OPT.TOOL ends generating a optimal FEM with 0.74% mass increased. The last consideration is about the processing time: because of the high complexity of the solution and the type of optimization task activated (in particular on the panel stiffeners), the analysis lasts almost five hours. This processing time doesn't last so long as the MSC.Nastran run. In order to validate the OPT.TOOL procedure a comparison with MSC.Nastran's optimization solver is performed. The comparison between the MSC.Nastran and OPT.TOOL solution on the same FE model is based. As the finite element software can not support a sizing optimization of the distance between the panel stiffeners, the theory of equivalent stiffness is used to define the properties of the panels in the model and the panel stiffeners optimization is omitted. In comparison with to the previous example a different definition of the requirements are report: (1) all the load cases are activated; (2) specific range defined on the cross section of stringers and frames and panel thickness (respectively 64 mm$^2 \leq A_i(d_i) \leq$ 2500 mm^2 and 8 mm $\leq t_p \leq$ 50 mm); (3) ultimate yield stress imposed to 270 MPa; (4) structural and stability security factors fixed respectively to $s_1 = s_2 = 1.0$: (5) because of the absence of the stiffener panel the local stability constraint is not taken into account on these elements. The comparison results show (Fig. 7) a small difference (about 20 kg) between the two global weights (Tab. 2b), with a long processing time for the FE software. This is why the MSC.Nastran time processing will be longer if the time for writing the execution (Sol 200) will be considered; OPT.TOOL doesn't need this kind of user's attention because of its FE interfacing algorithm. A different mass distribution between the groups is reached because of the stability conditions that increase the panels weights giving out the remaining groups mass.

[2] ALCATEL ALENIA Space is the owner of the FE model used for the analyses.

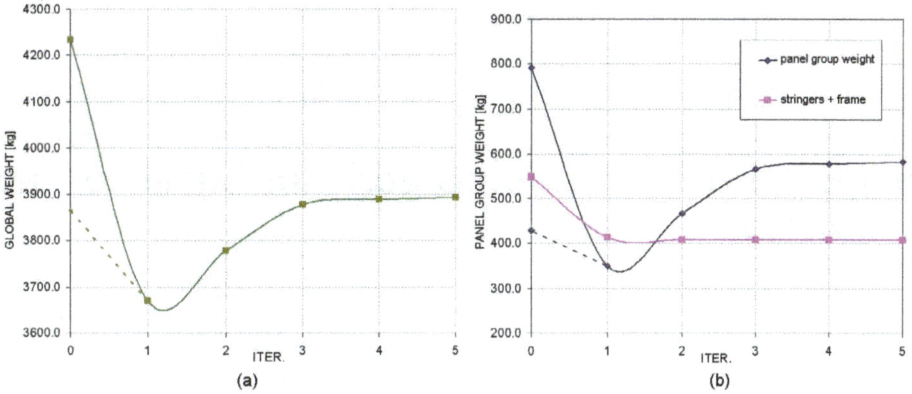

Fig. 8. OPT.TOOL space module analysis. Global (a) and local group weights (b) for the initial FE (dotted line) and the actual (continuous line) models.

Table 2. OPT.TOOL analysis (a) vs. MSC.Nastran comparison (b) results on a space structure.

	Start	OPT.TOOL(a)	OPT.TOOL(b)	MSC.Nastran(b)
global weight [kg]	3864.99	3891.82	3819.36	3798.69
panels [kg]	422.28	581.59	584.85	328.27
str., fra. [kg]	548, 64	416.16	340.08	576, 38
time	–	4^h48^m	2^h21^m	6^h05^m

4 Conclusions

The work presents a multilevel optimization method for the preliminary design based on the CO architecture. It has been showed how the OPT.TOOL procedure can be a good approach for the sizing optimization of realistic structures defined by a FE model and considering some structural and stability local effects. Some features have been presented in two examples on aeronautical and aerospace structures, where: (1) a good comparison with a commercial optimization solver has been demostrated (with better results, in a shorter processing time); (2) new structural local configurations (wing example) and lighter model (aerospace example) have been founded respect to the initial conditions; (3) the process is completely automatic and no FE modifications are requested; (4) the OPT.TOOL local results are entirely uploaded in a FE model and available for further analyses.

References

1. R. Braun, R.W. Powell, R.A. Lepsch, I.M. Kroo, in *Comparison of two multidisciplinary optimization strategies for launch-vehicle design,* Journal Of Space And Rocket, Vol. 52, Nov. 2, May (2006)
2. O. Kalden, U.M. Schottle, in *A software tool for analysis of future launch vehicle concepts,* international Astronautical Congress Paper, IAC -03-V.5.08
3. N. Durante, A. Dufour, V. Pain, in *Multidisciplinary analysis and optimization approach for the design of expendable launchers,* 10th AIAA/ISSMO Multidisciplinary Analysis and Optimization Conference, AIAA 2004-4441, 30 August–1 September, 2004
4. R. Braun, A. Moore, I. Kroo, in *Use of the collaborative optimization architecture for launch vehicle design,* AIAA Paper 96-4018, 1996
5. G.N. Vanderplaats, in *An efficient feasible direction algorithm for design synthesis,* AIAA Journal, Vol. 22, No. 11 (1984)
6. L. Guadagni, in *Development of multilevel procedures for the optimal design of aerospace vehicles,* Ph.D. Thesis, Politecnico di Torino, 2005
7. N.M. Alexandrov, R. M. Lewis, in *Analytical and computational aspects of collaborative optimization,* NASA Paper TM-2000-210104, 2000
8. N.M. Alexandrov, R.M. Lewis, In *Analytical and computational properties of distributed approaches to MDO,* 8th AIAA/USAF/NASA/ISSMO Symposium Of Multidisciplinary Analysis & Optimization, September 2000
9. P. Piperni, M. Abdo, F. Katyeke, in *The building blocks of a multi-disciplinary wing design method,* General Meeting and Conference (Toronto), April 26 and 27, 2005
10. S. Craig, P.E. Collier, in *Next generation structural optimization today,* Colier Research & Development Corporation, September 1997
11. N.F. Brown, J.R. Olds, in *Evaluation of multidisciplinary analysis tecniques applied to a reusable launch vehicle,* Journal Of Space And Rocket, Vol. 43, No. 6, (2006)
12. J. Wallace, J.R. Olds, A.C. Charania, G. Woodcock, in *A studi of arts: a dual-fuel reusable launch vehicle with launch assist,* 39th AIAA/ASME/SAE/ASEE Joint Propulsio Conference And Exibit, July 2003

9

An efficient global optimization algorithm based on augmented radial basis function*

Yun-Kang Sui[1,a], Shan-Po Li[1] and Ying-Qiao Guo[2]

[1] Centre of Numerical Simulation for Engineering, Beijing University of Technology, F100022 Beijing, P.R. China
[2] Laboratory of Mechanics, Materials & Structures, University of Reims Champagne-Ardenne, 51687 Reims, France

Abstract – In the structural optimization, the accuracy of approximation for the established mathematical model will directly affect the solution efficiency, even the convergence. The global optimization model based on the augmented Gaussian radial basis function h as a high approximation accuracy, but the solution efficiency will not be increased without a matched optimization algorithm. In this paper, we adopt the information at the interpolating points in large extent and the augmented Gaussian radial basis function to construct the approximate mathematical model. Using the explicit derivatives of the model for the sensitivities and sequential quadratic programming (SQP) algorithm for the optimization solving, an efficient algorithm of global optimization is proposed. It is simple to be realized and converges quickly. Two examples will illustrate the stability and efficiency of the present algorithm.

Key words: Structural optimization; approximate model; Gaussian radial basis function; global optimization algorithm

1 Introduction

In 1960, Schmit [1] brought forward the idea which introduced the mathematical programming into the structural optimization by using the systematic synthesis. At that time, he might not realize the importance of the establishment of mathematical models, so the optimal design was found by hundreds structural reanalyses. In fact, the objective function and constraint functions in an engineering problem can be rarely written as the explicit expressions in terms of design variables, the sensitivity analysis is very time-consuming.

In 1974, Schmit and Farshi [2] proposed some approximation concepts including the obtaintion of explicit approximate functions for constraint functions to improve the structural optimization efficiency and the iteration times largely decreased. Essentially, the establishment of an optimization model consists in constructing approximate functions for objective function or/and constraint conditions. From then on, the idea of establishing mathematical model has been accepted by many optimization researchers and is becoming a relative independent research domain. However, some researchers still use the methods without establishing any optimization model. Although the difficulty of sensitivity analysis can be avoided, the calculations are very time-consuming and the problem of converge often arises.

Therefore, the establishment of an approximate optimization model is a good choice for the structural optimization. The research works on the model establishment can be divided into three parts: the first one is to select suitable approximate functions, the second one is to provide the derivatives of these functions (sensitivity analysis), the third one is to choose an adapted algorithm to solve the optimization model. In the early used methods for mathematic model establishing, the Taylor's expansions of first order and/or second order are usually used, and analytical methods of sensitivity analysis are often adopted. The disadvantages of the sensitivity analysis are that the analytical derivatives is very difficult and the computation of derivatives is too expensive. This may be the reason that some scholars do not establish any optimization model.

Are there some methods to establish the mathematical model without sensitivity analysis? Recently, the Response Surface Methodology (RSM) (Mayer et al. [3]) is largely used for the structural optimization, a lot of research (Venter and Haftka [4], Hosder et al. [5], Jansson et al. [6], Liu et al. [7]) and improved works (Zheng and Das [8], Sui and Li [9]) have been done by using the RSM. The RSM is mainly a polynomial regression method, it usually uses the first order and/or second order polynomials and the approximation accuracy is good only in local domain. It is often combined with QP algorithm by

* Supported by National Natural Science Foundation of China (10472003) and Scientific Research Foundation of Doctoral Subjects in Chinese Universities (20060005010).

[a] Corresponding authors:
ysui@bjut.edu.cn; lsp102@emails.bjut.edu.cn; yq.guo@univ-reims.fr

sequential iterations, so an initial solution should be selected by user. Because this selection is very difficult in a structural optimization, the method can not promise to give the global optimal solution. Although high order polynomials can be used to obtain some global approximations, but a lot of response calculations are needed at the experiment points and this will lead to too expensive calculations.

In recent years, some researchers introduced an accurate global approximate method called Radial Basis Function method (RBF) (Mason and Cox [10], Krishnamurthy [11]). Fang and Horstemeyer [12] demonstrate that the approximation accuracy of the augmented Gaussian RBF is higher than that of other types of RBF and its approximation domain is also wider. How to solve this type of models with high approximation accuracy? From published papers, an optimization algorithm is often taken to solve the model without considering whether the algorithm is adapted to the characteristic of the model. In other word, it is deficient in considering the relation between an establishing approximate optimization model and an adapted algorithm. Consequently, the optimization computation is not efficient. In this paper, we introduce a new sequential algorithm corresponding to the approximation model based on the augmented RBF method. The numerical results illustrate the accuracy and efficiency of the present algorithm.

2 RBF and augmented RBF methods

The Radial Basis Function (RBF) can be expressed by the linear combination of the functions in terms of radial distance from the considered point to every interpolation point (or sample point or experiment design point) [12]:

$$s(\mathbf{x}) = \sum_{j=1}^{m} \lambda_j \varphi\left(\left\|\mathbf{x}-\mathbf{x}^j\right\|_2\right) \quad \mathbf{x} \in E^n \tag{1}$$

where n is the total number of design variables or the dimensional number of the design space, m is the total number of interpolating points, $\lambda_j (j = 1, \ldots, m)$ is the jth coefficient to be determined, $\mathbf{x}$ is the position vector of a design point, $\mathbf{x}^j$ is the position vector of jth interpolating point, $\left\|\mathbf{x}-\mathbf{x}^j\right\|_2$ is the jth radial distance between the two points:

$$r^{0j} = \left\|\mathbf{x}-\mathbf{x}^j\right\|_2 = \sqrt{\sum_{k=1}^{n} (x_k - x_k^j)^2}. \tag{2}$$

Different forms of the RBF $\varphi(r)$ are listed in Table 1.

Let F_i the real response value at the interpolating point $\mathbf{x}^i$, the formula (1) becomes:

$$\sum_{j=1}^{m} \lambda_j \varphi(r^{ij}) = F_i \quad (i = 1, \ldots, m) \tag{3}$$

$$r^{ij} = \left\|\mathbf{x}^i-\mathbf{x}^j\right\|_2 = \sqrt{\sum_{k=1}^{n} (x_k^i - x_k^j)^2}. \tag{4}$$

Table 1. Usually used radial basis functions.

Function name	Expression
Linear	$\phi(r) = cr$
Thrice	$\phi(r) = cr^3$
Thin plate spline	$\phi(r) = r^2 \ln(cr), 0 < c \leqslant 1$
Gaussian	$\phi(r) = e^{-cr^2}, 0 < c \leqslant 1$
Multi-quadratic	$\phi(r) = \sqrt{r^2+c^2}, 0 < c \leqslant 1$

where the constant c is usually taken as 1.

Then $\lambda_j (j = 1, \ldots, m)$ can be obtained by solving equations (3).

To increase the approximation accuracy and expand the approximation range of the RBF, a polynomial term can be added to the expression (1):

$$s(\mathbf{x}) = \sum_{j=1}^{m} \lambda_j \varphi(\mathbf{r}^{0j}) + \sum_{i=1}^{I} b_i P_i(\mathbf{x}) \tag{5}$$

where the polynomial functions $p_i(\mathbf{x}) (i = 1, ..., I)$ can be $1, x_1, \ldots, x_n, x_1^2, x_1x_2, \ldots, x_n^2, \ldots$ or a part of them. The above expression is called augmented RBF. Using m interpolating points and their corresponding real response values in expression (5), we obtain:

$$\begin{cases} \lambda_1 0 + \lambda_2 \varphi\left(\mathbf{r}^{12}\right) + \ldots + \lambda_m \varphi\left(\mathbf{r}^{1m}\right) + b_1 p_1\left(\mathbf{x}^1\right) + \ldots \\ \qquad + b_I p_I\left(\mathbf{x}^1\right) = F_1 \\ \lambda_1 \varphi\left(\mathbf{r}^{21}\right) + \lambda_2 0 + \ldots + \lambda_m \varphi\left(\mathbf{r}^{2m}\right) + b_1 p_1\left(\mathbf{x}^2\right) + \ldots \\ \qquad + b_I p_I\left(\mathbf{x}^2\right) = F_2 \\ \qquad \vdots \\ \lambda_1 \varphi\left(\mathbf{r}^{m1}\right) + \lambda_2 \varphi\left(\mathbf{r}^{m2}\right) + \ldots + \lambda_m 0 + b_1 p_1\left(\mathbf{x}^m\right) + \ldots \\ \qquad + b_I p_I\left(\mathbf{x}^m\right) = F_m. \end{cases} \tag{6}$$

In equations (6), there are m coefficients $\lambda_j (j = 1, \ldots, m)$ and I coefficients $b_i (i = 1, ..., I)$ as unknowns, so I equations have to be added. The added equations may be obtained according to the following orthogonal conditions:

$$\begin{cases} p_1\left(\mathbf{x}^1\right)\lambda_1 + \ldots + p_1\left(\mathbf{x}^m\right)\lambda_m = 0 \\ \qquad \vdots \\ p_I\left(\mathbf{x}^1\right)\lambda_1 + \ldots + p_I\left(\mathbf{x}^m\right)\lambda_m = 0. \end{cases} \tag{7}$$

The equations (6) and (7) can be rewritten in a matrix form:

$$\begin{bmatrix} \mathbf{A} & \mathbf{B} \\ \mathbf{B}^{\mathrm{T}} & \mathbf{0} \end{bmatrix} \begin{Bmatrix} \lambda \\ \mathbf{b} \end{Bmatrix} = \begin{Bmatrix} \mathbf{F} \\ \mathbf{0} \end{Bmatrix}$$

$$\text{with } \mathbf{A} = \begin{bmatrix} 0 & \varphi\left(\mathbf{r}^{12}\right) & \ldots & \varphi\left(\mathbf{r}^{1m}\right) \\ \varphi\left(\mathbf{r}^{21}\right) & 0 & \ldots & \varphi\left(\mathbf{r}^{2m}\right) \\ \vdots & \vdots & \vdots & \vdots \\ \varphi\left(\mathbf{r}^{m1}\right) & \varphi\left(\mathbf{r}^{m2}\right) & \ldots & 0 \end{bmatrix},$$

$$\mathbf{B} = \begin{bmatrix} p_1\left(\mathbf{x}^1\right) & \ldots & p_I\left(\mathbf{x}^1\right) \\ \vdots & \vdots & \vdots \\ p_1\left(\mathbf{x}^m\right) & \ldots & p_I\left(\mathbf{x}^m\right) \end{bmatrix}$$

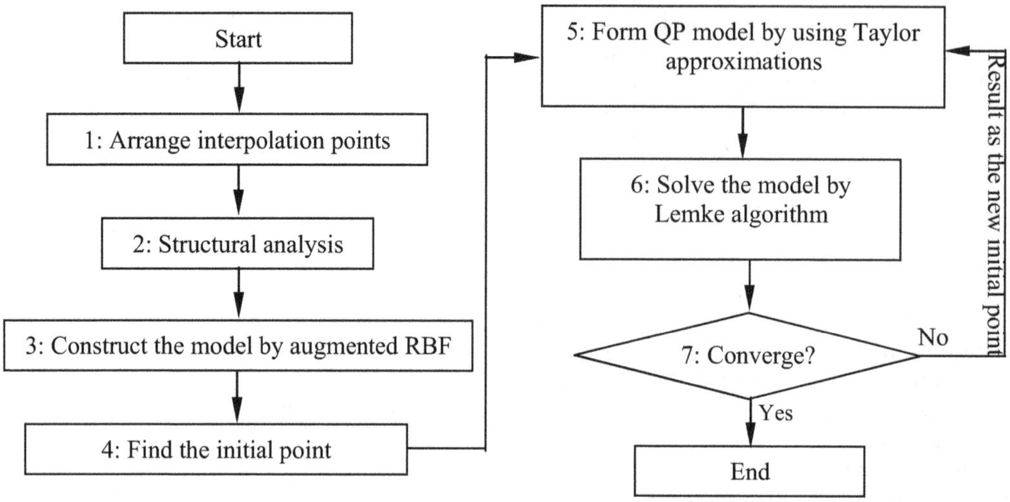

Fig. 1. Flow chart of present new algorithm.

$\lambda = (\lambda_1, \ldots, \lambda_m)^{\mathrm{T}}, \mathbf{b} = (b_1, \ldots, b_I)^{\mathrm{T}}, \mathbf{F} = (F_1, \ldots, F_m)^{\mathrm{T}}$.

The RBF and augmented RBF have no error at the interpolating points. In order to evaluate their approximation accuracy, K other points are selected in the design space. We calculate the approximate response $\overline{F}_i(i = 1, ..., K)$ using (5) and the real response $F_i^{real}(i = 1, ..., K)$ using the structural analysis solver, then the following root of mean square errors can be the evaluate index:

$$RMSE = \sqrt{\sum_{i=1}^{K} \left(\overline{F_i} - F_i^{\mathrm{real}}\right)^2 / K}. \tag{8}$$

3 Implementation of algorithms of modelling and optimization

The optimization problem of a mathematical programming or a structural optimization can be described by:

$$\begin{cases} \text{find} & \mathbf{x} \in E^n \\ \min & f(\mathbf{x}) \\ \text{s.t.} & c_k(\mathbf{x}) \leqslant b_k (k = 1, \ldots, K) \\ & \underline{x}_i \leqslant x_i \leqslant \bar{x}_i (i = 1, \ldots, n) \end{cases} \tag{9}$$

where K is the total number of behaviour constraints. The above mentioned Augmented RBF (Eq. (5)) can be used to approach the objective function and constraints functions in order to obtain an optimization model in global domain. But it is difficult to solve this mathematic model because of its high nonlinearity.

In order to find an efficient optimization algorithm matched with high order nonlinear characteristics of the model (9), the following solving strategy including the sequential quadratic programming (SQP) is proposed. It is carried out in 7 steps:

(1) Arrange a set of interpolating points evenly in the global domain representing the experiment design variables. Their feasible properties are not considered.

(2) A series of structural analysis at the interpolating points are carried out to obtain the corresponding response values.

(3) The objective function and constraint conditions based on the Augment Gaussian RBF are established and their corresponding coefficients λ_j and b_i are determined according to the response values.

(4) The vector of design variables giving the minimal value of the objective function at one of all feasible interpolating points is taken as an initial point for SQP.

(5) At the initial point, a standard QP model is formed by using the second order Taylor's expansion for the objective function and the first order Taylor's expansions for the constraint conditions.

(6) Combined with the rational move limits of design variables, the QP model is solved by Lemke algorithm.

(7) Checking the convergence criterion, if it is satisfied then exit, otherwise return to the 5th step and use the obtained result as a new initial point for the next step.

In 5th step, for the solution convenience, the objective function and the constraint conditions adopt different order Taylor's expansions (Sui et al. [13]). In 6th step, the rational move limits of design variables is a method considering accumulated information (Sui et al. [14]) and Lemke algorithm can efficiently solve a standard QP.

The flow chart in Figure 1 shows our new algorithm, named Augmented Gaussian RBF-SQP (AG-RBF-SQP) method.

In the QP model, the first order and second order Taylor's approximations of an arbitrary function $f(\mathbf{x})$ are given by the following expressions

$$f(\mathbf{x}) \approx f(\mathbf{x}^0) - \sum_{i=1}^{n} \frac{\partial f(\mathbf{x}^0)}{\partial x_i} x_i^0 + \sum_{i=1}^{n} \frac{\partial f(\mathbf{x}^0)}{\partial x_i} x_i \tag{10}$$

$$f(\mathbf{x}) \approx f(\mathbf{x}^0) - \sum_{i=1}^{n} \frac{\partial f(\mathbf{x}^0)}{\partial x_i} x_i^0 + \frac{1}{2}\sum_{j=1}^{n}\sum_{k=1}^{n} \frac{\partial^2 f(\mathbf{x}^0)}{\partial x_j \partial x_k} x_j^0 x_k^0 + \sum_{i=1}^{n}\left(\frac{\partial f(\mathbf{x}^0)}{\partial x_i} - \sum_{j=1}^{n} \frac{\partial^2 f(\mathbf{x}^0)}{\partial x_i \partial x_j} x_j^0\right) x_i + \frac{1}{2}\sum_{j=1}^{n}\sum_{k=1}^{n} \frac{\partial^2 f(\mathbf{x}^0)}{\partial x_j \partial x_k} x_j x_k. \tag{11}$$

For our algorithm, the Gaussian functions and linear functions are taken in the formula (5). In order to improve the computation efficiency, the analytical explicit derivatives are deduced. The partial derivatives of first order and second order are calculated as follows:

$$\partial s(\mathbf{x})/\partial x_l = -2\sum_{j=1}^{m} \lambda_j e^{-\sum_{k=1}^{n}(x_k - x_k^j)^2}(x_l - x_l^j) + b_l \tag{12}$$

$$\frac{\partial^2 s(\mathbf{x})}{\partial x_l \partial x_p} = \begin{cases} 4\sum_{j=1}^{m} \lambda_j e^{-\sum_{k=1}^{n}(x_k - x_k^j)^2}(x_l - x_l^j)(x_p - x_p^j) & \text{if } l \neq p \\ 2\sum_{j=1}^{m} \lambda_j e^{-\sum_{k=1}^{n}(x_k - x_k^j)^2}[2(x_l - x_l^j)^2 - 1] & \text{if } l = p. \end{cases} \tag{13}$$

Finally, the model (9) can be transformed to a standard QP model:

$$\begin{cases} \text{find} & \mathbf{x}^{(v)} \in E^n \\ \min & \mathbf{x}^{\mathrm{T}}\mathbf{H}^{(v)}\mathbf{x}/2 + \mathbf{C}^{(v)}\mathbf{x} \\ \text{s.t.} & \mathbf{A}^{(v)}\mathbf{x} \leqslant \mathbf{B} - \mathbf{A}_0^{(v)} \\ & \underline{x}_i^{(v)} \leqslant x_i \leqslant \bar{x}_i^{(v)} \end{cases} \tag{14}$$

where v denotes the vth iteration. The objective function is expanded into the second order Taylor expansion: $\mathbf{H}^{(v)}$ is the Hessian matrix, given by the fifth term in the expression (11), $\mathbf{C}^{(v)}$ is a vector, given by the fourth term in (11). The first three terms in (11) have no influence on the minimization of$f(\mathbf{x})$. The constraint functions in (9) are expanded into the first order Taylor expansion: $\mathbf{A}^{(v)}$ is a matrix given by the third term in (10), $\mathbf{A}_0^{(v)}$ is a vector given by the first two terms in (10), $\mathbf{B}$ represents b_k in (9).

4 Numerical examples

In order to evaluate the accuracy and efficiency of the present algorithm, a lot of mathematical functions and practical engineering examples have been treated, we just present two of them in this paper: a mathematical example, an engineering application, whose models are established by the Augmented Gaussian RBF and solved by the present algorithm. The optimal results of the first example is compared with the results obtained by using the Genetic Algorithm (GA) (Xuan et al. [15]) and Simulated Annealing Algorithm (SAA) (Yang and Gu [16]), implemented by the 7D-Soft High Technology Inc. and published in the software 1stOpt1.5 (Website [17]). The optimal results of the second example is compared with that of SRSM-QP (sequential response surface methodology and quadratic programming) algorithm (Yang et al. [18]).

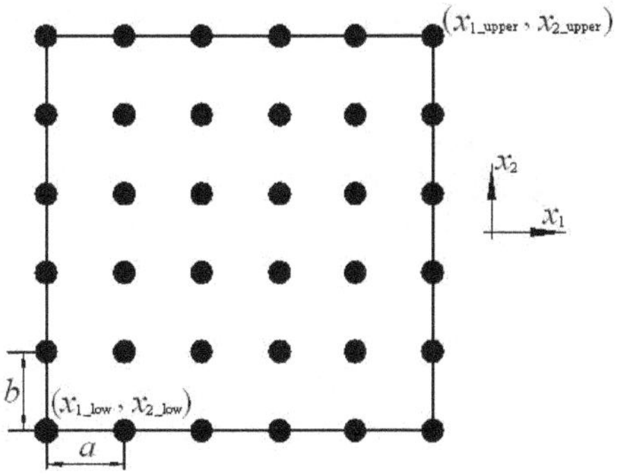

Fig. 2. 36 Interpolating points.

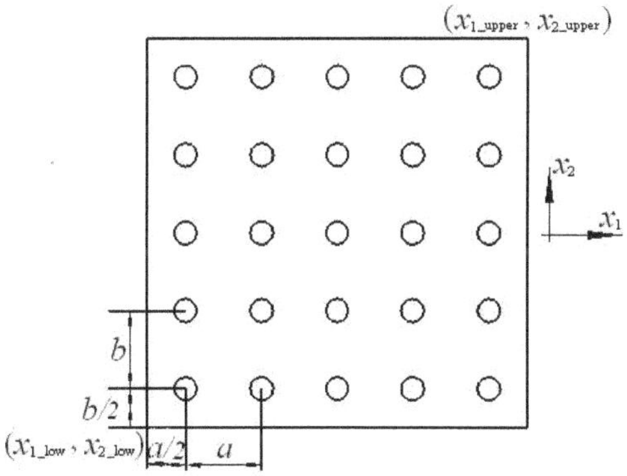

Fig. 3. 25 Non- interpolating points.

4.1 Maximization problem of a highly nonlinear objective function

$$\begin{cases} \text{find} & \mathbf{x} \in E^2 \\ \max & f(\mathbf{x}) \\ \text{s.t.} & 0.01 \leqslant x_1 \leqslant 1 \\ & 0 \leqslant x_2 \leqslant 1 \end{cases} \tag{15}$$

with $f(\mathbf{x}) = [0.8\sqrt{x_1^2 + x_2^2} + 0.35\sin(2.4\pi\sqrt{x_1^2 + x_2^2})/\sqrt{2}]$ $[1.5\sin(1.3\arctan(x_2/x_1))]$. The real response surface is illustrated in Figure 4. The exact solution gives the maximum value 1.6519 at the point (0.4276, 1.0). To establish an approximate model and evaluate the error, we evenly arrange 36 interpolating points and 25 non-interpolate points respectively in the global design space (Figs. 2 and 3, $a = 0.198$, $b = 0.2$). The approximate model obtained by using the Augmented Gaussian RBF and the given 36 sample points gives an approximate response surface (Fig. 5). The RMSE using the augmented RBF at the given 25 points is 0.0108 and that using RBF at the given 25 points is 0.0112. Table 2 gives the comparison of the results of different algorithms.

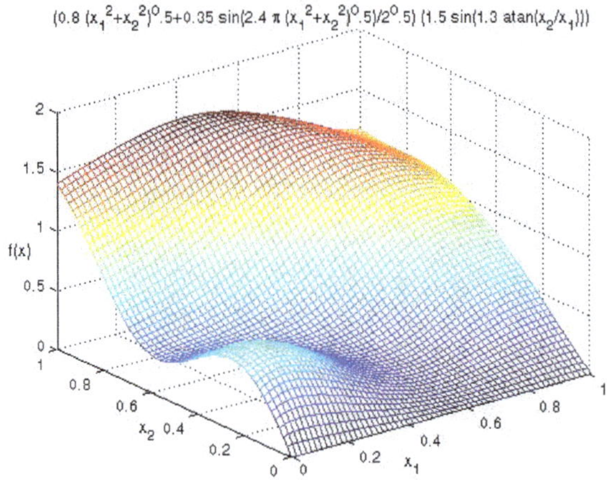

Fig. 4. Real function image of example 4.1.

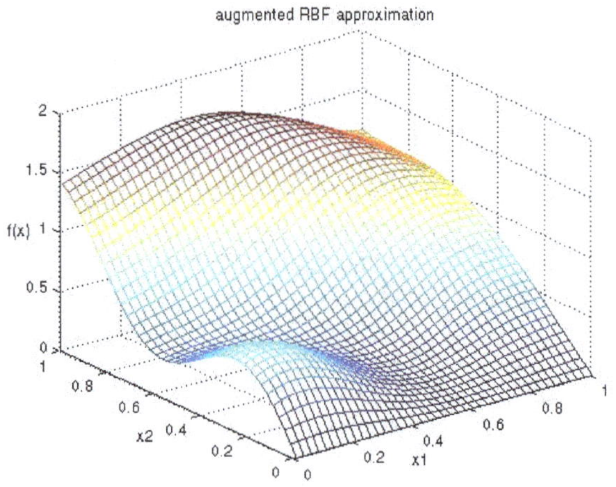

Fig. 5. Approximate function image of example 4.1.

Table 2. Comparison of the results of different algorithms.

Algorithm	Optimal point		Objective value	Number of
	x_1	x_2	$f(x)$	objective calculations
GA	0.4262	0.9999	1.6518	1304
SAA	0.4276	0.9999	1.6518	416
AG-RBF-SQP	0.4050	1.0000	1.6507	37

In Table 2, the right column shows the number of objective calculations with different algorithms. We note that the present algorithm is very fast with respect to the two others. The convergence is attained with only two times of model establishment and 37 calls of objective calculations. The first model is established by using the 36 calls at the 36 chosen interpolating points; in the second model, the previously calculated objective functions and constraint functions at the 36 chosen points are kept and one more call is carried out at the optimal point obtained with the first model. We can use all these 37 points or 36 points by removing the nearest point from the optimal point of the first model to establish the second model. Two SQP calculations are needed for the first and second models, respectively.

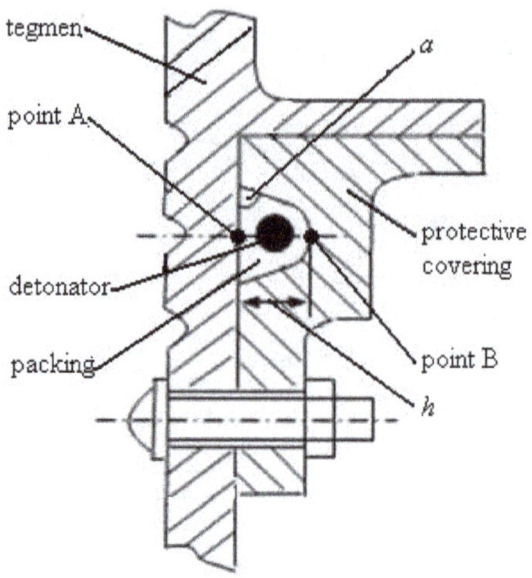

Fig. 6. 2D Section of axisymmetrical device for cutting off by explosion.

4.2 Cavity shape optimization charged detonator of equipment cutting off by explosion

An axisymmetrical device for cutting off by explosion is taken as an industrial application. Only its 2D section is illustrated (Fig. 6) and calculated. The cavity is charged with the detonator, so its shape and dimensions will determine the intensity and direction of the maximal cutting force. The cavity shape is defined by an arc and two segments, the height h of the cavity and the obliquity α of the segments are taken as the design variables. For the initial design, h is 0.36 cm, α is 1.56 rad. The structure is divided into 1635 elements. The aim of the optimization is to give the maximal effective stress at the point A (element No. 1377) to cut of the tegmen and a effective stress inferior to the allowable limit (310 MPa) at B (element No. 941) to preserve the protective covering. Therefore, the effective stress σ_{EA} at point A is taken as objective function and should be maximized, and the effective stress σ_{EB} at point B is taken as behaviour constraint. The following expression (16) describes this optimization problem:

$$\begin{cases} \text{find} & \alpha, h \\ \max & \sigma_{\text{EA}}(\alpha, h) \\ \text{s.t.} & \sigma_{\text{EB}}(\alpha, h) \leqslant 310 \\ & 0.8 \leqslant \alpha \leqslant 1.56 \\ & 0.33 \leqslant h \leqslant 0.43. \end{cases} \tag{16}$$

Two similar schemas of interpolating points (6×6, $a = 0.152$, $b = 0.02$ in Fig. 2) and non interpolating points (5×5 in Fig. 3) are used for this example. At the 25 non-interpolate design points, the RMSEs of objective function and constraint condition obtained by using augmented RBF are 2.658 and 2.349, respectively. They are a little better than those obtained by using RBF (2.734 and 2.408). Table 3 shows the comparison with SRSM-QP.

Similar to Example 4.1, this engineering application has two model establishing and two SQP calculations for every model. The AG-RBF-SQP algorithm use 37 structural analyses. From Table 3 and Figure 7, we note that

Table 3. Result comparison with SRSM-QP.

Algorithm	Optimal point		Objective value	Constraint value	Number of
	α/(rad)	h/(cm)	σ_{EA}/MPa	σ_{EB}/MPa	structural analyses
SRSM-QP [18]	0.9914	0.3300	262.2200	296.5100	85
AG-RBF-SQP	1.2371	0.3300	314.8250	295.8280	37

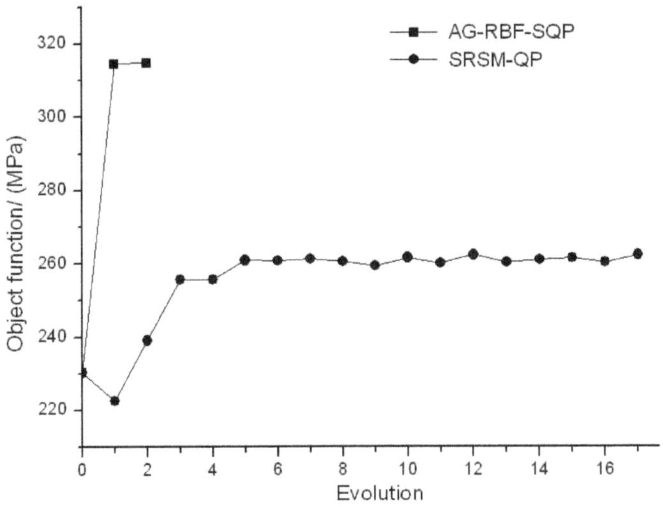

Fig. 7. Evolution of the objection function.

the global optimal objective value calculated by AG-RBF-SQP is 20.1% larger than the local one by SRSM-QP, and the number of structural analyses is much smaller than that by SRSM-QP.

The simulated explosive process lasts 21 μs. Figure 8 shows the curves of the effective stresses versus time before and after optimization, the maximum effective stress at point A is 314.825 MPa after optimization, it increase 36.7% compared with 230.31 MPa before optimization, the maximum effective stress at point B changes very little, from 298.11 MPa before optimization to 295.828 MPa after optimization, it also satisfies the constraint.

5 Conclusion

In this paper, the augmented Gaussian Radial Basis Function is used to establish an approximate model for a global optimization procedure. This approximate model expressed explicitly in terms of the design variables allows to easily obtain the derivatives of the objective function and constraint functions, required in many optimisation algorithms. Taking into account the characteristics of the augmented Gaussian RBF, the method of Sequential Quadratic Programming (SQP) is adopted as optimisation solver. Since the derivatives of the objective function and constraint functions are calculated analytically without structural reanalyses, the present optimisation procedure is very fast. Furthermore, it does not require the user to provide any initial solution. The numerical tests show that the optimal results given by the present algorithm are very close to those obtained by the usually used GA and SAA, but it is much more efficient than these two optimisation algorithms and SRSM-QP algorithm.

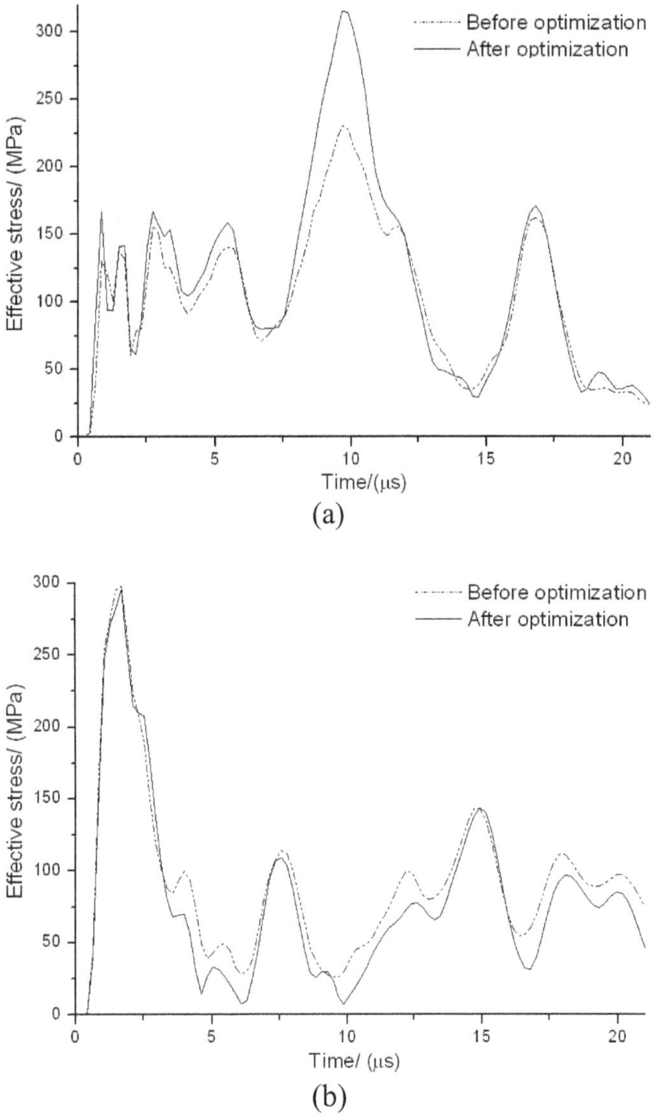

Fig. 8. Curves of the effective stress versus time. (a) Point A. (b) Point B.

References

1. L.A. Schmit, *Structural design by systematic synthesis*, in *Proc. 2nd Nat. Conf. Elect. Comput. ASCE*, 105–132 (1960)
2. L.A. Schimit, B. Farshi, Some approximation concepts for structural synthesis. AIAA J. **12**, 692–699 (1974)

3. R.H. Myers, D.C. Montgomery, *Response Surface Methodology*, Wiley and Sons New York (1995)
4. G. Venter, R.T. Haftka, Using response surface approximations in fuzzy set based design optimization. Struct. Multidisc. Optim. **18**, 218–227 (1999)
5. S. Hosder, L.T. Watson, B. Grossman, W.H. Mason, H. Kim, H. RT, Steven EC, Polynomial Response Surface Approximations for the Multidisciplinary Design Optimization of a High Speed Civil Transport. Optimization and Engineering **2**, 431–452 (2001)
6. T. Jansson, L. Nilsson, M. Redhe, Using Surrogate models and response surfaces in structural optimization-with application to crashworthiness design and sheet metal forming. Struct. Multidisc. Optim. **25**, 129–140 (2003)
7. B. Liu, R.T. Haftka, L.T. Watson, Global-local structural optimization using response surfaces of local optimization margins. Struct. Multidisc. Optim. **27**, 352–359 (2004)
8. Y. Zheng, P.K. Das, Improved response surface method and its application to stiffened plate reliability analysis. Engineering Structures **22**, 544–551 (2000)
9. Y.K. Sui, S.P. Li, *The application of improved RSM in shape optimization of two-dimension continuum*, The International Conference on Computational Methods, December 15–17, (2004), Singapore
10. J.C. Mason, M.G. Cox, *Radial basis functions for multivariable interpolation: a review in Algorithms for Approximation*, Oxford University Press, Oxford (1987)
11. T. Krishnamurthy, *Response surface approximation with augmented and compactly supported radial basis functions*, in *44th AIAA/ASME/ASCE/AHS/ASC structures, structural dynamics, and materials conference*, Norfolk, Virginia, April 7–10 (2003)
12. H.B. Fang, M.F. Horstemeyer, Global response approximation with radial basis functions. Eng. Optimiz. **38**, 407–424 (2006)
13. Y.K. Sui, in *Modeling, Transformation and Optimization: New Developments of Structural Synthesis Method*, Dalian University of Technology Press (1996) (in Chinese)
14. Y.K. Sui, A.Q. Zhang, L.C. Long, Rational move limits and its application in optimization algorithm. J. Basic Science and Engineering (to be published, in Chinese)
15. G.N. Xuan, R.W. Cheng, X.J. Yu, G.G. Zhou, in *Genetic algorithm and engineering optimization*, Qinghua University Press Beijing, (2004) (in Chinese)
16. R.L. Yang, J.F. Gu, An efficient simulated annealing algorithm for global optimization. Systems Engineering-Theory Practice **17**, 30–33 (1997) (in Chinese)
17. http://7d-soft.com/cn/download.htm
18. Z.G. Yang, M. Chen, Y.K. Sui, Application of response surface methodology in optimum design of cylindrical shell structure. J. Projectiles, Rockets, Missiles and Guidance (to be published, in Chinese)

10

A model for defining evacuation policies for emergency escape from buildings

Belarmino Adenso-Díaz[1,a], Pilar González-Torre[1], Verónica Ordóñez[1] and Juan José del Coz[2]

[1] Operations Management Department, University of Oviedo, Gijón, Spain
[2] Civil Engineering Department, University of Oviedo, Gijón, Spain

Abstract – Over the last 10 years, the need to guarantee safety in the habitability of buildings has led to the development of different models (which are usually classified either as optimization, simulation, or risk assessment approaches) that help define how urgent evacuation of the building should take place in the event of a fire or any other circumstance that requires such a measure. This paper presents an optimization model embedded in a simulation model that enables the user to define which evacuation routes should be used to minimize the time needed to evacuate all occupants. The system is presented here along with an experimental framework designed to assess the quality of the proposed solutions and the system's performance with respect to different types of buildings. Results analysis show improvements around 9% of the time needed to evacuate a building, using an experimental framework that considers three main factors: type of building, number of exit doors and capacity flow in the corridors.

Key words: Simulation; evacuation policies; optimization; emergency planning.

1 Introduction

The events that periodically appear in the world news highlight the importance that must be given to safety in buildings, as well as the difficulty of conducting an efficient evacuation should an emergency arise. Different computer assisted tools have appeared in recent years to help in the analysis of how to evacuate a building. These models permit the assessment of what will happen in a particular building when faced with the event of a fire or a forced evacuation due to any other cause.

These computer tools mostly address the problem from three distinct approaches [1]: simulation (see what occurs and test the alternatives), optimization (seek the best way of carrying out the evacuation), and risk assessment approaches (based on computer assisted tools, the main objective being only to assess exit conditions [2] or to obtain the total time need to evacuate the building [3]).

The research study presented here employs a mixed approach that simultaneously uses the first two paradigms, attempting to optimize the design of evacuation routes by the joint use of a heuristically determined simulation in order to fine tune egress parameters. The overall aim of this research study consists in providing those responsible for projecting and planning buildings with a simulation tool that allows them to study the egress movement of the occupants of a building during the process of evacuating people from a particular building, with the aim of assessing and comparing the feasibility and effectiveness of different evacuations plans. With this purpose in mind, a system was designed that is capable of automatically directing the search for the best parameters for defining the evacuation plan for a given building by means of iterative interaction with the defined simulator. The computer tool simulates the movements of the occupants in the building under analysis by means of the definition of rules and automatically defines the egress route that permits the highest speed of evacuation.

1.1 Review of some previous models

Two approaches have traditionally been used in designing evacuations in the event of an emergency:

a. Full/scale evacuation drills. This means using real people with similar characteristics to those who usually occupy the building and experimentally testing what will occur in the event of an emergency. This methodology presents several inconveniences, including those of an ethical nature (there may be casualties if an attempt is made to make the experiment realistic), as well as being very costly, since it is necessary to employ many people (if the aim is to obtain statistically significant results).
b. *Strict application of predefined regulations.* This consists in accepting or rejecting the proposed design on

[a] Corresponding author: adenso@epsig.uniovi.es

Table 1. Summary of the features of the reviewed evacuation models.

Model	References	Objetive	Evacuation route	Simulation conditions	Influence egress	Decisional capacity
1. Allsafe	[1,2]	O.1	R.3	C.4	E.2	D.2
2. Aseri	[1,2]	O.3	R.1/R.4	C.4	E.1	D.1
3. Assisted Evacuation Simulation	[4]	O.1	R.4	C.4	---	---
4. Bfires	[2,5]	O.2	R.5	C.4	E.1	D.1
5. Bgraf	[2]	O.4	R.5	C.4	E.1	D.1
6. Crisp3	[2,6]	O.3	R.1.a	C.1/C.4	E.1	D.1
7. Eescape	[1,2]	O.5	R.3	C.1	E.2	D.2
8. Egress	[1,2,7,8]	O.3	R.1.b	C.4	E.1	D.1
9. Egress Complexity model	[2,3]	O.5	R.6	C.1	E.2	D.2
10. Egress Pro	[1,2]	O.5	R.3	C.3	E.2	D.2
11. E-Scape	[2]	O.3	R.1.b	C.4	E.1	D.1
12. Evacnet4	[1,2,8,9]	O.6	R.2	C.1	E.1	D.2
13. Evacs	[1]	O.1	R.4	C.1	---	D.2
14. Evacsim	[2]	O.3	R.1.b	C.4	E.1	D.1
15. Exit89	[1,2]	O.2	R.1/R.4	C.3	E.1	D.2
16. Exitt	[1,2,10]	O.4	R.1.b	C.4	E.2	D.1
17. Exodus	[1,2,4,8,11], [12–14]	O.3	R.1.b	C.4	E.1	D.1
18. Firecam	[2]	O.7	R.4	C.3	---	D.2
19. Fpetool	[1]	O.1	R.4	C.3	E.2	D.2
20. Gridflow	[1,2]	O.3	R.1/R.4/R.6	C.3	E.1	D.2
21. GSpaces		O.3	R.5	C.2	E.1	D.1
22. Legion	[2]	O.3	R.5.a	C.4	E.1	D.1
23. Magnetic Model	[2]	O.3	R.1/R.4/R.5	C.2	E.1	D.1
24. Myriad	[2]	O.5	R.1	C.1	E.1	D.2
25. Pathfinder	[1,2]	O.3	R.1/R.4	C.1	E.1	D.2
26. Pedgo	[2,4]	O.3	R.4	C.2	E.1	D.1
27. Pedroute y Paxport	[2]	O.3	R.1/R.2/R.3	C.2	E.1	D.1
28. Sgem	[2,15]	O.3	R.1.b	C.2	E.1	D.1
29. Simulex	[1,2,8,16,17]	O.3	R.1/R.4	C.2	E.1	D.2
30. Steps	[1,2]	O.3	R.1.a/R.4	C.1	E.1	D.2
31. Takahashi's Fluid Model	[2]	O.3	R.2	C.1	E.1	D.2
32. Timtex	[2]	O.1	R.1/R.4	C.1	E.2	D.2
33. Vegas	[2]	---	R.1.b	C.4	E.1	D.1
34. Wayout	[1,2]	---	R.3	C.3	E.1	D.1

the basis of its adaptation to a set of predefined regulations. In Spain, for example, this takes the form of a mandatory code (NBE-CPI-96, *Norma Básica de Edificación, Condiciones de Protección contra Incendios*, i.e., Basic Building Code, Fire Protection Conditions). However, it is unlikely that this procedure can take into account all the characteristic features of each building which condition the efficiency of its evacuation plan. These features may be classified under 4 categories: Configurational factors (layout, number of exits, width of exits, interior distances, etc.); Environmental factors (the debilitating effect of smoke, heat, gases, etc.); Procedural factors (the actions of the occupants of the building, their training, emergency signs, etc.); and Behavioral factors (the existence of families or groups in the building, initial reaction time, speed of movement, etc.).

To be able to consider all these factors that need to be taken into account in an evacuation process, the usual procedure is to develop computer assisted evacuation models. Following a review of the literature, 34 fire evacuation systems were identified (Tab. 1), which may be classified according to the following features:

- According to the objective of the program:

  O.1. To calculate the total time need to evacuate the building.
  O.2. To simulate the egress movement of evacuees, without being able to view the process. Here, only information relative to the positions occupied by the evacuees during the process and to the chosen evacuation route is obtained.
  O.3. To simulate the egress movement of evacuees, with the possibility of viewing the process. Information is also obtained on the bottlenecks and jams produced during evacuation.
  O.4. To study the decision-taking process of evacuees.
  O.5. To evaluate the exit conditions.

- According to the evacuation route chosen by the occupants of the building:

  R.1. The shortest route. Two, more specific options may exist in this case:

    R.1.a. The shortest route modified according to the density of people.

    R.1.b. The shortest route modified according to the behavior of the evacuees.

  R.2. The route that minimizes evacuation time.

  R.3. The only route possible given the geometry of the building.

  R.4. The route defined by the user.

  R.5. The route chosen according to the behavior of the evacuees. A more specific application is:

    R.5.a. The route dictated by the behavior of the evacuees modified by the density of people.

  R.6. Assigning a probability to each of the possible exits. The occupants will choose a particular exit on the basis of this probability.

- According to simulation conditions:

  C.1. Does not consider the behavior of evacuees or the effects of the fire on the evacuation.

  C.2. Considers the behavior of evacuees, but not the effects of the fire on the evacuation.

  C.3. Considers the effects of the fire, but not the behavior of evacuees on the evacuation.

  C.4. Considers the behavior of evacuees and the effects of the fire on the evacuation.

- According to the influence on egress movement owing to the presence of other occupants:

  E.1. The egress movement of occupants affects the rest of the evacuees.

  E.2. The egress movement of occupants does not affect the rest of the evacuees.

- According to the decisional capacity of occupants:

  D.1. The occupants take decisions during the evacuation.

  D.2. The occupants do not take decisions during the evacuation.

These programs may be divided into two major types: models that study how the movement of people is produced under normal non-emergency conditions (the precursors of which were Predtechenski and Milinksii [18], and subsequently Fruin [19], which, applied to areas with a large number of people and on stairs, gave rise to movement models such as PEDROUTE by Buckmann and Leather [20]); and evacuation models (which are more recent, some of the first studies being those by Stahl [5]). The latter type of model may in turn be classified into those that only consider deterministic movements (people are assumed not to think in a personal way, but react automatically to external stimuli, exiting as quickly as possible when the event occurs, at a specific speed, etc., which is how the model by Takashi et al. functions), and models which also consider physical features, those relating to the behavior of people, treating them as individual agents with different reaction times, preference for certain exits, etc. (the most paradigmatic example being the EXODUS software [10,21]). At the same time, several studies have shown how psychological factors have a major influence on the results of the evacuation [22].

In our model the goal pursued is to obtain the most appropriate evacuation policies for a given building, the result of the program being the route that the evacuees should follow in order to minimize the total time taken to completely evacuate the building. A deterministic approach is considered, the egress routes being defined by the use of heuristic algorithms linked to a simulation model for assessing the quality of the explored routes.

2 Proposed model

Figure 1 represents the scheme of the proposed system[1]. As can be seen, the building layout will be the starting point of the entire system. A graph including all the geometrical information on the building will be constructed on the basis of this layout (generated by module T1) in order to model all possible movements (see Fig. 2 for an example of a hotel floor). The graph will have 4 different types of nodes (access to rooms, widening doors inside a corridor, crossing of three or more corridors, and exit doors), the corridors being represented by the edges (see more details of implementations in Ordoñez et al. [23]). Additional data on the maximum capacity of corridors and maximum speed along them (as a function of the people in the corridor) and maximum flow crossing the doors and crossing are also needed for the simulation. In addition, the time for people to reach room doors (i.e., second by second, how many people will be passing through their room doors and trying to get into the corridor) is calculated by the room simulator (module T2 in Fig. 1), which will complete the input data for the simulation.

Note that an evacuation policy using these data will simply be an n-tuple indicating by means of each of the n nodes of type "crossing" which direction should be followed when arriving at that point. For instance, in Figure 2, given that crossing nodes are ⟨18, 19, 21, 23, 24, 26, 27, 28, 29, 30, 31, 32, 33⟩, a solution would be the n-tuple ⟨19, 21, 3, 24, 26, 27, 6, 27, 28, 29, 30, 31, 32⟩ which determines for each cross the "next corridor" to follow. Given this information and assuming determinism in the behavior of the evacuees, simulation (module T4) will consists in moving each person through the graph, second by second, considering the permissible speed and directions when arriving at a crossing, until everybody has exited the building, providing as output the time needed for its total evacuation.

The initial policy generator (T4) will be responsible for providing the simulator with the first strategy to

[1] A freeware version of the system called SIMUDRILL can be downloaded from http://coruxa.epsig.uniovi.es/ $\sim$adenso/file_d.html

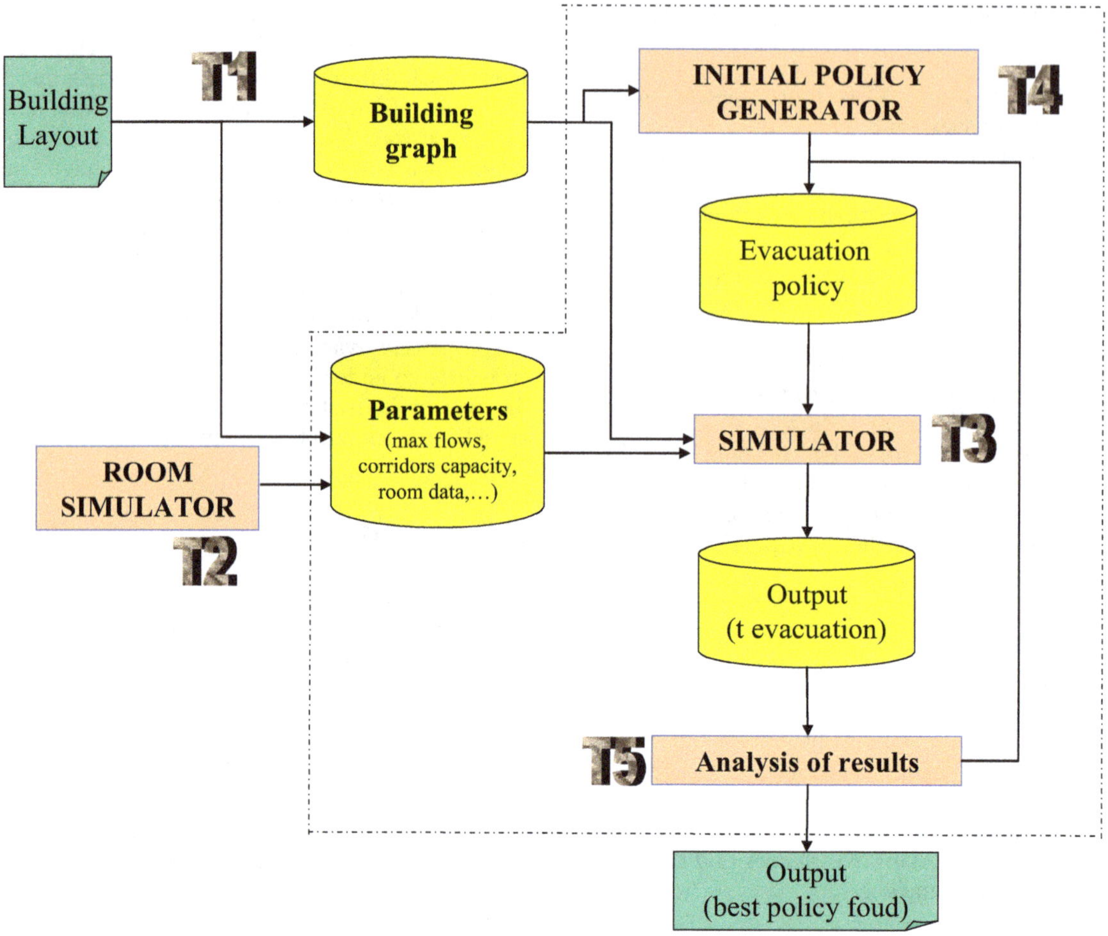

Fig. 1. Interrelationships among the modules of the designed system.

follow during evacuation (initial evacuation plan), which will be iteratively improved by considering the results of previous simulations (analyzed by module T5, which will be in charge of trying to find a solution that would minimize the required evacuation time).

As regards the logic of modules T4 and T5, the initial solution is generated using a heuristic procedure which has been proven to offer a good initial solution: from each room door in the building, we calculate the shortest path (using Dijkstra's algorithm) to any of the exit doors. The next corridor chosen at each crossing node defines the initial evacuation policy (in the case of a certain crossing node not being used by the shortest path from any room, Dijkstra's algorithm is used to find the shortest path from that crossing node to any exit door. This path then defines the next corridor for that crossing node).

The logic for optimization module T5 needs to be more complex. It comprises a local search procedure that explores the neighborhood of the current solution trying to find a policy with a shorter evacuation time. The algorithm starts with the current solution, and explores for each node of the type "crossing", which is the node with the slowest corridor that is in that solution. By forcing the escape from that node via a different corridor, the neighborhood of the current solution is defined, continuing the search until no further improvement is found. Figure 3 shows a pseudocode for the algorithm.

3 System performance analysis

An experimental framework was defined to analyze the performance of the proposed system, studying its behavior with respect to different factors as well as the improvement in quality achieved by means of the optimization phase. After a review of the system and its performance in preliminary experiments, three factors were identified that are considered likely to influence the quality of the obtained solutions: the type of building, the number of existing exits in the building, and the capacity (width) of its corridors.

The type of building refers to the fact that the rooms and corridors are all close to one another with a compact building design, or, to the contrary, that the rooms present a linear layout due to a longitudinal building design. The latter alternative may make it more difficult to evacuate the building in the case of an emergency if emergency exits do not also linearly exist along the length of the building. Two different types were considered in the

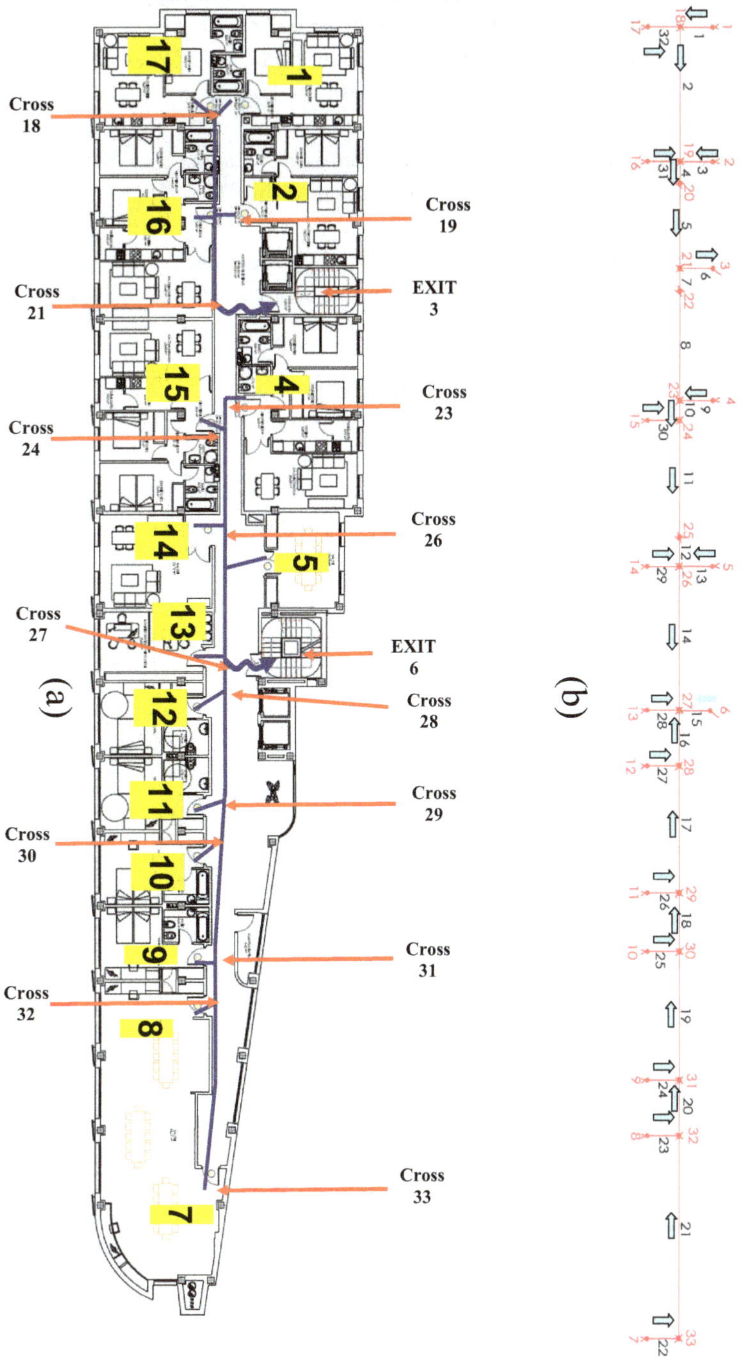

Fig. 2. Example of the layout of a hotel floor (a) and corresponding graph (b) input to the system. In the layout, the number of rooms and crossing nodes are indicated. In the graph, an example of evacuation policy is depicted.

experiments: a compact type building and a linear building (Fig. 4).

In the experimentation, all the buildings corridors have one of these lengths: 5, 7, 8 or 25 m in the compact case, and 5 or 10 m in the linear one.

The maximum distance from any point in the building to the emergency exit is regulated in the legislation on the construction of buildings. However, a greater number of exits facilitates evacuation and therefore may also be a factor to take into account. Two levels were considered for this factor in the experiments: a building with few exits (only two) and one with many exits (four).

The graph that defines the layout of the building is fully determined with these two factors. It now only remains to define aspects such as: the maximum flow crossing doors and corridors crosses (in all cases a constant flow of 5 people per second in crosses, 4 people per second in exists when only two exists and 2 people per second in case of 4 exits were defined); the time for people reaching rooms doors (in all cases, a schedule of 2 people

1. Let S_{cur} be the current solution and T_{Scur} its evacuation time.
2. $\forall$ crossing node not explored so far, let C' the node with slowest "next corridor" in S_{cur}.
3. Let $\Omega_{C'}=\{P_{C',1},P_{C',2},\ldots,P_{C',k}\}$ be the set of k corridors adjacent to node C', excluding the "next corridor" in S_{cur}.
4. $\forall$ $P\in\Omega_{C'}$, we build the solution S_P obtained from S_{cur}, by forcing P as "next corridor" in node C'. Calculate by calling simulation module its evacuation time T_{Sp}.
5. Let $P'\in\Omega_{Cr'}$ the corridor with smallest $T_{Sp'}$.
   IF $T_{Sp'}<T_{Scur}$ ➔GOTO **6** (*a better solution than S_{cur} was found*)
   ELSE ➔ GOTO **7** (*continue exploration from S_{cur} or a solution worse than S_{cur}*)
6. Update $S_{cur}:=S_{p'}$;
   GOTO **2**
7. IF all crossing nodes have been already explored ➔
   IF $T_{Sp'}>1.10*T_{best\_solution\_found\_so\_far}$) ➔GOTO **8** (* end of exploration*)
   ELSE ➔ {$S_{cur}:=S_{p'}$;
   GOTO **2**} (*continue exploration from the new S_{cur}*)
   ELSE ➔ GOTO **2** (*continue exploration of neighborhood of original S_{cur}*)
8. Output the best solution found so far; END.

Fig. 3. Pseudocode for the optimization phase of the system.

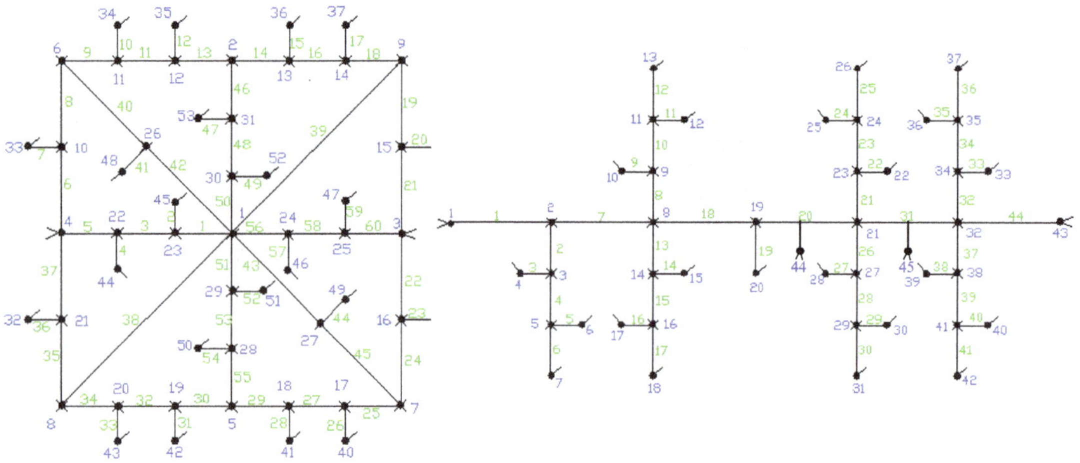

(a) (b)

Fig. 4. Basic structure of the (a) compact and (b) linear buildings used in the experiments, indicating each of their nodes and edges. Exit nodes in compact case are number 3, 4, 8 and 9 (3 and 4 when only two exits), and in the linear case nodes 1, 43, 44 and 45 (1 and 43 when only two exits).

reaching the door in second 5 was defined for rooms, and 3 people in second 5, 4 in second 6, 5 in second 7, 4 in second 8, 3 in second 9 and 1 in second 10 for all meeting rooms). Regarding the speed of evacuation in the corridors or their maximum capacity, these factors will be defined on the basis of the number of people who fit into a linear meter of corridor (corridor width). Five levels were considered for this factor: 1, 2, 3, 4, or 5 people per linear meter of corridor. In this way, the maximum capacity is calculated as the product of the corridor length times this factor, and the maximum running speed in the corridors, as some constants depending in the number of people in the corridor and its length (see Table 2).

Experiments were carried out for $2 \times 2 \times 5 = 20$ different buildings and scenarios, varying the levels of the 3 factors under consideration. The descriptive results of this experimentation (the evacuation time given by the initial solution and that given by the optimized solution) are presented in Table 3.

3.1 Analysis of the results

To verify whether the results obtained after the optimization are better than those obtained initially, the type of distribution of the variable under study were previously studied. By means of the Kolmogorov-Smirnov test, it was shown that both the data from the initial solution ($Z = 0.943$, bilateral significance $= 0.336$) and those corresponding to the optimized solution ($Z = 0.987$, bilateral significance $= 0.284$) are normally distributed,

Table 2. Evacuation speed (m/s) as a function of the number of people in a corridor and its length.

Number of people in the corridor	5 m corridor	7 m corridor	8 m corridor	10 m corridor	25 m corridor
1 person – 1/4 of maximum capacity of the corridor	5	6	7	6	8
1/4 of maximum capacity of the corridor – 1/2 of maximum capacity of the corridor	2	3	4	4	5
1/2 of maximum capacity of the corridor – corridor full of people	1	1	1	1	1

Table 3. Descriptive statistics of the variables involved in the experiments.

Type of solution	Factor/Levels	Evacuation time Average (s)	Evacuation time Standard deviation (s)
Initial		38.15	18.88
	Layout type:		
	Compact	32.60	12.01
	Linear	43.70	23.23
	Number of exits:		
	Two	42.20	19.48
	Four	34.10	18.34
	Corridor capacity:		
	1 people/m	69.50	19.33
	2 people/m	36.75	8.85
	3 people/m	30.75	6.34
	4 people/m	27.50	5.20
	5 people/m	26.25	3.40
After optimization		34.80	16.35
	Layout type:		
	Compact	32.20	11.70
	Linear	37.40	20.31
	Number of exits:		
	Two	40.10	19.20
	Four	29.50	11.55
	Corridor capacity:		
	1 people/m	61.00	18.81
	2 people/m	34.50	7.42
	3 people/m	29.00	4.55
	4 people/m	25.00	4.08
	5 people/m	24.50	4.12

thus allowing the use of parametric tests in the statistical analyses.

To compare the results given by the initial and optimized solutions, an analysis of the initial correlation shows a strong positive relation between the two crossed variables ($R = 0.944$), while the t-test for paired data ($t = 2.340$, bilateral significance $= 0.030$) confirms the existence of statistically significant differences between the average values of the two variables. From Table 3, it can be seen that an average improvement in evacuation time of 8.78% is obtained, thus verifying the effectiveness of the optimization algorithm. Computational times for the experimentation are nearly negligible (less than one second).

To test the effects of the 3 considered factors on the performance of the optimization phase, a Mann-Whitney's test was carried out for the case of the type of building and the number of exits (which only have two levels), considering as the analysis variable the ratio between the optimized solution an the initial one. Test results show significative differences between the results obtained after the optimization phase depending on the type of layout ($Z = -3.538$, bilateral significance near 0), while no differences are shown for the factor number of exits ($Z = -0.115$, bilateral significance 0.908). The appropriate number of exit doors on the floor depends on the scale of the floor and the total length of routes [24], although their widths also affect the survival rates [17]. According to the literature, other factors having influence on the successful evacuation are the good knowledge by the people of the building layout, the existence of good instructions and signals to follow the evacuation route [18], the queues length [17] or the panic syndrome [24].

A Kruskal Wallis' test, again for the ratio of improvement after optimization phase, was used to analyze the effects of the capacity of the corridors on evacuation time. The results obtained ($\chi^2 = 0.884$, signification 0.927) show no significant differences among the five levels taken into consideration. However, Table 3 shows that, without any other variation in the data, the greater the capacity of the corridor, the lower the time needed to evacuate the building, obtaining an improvement in evacuation time of almost 60% between the two simulated extremes (1 person per meter and 5 people per meter).

4 Conclusions

To date, several approaches have been used to define the policies for evacuating a building. Some of them studied the fire advance [25, 26], while some other [9, 11, 15, 17, 21, 24, 27] studied the people behavior when evacuating a building. However in this paper we propose a new approach to define the best evacuation route, after analyzing, from an optimization point of view, the different alternative routes from each corridor cross to the different exit doors. This paper presents a model based on the definition of an initial policy that is subsequently improved iteratively by means of a heuristic algorithm linked to the simulator that assesses the quality of the solutions that are being explored. The proposed model uses a network to represent the building that consists of a set of nodes (rooms) connected by arcs (corridors). The progressive motion and behavior of each individual is determined by a set of heuristics rules.

This model was tested using an experimental framework which considers 3 factors (type of building, abundance of exit doors from the building, and the capacity of the corridors). Simulations were carried out on 20 different scenarios which allowed us to confirm that the optimization phase achieves a statistically significant average improvement of 8.72% in the evacuation time offered by the initial starting-out solutions, in addition to the influence of the 3 factors on the performance of the algorithm. Future research should analyze the goodness of the generated solutions versus other heuristic procedures for generating initial solutions.

Acknowledgements

This research was funded by the Spanish Ministry of Infrastructures, contract number MFOM-06-2.

References

1. E. Kuligowski, R. Peacock, *A review of building evacuation models*, National Institute of Satndards and Technology, Technical Note 1471, 2005
2. S. Olenick, D. Carpenter, Journal of Fire Protection Engineering **13**, 87 (2003)
3. H.A. Donegan, A.J. Pollock, I.R. Taylor, 1994, Egress Complexity of a Building, in *Proceedings of Fourth International Symposium on Fire Safety Science*, edited by T.l. Kashiwagi, 601-12
4. M. Jafari, I. Bakhadyrov, A. Maher, *Technological advances in evacuation planning and emergency management: current state of the art.* U.S. Department of Transportation. Center for Advanced Infrastructure and Transportation, Report n° EVAC-RU4474, 2003
5. F. Stahl, Fire Technology **18**, 49 (1982)
6. J.N. Fraser-Mitchell, An Object-Oriented Simulation (CrispII) for Fire Risk Assessment, in *Proceedings of the Fourth International Symposium of Fire Safety Science*, edited by T. Kashiwaga, 1994, pp. 793–804
7. H.W. Hamacher, S.A. Tjandra, Mathematical Modelling of Evacuation Problems: A State of the Art. Berichte des Fraunhofer ITWM, 2001, No. 24
8. G. Santos, B.E. Aguirre, *A critical review of emergency evacuation simulation models.* University of Delaware. Disaster Research Center, Preliminary Paper 339, 2004
9. T.M. Kisko, R.L. Francis, Fire Safety Journal **9**, 211 (1985)
10. B. Levin, Exitt, a simulation model of occupant decisions and actions in residential fires, in *Proceedings of the Second International Symposium on Fire Safety Science*, edited by T. Wakamatsu, Y. Hasemi, A. Seizawa, P. Seeger, P. Pagni, C. Grant, 1989, pp. 561–570
11. E.R. Galea, J.M.P. Galparsoro, Fire Safety Journal **22**, 341 (1994)
12. E.R. Galea, M. Owen, Emergency Egress From Large Buildings Under FIRE conditions Simulated Using The Exodus Evacuation Model, in *Proceedings of the Seventh International Fire Science and Engineering Conference Intertlam*'96, edited by C. Franks, S. Grayson, 1996, 711–720
13. S. Gwynne, E.R. Galea, P.J. Lawrence, M. Owen, L. Filipidis, *Further Validation of the building Exodus Evacuation Model Using the Tsukuba Dataset*, CMS Press, 1998, No. 9, XIM:31
14. S. Gwynne, E.R. Galea, P.J. Lawrence, L. Filipidis, Fire Safety Journal **36**, 327 (2001)
15. S.M. Lo, Z. Fang, P. Lin, G.S. Zhi, Fire Safety Journal **39**, 169 (2004)
16. P. Thompson, J. Wu, E.W. Marchant, Fire Engineers Journal **56**, 6 (1996)
17. P. Thompson, E.W. Marchant, Fire Safety Journal **24**, 149 (1995)
18. V.M. Predtechenskii, A.I. Milinskii, *Planning for Foot Traffic in Buildings*, New Delhi: Amerind Publishing Co. Pvt. Ltd, 1978
19. J.J. Fruin, *Pedestrain Planning and Design*, Mobile, AL: Elevator World, Inc., 1987
20. L.T. Buckmann, J. Leather, Traffic Engineering and Control **35**, 373 (1994)
21. M. Owen, E.R. Galea, P.J. Lawrence, Journal of Fire Protection Engineering **8**, 65 (1996)
22. N.R. Johnson, W.E. Feinberg, Journal of Environmental Psychology **17**, 123 (1997)
23. V. Ordoñez, B. Adenso-Díaz, P. González-Torre, J.J. del Coz, Simulation of performance of evacuation policies in a building, *Proceedings of the First International Conference on Multidisciplinary Design Optimization and Applications*, ASMDO, 17–20 April, 2007 (Besancon, France)
24. H. Notake, M. Ebihara, Y. Yashiro, Safety Science **38**, 127 (2001)
25. G. Hadjisophocleous, Z. Fu, S. Fu, Ch. Durcher, Journal of Fire Protection Engineering **17**, 185 (2007)
26. K. Remesh, K.H. Tan, Journal of Fire Sciences **25**, 321 (2007)
27. T.T. Pires, Fire Safety Journal **40**, 177 (2005)

11

Optimal design and analysis of a new thermally actuated microscanner of high precision

W.L. Gambin[a] and A. Zarzycki

Warsaw University of Technology, ul. Sw. A. Boboli 8, 02-525 Warszawa, Poland

Abstract – A design of a precise, thermally actuated microscanner is proposed, and next, its thermal and mechanical behaviour is considered. The device consists of a micromirror and four thermo-bimorph cantilevers with electric resistors. After the forming process, the mirror assumes an out-of-plane rest position. The process is very simple and compatible with the IC fabrication technique. The mirror is capable of two-dimensional (2D) scans for optical raster imaging. The scanner works both in a non-resonance (for the frame scanning) and resonance mode (for the raster scanning). The high precision of scanning action is achieved due to a special position of the mirror rotation axes with respect to the cantilever beams. The above position assures, that the distance of the mirror centre from the light source is the same during the whole scanning process and the inertial moments of movable parts, as well as the influence of air damping, are minimized. To find the optical angle amplitudes of the mirror, deflections of the cantilevers caused by changes of temperature are determined. The analysis of dynamic temperature distribution enables to determine the thermal cut-off frequency, below which the amplitude of mirror deflections is frequency-independent. To find the resonance frequencies of the device, the dynamical analysis of the scanner is performed and free vibrations of the considered system are examined.

Key words: Microscanners; thermal bimorph actuators; raster scanning system; silicon etching technology; FEM analysis; resonance frequencies; thermal cut-off frequency.

1 Introduction

Scanning micromirrors have a wide range of applications. One can find them in devices for imaging, bar-code reading, laser surgery, laser machining, etc. Modern MEMS and MOEMS technologies [1] enable to produce those microscanners smaller and smaller. One can classify microscanners according to their actuation principles. The most common are: electrostatic, piezoelectric, electromagnetic and thermally activated devices.

Microscanners based on *electrostatic actuation* make the most extensive group of microscanners, because electrostatic actuation is the simplest one [2]. Moreover, these scanners are characterized by very high work frequency – advantageous for such devices. Unfortunately, the deflection angle of the mirror actuator is mostly limited. In addition, electrostatic actuation needs quite large voltage. Moreover, it needs a nonlinear control voltage.

Microscanners with *piezoelectric actuation* have similar properties, although the most important feature of piezoelectric actuation is an almost proportional dependence between the induced strain and the applied electric field [3]. Additional advantages are: high pressure, high power and lower operating voltage during electrostatic actuation. Disadvantages of piezoelectric microscanners are much smaller mirror rotations and more complicated output process.

Micromirrors moved by *electromagnetic actuators* can realize large scan angle [4]. Movement of the mirror is due to large attraction and repulsion forces generated by the actuator. However, electromagnetic scanners are large because magnetic actuation requires large electromagnets. The large size of the devices leads to great weight of movable parts and, in result, to small operating frequency.

Micromirrors activated thermally, used in the proposed project, have some interesting properties. *Thermal actuators*, formed as thermo-bimorphic beams, provide large scan angle, nearly linear deflection-power relationship and moderate power consumption. Moreover, fabrication of such microscanners is based on very simple technology. Thermal actuators do not operate with the frequency as high as the electrostatic or piezoelectric actuators, but high enough for some optical applications. Limited frequency is mainly determined by thermal time constants, but it can be improved by using materials with high thermal conductivity. Finally, micromirrors

[a] Corresponding author:
`Wiktor.Gambin@mchtr.pw.edu.pl`

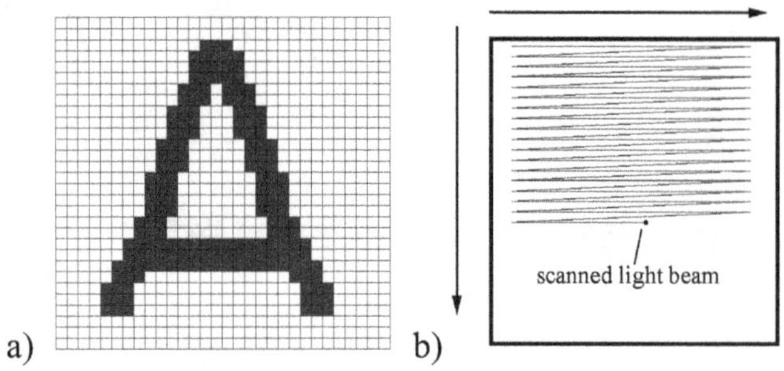

Fig. 1. Display method by raster scanning: a) discrete pixel array, b) light beam motion.

with thermo-bimorphic actuation seem to be very attractive.

2 Display technique

Nowadays there are two main display technologies: *the two-dimensional discrete pixel arrays* [5] and *the raster scanning system* [6]. There is also the third technique, which combines the above ones [7]. The first technology is based on a micromirror array, where each physical pixel on it responds to a small part of the displayed image. Examples of display technologies using two-dimensional panel include liquid crystal display (LCD) and digital micromirror array, popularly called the digital micromirror display (DMD[1]). The LCD technique is used mostly in different kinds of displays, such as modern television sets, computer monitors etc. The DMD technique is a very popular method of displaying coloured image, and for that reason the producers of multimedia projectors often use it.

The raster scanning system creates a light beam that produces a single bright pixel. The last one is scanned in two dimensions to create an image, in the same manner as a cathode ray tube (CRT) is scanned (Fig. 1). The light beam starts its movement from the left top corner of a screen and moves horizontally (from the left to right). When it reaches the end of the scanned line, the light beam is turned off and moves back to the beginning of the second line. Next, the beam is moving along the second line. That combination of two movements of the beam is repeated till it reaches the end of the screen. After the whole cycle, an image is displayed on the screen. The movement of the light beam along the horizontal line (with a high speed) is called *raster scanning*, whereas the movement between the lines (with a low speed) is called *frame scanning*. Both movements are precisely correlated to each other. Raster scanning can be realized in two ways, either by two mirrors scanning in two orthogonal directions or, like in our design, by one two-dimensional scanning mirror. Using only one mirror, the size of the whole device is much more reduced.

As it has been mentioned above, there is the third display technology – the *linear-scanning system*. It combines features of the pixel-array and raster techniques using a one-dimensional array of micromirrors. The micromirror array corresponds to the one-dimensional linear (horizontal or vertical) pixel array of displayed image. In order to display the whole image, the linear pixel array is scanned across the screen, once per frame. The linear-scanning system is widely explored by Silicon Light Machine Inc. The system proposed by them has been called Grating Light Modulator (GLV).

3 Device design

Interesting solutions of one-dimensional and two-dimensional scanning devices have been described in the book [8]. However, in the case of a two-dimensional microscanner, a certain imperfection is lack of thermal and mechanical separation of perpendicular actuators. It causes undesired twisting and bending of the actuators, and as a consequence, a global distortion of the displayed image. More important is the fact that the distance of the mirror centre from the light source changes during the scanning process. It causes nonuniform resolution of the projected image. Below we propose a solution free from the above drawbacks.

The proposed microscanner is composed of a round micromirror and four thermal-bimorph beams. The beams play the role of actuators, revolving the mirror around two perpendicular axes (Fig. 2a) in two orthogonal planes. One pair of parallel actuators (moving with a high speed and designed for raster scanning) is fixed directly to the mirror, and to a rigid movable frame. Second pair of the bimorph beams (moving at a low speed and designed for frame scanning) is situated perpendicularly to the first one and connects the frame with a stationary silicon substrate.

The actuators are composed of two layers with different coefficients of thermal expansion (CTE) and a thin insulator layer located between them (Fig. 2b). The bottom layer (the *passive* one) has a lower CTE, whereas the top layer (the *active* one) has a higher CTE. In our project, the bottom layer has the properties of an electric heater resistor connected with an electric supply. Due to mismatch between the CTE of the materials, the bimorphs after the forming process curl out of plane and make the whole device a 3D-structure.

It is necessary to point out that the described microscanner may be produced using very cheap silicon etching technology [1]. In fact, only two materials are used: silicon and aluminium. As the insulating layer, nitride silicon is sputtered on the bottom layer (with lower CTE) of the actuator, and as a heat resistor is used the same bottom layer made of doped silicon. Due to the special position of the mirror centre with respect to its rotation axes, mirror diameter may be relatively large, what assures a large precision of the displayed image. In the considered project it is assumed as 400 μm. Thickness of the mirror and its supporting frame is equal to 15 μm. Total thickness of the bimorph cantilevers is equal to 1.85 μm, and succeeding layers have the following thickness: doped

[1] DMD is a registered trade mark of Texas Instruments Inc.

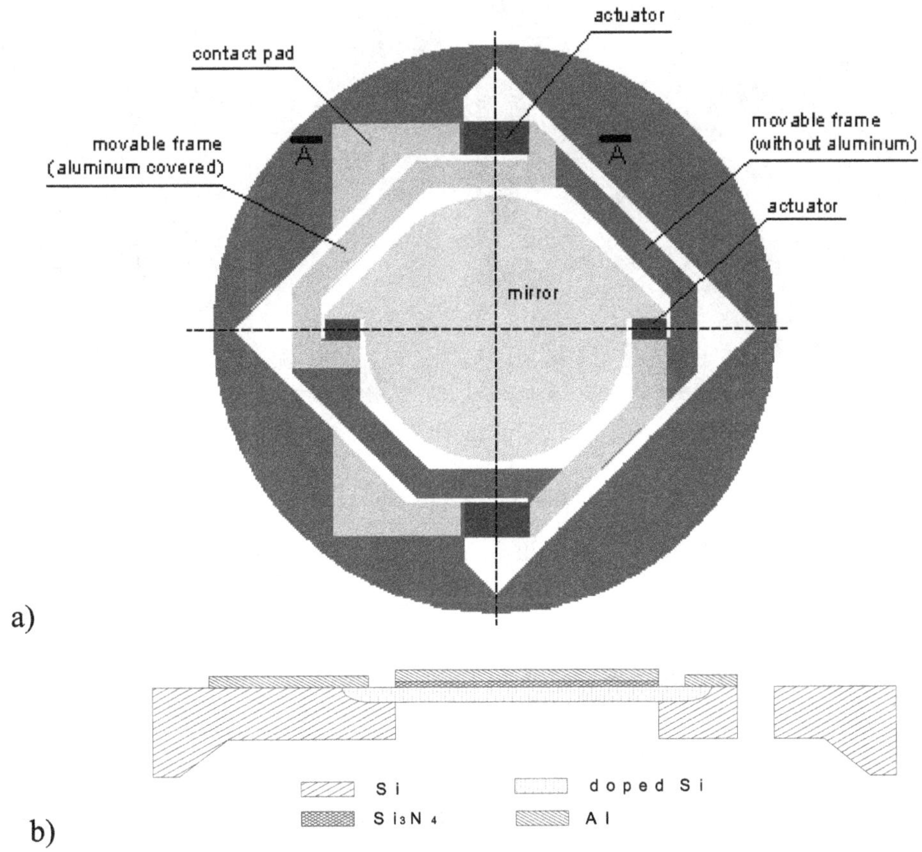

Fig. 2. Proposed microscanner a) planar projection, b) cross-section **A-A** through the frame and the scanning actuators; materials and layers thickness: Si (frame and mirror) – 15 μm, doped Si (passive layer) – 1 μm, silicon nitride (insulator) – 0.15 μm, Al (active layer) – 0.7 μm.

Si (passive layer) – 1 μm; silicon nitride (insulator) – 0.15 μm; Al (active layer) – 0.7 μm. Width of both types of cantilevers is equal to 50 μm, and its length is 30 μm (for raster scanning) and 100 μm (for frame scanning), respectively.

Short description of the microscanner fabrication process is the following. In the first step, the silicon substrate is doped in selected areas. The obtained semi-conductive layer is used, both as an electric heater resistor and a passive layer of actuators. Next, the upper surfaces of the doped areas are covered by 0.15 μm insulating layer of Si_3N_4. Afterwards, on the insulating layer, on the upper surface of the mirror and some parts of the moving frame, a thin (0.7 μm) aluminium layer is deposited (Fig. 3a). After the cooling process, shrinkage of the aluminium layers of relatively thin actuators causes initial rotation of the mirror (Fig. 3b).

In the room temperature, when electric current is passing through the heater resistor, the temperature of the actuators increases and the structure, initially deflected out of plane, deflects downwards to the substrate plane. Actuators directly connected with the mirror are short and move together with a high resonant frequency, whilst the two others are longer and move together with a low, non-resonant frequency. Due to the fact that the pairs of actuators are perpendicular to each other, and that they move with different frequency, the laser beam reflected by the mirror draws a 2D raster image on a screen.

As we can see, due to the fact that the cantilevers acting in two different planes are thermally and mechanically separated, one can avoid a global distortion of displayed image. However, the high precision of scanning action is achieved due to a *special position of the mirror centre with respect to its rotation axes*. In our case, the mirror centre lies on the cross-section of two orthogonal lines joining the centres of cantilevers, (Fig. 2a). The above position assures that the distance of the mirror from the light source is the same during the whole scanning process. Additional advantage is the fact that the inertia moments of movable parts, as well as the influence of air damping, are minimized. It allows to achieve a higher frequency for the frame scanning and a higher resonance frequency for the raster scanning. Higher frequencies result in more accurate projecting image because of greater image refreshing. On the other hand, more precise motion of the mirror causes smaller image distortion. Therefore, keeping a fixed distance between the mirror and the light source during the whole scanning process may significantly improve the displayed image.

To specify the work conditions of the microscanner, it is necessary to determine the dimensions, mechanical

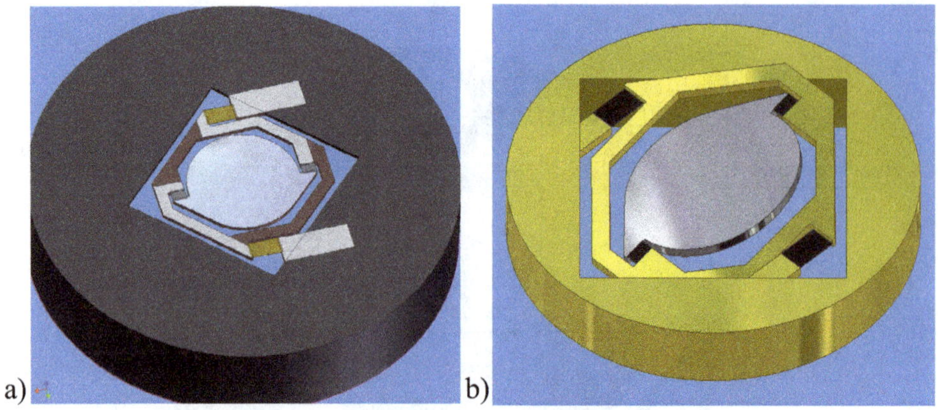

Fig. 3. General view of the microscanner: a) during the forming process, b) after the forming process.

Table 1. Material data.

Material	E [GPa]	CTE [K^{-1}]	λ[W/m K]	C [J/kg K]	ρ[kg/m^3]	R [Ω m]	ν
Si	190	2.6×10^{-6}	168	678	2.33×10^3	2.3×10^3	0.22
Doped Si	190	2.6×10^{-6}	168	678	2.33×10^3	85×10^{-6}	0.22
Si_3N_4	270	3.3×10^{-6}	30	170	3.10×10^3	10^{14}	0.24
Al	69	23×10^{-6}	235	879	2.60×10^3	2.65×10^{-8}	0.33

and thermal parameters of the materials, as well as the quantities of exploitation parameters: current, voltage, temperature, resonance and non-resonance frequency and finally the stresses in deformed elements. The assumed mechanical and thermal parameters are presented in Table 1. There are: Young's modulus (E), coefficient of thermal expansion (CTE), thermal conductivity (λ), heat capacity (C), density (ρ), resistivity (R) and Poisson's coefficient (ν).

During the mechanical analysis of the whole device, it is necessary to distinguish the mass and geometrical parameters of the mirror element – working in the resonance mode (Fig. 4a), as well as the mass and geometrical parameters of the element composed of the mirror and a movable frame – working in non-resonance mode (Fig. 4b). Let us introduce a Cartesian system of coordinates, with the centre fixed at the geometrical centre of the mirror. The first element rotating with high speed around the horizontal x-axis is divided into two parts, A and B, of the same weight. The mass centres of these parts have coordinates $\{x_A, y_A\}$ and $(x_B, y_B\}$. In the same way, the second element rotating with lower speed around the vertical y-axis is divided into parts C and D of the same weight. The corresponding mass centres have coordinates $\{x_C, y_C\}$ and (x_D, y_D). Notice that the mass centre of the whole device (mirror + frame) lies in the geometrical centre of the mirror.

Owing the above mass distribution, the gravity forces acting on the ends of moving actuators are negligibly small compared with the inertia moments. It considerably reduces the supply of electric power and increases the fundamental resonance frequency of the device. Moreover, the pure bending process of actuators assures a linear dependence between the current voltage and mirror deflection.

4 Principle of working

4.1 Analysis of the "actuator + mirror" system

To show that distance of the mirror from the light source is the same during the scanning process, let us make the following simple geometrical analysis. Consider a mirror attached to one pair of actuators. Its side view, in the coordinate system $\{x, y\}$, is shown in Figure 5.

Let the mirror and the cantilevers be located in the plane $y = 0$. The beginning and the end of the mirror are at the points A_0 and B_0, and the geometrical centre of the mirror - at the point C_0. Actuators, with the length L each one, are fastened at the point X_0 and attached to the mirror at the point Y_0. The distance between the fixing point of the actuators and the centre of the mirror is $a_0 = L/2$.

After the forming process, the actuators bend out-of-plane and the characteristic points mentioned above: A_0, B_0, C_0, X_0, Y_0, take positions of the points: A, B, C, $X = X_0$, Y. It is assumed that the pure bending process is due to thermal deformations only, and a new shape of the actuators is a circular arc with the radius R and the centre at the point O. The angle ϕ (deflection angle) describes the bend of the actuators and the slope of tangent to the arc at the point Y.

Below, we will show that the distance between the point C and the plane $y = 0$ is smaller than 2% of the cantilever length L, if the angle ϕ is smaller than 45°. To do it, denote by S the point of intersection of the moving mirror with the x-axis of the coordinate system $\{x, y\}$, and by a_S – the distance between $S$and the fixing point $X = X_0$. Let δ_C denote the distance between the centre C of the moving mirror and the point S. Notice, that *due to symmetry with respect to the O-S axis, the distance* δ_C *is*

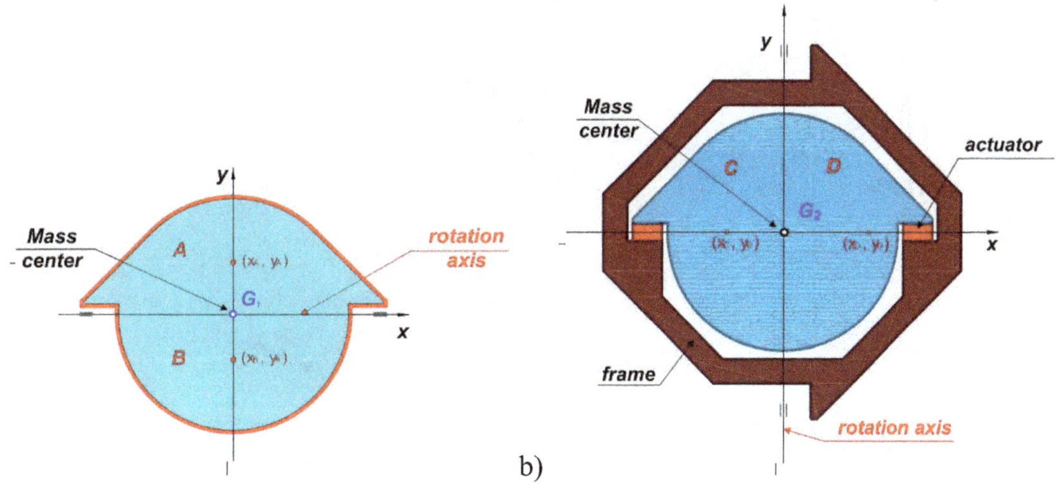

Fig. 4. Mass and geometrical parameters of two elements: a) resonance element (mirror), b) non-resonance element (mirror + frame).

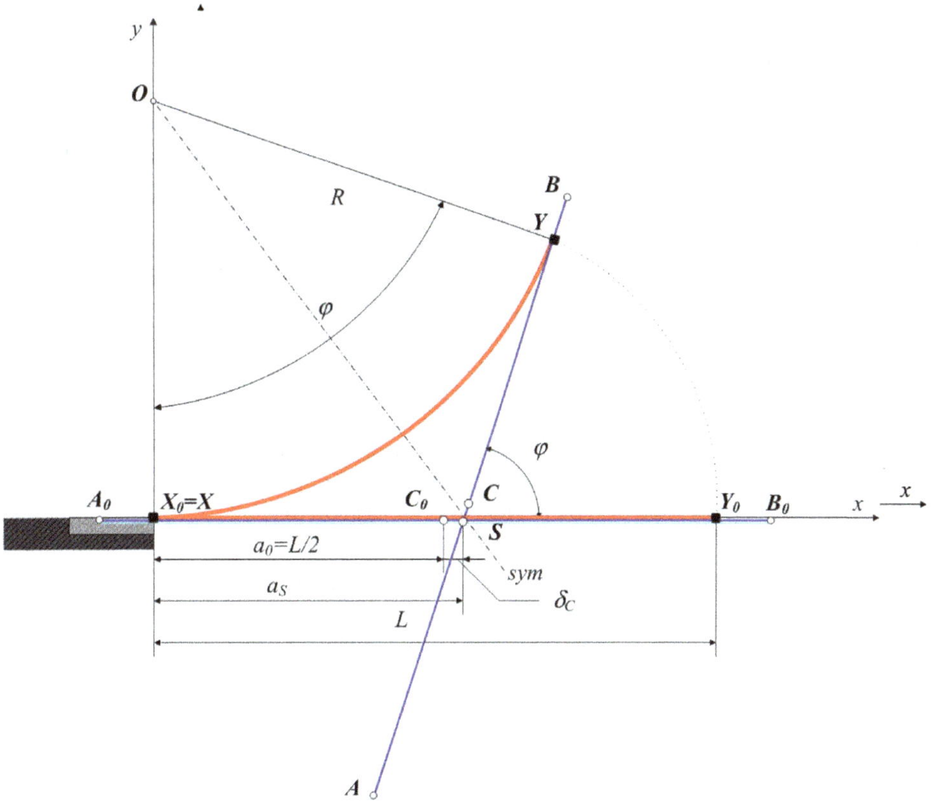

Fig. 5. Geometrical model of the mirror.

the same as the distance between C_0 *and* S. During the mirror motion, the quantity δ_C changes with the angle ϕ. Denote by δ_C^y the distance between the current position of the mirror centre C and the plane $y = 0$. Then, one can write

$$\delta_C^y = \delta_C \sin\varphi. \tag{1}$$

Assuming that $0 \leqslant \varphi \leqslant \pi/2$, one can try to estimate the value of δ_C^y. Because the coordinates of the moving end of the actuator are given by the relations

$$\begin{cases} x_Y = R\sin\varphi = \frac{L}{\varphi}\sin\varphi, \\ y_Y = R - R\cos\varphi = \frac{L}{\varphi}(1-\cos\varphi), \end{cases} \tag{2}$$

the line tangent to the circle at point Y may be described by one of the following equations:

$$\frac{y - y_Y}{x - x_Y} = \mathrm{tg}\varphi \ \text{ or } \ y = x\,\mathrm{tg}\varphi + \frac{L}{\varphi}(1 - \cos\varphi - \sin\varphi\ \mathrm{tg}\varphi). \tag{3}$$

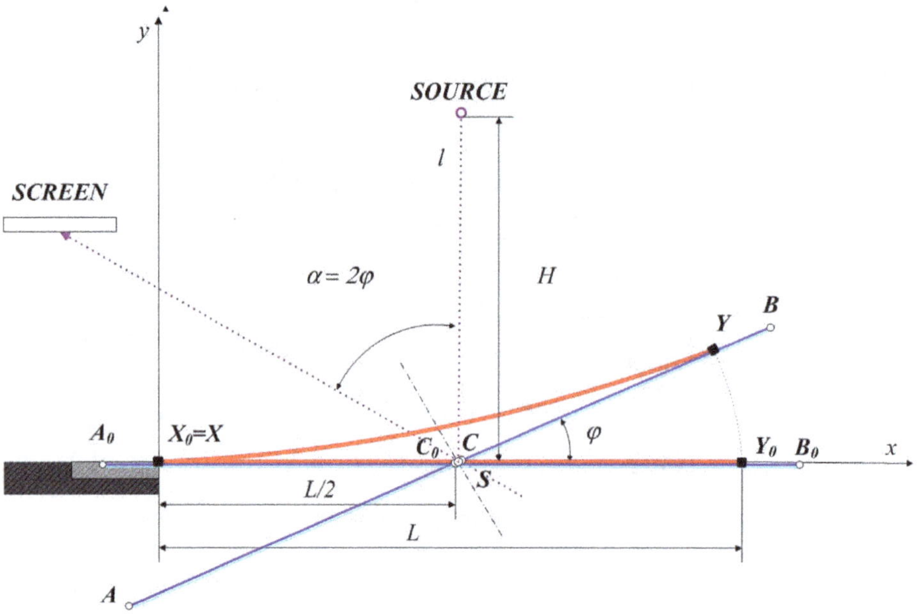

Fig. 6. Geometrical model of the mirror with light source.

Then, the point S, which moves along the line $y = 0$ during the mirror motion, has the coordinates

$$\begin{cases} x_S = a_S, \\ y_S = \frac{L}{\varphi}(1 - \cos\varphi - \sin\varphi \operatorname{tg}\varphi) + a_S \operatorname{tg}\varphi = 0. \end{cases} \tag{4}$$

The last relation leads to the formula

$$a_S = \frac{L}{\varphi} \cdot \frac{1 - \cos\varphi}{\sin\varphi}, \quad \text{for } 0 < \varphi \leqslant \pi/2. \tag{5}$$

Because $\delta_C(\varphi) = a_S(\varphi) - \frac{1}{2}L$, relation (1) takes the form

$$\delta_C^y = L\left(\frac{1 - \cos\varphi}{\varphi} - \frac{\sin\varphi}{2}\right). \tag{6}$$

From the above rule, one can calculate values of the function $\Delta = 100\% \times \delta_C^y/L$ for $0° \leqslant \varphi \leqslant 45°$.

It appears that for φ equal to 10°, 15°, 20°, 25°, 30°, 35°, 40°, 45° – the values of Δ are equal to: 0.0209%, 0.0744%, 0.1795%, 0.3418%, 0.5873%, 0.9264%, 1.3720%, 1.9370%, respectively.

Conclusion. For deflection angles from the interval $0 \leqslant \varphi \leqslant 45°$, the distance of the mirror centre from the substrate is smaller than 2% of the actuator length.

4.2 Analysis of the "actuator + mirror + light source" system

According to the previous considerations, one can assume that the points C_0, C and S at Figure 5 take the same position during a motion of the mirror. It means that the centre of the mirror is at rest when the mirror rotates. A scheme of the considered model supplemented by light source is presented at Figure 6. Light source is stationary and lies on a line l that is parallel to the y-axis of the coordinate system and passes through the point S. Because the position of the point S practically does not change during the mirror motion, one can assume that the distance from the light source to the point Sdenoted by H is constant. Denote by α the angle between the light beam incident on the mirror and the reflected one. It is equal twice ϕ and is called an optical angle. A light beam incident on the mirror at the centre S is reflected and appears on a screen at the point determined by the optical angle α. During the scanning process, the angle α is changing in certain interval. Due to bending of actuators and mirror rotations, optical signals transmitted by the light beam create an image on the screen.

5 Thermal analysis

5.1 Thermal-bimorph beam deflections

The initial flexure of actuators appears during their *forming process*. When flat bimorph beams are cooled from the temperature 300 °C to the room temperature 20 °C, its take the shape of arc with the initial deflexion angle φ_0. To find φ_0, we may use the following rule connecting the change of deflexion angle $\Delta\varphi$ with the change of temperature ΔT [9]:

$$\begin{aligned} \Delta\varphi &= \frac{6E_1E_2t_1t_2(t_1+t_2)(\alpha_1-\alpha_2)}{4E_1E_2t_1t_2(t_1+t_2)^2 + (E_1t_1^2 - E_2t_2^2)^2} \cdot L \cdot \Delta T \\ &= k_t \cdot L \cdot \Delta T, \end{aligned} \tag{7}$$

where k_t [μm^{-1} K^{-1}] is the thermal sensitivity of the bimorph, and for i = 1, 2, E_i– the Young modulus, t_i – thickness of the layer, α_ι – coefficient of thermal expansion, and L – the length of cantilever. Here, $i = 1$ for the active layer (with larger CTE), and $i = 2$, for the passive

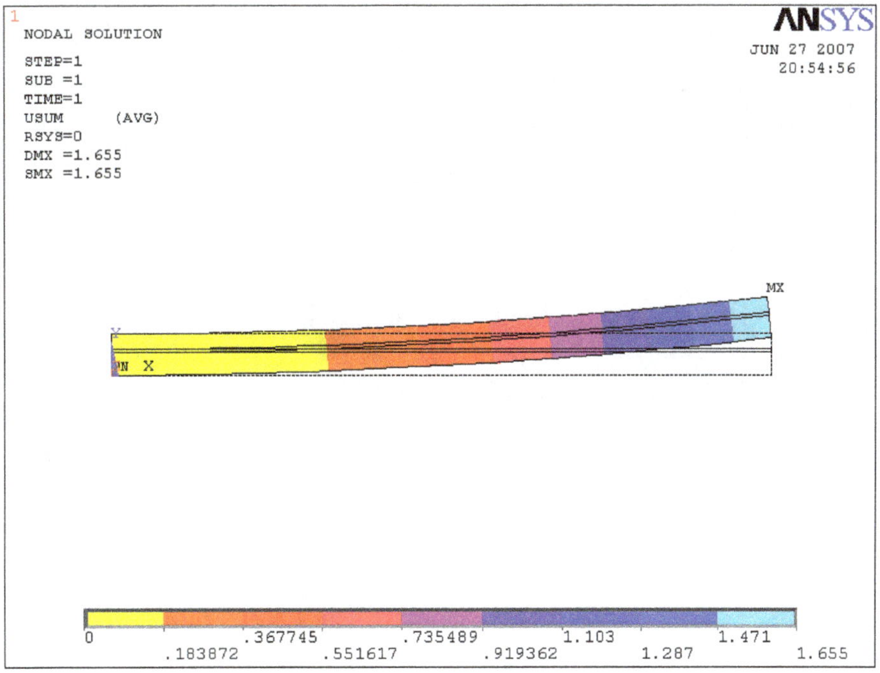

Fig. 7. Initial deflexion of raster scanning actuator.

one (with smaller CTE). To simplify the calculations, the passive layer is taken as the material composed of doped Si and silicon nitride with thickness $t_2 = 1.15$ μm, with a mean volumetric material parameters. Using the data given in the Table 2, one can obtain the thermal sensitivity of the actuators $k_t = 12.19$ μm^{-1} K^{-1}. Finally, taking $\Delta T = 280$ °C, we have - for raster scanning actuators ($L = 30$ μm) – the initial deflexion angle $\varphi_0 = 5.87°$, and for frame scanning actuators ($L = 100$ μm) – the angle $\varphi_0 = 19.56°$.

In the above analytical model, actuators were considered as beams composed of two layers with different mechanical properties. Numerical FEM simulations of the shorter actuator, as a beam composed of three layers indicate that insertion of the insulating layer reduces the initial deflexion angle φ_0. Instead of $\varphi_0 = 5.87°$, we obtain $\varphi_0 = 5.76°$ (Fig. 7). In fact, we have to do with a plate composed of three different layers. During the cooling process such a plate buckles in two perpendicular directions. To verify the influence of the transversal buckling on values of $\Delta\varphi$, numerical simulations of the shorter actuator were performed. In Figure 8, a transversal shrinkage of the plate is shown using the contour map. The corresponding value of the initial deflexion angle φ_0 higher, it is equal to 7.33°.

During the *exploitation process*, when a driving voltage is applied to the device, the free ends of the actuators oscillate, bending back from their initial positions determined by the angles φ_0, to smaller angles $\varphi = \varphi_0 - \Delta\varphi$. Using the procedure described above, one can find the values of $\Delta\varphi$.

The values of mechanical angles $\Delta\varphi$ presented in Table 2 are calculated for shorter actuators, which are working in resonance mode with much higher amplitude. The corresponding values $\Delta\varphi$ obtained for longer actuators are four times higher. Equivalent values of optical angles $\Delta\alpha = 2\Delta\varphi$ take magnitudes order of 40° and are large enough to use the device for image display, in the way shown in Figure 4.

Table 2. Changes of the deflexion angle for the raster scanning actuators due to changes of temperature.

ΔT =	50 °C	100 °C	150 °C	200 °C	250 °C
$\Delta\varphi$ =	1.07°	2.01°	3.06°	4.11°	5.16°

5.2 Thermal cut-off frequency

Motion of the microscanner is closely connected with the heat flow and temperature distribution during the actuation process. For frame scanning, one can assume that the temperature in the device is stable and we have to do with a free convection state corresponding to DC drive of the actuators. For raster scanning, when the microscanner is actuated at the resonance frequency, we have to do with a forced heat convection corresponding to AC mode of beams actuation. In the first case, if changes of the temperature are too fast, the amplitude of the scanner may be frequency-dependent. For that reason, it is necessary to determine the highest frequency exciting the non-resonance vibrations of the scanner. The time constant τ determining the thermal cut-off frequency of cantilever may be obtained from the following heat transport equation [10]:

$$\rho \cdot C \cdot \frac{\partial T}{\partial t} = \lambda \cdot \frac{\partial^2 T}{\partial x^2} + P, \tag{8}$$

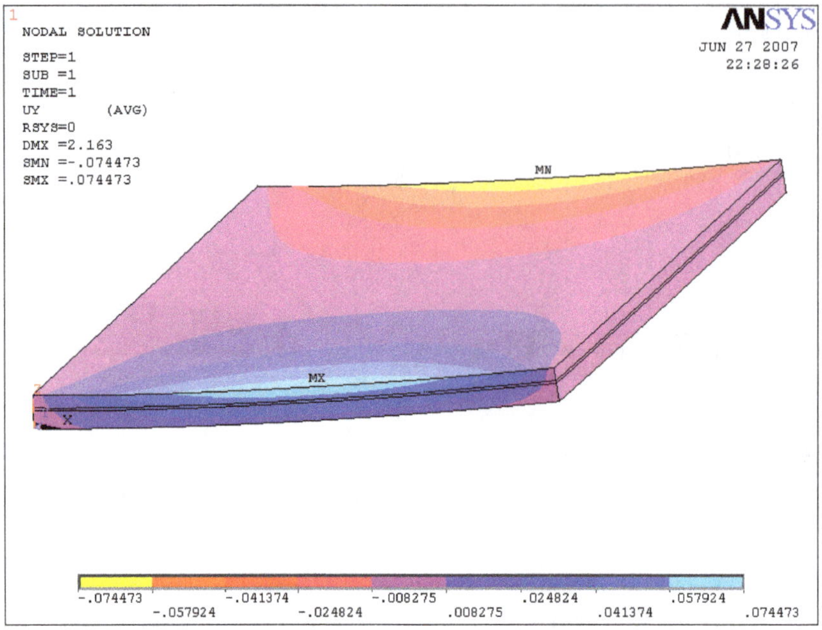

Fig. 8. Transversal shrinkage of raster scanning actuator.

where x is the distance from the fixed point of the cantilever, ρ, C, λ – mean parameters of the beam materials, given in Table 1, and P is the mean electric power dissipated in the heating resistor. At the free end of the beam, when $x = L$ and the maximal stable temperature $T = T_{max}$, we have

$$\tau \frac{\partial T}{\partial t}|_{x=L} = T_{max}, \text{ where } \tau = \frac{\rho \cdot C}{2\lambda} L^2. \quad (9)$$

The quantity τ is the *thermal constant time.* In our case, mean values of material parameters are the following: $\rho = 2.49 \times 10^3$ kg/m^3, $C = 713$ J/kg K, $\lambda = 182$ W/m K, and length of the cantilever is 100 μm. Then, $\tau = 0.049$ ms. Theoretical value of the *thermal cut-off frequency*

$$f_{cut-off} = \frac{1}{2\pi\tau} \quad (10)$$

gives quite large value 3250 Hz.

6 Resonance frequency calculations

6.1 Analytical model

To find the fundamental resonance frequency analytically, it is necessary to make some simplifications. Analytical model does not take into account the large deflections of moving cantilevers, and its initial curvature is not taken into account. Also the air dumping and changes of temperature are neglected. Calculations will be performed for the resonance element shown in Figure 4a.

Let us consider, in the coordinate system $\{x, y\}$, a cantilever (actuator), fixed at the point A, with a *rigid* flattened brick element (mirror), attached to its free end B (Fig. 9). Length, width and thickness of the cantilever are denoted by: l_a,b_a, and t_a, respectively. Length of the loading element is l_m, and mass is m_m. Moreover, denote by J the moment of inertia of the brick element with respect to its rotation axis (orthogonal to $\{x, y\}$-plane at the point B). When the brick element oscillates with rotating motion around its center C_0, at the end of the cantilever act the force F and the moment M shown in Figure 9. Because the point C_0 is practically motionless, we have to do with the case of pure bending considered in the Section 4. Then, one can assume that the force F is equal to zero. If we denote by φ the angle between the tangent to the end of the falling beam and the x-axis, then

$$F = 0 \text{ and } M = J\ddot{\varphi}, \quad (11)$$

where $J = \frac{1}{12} m_m l_m^2$, and $\ddot{\varphi}$ denotes the second time-derivative of the function φ.

To find the fundamental resonance frequency, one can use the general equation of harmonic vibrations: $\boldsymbol{F} + \boldsymbol{K} \cdot \boldsymbol{y} = \boldsymbol{0}$, ($\boldsymbol{F}$ is the generalized force vector, $\boldsymbol{y}$ – the generalized displacement vector, and $\boldsymbol{K}$ – the stiffness matrix). In our case, the free vibrations are due to the bending moment M only, and the beam stiffness is equal to EI/l_a. Denoting by b the width of the bimorph cantilever composed of two different materials, we have [9]:

$$EI = \frac{b}{12} \cdot \frac{E_1^2 t_1^4 + E_2^2 t_2^4 + E_1 E_2 (4t_1^2 + 6t_1 t_2 + 4t_2^2)}{E_1 t + E_2 t_2}. \quad (12)$$

The equation of free harmonic vibrations, takes the form of the following differential equation:

$$J\ddot{\varphi} + \frac{EI}{l_a}\varphi = 0. \quad (13)$$

Looking for non-vanishing solution of the form: $\varphi = \varphi_0 e^{i\omega t}$, we obtain the following form of the

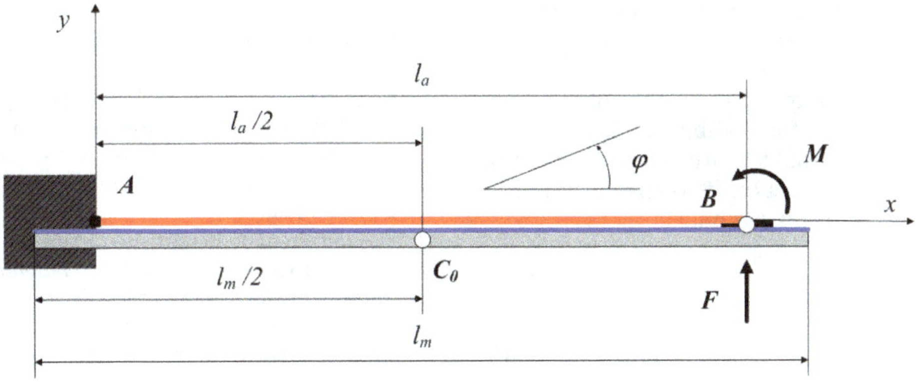

Fig. 9. Cantilever loaded by a mirror attached at point B.

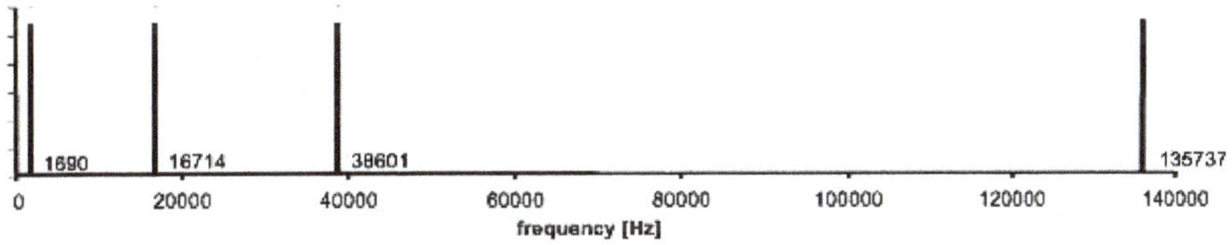

Fig. 10. The first four modes of the resonance frequency of the mirror.

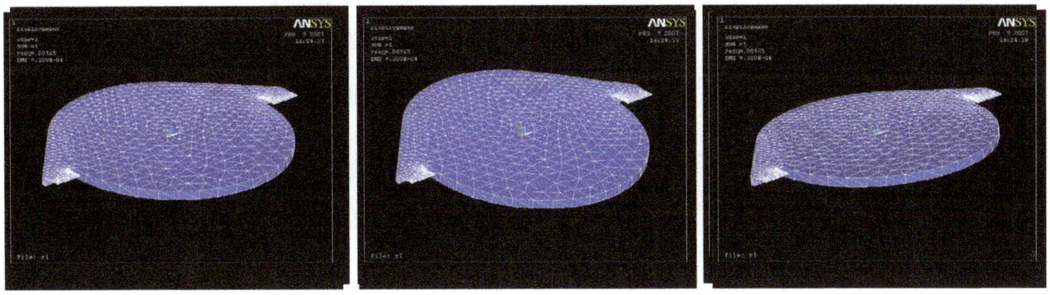

Fig. 11. First mode of the resonance frequency of the mirror.

fundamental natural frequency ω_0:

$$\omega_0 = \eta \sqrt{\frac{12\,EI}{m_m\, l_a^3}} \quad \text{where} \quad \eta \equiv \frac{l_a}{l_m}. \tag{14}$$

Using the dimensions and mass density describing the mirror and cantilever, we obtain the fundamental natural frequency $\omega_0 = 9266$ rad/s. It corresponds to the frequency $f = 1475$ Hz. Because the model used is simplified (2D deflections, the rigid mirror without deformations), the real value of f may appear to be smaller. To obtain a more realistic estimation, it is necessary to perform numerical simulations using 3D model with deformable mirror element.

6.2 Numerical simulations

For estimation of the resonance modes of the mirror suspended on the shorter thermo-bimorph actuators, finite element simulations were performed. The obtained five succeeding resonance modes with associated frequencies are: I mode: 1690 Hz, II mode: 16 719 Hz, III mode: 38 601 Hz, IV mode: 135 137 Hz and V mode: 135 760 Hz.

The first mode of oscillations is due to rotations of the mirror around the axis crossing the free end of the cantilever (Fig. 11). Quite interesting is the second mode, when the mirror is moving without rotations, remaining in parallel horizontal planes. The higher modes are connected with twisted forms of the mirror plate. Notice that the obtained frequency $\omega_0 = 1052$ Hz, described by mode I, is large enough for the scanning process.

7 Conclusions

Detailed analysis of the solutions presented in the book [8] leads to a new solution of the thermally actuated microscanner of high precision. The most important conclusions are the following:

- thermal and mechanical separation of actuators working in different planes enables to avoid their twisting and, in consequence, it is possible to avoid global distortions of the displayed image;
- locations of mass centres of the moving parts of the device in the geometrical centre of the mirror, suppress the inertial forces acting on the rotating elements and, in consequence, it considerably diminishes the

power supply, increases fundamental resonance frequency and assures a linear dependence between the current voltage and mirror deflection;
- special location of the geometrical center of the mirror with respect to actuators enables to keep a constant distance from the light source to the center of the mirror and, in consequence, it assures a uniform, high resolution of the displayed image.

References

1. M. Madou, *Fundamentals of Microfabrication* (CRC Press, Boca Radon, Florida, USA 1997)
2. K.E. Petersen, IBMJ. Res. Develop. **24**, 631 (1980)
3. K. Yamada, T. Kuriyuma, A novel asymmetric silicon micro-mirror for optical beam scanning display, in *Proceedings of MEMS'98* (Heidelberg, Germany, pp. 110–115, 1998)
4. R.A. Miller, W.G. Burr, SPIE **2687**, 47 (1996)
5. K.E. Petersen, Appl. Phys. Lett. **31**, 521 (1977)
6. M.-H. Kiang, O. Solgaard, K.Y. Lau, J. Microelectromechanical Systems **7**, 27 (1998)
7. R.W. Corrigan, *Scanned linear architecture improves laser projectors*, Laser Focus World (1999), pp. 169–172
8. G. Lammel, S. Schweizer, Ph. Renaud, *Optical Microscanners and Microspectrometers using Thermal Bimorph Actuators* (Kluwer Academic Publishers, Boston, 2002)
9. S. Timoshenko, J. Opt. Soc. America **11**, 233 (1925)
10. F.P. Incropera, D.P. DeWitt, *Fundamentals of Heat and Mass Transfer* (John Wiley & Sons, New York, 1996)
11. W.L. Gambin, A. Zarzycki, Optimal design of a new thermally actuated microscanner of high precision, *First International Conference on Multidisciplinary Design Optimisation and Application*, April 17–20, (2007) EDP Sciences, ISBN 978-27598-0023-0
12. W.L. Gambin, A. Zarzycki, Thermal and dynamical analysis of new microscanner behaviour, *First International Conference on Multidisciplinary Design Optimisation and Application*, April 17–20, (2007) EDP Sciences, ISBN 978-27598-0023-0

12

Identification of a spatial linear model based on earthquake-induced data and genetic algorithm with parallel selection

David Hicham Bassir[1,a], José Luis Zapico[2], María Placeres González[2] and Rodolfo Alonso[2]

[1] Institute of FEMTO-ST, UMR / CNRS 6174, Dept. LMARC, 24, rue de l'Epitaphe, 25000 Besançon, France
[2] Universidad de Oviedo, Departamento de Construcción e Ingeniería de Fabricación, Campus de Gijón, 33203 Gijón, Spain

Abstract – This paper deals with the parametric identification of a small-scale bridge model that was intended to approximately reproduce the transversal dynamic behaviour of its corresponding prototype. This is representative of a typical multi-span continuous-deck irregular bridge. A linear model with viscous damping is proposed to reproduce its dynamic response. The identification is carried out in the time domain using experimental earthquake-induced data and assuming the mass is known. An identification procedure including a genetic algorithm with parallel selection has been developed. In the studied case, the procedure has demonstrated to be robust. It is shown that the model could be improved by using a non linear approach for the dissipative forces.

Key words: Model identification; linear model; bridge; genetic algorithms; parallel selection; signal processing

1 Introduction

This paper is an updated and revised version of the conference paper [1]. The experimental model studied in this article was originally intended to reproduce approximately the transversal dynamic response of its prototype under earthquake excitation. Both the elastic response originated by minor earthquakes and the inelastic response due to strong motions was considered in the design of the model. The prototype is representative of a typical multi-span continuous-deck irregular bridge. The bridge had four identical straight spans of 50 m each. The three piers were respectively 14, 7 and 21 m high, which corresponds to a rather irregular distribution.

The model was fabricated and dynamically tested on a shaking table at the facilities of the Earthquake Engineering Research Centre of the University of Bristol. The experiments were framed within the PREC8 (Prenormative Research for Eurocode 8) project, which was aimed to update the Eurocode 8 [2]. Several large-scale models of the same prototype were also developed and tested with equivalent earthquakes in different European laboratories, so as to allow the results to be compared. More details of this testing campaign can be found in reference[2–5]. Later on, the authors have developed a Finite Element (FE) model of the bridge. This FE model was updated on the basis of its natural frequencies, which were identified from modal tests carried out before the seismic ones [5].

Studies on the effects of past strong earthquakes (Northridge 1994, and Kobe 1995) on steel structures have revealed some deficiencies in the modern seismic codes. The response of the structures in these earthquakes was good in terms of safety; only a few of them collapsed. The level of damage, however, was higher than that predicted by the codes in many cases [6–8]. As a consequence, the so-called performance-based approach has been recently proposed for the seismic design [9]. Several performance levels are considered in this new approach. At the Immediate Occupancy Level, only limited overall structural damage is allowed. The evaluation of this performance level is based on the structural seismic response, which is obtained through mathematical models, the structural displacements being the essential evaluation parameters. As only light structural damage is allowed at this level a linear elastic model with viscous damping is recommended for the structural analysis [9]. Recent studies [10], however, have proved that the calculated seismic response strongly depends on the modelling of the damping. This is due to the fact that only a small part of the energy is dissipated by plastic deformation at this level, most of it being dissipated by damping. The aforementioned reference only includes analytical simulation on single-degree-of-freedom models with different kinds of damping, and it is concluded that further research on this issue is needed.

The present article incorporates the experimental aspect to this research field taking advantage of the previous

[a] Corresponding author: david.bassir@utbm.fr

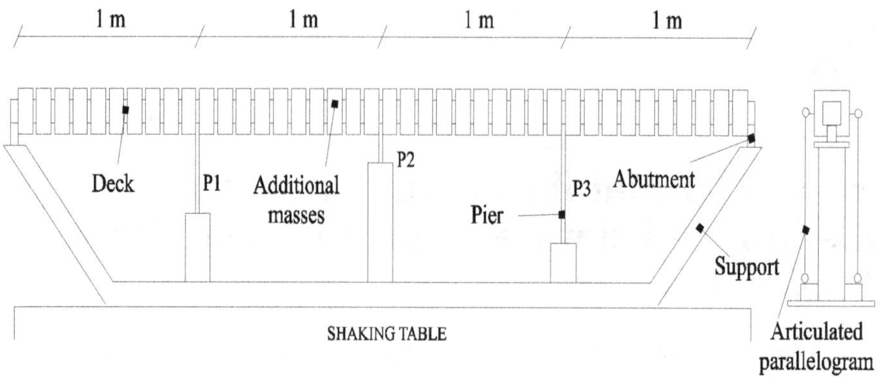

Fig. 1. Elevation of the experimental model.

tests on the small-scale bridge. The aim is to evaluate the accuracy of a linear viscous model to reproduce the pre-yielding seismic response of the actual bridge. For this purpose, an analytical model is proposed instead of the of the FE one, and fitted to the available seismic data in the time domain. Namely, the terms of the stiffness and damping matrices are directly obtained by minimizing the discrepancies between both the analytical model and the experimental model. As there is not a previous calibration of the damping coefficients in this case, and they can have a large interval of variation, a Genetic Algorithm (GA) has been selected for solving the optimization problem, so as to increase the probability of finding the global minimum solution. The seismic response of this calibrated analytical model is finally compared with the measured one. Besides, the terms of the stiffness matrix obtained in this process are likened to those of the previous FE model in order to evaluate the agreement between both approaches.

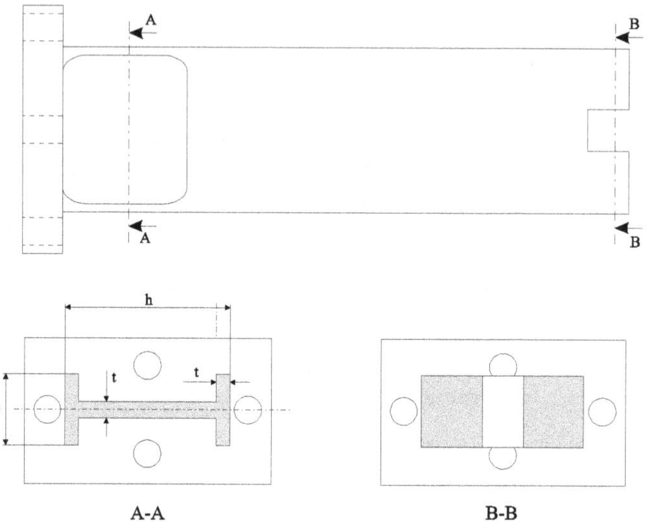

Fig. 2. Details of the piers.

2 Experimental model

The model scaling was mainly conditioned by the characteristics of the earthquake simulator. In order not to exceed the maximum available length of the shaking table, a geometric scale factor of 50 was adopted. An artificial mass simulation model was chosen, with the acceleration scale factor set equal to one [5]. The adopted geometric scale meant that the materials used for the model had to be different from those of the prototype. Structural steel BS 4368 grade 43c was chosen instead of post-tensioned concrete for the model deck, and aluminium alloy grade 6082 T6, which was annealed before machining, was chosen instead of reinforced concrete for the model piers. Three models labelled as A, B, C, which correspond to different configurations of the piers, were used in the initial project. Configuration A was designed according to Eurocode 8, while configurations B and C were possible alternatives to the irregular issue. The former consist on reinforcing the shortest pier in order to reduce its ductility demand. In the latter configuration the taller piers were reinforced, so as to reduce the forces acting in the shortest pier. Only Configuration A has been considered for identification purposes herein.

2.1 Deck

The model deck was designed with a continuous square hollow section. A 60 × 60 mm, 3.2 mm thick section was used, that had a similar second moment of area to that required for similarity between the prototype and model. Supplementary masses were added to the model, so as to attain mass similarity. These additional masses are distributed along the deck, and they consist of steel blocks bolted to it. In order to transmit the extra weight of the additional masses and to constrain the deck to translate only in the horizontal direction, articulated parallelograms are attached to the deck at the pier locations (Fig. 1).

2.2 Piers

The piers were designed with an I-beam section at the bottom, and a rectangular section for the remainder. Both sections had the same depth and width for each pier (Fig. 2). This design was intended to approximately reproduce the response of the prototype piers under vertical and transverse loads, acting simultaneously.

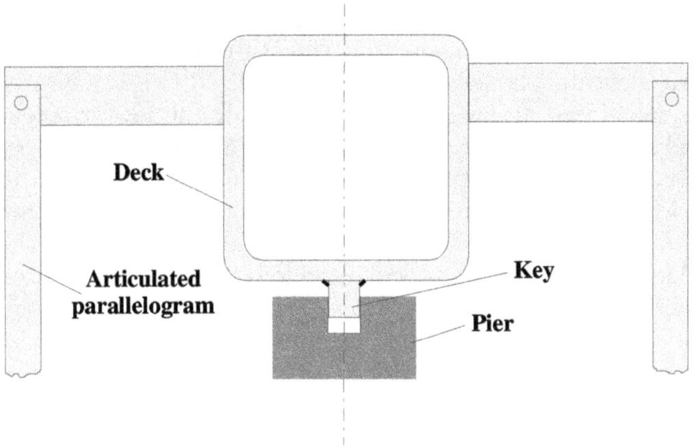

Fig. 3. Connection of the piers to the deck.

2.3 Connections

The deck ends hinged on the abutments through two vertical pins 20 mm in diameter. All of the piers had the same type of connection to the support including four 6 mm diameter bolts. The connections to the deck were also the same for all piers and consisted of square keys welded to the deck that fitted into slots in the piers. This ensured that the articulated parallelograms rather than the piers took the weight of the deck (Fig. 3).

2.4 Testing

In each configuration of the bridge, an initial modal test followed by several seismic tests was carried out. In the modal tests, the model was directly excited in the transverse direction by the shaking table with a low intensity random vibration. The response of the model was measured by an accelerometer that was placed at different positions on the deck and support using a magnetic base.

The model was also shaken in the transverse direction in the seismic tests. They consist of a series of earthquakes with the same time-history but increasing intensity, namely: 0.5, 0.8, 1.0, 1.2 and 2.0 times the design intensity. The reference earthquake was a synthetic one fitting the Eurocode 8 elastic response spectrum for medium soil conditions, had a maximum acceleration of 0.35 g and duration of 13 seconds at full scale.

During these tests, the absolute acceleration of the deck at the connection to the piers was measured by means of linear accelerometers. The absolute acceleration of the table was also recorded in all the configurations and tests. More details of the experimental model and testing can be found in reference [5].

2.5 Signal pre-processing

The analogue signals were converted into digital ones at sampling frequency of 1035 Hz. An initial interval previous to loading was also recorded in order to check the quality of the signals. All the measurements of this initial part exhibit similar shape; an example is shown in

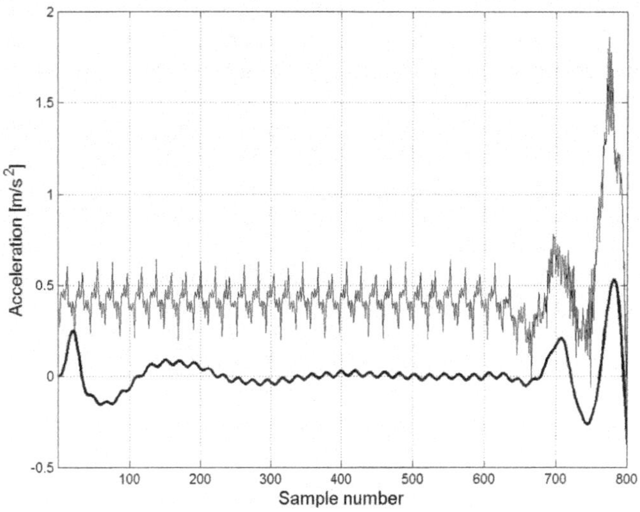

Fig. 4. Initial part of the raw signal (fine line) and the corresponding filtered one (thick line).

Figure 4. As it can be seen, the signal contains an offset and a sort of periodic high-frequency noise. As the offset is quite constant in this part, it was initially removed by subtracting the mean value of the signal in the complete sampling interval. The further integration of this acceleration signal, however, showed trends of the displacements with magnitude even higher than the pick-to-pick displacements. This means that the offset is not constant, but it changes during each test.

A band-pass filter, which is intended to remove both the low-frequency signal trend and the high-frequency noise, was eventually used. It was a four-order Butterworth filter with cut-off frequencies at 4.14 and 36.22 Hz, which was applied to both the input and the output signals. These cut-off frequencies were selected as far as possible from the first natural frequencies of the model (11.6, 13.6 and 25.75 Hz), which have the more significant influence in the seismic response of the bridge, seeking for a clean signal but retaining all the significant dynamic information of the raw one.

The filtered signal is shown in Figure 4 along with the raw one. It is visually evident that both the offset and the high-frequency noise are reduced. Nevertheless, the components of the noise within the filtering interval still remain in the pre-processed signal.

3 Definition of the model and criterion of fitting

3.1 Analytical model

The identification is based on the available experimental data induced by the lowest intensity earthquake (0.5×design intensity) carried out in the experiments. Under these conditions the bridge is expected to behave quasi elastically and with low displacements. Hence, an analytical linear model with viscous damping has been selected as a first approximation to reproduce the dynamic response of the experimental one. The mathematical formulation of the earthquake-induced response of the model is

$$[M]\ \{\ddot{x}_a(t)\} + [C]\ \{\dot{x}(t)\} + [K]\ \{x(t)\} = 0, \qquad (1)$$

where the vector $\{x\}$ represents the transversal displacements of the deck relative to the table at the connections to the piers. $\{\ddot{x}_a\}$ is the vector of absolute accelerations at the same Degrees of Freedom (dofs). $[M]$, $[C]$, $[K]$ are respectively the corresponding mass, damping and stiffness matrices.

It is assumed that the damping matrix is diagonal, and the stiffness matrix is symmetric according to the Betti's law. In order to avoid the identification process to be ill-conditioned, the mass matrix is set constant and equal to that obtained by dynamic condensation of the updated FE model of the bridge developed previously [5]. Summing up, the parameters of the model to be identified are the six independent terms of the stiffness matrix and the three terms of the leading diagonal of the damping matrix.

3.2 Objective and penalty functions

The analytical model was calibrated by fitting its response to the experimental data in time domain. The available experimental data are the absolute acceleration of both the table a_g and the deck $\{\ddot{x}_a\}$. From these, the acceleration of the deck relative to the table is obtained as follows

$$\{\ddot{x}(t)\} = \{\ddot{x}_a(t)\} - \{1\}\ a_g(t). \qquad (2)$$

Then, the velocity $\{\dot{x}\}$ and the displacement $\{x\}$ are in turn computed by numerical integration through the trapezium rule. The displacements predicted by the model $\{x'\}$ are calculated from (1) on the basis of the measured absolute acceleration $\{\ddot{x}_a\}$ and the computed relative velocity $\{\dot{x}\}$

$$\{x'(t)\} = -\,[K]^{-1}\left([M]\ \{\ddot{x}_a(t)\} + [C]\ \{\dot{x}(t)\}\ \right). \qquad (3)$$

A Normalized Mean Square Error (NMSE) is defined to quantify the discrepancies between the computed relative displacements $\{x\}$ and the predictions of the model $\{x'\}$

$$NMSE = \frac{1}{3}\sum_{i=1}^{3}\frac{\sum_{t=1}^{N}\left(x_i'(t) - x_i(t)\right)^2}{\sum_{t=1}^{N}\left(x_i(t)\right)^2}, \qquad (4)$$

which constitutes the objective function to be minimized.

The stiffness matrix is constrained to be positive definite during the minimization process, so as to obtain solutions with physical meaning. This is achieved by penalizing the non positive definite solutions. For this end, the eigenvalues of the stiffness matrix are computed at each stage of the minimization process. If the lowest eigenvalue is positive, which means that the stiffness matrix is positive definite, the solution is not penalized and the NMSE is calculated according equation (4). If the lowest eigenvalue is less than or equal to zero, then the solution is penalized by setting the NMSE equal to the summation of NMSE_{max} (maximum admissible value obtained through the optimization process) and the absolute values of the eigenvalues (only the negative eigenvalues are considered in the summation).

A problem arising from the numerical integration of the relative acceleration (Eq. (2)) is the presence of linear and quadratic drifts in the calculated velocity and displacement, respectively. The magnitude of these drifts can be significant, and they might cause the identification process to fail. The drifts are due to unknown constant values of the acceleration and velocity that are not present into the signals. This problem is circumvented herein including these constant or initial values as parameters to be estimated in the minimization process.

4 Minimization method

As the objective of the identification consists on minimizing an error function that is highly multimodal, it's difficult to find one global solution. To overcome this problem, a robust identification procedure based on Genetic Algorithm with Parallel Selection (GAPS) coupled with a local search method has been developed [11]. Next subsections include a detailed description of such a procedure.

4.1 Genetic algorithm with parallel selection

Based on the theory of natural selection, the Genetic Algorithms (GAs) provide an alternative to traditional optimization techniques to locate the optimal solutions in a complex landscape. The theoretical foundations were first led by Holland in 1975 [12] and since that time, the number of applications and publications concerning GAs has increased with exponential manner. In 1989, Goldberg [13] gave to GAs their signs of nobility as an efficient and general method to overcome the complex optimization problems [14–20].

The principle of GAs is to simulate the evolution of one population of individuals to which different production operators (*selection, crossover and mutation*) are applied. As GAs start searching from different initial solutions, this gives them a global view of the problem. This global perspective prevents them to be trapped locally and allow them to explore all the search landscape. Such algorithms know the problem only through the value of the cost function and the values of the constraints. The behaviour of this algorithm is similar to a black box with several entries and one exit as described below in Figure 5.

The functioning of GAs can be carved in three main parts [21]: coding of parameters, genetic operators and choice of the objective function.

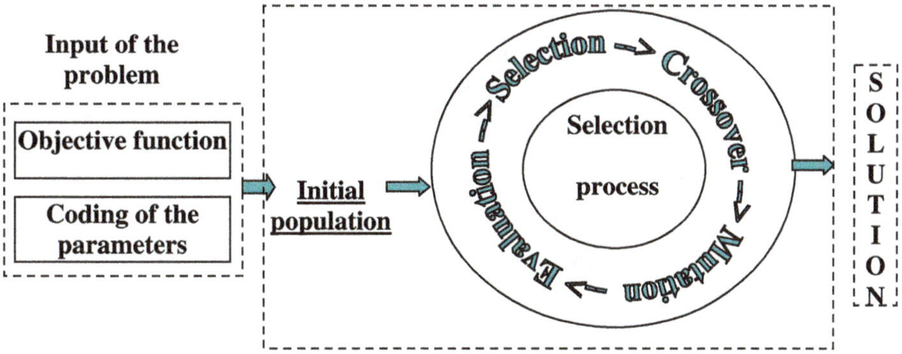

Fig. 5. Principle of a genetic algorithm.

The first problem to be faced during the use of the GAs is the representation of the individuals (coding of parameters). It is the manner each variable of the optimization problem is coded. The coding can be binary [17] coding or real coding [17, 19]. The binary coding is the common representation that is most implemented in the GAs. For continuum variables, many authors [17, 20, 22] prefer using a real coding for their simplicity and efficiency for real problems. The real coding eliminate the difficulties of achieving arbitrary precision in decision variables and the Hamming cliff problem associated with binary string representation of real number [13, 17, 19]. In this article, the real coding [22] was chosen to be implemented in our program.

As concern the objective function, its definition is very important in the evolution process because GAs search in the landscapes by using only the discrete values of the objective function. If this function is not well defined, GAs cannot guaranty the location of the global solution.

Finally, the different operators used in GAs are the selection, the crossover and the mutation. These operators are widely described in the literature [13, 15, 19, 27].

In the following, we will introduce the basic ideas of the common operators with references to the literatures. Then, we will present the parallel selection introduced in our GAPS that is based on the advantage of the parallel processes in computing. The common selection method will be extended to the case of parallel calculation.

4.2 Selection

After creating randomly the first generation and evaluating the objective function for all individuals, the process of selection on the population is applied. There exist a number of selection operators in the literature of GAs. The main idea consists on choosing within a population P a number N of individuals (to create a mating pool) that are well adapted to survive and to follow their evolution to access to the next generation. The choice of keeping one individual is made by comparing its fitness function. One of the characteristics of the selection process is the selective pressure applied on the population. If this pressure is too strong, the diversity within the population is lost; whereas, if this pressure is too small, the behaviour of GAs becomes random.

In the following, focus on the tournament selection will be done. This selection method starts by choosing randomly R individuals (*R is the size of the tournament*) of the population of N individuals. The individual with a greatest fitness value wins the tournament and is saved for the next generation. This operation is repeated until N/2 individuals are obtained. It is entirely possible that some individuals participate in several tournaments: if they earn several times, they will have therefore straight to be copied several times. In this approach, the selection is strong enough, that is why one often reduce this pressure by decreasing the probability of selection in the tournament. This method has been integrated in GAPS as a pre-selection method.

4.3 Crossover

Crossover operator is applied next to the individuals of the mating pool. When the intermediate generation is half filled, it chooses randomly couples of individuals (parents) with a probability Pc of participation to the crossover. From a couple of parent, two children are created with the genetic patrimony of the parents, and then introduced in the population. In more concrete terms, the crossover operator favours the exploration of the research landscape. This exploration can be made by binary crossover (case of the binary coding) or by real crossover (case of the real coding).

In the case of real crossover, the same notion can be used either as for binary crossover or arithmetical approach. Thus the notion of point cut apparent more to a simple permutation of variables component without modifying the variable itself. In the case of arithmetical approach, the children R_1' and R_2' are generated by linear combination of the parents R_1 and R_2. This solution is created in the range of $[R_1 - \alpha(R2 - R_1), \mathrm{R2} + \alpha(R2 - R_1)]$ as follows:

$$\begin{cases} R_1' = [\lambda R_1 + (1-\lambda) R_2] \\ R_2' = [\lambda R_2 + (1-\lambda) R_1] \end{cases}, \quad (5)$$

where $\lambda = ((1+2\alpha)P_u - \alpha)$ is the dispersion parameter, α define the range of the parents interval and P_u a random value between 0 and 1.

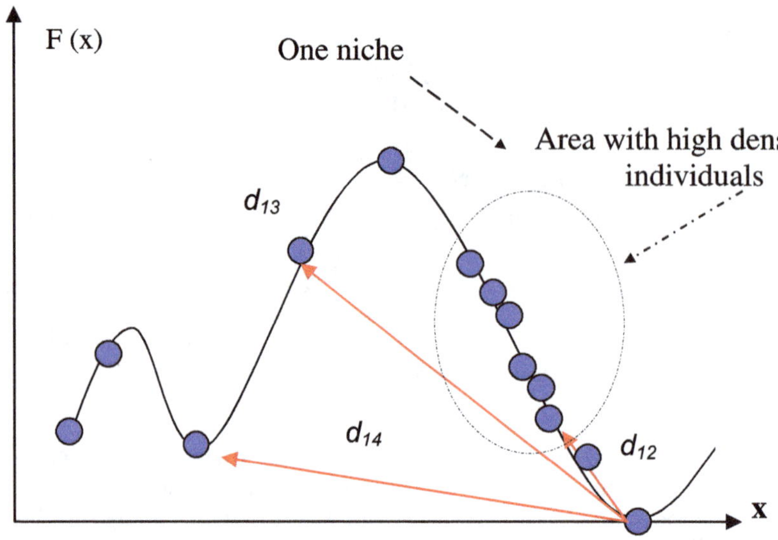

Fig. 6. Description of the relative distance with the sharing strategy.

A lot of studies about the dispersion parameter have been done specially by Deb [22–24] to understand and to control the dispersion of the individuals (children) in comparison with their parents. In the GAPS we have included the Deb approach in order to control the spread of the children. The children R_1' and R_2' are calculated as follows:

We choose one random value P_u between 0 and 1, then we evaluate the dispersion parameter β (Eq. (6)) as a function of the parameter n that define the probability of creating the new individuals close or far from the original individuals. If n is small value (respectively if n is a great value) we have a great probability to obtain the new individuals far from their parents (respectively close to the parents).

$$\beta = \begin{cases} (2P_u)^{\frac{1}{n+1}} & if\ P_u \leqslant 0.5, \\ \left(\frac{0.5}{(1-P_u)}\right)^{\frac{1}{n+1}} & otherwise. \end{cases} \tag{6}$$

Finally, the new solution $R_1'$and R_2' are calculated as follows:

$$R_1' = 0.5[(1+\beta)R_1 + (1-\beta)R_2],$$
$$R_2' = 0.5[(1+\beta)R_1 + (1-\beta)R_2]. \tag{7}$$

We can find more details about the influence and the sensitivity of the individuals compared to their parent in the following articles [22] and [24].

4.4 Mutation

This operator intervenes in a population by modifying the genetic code of an individual. The need of mutation is to introduce diversity in the population, because the perpetual selection decreases gradually this diversity. Similarly to crossover operator, the mutation can be applied on a binary coding or on a real coding. In the case of a binary coding, the mutation operator inverts one bit of a chromosome from 1 to a 0 and vice versa with a small probability P_m. An individual can move depending on the importance of the bit inverted. In the case of a real coding, the mutation of an individual R_1 in a range of $[R_{1inf}, R_{1sup}]$ is done in general as follows:

$$R_1' = R_1 + \delta(R_{1sup} - R_{1inf}), \tag{8}$$

where δ is the mutation parameter that defines the importance of the perturbation.

4.5 Sharing method

The concept of sharing is inspired by the nature. In a domain, two species can coexist and share the same resources. This idea [23] is introduced artificially in GAs through a sharing function S. This function consists in expelling some individuals from a zone of research that is reducing (*example of the niches or zones of local minima*) while the density of the population increases. The sharing function compares the relative distance d_{ij} between two individuals R_i and R_j (see Fig. 6).

If d_{ij} is greater or equal to d (d namely niche radius), then S is set equal to 0. Whereas if d_{ij} is less than d, then S is set equal to a constant c in the range [0, 1]. Once this evaluation $S(R_i, R_j)$ is done for all individuals, a new fitness function F is attributed to each one such as:

$$F'(R_i) = \frac{F(R_i)}{\sum_j S(R_i, R_j)}. \tag{9}$$

Even if this idea increases the calculation time of GAs, it is very suitable in a lot concrete cases with many local optima because it makes stronger the exploration capabilities of GAs. That is why it has been introduced in the GAPS.

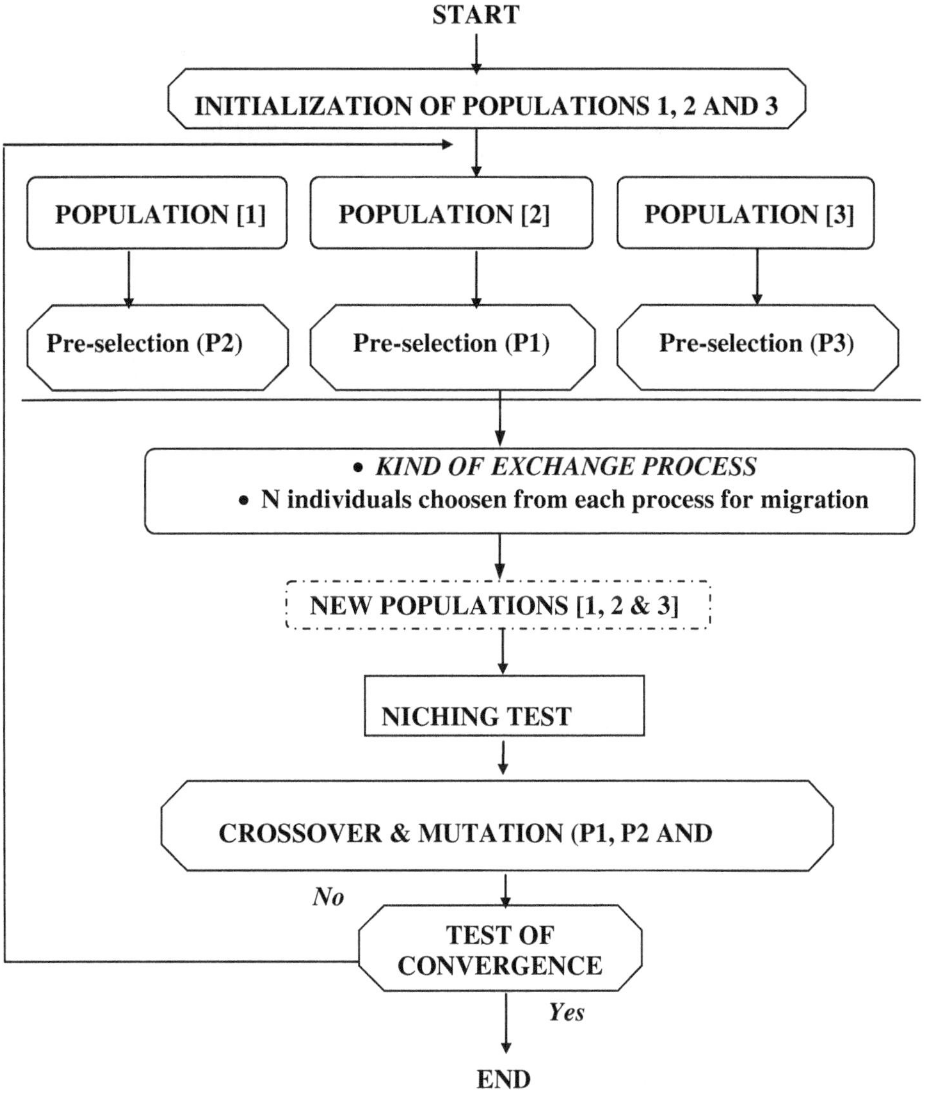

Fig. 7. Flow chart of the GAPS applied on 3 processes with an exchange of 2N individuals.

4.6 Parallel selection

The Parallel Selection that is implemented in the GAPS is based on several populations that evaluate independently in different processes. The operator of selection occurs in a population by taking into account the other selections to introduce the diversity in each process. This diversity allows the pressure of selection to decrease and gives a chance to the other individuals to transmit their genetic patrimony. The first works related to parallelism of GAs consisted in simply running several GA programs in different process in parallel and waiting for the convergence of each process. Then, some researchers have leaned on the idea to establish parallelism in the operators of GAs themselves [25,26]. This parallelism consists on subdivide a population in several sub-populations that evolve not in the same global area, but in a reduced one. Selection, crossover and mutation operators are then applied to these sub-populations.

The parallel selection method (see Fig. 7) consists in introducing in a selection one or several individuals coming from other selection processes. This allows the different process to exchange information related to the area of research for a better exploration in all the landscape. This idea presents the following advantages. First, it decreases the number of evaluations of the objective function. Second, all the processes converge in the same area. In this operator of selection, the pressure of selection can be increased or decreased by increasing or by decreasing the number of individual to exchange between the different processes. Similarly, the choice of individuals that migrate to the other processes governs the rank of the diversity in the population.

5 Application and analysis of results

To apply the above identification strategy to the bridge, we have started by defining a large domain for each

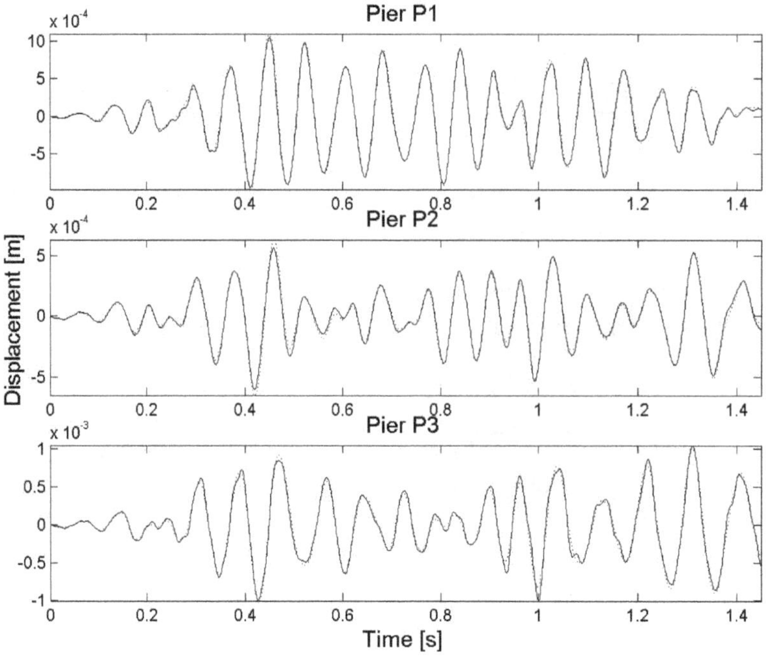

Fig. 8. Comparison between the displacements obtained by numerical integration of the measured accelerations (dotted line) and those predicted by the model (solid line).

parameter to be identified. For the coefficients C_{ii} of the damping matrix a range of variation between 0 and 10^4 have been used. For the coefficients K_{ij} of the stiffness matrix a range of variation between -10^7 and 10^7 have been used. Finally, as concern to the initial conditions of the velocity and the acceleration a range between -1 and 1 has been considered.

From the first GAPS tests, we have noted that the initial acceleration does not have much influence on the system response in this case. From this fact, then the number of parameters was reduced from fifteen to only twelve, the initial acceleration values being set equal to zero. The value of the objective function obtained with the GAPS at the convergence was 0.796% with a maximum number of iterations equals to 1200 generations and the time consuming equals to about 120 hours. The time of convergence can be explained by the initial intervals defined for the stiffness values K_{ij} that was too wide. The corresponding optimal solutions of the parameters are given below.

Stiffness matrix:

$$[K] = \begin{bmatrix} 1251132.51 & -987862.71 & 400913.73 \\ -987862.71 & 3164876.15 & -967343.67 \\ 400913.73 & -967343.67 & 979652.18 \end{bmatrix}. \quad (10)$$

Damping matrix:

$$[C] = \begin{bmatrix} 397.31 & 0 & 0 \\ 0 & 1387.57 & 0 \\ 0 & 0 & 121.56 \end{bmatrix}. \quad (11)$$

Initial velocity:

$$[\dot{x}_0] = \begin{bmatrix} -1.03e-3 \\ -8.34e-3 \\ -1.22e-3 \end{bmatrix}. \quad (12)$$

A comparison between the responses predicted by the model and those obtained directly by numerical integration of the measured accelerations are shown in Figure 8. In general, the predictions of the model are lower than those obtained by numerical integration for low displacements, and vice versa in the case of large displacements. Additionally, the absolute values of all the terms of the obtained stiffness matrix are lower than those obtained by FE model updating; the discrepancies being between 3% and 21% (see Fig. 9).

All these results indicate that the values of the "true" stiffness matrix diminish when the displacements increase, i.e., the behaviour of the bridge is not exactly linear. Thus, the response of the bridge could be more precisely described by a non linear model.

6 Conclusions

In this article, the properties of a small-scale irregular bridge and its dynamic tests have been described. An analytical linear model with viscous model is proposed to reproduce the response of the model to medium intensity seismic excitation. The parameters of the model are identified by minimizing the discrepancies between the response predicted by the model and that obtained in the

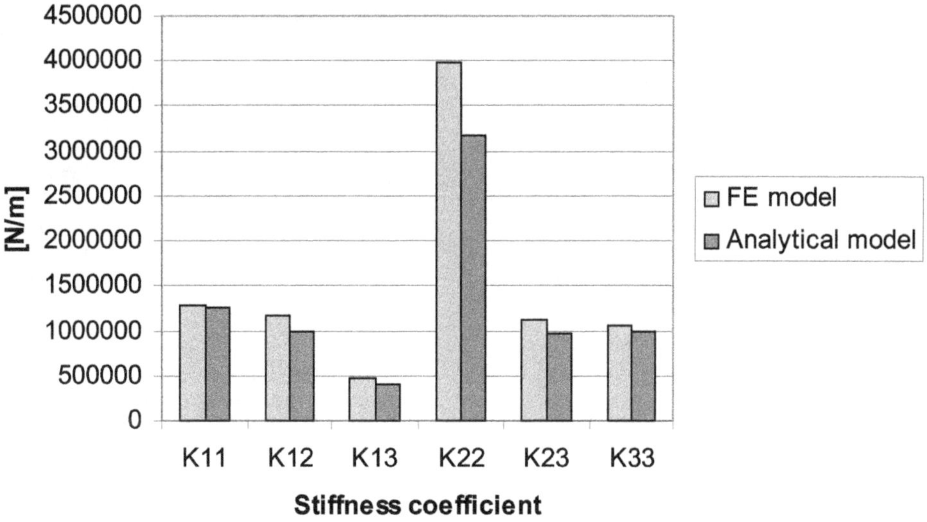

Fig. 9. Comparison between the stiffness coefficients corresponding to the updated FE model (Ref. [5]) and the analytical one

experiments in the time domain. A procedure based on GAPS has been developed for minimizing such discrepancies.

The application of the identification process shows that the strategies adopted for the signal pre-processing and GAPS are robust and efficient. The proposed model gives good dynamic predictions. The analysis of results, however, reveals that the model could be improved by using a non linear hysteretic scheme for the dissipative forces. This will be the matter of future studies on the bridge model.

Acknowledgements. The experimental part of this work was developed thanks to the ECOEST2 network project under the Human Capital and Mobility programme of the European Commission DGXII. The authors are grateful to it for their financial support. The tests took place in the facilities of the Earthquake Engineering Research Centre of the University of Bristol. The collaboration of the EERC staff, particularity Professor Severn, is greatly acknowledged. A part of this work was carried out under the AMERICO project (Multiscale Analysis Innovating Research for CFRP) directed by ONERA (French Aeronautics and Space Research Centre) and funded by the DGA/STTC (French Ministry of Defence), which is gratefully acknowledged.

References

1. J.L. Zapico, M.P. Gonzalez, D.H. Bassir, "Model identification of small-scale bridge using a genetic algorithm with parallel selection", in Proceedings of The Tenth International Conference on Civil, Structural and Environmental Engineering Computing, B.H.V. Topping, (Editor), Civil-Comp Press, Stirling, United Kingdom, paper 248, 2005
2. R.T. Severn, Proceedings of the Institution of Civil Engineers Structures & Bridges, European experimental research in earthquake engineering for Eurocode 8 (1999) 134: 205–217
3. V.A. Pinto, Pseudo-dynamic and shaking table tests on RC bridges. Report No 5, ECOEST & PREC8, 1996
4. V.A. Pinto, G. Verzeletti, P. Pegon, G. Magonette, P. Negro, J. Guedes, Pseudo-dynamic testing of large-scale R/C bridges, Report EUR 16378 EN, ELSA, Ispra, Italy, 1996
5. J.L. Zapico, M.P. González, M.I. Friswell, C.A. Taylor, A.J. Crewe, Finite Element Model Updating of a Small Scale Bridge. Journal of Sound and Vibration **268**, 993–1012 (2003)
6. FEMA-355E., State of the art report on past performance of steel moment-frame buildings in earthquakes. Federal Emergency Management Agency, Washington, DC, 2000
7. M. Nakashima, K. Inoue, M. Tada, Classification of damage to steel buildings observer in the 1995 Hyogoken-Nambu earthquake. Engineering Structures **20**, 271–281 (1998)
8. S.A. Mahin, Lessons from damage to steel buildings during the Northridge earthquake. Engineering Structures **20**, 261–270 (1998)
9. FEMA-350., Recommended seismic design criteria for new steel moment-frame buildings, Federal Emergency Management Agency, Washington, DC, 2000
10. D.V. Val, F. Segal, Effect of damping model on pre-yielding response of structures. Engineering Structures **27**, 1968–1980 (2005)
11. D.H. Bassir, S. Carbillet, L.M. Boubakar, Algorithme génétique à sélection parallèle, Revue des Composites et Matériaux avancés, **15**, 53–70 (2005)
12. J.H. Holland, Adaptation in natural and artificial systems, Ann Arbor: University of Michigan Press (1975)
13. D.E. Goldberg, Genetic algorithms in search, optimisation, and machine learning, New York: Addison-Wesley (1989)
14. L. Ljung, System Identification. New Jersey: Prentice Hall PTR (1999)
15. Z. Michalewicz, Genetic Algorithms + Data Structures = Evolution, Springer-Verlag, Heidelberg (1994)
16. C.A.C. Coello, Use of the self-adaptive penalty approach for engineering optimization problems. Comput. Ind. **41**, 113–127 (2000)

17. C.Z. Janikow, Z. Michalewicz, An experimental comparison of binary and floating point representation in genetic algorithms, Proc. of the Fourth Int. Conference on genetic algorithms, R. Belew, L.B. Booker (Eds), (Morgan Kaufmman, San Mateo), 31-36, 1991.
18. K.A. DeJong, WM. Spears, An analysis of the interacting roles of population size and crossover in genetic algorithms, Proc., First workshop parallel problem solving from nature, Springer-Verlag, Berling, 38–47, 1990
19. A. Wright, Genetic algorithms for real parameter optimization. Foundations of genetic algorithms 1, G.J.E. Rawlin (Ed.) (Morgan Kaufmman, San Mateo), 205–218 (1991)
20. K. Deb, S. Gulati, Design of truss-structures for minimum weight using genetic algorithms. Finite Elem. Anal. Des. **37**, 447–465 (2001)
21. K. Sastry, U.M. O'Reilly, D.E. Goldberg, Population sizing for genetic programming based upon decision making, IlliGAL, Report N. 2004028, April 2004
22. K. Deb, R.B. Agrawal, Simulated binary crossover for continuous search space. Complex Systems, 115–148 (1995)
23. K. Deb, D.E. Goldberg, An investigation of niche and species formation in genetic function optimization, in: J. D. Schaffer, ed., Proceedings of the Third International Conference on Genetic Algorithms (Morgan Kauffman, San Mateo), 42–50, 1989
24. K. Deb, H.G. Beyer, Self-Adaptive genetic algorithms with simulated binary crossover, Technical Report No. CI-61/99, March 1999, Department of Computer Science/XI University of Dortmund, 44221 Dortmund, Germany
25. E.C. Paz, D.E. Goldberg, Efficient parallel genetic algorithms: theory and practice. Comput. Method. Appl. M. **186**, 221–238 (2000)
26. N.J. Roudcliffe, Non linear genetic representations parallel problem solving from nature 2, R. Männer, B. Manderick (Ed.) Elsevier Science Publishers, Amsterdam, 259–268 (1992)
27. C.A.C. Coello, E.M. Montes, Constraint handling in genetic algorithms through the use of dominance-based tournament selection. Adv. Informatics **16**, 193–203 (2002)

13

Multi-objective optimization on mix proportions of HSHPC applied to SRC composite structures

Shan-Suo Zheng[1,a], Lei Zeng[1], Wei-Hong Zhang[2], Jie Zheng[1], Bin Wang[1] and Lei Li[1]

[1] School of Civil Engineering, Xi'an University of Architecture and Technology, Xi'an, 710055, P.R. China
[2] Northwestern Polytechnical University, Xi'an, P.R. China

Abstract – The fine bond behaviour between steel shape and concrete is the base of their cooperation. For designing high strength and high performance concrete (HSHPC) applied to steel reinforced concrete (SRC) structures, a mathematical model of multi-objective and nonlinear optimization is established based on the bond-slip theory between steel shape and concrete, which takes the materials' cost and bond strength as objective functions, and takes the other performance indexes of HSHPC as constraint conditions. Optimal mix design is implemented with an application of convergent sequential quadratic programming and based on Matlab language. The optimized concrete mix proportions can satisfy all the specifications, and especially can raise the bond strength between steel shape and concrete and reduce unit cost of HSHPC.

Key words: SRC structure; HSHPC; mix proportion; optimization; multi-objective.

1 Introduction

Modern engineering structure advanced along the trend of long-span, heavy-load, high-rise. Attention to the durability of concrete structures promotes tremendous development in the technology of high strength and high performance concrete (HSHPC) [1,2]. The design of concrete mix proportions is changing from traditional strength design to multi-performance design that includes strength, durability and working performance etc. Traditional design, which depends on a method of experiential semi-quantitative concrete mix proportions, will be renovated from feasibility design to optimization design. Steel reinforced high strength and high performance concrete structure is a kind of steel and concrete composite structure that applies HSHPC to steel reinforced concrete (SRC) structure. The main idea is using the good mechanical property and durable property of HSHPC to improve the cooperative performance between steel shape and concrete and increase security and durability of structures during service time [3].

This paper discusses multi-objective and nonlinear optimization design on mix proportions of HSHPC applied to SRC structures to improve synthetically performance of the structure. Design objective is transformed from traditional design of single strength target to multi-objective and nonlinear design considering working property, strength, the cooperative performance between steel shape and concrete, economic cost and endurance. For designing HSHPC applied to SRC structures, a multi-objective and nonlinear optimal mathematic model is established based on the bond-slip theory between steel shape and concrete, which takes the materials' cost and ultimate bond strength as objective functions, and takes the other performance indexes of HSHPC as constraint conditions. The optimum problem is solved by convergent sequence quadratic programming. As a result, the choice of raw material parameters and the control of construction quality will be controlled by computer automatically.

2 Key performance indexes of HSHPC

2.1 Pore structure and interface transition layer

On mesoscopic level, concrete can be regarded as composition of coarse aggregate, hardened sanded cement grout and transition region between them. Certain of concrete material's performance indexes such as strength, deformation ability, permeability and endurance are influenced by its pore structure and interface transition layer [4].

In hardened gelled material pores distribute in network shape, whose character strongly affect concrete performance. The following is a model to calculate porosity of hardened sanded cement grout, and proportion of capillary pore and gelled pore in whole pore is given as follows For capillary pore:

$$[(W/C - \alpha(W/C^*))]/[(D_W \times V_c) + W/C]. \quad (1)$$

For gelled pore:

$$[mgV_g \times \alpha]/[V_C + (W/C)/D_W] \quad (2)$$

[a] Corresponding author: zhengshansuo@263.net

where W/C^* is critical water cement ratio when cement is hydrated wholly; D_W is density of pore liquid in freshly mixed slurry; V_C is specific volume of unhydrated cement; V_g is average specific volume of gelatum water; α is percentage of hydrated cement; mg is weight of gelatum water in a unit weight cement grout hydrated wholly.

According to the above model, volume of capillary pore in a unit weight cement grout is the difference between total volume and the sum volume of unhydrated cement and cement hydrate, which is denoted as $[W/C - (W/C^*)\alpha]D_W$. Volume of gelled pore in a unit weight cement grout is $mg \cdot V_g \cdot \alpha$.

According to the relation of water cement ratio and porosity mentioned above, it is clear that porosity of cement grout increases with increasing of water cement ratio.

Based on experiments the relation of compression strength and porosity of concrete is expressed as the following equation, which illustrates that concrete strength and Young's modulus drop along with increase of porosity.

$$f_c = a(1-\rho)^3 \tag{3}$$

where f_c is compression strength of concrete; ρ is porosity; a is sanded cement grout strength when porosity equals to zero.

The relation of permeation ratio and pore structure can be expressed by an empirical equation as follows

$$K = 1.684 r_\infty^{3.284} \times 10^{-22} \tag{4}$$

where K denotes permeation ratio; r_∞ denotes maximum continuous radius, which is defined by the point of greatest gradient on mercury intrusion experiment curve.

The relation of permeation ratio, porosity and hydraulic radius is as follows

$$\log K = 38.45 + 4.08 \log\left(\varepsilon r_a^2\right) \tag{5}$$

where ε denotes porosity; r_a denotes hydraulic radius.

Mchta and Mamohan adopted four pore parameters to establish relation with permeation ratio

$$K = \exp\left(3.84V_1 + 0.2V_2 + 0.56TD + 8.09MTP - 2.53\right) \tag{6}$$

where V_1 is volume ratio of pore greater than 1320 Å and total pore; V_2 is volume ratio of pore between 290 Å to 1320 Å and total pore; TD is critical aperture, MTP is ratio of total volume and hydrating degree.

2.2 Simplified model and method to improve concrete strength

Concrete can be simplified as composition of hardened sanded cement grout and coarse aggregate, which link together through bonding effect. The simplified model is shown in Figure 1. When bearing external force, the two portions work together and have the same stress. Because Young's modulus of them is not equal, the two portion's strain is not the same, total strain is sum of the two portions, i.e.

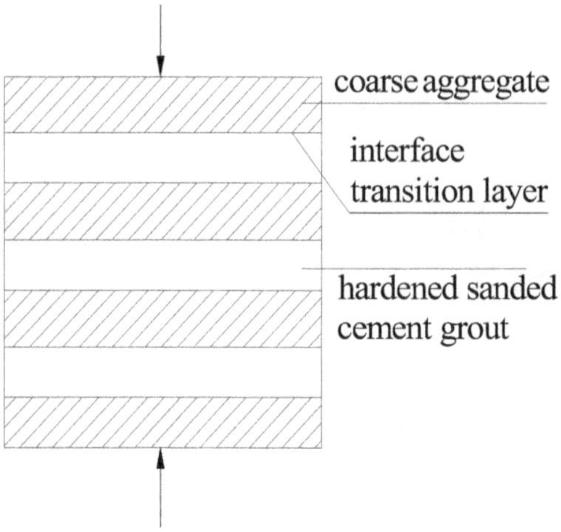

Fig. 1. Simplified concrete model.

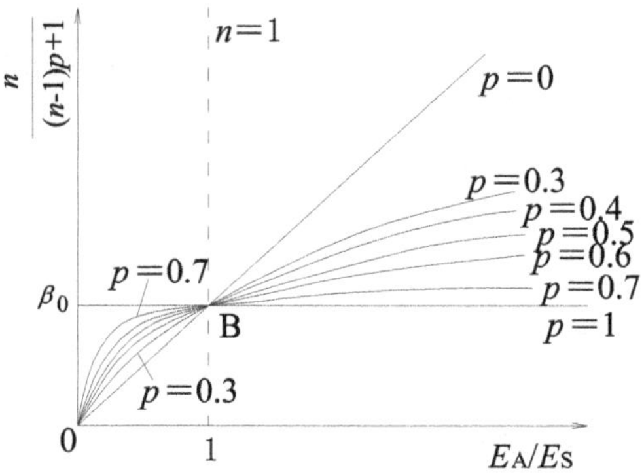

Fig. 2. Young's modulus ratio of concrete, aggregate and grout.

$$\begin{cases} \dfrac{R_c}{E_c} = \dfrac{R_A}{E_A}(1-P) + \dfrac{R_S}{E_S} \cdot P \\ R_C = R_A = R_S. \end{cases} \tag{7}$$

Defining $n = \frac{E_A}{E_S}$, then Young's modulus of concrete is given by

$$E_C = \frac{nE_s}{(n-1)p+1}. \tag{8}$$

Defining $\beta = \frac{n}{(n-1)p+1}$, we can write

$$E_C = \beta E_S \tag{9}$$

where E_C, E_S, E_A denote Young's modulus of concrete, coarse aggregate and hardened sanded cement grout respectively; P denotes grout's volume content; n denotes Young's modulus ratio of aggregate and grout; R_C, R_S, R_A denote concrete's total stress, aggregate's stress and grout's stress respectively.

The above analysis indicates that cement grout and aggregate both affect concrete strength. Equations (7)–(9) and Figure 2 reflect the relationship of P and n. In Figure 2, area $\Delta_{\alpha_0 o B}$ represents high strength concrete, in which cement grout's strength is higher than aggregate. The following are methods to improve concrete strength.

a. To improve P

In equations (7)–(9), n is definitive after materials to produce concrete are picked up. Within $\Delta_{\alpha_0 oB}$ area shown in Figure 2, raising cement quantity can improve P. On the other hand, the improvement is not so effective. When n equals to 0.5, P increases to 0.6 from 0.3, accordingly β increases to 0.701 from 0.588, and cement quantity is doubled but strength is raised only 17%, which means materials waste. This method is infeasible in actual engineering.

b. To improve n

Improving n means improvement of aggregate strength. In like manner, when $P = 0.5$, n increases to 0.4 from 0.2, accordingly α increases to 0.571 from 0.333, and improvement degree is 71.5%, which means applying high strength and high quality aggregate is the key point in producing high strength and high performance concrete.

c. To improve E_S

With regard to high strength and high performance concrete, coarse aggregate strength is lower than cement grout. Concrete damage lies on aggregate strength, when cement grout strength increases to a certain extent concrete strength will not be improved any more.

2.3 Bond strength between steel and concrete

In SRC structures there exists bonding action between steel shape and concrete, which are two kinds of materials with different performance. Bonding action enables stress to transfer effectively each other on the interface of steel and concrete and establishes working stress needed to fulfill loading capacity of structures in steel and concrete [5]. Bond stress is a kind of shear force from macroscopic effect, the experiments showed that the distribution of bond stress on the interface of steel and concrete.

Based on the theory of mechanics of materials, the distribution of bond stress along steel surface is given as follows

$$\tau(x) = \frac{A_s}{u}\frac{d\sigma_x}{dx} = \frac{A_s}{u}E_s\frac{d\varepsilon_x}{dx} = \frac{-k_1 A_s \varepsilon_{max}}{u}E_s e^{-k_1 x} \tag{10}$$

where E_s is elastic modulus of steel; $d\sigma_x/dx$, $d\varepsilon_x/dx$ are steel cross-section normal stress and normal strain increment along anchorage length respectively; σ_x, A_s and u, for full section, are average stress, area and perimeter of steel section respectively, for flange or web, are average stress, area, perimeter of flange or web section respectively.

The distribution of slip between the interface of steel and concrete is given as follows

$$s(x) = s_{max}e^{-k_2 x} \tag{11}$$

where $s(x)$ is the slip quantity at different part of cross-section along anchorage length; s_{max} is the maximum slip quantity along anchorage length; k_2 is characteristic exponent of slip distribution.

The bond-slip pull-out test of SRC members indicates that on the interface of steel and concrete the relationship of bond strength and concrete cube compression strength is as follows [6, 7]:

$$\tau_u = 0.0130 f_{cu} + 1.2150 C_a/h_a - 0.0640 l_a/h_a + 1.8682 \tag{12}$$

where τ_u is ultimate bond strength; f_{cu} is concrete cube compression strength; C_a is thickness of covering layer; h_a is sectional height; l_a is the embedded depth of steel shape.

3 Multi-objective optimization design of concrete mix proportions

3.1 Mathematical model

Mix proportion optimization design of HSHPC applied to SRC structures is a nonlinear multi-objective problem. In order to establish the relationship of every objective function and constraint conditions and finish optimum solution, multi-objective optimization technique is applied. The mathematical model of multi-objective optimization for concrete mix proportions is shown as follows [8]:

$$\begin{cases} \min \boldsymbol{F(x)}, \\ \min \boldsymbol{G(x)}; \\ \text{s.t. } \boldsymbol{s}_i(\boldsymbol{x}) \leqslant 0, \quad i = 1, 2 \ldots, m \\ \qquad \boldsymbol{h}_j(\boldsymbol{x}) = 0, \quad j = 1, 2 \ldots, l (l < n) \\ \qquad \boldsymbol{x}_i \leqslant \boldsymbol{x} \leqslant \boldsymbol{x}_u \end{cases} \tag{13}$$

where $\boldsymbol{x} = [x_1, x_2, \ldots, x_n]^T$ is decision-making vector including n components; $\boldsymbol{F(x)}$, $\boldsymbol{G(x)}$ is objective function vector; $\boldsymbol{s}_i(\boldsymbol{x}) \leqslant 0$, $\boldsymbol{h}_j(\boldsymbol{x}) = 0$ is constraint conditions. The decision-making Vector satisfying all restraint conditions is called feasible solution or feasible point, and the aggregate of all feasible points is called feasible aggregate. All optimization solution procedure is to find one point $\boldsymbol{x}^*$ to make objective function obtain extremum in this point. $\boldsymbol{x}^*$ is the optimal solution to the problem of concrete mix proportions accordingly.

3.2 Constraint conditions

In this paper concrete raw material is composed of cement, sand, macadam, mineral admixture, high efficiency water reducing agent and water, which are calling six components of concrete. Mix proportion design is to determine the consumption of each component, the consumption of these six kinds of components are expressed as x_1, x_2, $\ldots x_6$ individually [9].

a. The limits of each component consumption

$$x_{il} \leqslant x_i \leqslant x_{iu}, \ i = 1, 2, \ldots, 6 \tag{14}$$

where x_{il}, x_{iu} are lowest limit and maximal limit for x_i respetively.

b. The restriction of water gel ratio

$$k_l \leqslant x_6/(x_1 + x_4) \leqslant k_u \tag{15}$$

where, $x_6/(x_1 + x_4)$ is the ratio of water and gel material (the sum of cement and mineral admixture); k_l, k_u are lowest limit and maximal limit for water gel ratio respetively.

c. Sand ratio

$$S_l \leqslant x_2/(x_2 + x_3) \leqslant S_u \tag{16}$$

where, $x_2/(x_2 + x_3)$ is the ratio of sand and coarse aggregate; S_l, S_u are lowest limit and maximal limit for Sand ratio respetively.

d. The total consumption of gel material

$$C_l \leqslant x_1 + x_4 \leqslant C_u. \tag{17}$$

e. Additive consumption

$$R_l \leqslant x_5/(x_1 + x_4) \leqslant R_u \tag{18}$$

where, $x_5/(x_1 + x_4)$ is the ratio of water reducing agent blend dosage and gel material dosage, and R_l, R_u are its lowest limit and maximal limit respetively.

f. The restraint of material volume

Assumed that the volume of concrete equals the sum of the absolute volume of each component and the volume of air contained in concrete, the amount of raw materials in every cube concrete requires to satisfy the following equation [10]

$$\sum_{i=1}^{6} x_i/\rho_i + 10\alpha - 1000 = 0 \tag{19}$$

where, ρ_i is the density of ith component; α is the percentage of air in concrete and it equals 1 when air entraining agent not be used.

g. Cube compression strength

Admixture activity index was introduced to the calculation of mix proportions. Experimental investigation and the results of theoretical analysis show that the relationship of water gel ratio and concrete formulation strength presented by reference [11] possesses definite physical concept and better applicability, and can be applied to this optimum design. The relationship of water gel ratio and concrete formulation strength is as follows:

$$A \cdot \left(-0.4952 + \frac{5.514 \times x_6}{x_1 + x_4}\right) \cdot \alpha_a \cdot f_{ce} \cdot \left(\frac{x_1 + x_4}{x_6} - \alpha_b\right) - f_{cu,k} + 1.645\sigma \geqslant 0 \tag{20}$$

where A is mineral admixture activity index; $f_{cu,k}$ is concrete cube compression strength; f_{ce} is cement strength; σ is standard deviation of concrete strength; α_a, α_b are regression coefficients.

3.3 Objective functions

The cooperating working ability of concrete and steel shape and the economic cost of concrete are chosen as the target of mix proportion optimization.

τ_u, which is ultimate bond strength on the interface of steel shape and concrete, characters the cooperating working ability of concrete and steel shape. According to formula (3) the objective function of ultimate bond strength can be established as follows

$$\tau_u = \sum_{i=1}^{6} F_i(x_i)\, x_i \tag{21}$$

where, x_i is the material factor of cement, aggregates, mineral admixture, etc.; $F_i(x_i)$ is effect of material factor on ultimate bond strength, and the function relationship can be defined by equations (12) and (20).

The unit price of each material component are expressed as y_i, the objective function of unit cost of concrete is as follows

$$c = \sum_{i=1}^{6} y_i x_i. \tag{22}$$

3.4 Solving process

The mix proportion optimization problem of HSHPC applied to SRC structures can be solved by using the method of sequential quadratic programming. It is the best means to solve the optimization problem with nonlinear objective function and nonlinear constrain function. Under the support of computer calculation speed, the means constructs quadratic programming sub-problem at every iteration point, and uses the solution of this sub-problem as iterative search direction to approach the solution of constrain optimization problem [12].

Typical inequality constrained optimization question is shown as

$$\begin{aligned} &\min c_0(\boldsymbol{x}) \\ &\text{s.t.} \ \ c_i(\boldsymbol{x}) \leqslant 0, \ i \in I = 1, 2, \ldots, m \end{aligned} \tag{23}$$

where $c_0(\boldsymbol{x})$ is objective function; $\boldsymbol{x}$ is design variables; $c_i(\boldsymbol{x})$ is constraint function of state variables.

The following notation and definition are introduced for the current iterative point $\boldsymbol{x}^{(k)} \in \boldsymbol{R}^n$

$$\left.\begin{aligned} \boldsymbol{I}_k^- &= \boldsymbol{I}^-\left(\boldsymbol{x}^{(k)}\right) = \left\{i \in \boldsymbol{I} \,\middle|\, c_i\left(\boldsymbol{x}^{(k)}\right) \leqslant 0\right\} \\ \boldsymbol{I}_k^+ &= \boldsymbol{I}^+\left(\boldsymbol{x}^{(k)}\right) = \left\{i \in \boldsymbol{I} \,\middle|\, c_i\left(\boldsymbol{x}^{(k)}\right) > 0\right\} \end{aligned}\right\} \tag{24}$$

$$\begin{aligned} \varphi_k = \varphi\left(\boldsymbol{x}^{(k)}\right) &= \max\left\{0, c_i\left(\boldsymbol{x}^{(k)}\right) | i \in \boldsymbol{I}\right\} \\ &= \max\left\{0, c_i\left(\boldsymbol{x}^{(k)}\right) | i \in \boldsymbol{I}_k^+\right\} \end{aligned} \tag{25}$$

$$D_i\left(\boldsymbol{x}^{(k)}\right) = \begin{cases} c_i\left(\boldsymbol{x}^{(k)}\right)^2, & i \in \boldsymbol{I}_k^- \\ \left(\varphi\left(\boldsymbol{x}^{(k)}\right) - c_i\left(\boldsymbol{x}^{(k)}\right)\right)^2, & i \in \boldsymbol{I}_k^+ \end{cases} \tag{26}$$

$$\boldsymbol{I}\left(\boldsymbol{x}^{(k)}\right) = \left\{i \in \boldsymbol{I} \,\middle|\, D_i\left(\boldsymbol{x}^{(k)}\right) = 0\right\}. \tag{27}$$

Accordingly the inequality constrained optimization can be translated into quadratic programming question.

$$\left.\begin{aligned} &\min \boldsymbol{d}^T \nabla c_0\left(\boldsymbol{x}^{(k)}\right) + \frac{1}{2}\boldsymbol{d}^T \boldsymbol{B}_k \boldsymbol{d} \\ &\text{s.t. } c_i\left(\boldsymbol{x}^{(k)}\right) + \boldsymbol{d}^T \nabla c_i\left(\boldsymbol{x}^{(k)}\right) \leqslant 0, \quad i \in \boldsymbol{I}_k^- \\ &\quad\;\; c_i\left(\boldsymbol{x}^{(k)}\right) + \boldsymbol{d}^T \nabla c_i\left(\boldsymbol{x}^{(k)}\right) \leqslant \varphi_k, \quad i \in \boldsymbol{I}_k^+ \end{aligned}\right\} \tag{28}$$

where matrix $\boldsymbol{B}_k$ produces convergent descending direction, and $\boldsymbol{d}$ is principal searching direction, then the solution to question (28) will approach the solution to question (23).

Quadratic programming question has its feasible solution $\boldsymbol{d} = \boldsymbol{0}$, assuming $\left(\boldsymbol{d}_0^{(k)}, \lambda^{(k+1)}\right)$ is its optimal solution.

Giving the following definition

$$\left.\begin{aligned} \boldsymbol{A}_k &= \boldsymbol{A}\left(\boldsymbol{x}^{(k)}\right) = \left(\nabla c_i\left(\boldsymbol{x}^{(k)}\right), i \in \boldsymbol{I}\right) \\ \boldsymbol{D}_k &= \boldsymbol{D}\left(\boldsymbol{x}^{(k)}\right) = \left(\text{diag }\left(\boldsymbol{D}_i\left(\boldsymbol{x}^{(k)}\right)\right), i \in \boldsymbol{I}\right) \end{aligned}\right\} \tag{29}$$

$$\left.\begin{aligned} \boldsymbol{Q}_k &= \boldsymbol{Q}\left(\boldsymbol{x}^{(k)}\right) = \left(\boldsymbol{A}_k^T \boldsymbol{A}_k + \boldsymbol{D}_k\right)^{-1} \boldsymbol{A}_k^T \\ \boldsymbol{P}_k &= \boldsymbol{P}\left(\boldsymbol{x}^{(k)}\right) = \boldsymbol{E} - \boldsymbol{A}_k \boldsymbol{Q}_k \end{aligned}\right\} \tag{30}$$

where $\boldsymbol{E}$ is identity matrix.

Constraint index set of quadratic programming question is $\boldsymbol{L}(k)$, and

$$\boldsymbol{L}_k = \boldsymbol{L}\left(\boldsymbol{x}^{(k)}\right) = \boldsymbol{L}_k^+ \cup \boldsymbol{L}_k^- \tag{31}$$

$$\begin{aligned} \boldsymbol{L}_k^- &= \boldsymbol{L}^-\left(\boldsymbol{x}^{(k)}\right) \\ &= \left\{i \in \boldsymbol{I}_k^- \,\middle|\, c_i\left(\boldsymbol{x}^{(k)}\right) + \left[\nabla c_i\left(\boldsymbol{x}^{(k)}\right)\right]^T \boldsymbol{d}_0^{(k)}\right\} \end{aligned} \tag{32}$$

$$\begin{aligned} \boldsymbol{L}_k^+ &= \boldsymbol{L}^+\left(\boldsymbol{x}^{(k)}\right) \\ &= \left\{i \in \boldsymbol{I}_k^+ \,\middle|\, c_i\left(\boldsymbol{x}^{(k)}\right) + \left[\nabla c_i\left(\boldsymbol{x}^{(k)}\right)\right]^T \boldsymbol{d}_0^{(k)} \right. \\ &\quad \left. = \varphi\left(\boldsymbol{x}^{(k)}\right)\right\}. \end{aligned} \tag{33}$$

Quadratic programming question (28) can be generally solved in proper neighborhood of solution of inequality constrained optimization question (23), but the latter does not always have finite solution because $\boldsymbol{B}_k$ is not limited positive definite form, it is necessary to introduce a convergent subsidiary searching direction $q^{(k)}$ [13–15].

$$q^{(k)} = \rho_k^\theta\left[-\boldsymbol{P}_k \nabla c_0\left(\boldsymbol{x}^{(k)}\right) + \boldsymbol{Q}_k^T v^{(k)}\right], \quad \theta > 0 \tag{34}$$

$$\left.\begin{aligned} \rho_k = \rho\left(\boldsymbol{x}^{(k)}\right) &= \frac{\left\|\boldsymbol{P}_k \nabla c_0\left(\boldsymbol{x}^{(k)}\right)\right\|^2 + \omega_k + \varphi\left(\boldsymbol{x}^{(k)}\right)}{1 + \left|e^T \pi^{(k)}\right|} \\ v_j^{(k)} &= \begin{cases} -1 - \rho_k, & \pi_j^{(k)} < 0 \\ D_j\left(\boldsymbol{x}^{(k)}\right) - \rho_k, & \pi_j^{(k)} \geqslant 0 \end{cases} \end{aligned}\right\} \tag{35}$$

$$\left.\begin{aligned} \pi^{(k)} &= \left(\pi_j^{(k)}, j \in \boldsymbol{I}\right) = -\boldsymbol{Q}_k \nabla c_0\left(\boldsymbol{x}^{(k)}\right) \\ \omega_k &= \omega\left(\boldsymbol{x}^{(k)}\right) = \sum_{j \in I} \max\left\{-\pi_j^{(k)}, \pi_j^{(k)} D_j\left(\boldsymbol{x}^{(k)}\right)\right\} \end{aligned}\right\}. \tag{36}$$

Because matrix $\boldsymbol{A}_k^T \boldsymbol{A}_k + \boldsymbol{D}_k$ is positive definite form, the mentioned definition is of significance. If and only $\rho_k = 0$, $\boldsymbol{x}^{(k)}$ is solution to optimization question (23). If $\left(\boldsymbol{d}_0^{(k)}, \varphi_k\right) = (\boldsymbol{0}, 0)$, then $\left(\boldsymbol{x}^{(k)}, \lambda^{(k+1)}\right)$ is solution to quadratic programming question (28).

4 Example of mix proportion design

Design HSHPC applied to the SRC concrete structures. The strength grade range is from C60 to C100. Preferred cement is Chinese Qingling brand P.O 52.5R normal silicate cement. Preferred coarse aggregate is from Jingyang city, Shaanxi province, China; the particle size is $5 \sim 20$ mm, and the density is 3.2 kg/m^3. Fine aggregate is coarse river sand from Ba river in Shaanxi province of China; the modulus of fineness is 2.8, and the density is 2.61 g/m^3. Admixture chooses polycarboxylate high efficiency water reducing agent, and the test of cement paste fluidity shows that it is better consistency with Qingling brand cement. Choosing fly ash and silicon ash as mineral additive, which are superfine particle and own high activity. Mixing water is drinking water.

In addition, Pull-out test is made to determine/compare ultimate bond strength of optimized

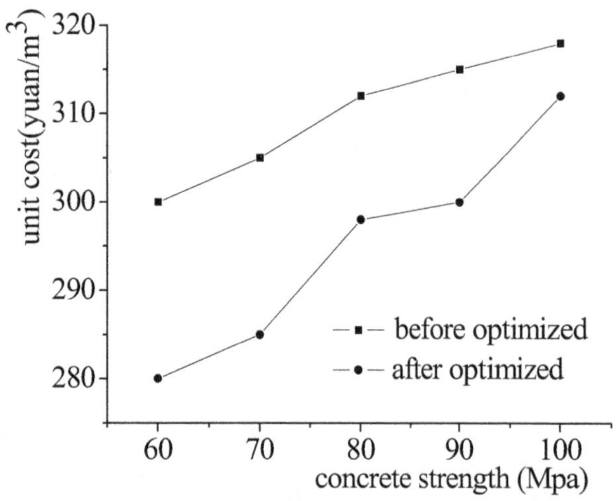

Fig. 3. Comparative analysis of concrete unit cost.

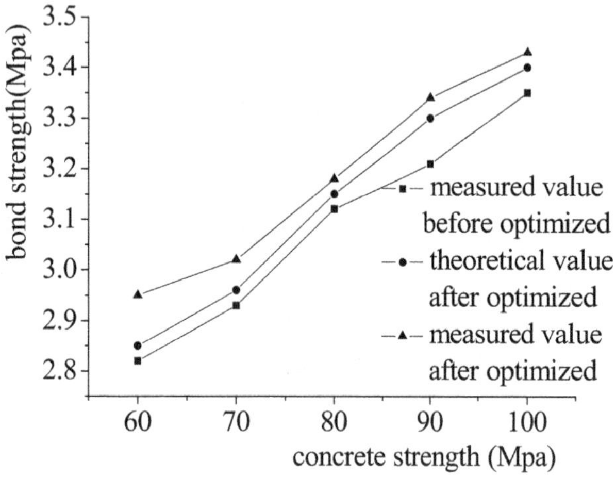

Fig. 4. Comparative analysis of ultimate bond strength.

HSHPC and/with normal HSHPC (with steel shape). I-shaped cross-section steel in the specimens is composed of two number 10 channel and 6 mm thick steel plate. The thickness of covering layer is 60 mm, the anchorage length is 740 mm.

The optimum problem is solved by convergent sequence quadratic programming and the design program is compiled based on Matlab language [16, 17].

The comparison analysis of concrete unit cost and ultimate bond strength of optimized HSHPC with normal HSHPC is shown in Figures 3 and 4, and the test results of ultimate bond strength are given in Figure 4. It can be seen that the bond strength of optimized HSHPC and steel shape increases about 12%, and the results of theoretical analysis with experiment is approximately consistent; cost per cubic meter can be saved about 10% through optimization.

5 Conclusions

HSHPC mix proportion design discussed in this paper is a kind of method of multi-objective and nonlinear optimization. Using convergent sequence quadratic programming and based on Matlab language, the optimum problem can be solved satisfactorily. Optimized HSHPC can meet all performance requirements of SRC structures and have an obvious economic benefit.

Acknowledgements

The authors would like to thank National Natural Science Foundation of China and the Educational Office of Shaan'xi Province in China for their support throughout this research.

References

1. P.K. Mehta, Greening of the concrete industry for sustainable Development, *Concrete International* (2000)
2. Feng Naiqian, High performance concrete structure (Beijing: Metallurgical Industry Press, 2004)
3. A. Ghezal, K.H. Khayat, J. ACI Materials **99**, 264 (2002)
4. O. Kontani, S.P. Shah, Pore pressure in sealed concrete at sustained high temperatures. *Proceeding Concrete Under Severe Conditions: Environment and Loading* (Tokyo, Japan, 1995)
5. C.W. Roeder, R. Chmielowski, J. Struct. Eng. **125**, 142 (1999)
6. S.S. Zheng, G.Z. Deng, Y. Yang, M.H. Yu, Eng. Mech. **20**, 63 (2003)
7. Zheng Shansuo, Study on the bond performance between steel shape and concrete in SRC structures, *Proc. of 7th International Symposium on Structural Engineering for Young Experts* (2002)
8. Zhang Wei, Xu JIaqing, *Optimization Method* (Shengyang: Northeast University Press)
9. J.P. Cannon, G.R. Krishna Murti, Cement and Concrete Research **1**, 353 (1971)
10. Narita Takeshi, Sekino Kazuo, Concrete Research and Technology **14**, 43 (2003)
11. Wan Chaojun, Concrete 41 (2002)
12. Jian Jinbao, J. Acta Math. Sci. **21**, 268 (2001)
13. Chao Weihua, Guo Zheng, *Optimization technique and implementation by Matlab* (Beijing: Chemical Industry Press)
14. Xu Jinming, *Practical tutorial of Matlab* (Beijing: Tinghua University Press, 2005)
15. Xu Chengxian et al. Modern method of optimization. (Beijing: Science Press, 2002)
16. R.H. Byrd, J. Nocedal, J. Numer. Anal. **26** 727 (1989)
17. A. Ghezal, K.H. Khayat, Optimizing self-consolidating concrete with limestone filler by using statistical factorial design methods. *ACI Mate* (2002)

14

Multi-component layout design with coupled shape and topology optimization

J.H. Zhu[1,2,a], W.H. Zhang[1] and P. Beckers[2]

[1] Sino-French Laboratory of Concurrent Engineering, Northwestern Polytechnical University 710072 Xi'an, Shaanxi, P.R. China
[2] LTAS – Infographie, Université de Liège, 4000 Liège, Belgium

Abstract – A Coupled Shape and Topology Optimization (CSTO) method is proposed here to deal with the layout design of the multi-component system. Considering a complex packing system for which several components will be placed in a container of specific shape, the aim of the design procedure is to find the optimal location and orientation of each component as well as the configuration of the structure that supports and interconnects the components. Compared with existing packing design approaches, two significant improvements are made in the CSTO method. On the one hand, a new Finite-circle Method (FCM) is used here to discretize boundaries of all components and the container into a number of circum-circles. Hence geometric constraints can be suitably modelled to avoid the overlap among the components as well as the overlap between the components and the container contour. Besides, the FCM approximation is also convenient to deal with components and the container with concave or some other complex shapes. On the other hand, the design procedure is able to take into account the mechanical performances of the structural system. Here, the location and the orientation of the components will be updated to improve the system rigidity by using shape optimization procedure. Meanwhile, the optimal material layout of the supporting structure in the design domain is designed by topology optimization. Due to the iterative movement of components, the technique of the embedded mesh is used to update the local FE mesh around each component in the design domain and pseudo density design variables assigned to density points instead of finite elements will be used to follow such a FE mesh variation. Several design problems are tested in this paper, and numerical results show the proposed CSTO method extends the actual concept of topology optimization and is efficient to generate reasonable design patterns.

Key words: Multi-component system; shape and topology optimization; finite-circle method; concave shape; density points; embedded mesh.

1 Introduction

Structural optimization and its industrial applications are always active research topics in engineering community covering aerospace, marine and mechanical systems, etc. Among others, structural topology optimization is recognized increasingly as one of the powerful tools supporting practical structural designs. As we can see, most of the industrial products can be considered to be multi-component systems made up of a container, i.e., a design domain and a number of components and structures to support and interconnect the container and components. In this paper, two basic problems involved in the design of a multi-component system are discussed.

On the one hand, the position and orientation of each component in the design domain should be designed. This is a kind of packing optimization problem. Packing problem or configuration design is still a CAD-based design with the compactness, position of gravity centre, etc. to be the design objective (see [1,2]). The components placement in the design domain can be treated as a shape optimization problem in which component locations and orientations are defined as the design variables. Among others, one of the key difficulties lies in that the geometry constraints have to be properly specified in order to avoid the components overlap and their overlap with the design domain boundary. However, varieties of component shapes and design domain boundaries will lead to nonlinearity and even discontinuity of the constraint functions, especially when the components or the design domain boundary have complex, concave or even 3D shapes, that limit the application of the gradient based optimization algorithms in the traditional formulation of packing problems. Therefore, non-deterministic computational approaches such as Genetic Algorithms (GA), Simulated Annealing and some extended patterns are mostly used in packing optimization problem. Recently, Zhang and Zhu [3] propose a new Finite-Circle Method (FCM). The idea is to replace approximately the exact shape of

[a] Corresponding author: jh.zhu_fea@hotmail.com

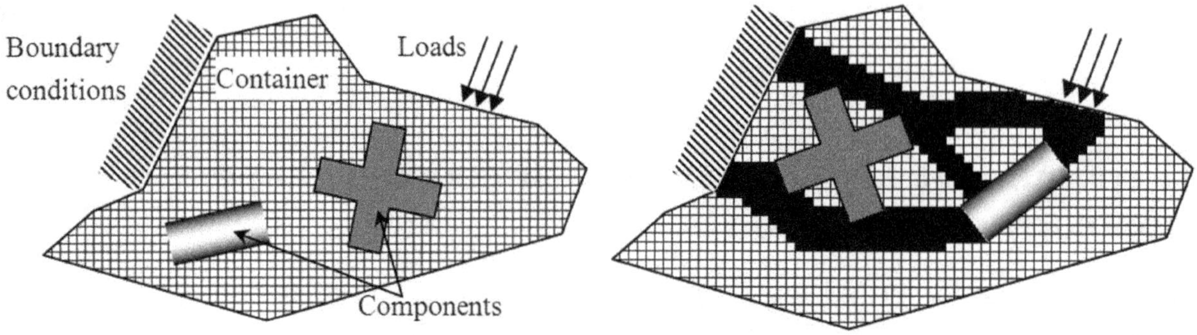

Fig. 1. An illustrative multi-component design problem.

each component with a family of circum-circles. Thus, the complex overlap constraints between components can be approximated into distance constraints between circum-circle centres themselves. As a result, constraint functions are kept to be continuous, differentiable and beneficial in sensitivity analysis and application of gradient based algorithms.

On the other hand, the system performances of embedded objects are the primary design objectives to be optimized. With this idea in mind, the design of the supporting structures connecting the components under specific loads and boundary conditions can be stated as a topology optimization problem. In the past one or two decades, standard topology optimization methods are developed to find optimal material layout for a single prescribed design domain with predefined loads and boundary conditions (see [4]). Meanwhile, effective approaches like homogenization based method [5, 6], SIMP model (Solid Isotropic Material with Penalization) [7, 8], evolutionary method [9], etc. have been proposed so far. Later on, more extended approaches are proposed to solve some complex problems such as microstructures design (see [10, 11]) and design-dependent loads [12, 13]. Moreover, the concept of topology design domain is also extended to design structural supports and joints modeled with spring elements. Some numerical examples of the latter were presented by Jiang and Chirehdast [14], Buhl [15], Zhu and Zhang [16] to solve structural stiffness, compliant mechanism and natural frequency problems. By constraining the material volume and the number of interconnections, Chickermane and Gea [17], Qian and Ananthasuresh [18] assigned each component and their joints as sub-design domains in topology optimization, respectively.

In fact, both of the above two aspects have to be carried out simultaneously in a multi-component system design. Compared to the standard topology optimization procedure in which topology design variables are assigned to elements of a fixed mesh, difficulties raised here are how to define topology design variables for a moving mesh due to the components placement in shape design iteration. Qian and Ananthasuresh [18] presented a compromised approach, in which the exact geometry of each component is not included but simulated with a predefined material interpolation model, i.e., the geometrical movement of a component is actually simulated as a physical variation of the material properties, while the supporting structure is designed with the standard topology optimization.

In this paper, an alternative method named CSTO is proposed. The exact geometry contour of each component will be modeled and embedded in the design domain. By means of the FCM method, the components layout is optimized with standard sensitivity-based shape optimization procedure while the supporting structures are designed with a new topology optimization approach. An illustrative multi-component system design problem is shown in Figure 1. By assigning the density variables to the predefined density points rather than finite elements, this approach doesn't require a fixed FE mesh. Relevant examples with components embedded in the design domain will be solved using the CSTO in the present paper.

2 Formulation of finite-circle method

In existing packing optimization approaches, different functions are used to describe each particular component with different shapes. Few of these functions are however differentiable and lead to the difficulties in sensitivity analysis. The FCM proposed by Zhang and Zhu [3] is a general method for packing optimization. The complex geometry of each component is transformed into a family of circum-circles so that the overlap between two components is avoided by constraining the distance between two related circles, which is a simple and differentiable function. Here, this method is improved to take into account more complex situation where the boundary of the design domain is concave.

2.1 Components approximation

As shown in Figure 2, Γ_1 and Γ_2 are the domains occupied by the 1st and 2nd components respectively. One design constraint should be defined to avoid the overlap of them

$$\Gamma_1(x_1, y_1, \theta_1) \cap \Gamma_2(x_2, y_2, \theta_2) = \emptyset. \tag{1}$$

Unfortunately, this kind of description is unable to directly be applied in our optimization model. By FCM,

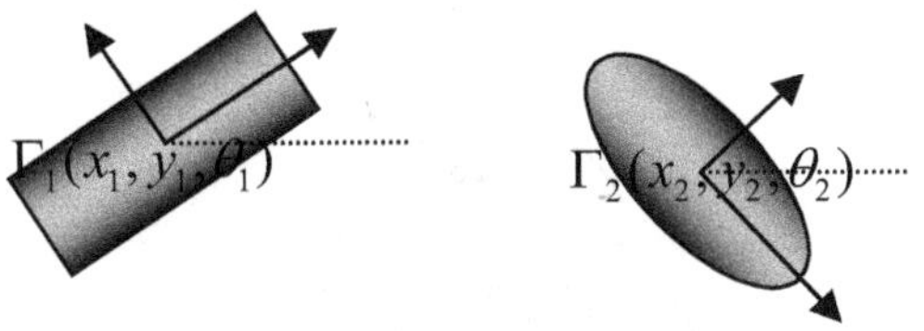

Fig. 2. Two components without any approximation.

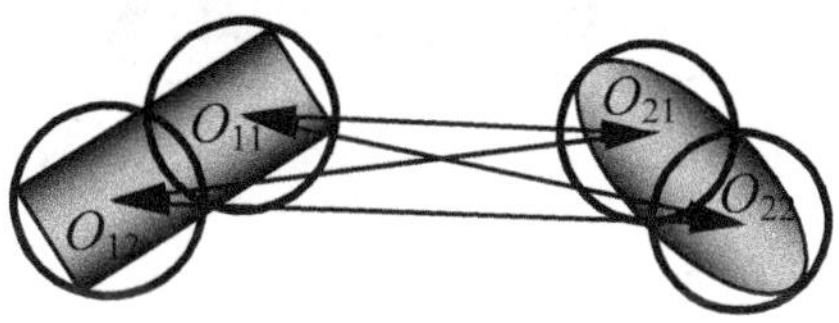

Fig. 3. Simplified geometry constraints with FCM.

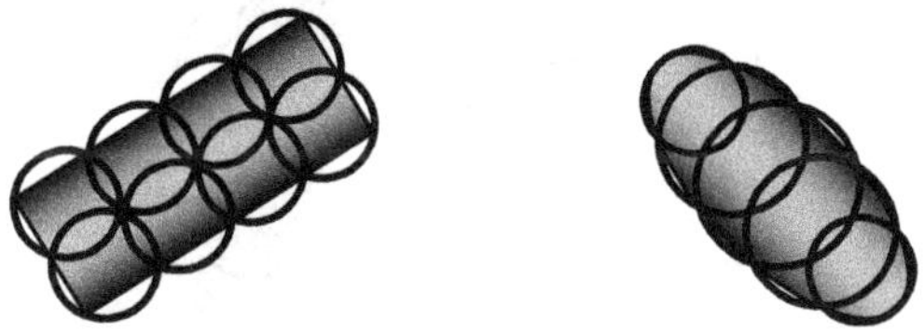

Fig. 4. A refined approximation with more circum-circles.

circum-circles are used to approximate each component as shown in Figure 3.

To avoid the overlap of the components, circles belonging to one component are constrained not to overlap with any circle of the other component. In Figure 3, two circles are attached to each component and the design constraints can be written as

$$\begin{array}{l} \|O_{11}O_{21}\| \geqslant R_{11}+R_{21}\ ,\ \ \|O_{11}O_{22}\| \geqslant R_{11}+R_{22} \\ \|O_{12}O_{21}\| \geqslant R_{12}+R_{21}\ ,\ \ \|O_{12}O_{22}\| \geqslant R_{12}+R_{22} \end{array} \tag{2}$$

where O_{11} denotes the centre of the 1st circle belonging to the 1st component, whose coordinates can be calculated by the location and orientation of the component. R_{11} is the radius of this circle. Now, the complex geometry constraint of equation (1) is expanded to several constraints of simple form shown in equation (2) in a systematic way.

In most cases except that the component is originally circular or holds some other particular shapes, the precision of the approximation can be improved by increase the number of circum-circles as shown in Figure 4. This is just like the refinement scheme of Finite Element Method. In Figure 4, we can find the approximation is more precise but it leads to 40 design constraints of equation (2). Similarly, for a 3D problem, the circum-circles will be upgraded to spheres for the components approximation.

2.2 Design domain boundary approximation

Although the components have been described by families of circum-circles, additional geometry constraints should be defined to ensure all components are completely lo-

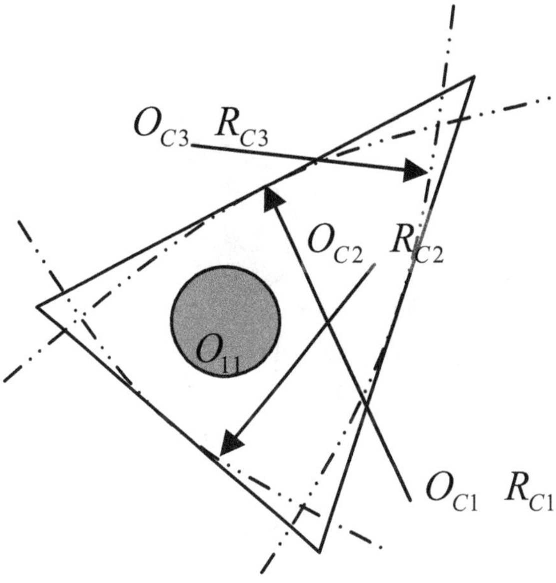

Fig. 5. Approximation of a convex design domain.

cated inside the design domain. Namely,

$$\begin{array}{l} \Gamma_1(x_1, y_1, \theta_1) \subset \Gamma_D \\ \Gamma_2(x_2, y_2, \theta_2) \subset \Gamma_D \end{array} \tag{3}$$

where Γ_D is the area occupied by the global design domain Regarding components shown in Figure 2 as an example, equation (3) can be transformed to

$$\begin{array}{l} \Gamma_{11} \subset \Gamma_D, \Gamma_{12} \subset \Gamma_D \\ \Gamma_{21} \subset \Gamma_D, \Gamma_{22} \subset \Gamma_D \end{array} \tag{4}$$

where Γ_{11} denotes the area taken by the circle O_{11}. Without the loss of generality, let us consider the circle O_{11}.

As shown in Figure 5, suppose the circle $O_{11}$is located in a triangle design domain whose contour is modelled as several circular arcs. Now the constraint becomes

$$\begin{array}{l} \|O_{11}O_{C1}\| \geqslant R_{C1} - R_{11} \\ \|O_{11}O_{C2}\| \geqslant R_{C2} - R_{11} \\ \|O_{11}O_{C3}\| \geqslant R_{C3} - R_{11}. \end{array} \tag{5}$$

The precision of the contour approximation can also be improved by adjusting centre positions and the radiuses of big circular arcs.

However, for a concave design domain shown in Figure 6, an improvement of FCM is needed. To this end, the concave part of the design domain is filled and considered as a fictive fixed component Γ_F. In this way, the design domain is transformed into a convex one and this method can be directly applied for general cases. Only more overlap constraints related to equation (6) are added to equation (5).

$$\Gamma_{11} \cap \Gamma_F = \emptyset. \tag{6}$$

2.3 A packing example

A compact packing design example is tested here to show how the FCM works. In Figure 7, 6 equilateral triangles

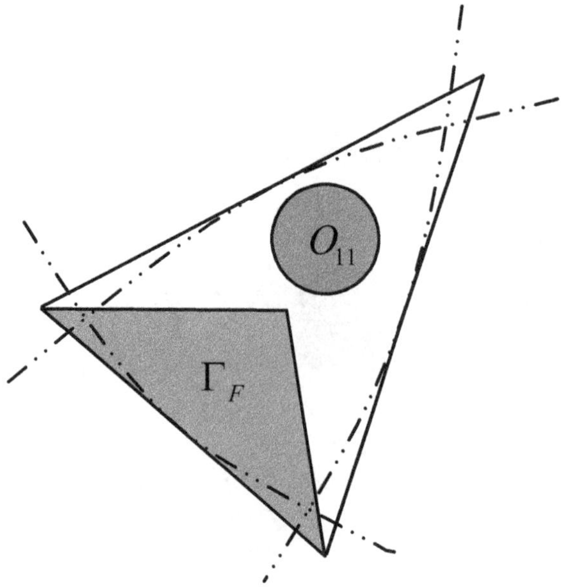

Fig. 6. Approximation of a concave design domain.

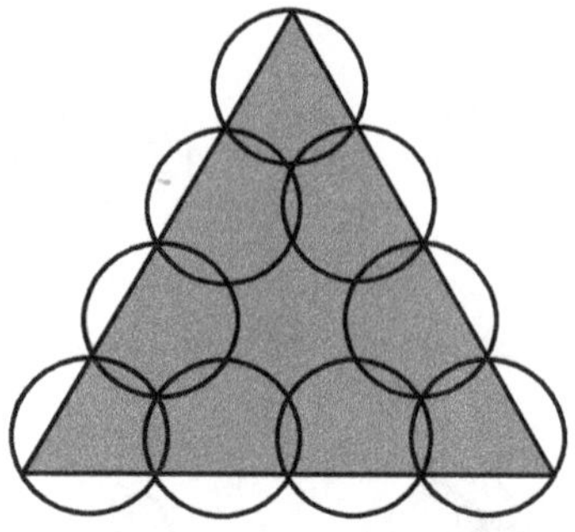

Fig. 7. Equilateral triangle component approximated with 9 circles.

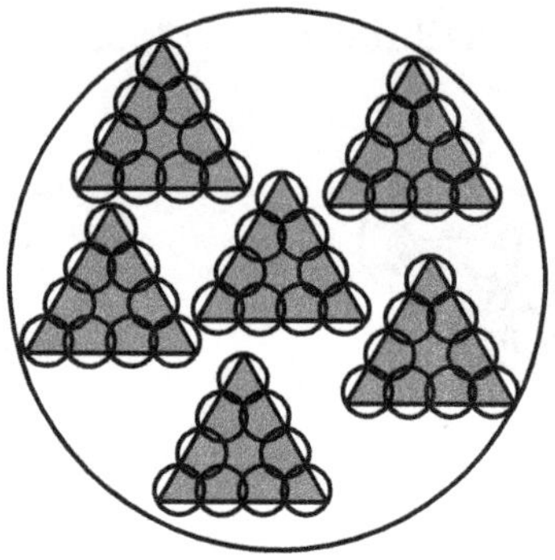

Fig. 8. The original configuration.

are considered to be the components. And one circle is defined as the boundary of packing domain. Each triangle will be approximated with 9 small circles.

The design purpose is to find the layout of 6 components and a packing domain of minimum area being able to contain them without overlap. The original configuration is shown in Figure 8. By using the gradient based GCM algorithm of BOSS-Quattro^{TM} software [19], this problem costs 9 iterations to reach the convergence. The design iteration and the final configuration are shown in Figure 9.

In the final configuration, 6 equilateral triangles are arranged compactly as an equilateral hexagon. It should be mentioned that this configuration is a feasible compact design because of the conservative circum-circle approximation.

3 Coupled shape and topology optimization

As the mechanical performances are included in the multi-component system optimization, the design procedure is much more complicated than a pure packing design. It is concerned with an integrated design of the packing and supporting structures and can be formulated as a CSTO problem.

3.1 Density points and embedded mesh technique

For a CSTO problem, the difficulty lies in that when the shape parameters are changed, i.e., the components translate or rotate in the design domain, the FE mesh should be updated simultaneously. As a result, the pseudo-densities as used before in topology optimization can no longer be assigned to fixed elements. Here, the concept of the density points is presented. As shown in Figure 10a, density points are prescribed in a design domain. Pseudo-density values (0-1) of design variables are attached to these points that remain unchanged during the optimization. For each finite element discretizing the design domain, the nearest density point will be identified and the pseudo-density value will be attributed correspondingly to the element. As illustrated in Figure 10b, the pseudo-density value of density point 1 will be attributed to the considered element as the distance from this density point to the centroid of the element is the shortest. Hence, one density point may dominate a group of local elements.

In summary, the procedure is as follows. Firstly, the whole design domain is discretized into finite elements. Such a mesh is here referred to as basic mesh whose centroids are used to define positions of density points as shown in Figure 11a. Suppose components are meshed independently. As they are placed in the design domain, an embedded mesh consisting of the basic mesh and component meshes is then created automatically by local mesh merging and smoothing as shown in Figure 11b. At this rate, when positions and orientations of the components change in the optimization process, only local mesh merging and smoothing are needed while most elements of the basic mesh remain unchanged.

3.2 Optimization problem formulation

Consider the stiffness of a system of m components as the objective function. The simultaneous design of the

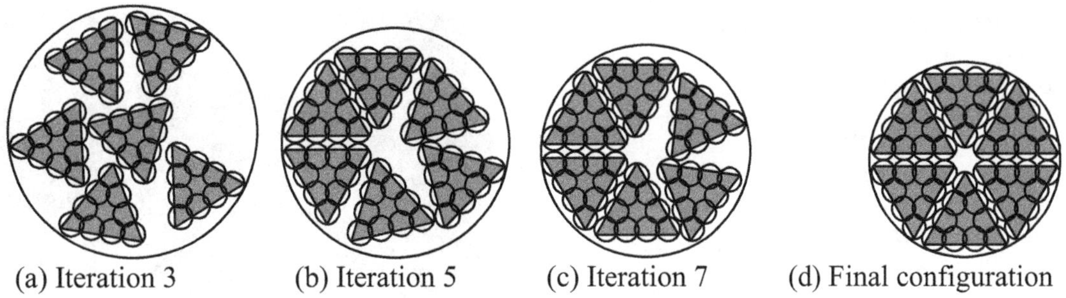

Fig. 9. Design iteration and the packing result.

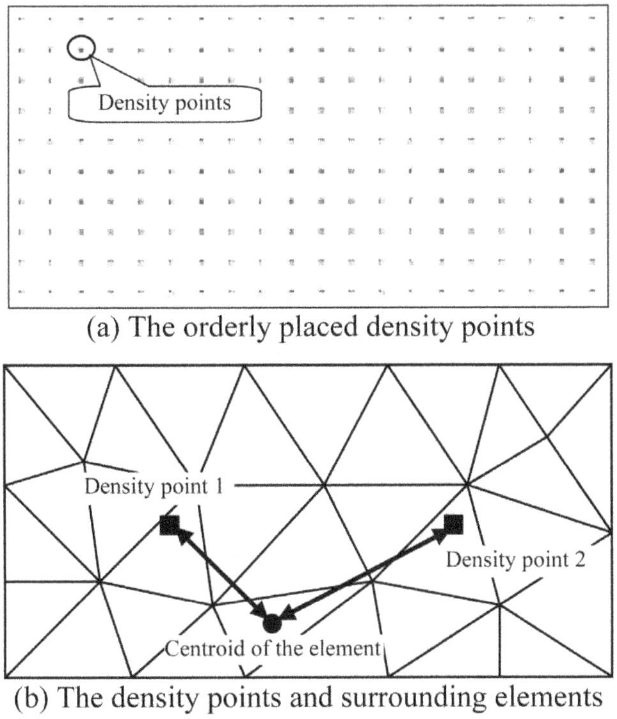

Fig. 10. Introduction of the density points.

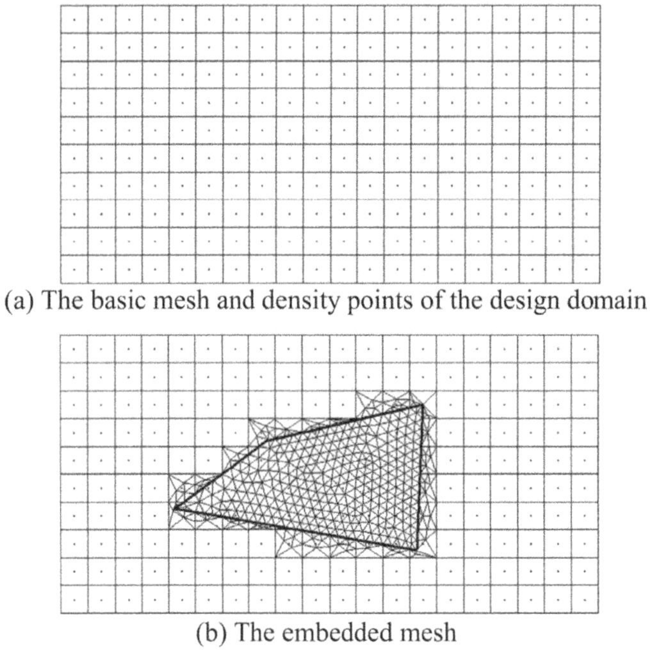

Fig. 11. Modeling of the basic mesh and embedded mesh.

components packing and supporting structures subject to the volume and geometry constraints can be expressed as

$$\begin{array}{ll} \text{Find}: & \boldsymbol{\eta}=\{\eta_1,\eta_2,\ldots,\eta_n\}\ \ \boldsymbol{S}=\begin{Bmatrix} x_1 & y_1 & \theta_1 \\ x_2 & y_2 & \theta_2 \\ & \vdots & \\ x_m & y_m & \theta_m \end{Bmatrix} \\ \min: & C=\frac{1}{2}\boldsymbol{u}^T\boldsymbol{K}\boldsymbol{u} \\ \text{s.t.}: & V\leqslant V^u \\ & 0<\eta^l\leqslant\eta_i\leqslant 1\ \ (i=1,2,\ldots,n) \\ & \Gamma_j(x_j,y_j,\theta_j)\cap\Gamma_k(x_k,y_k,\theta_k)=\emptyset \\ & \Gamma_j(x_j,y_j,\theta_j)\subset\Gamma_D \\ & j\neq k\ \ (j=1,2,\ldots,m\ \ k=1,2,\ldots,m) \end{array} \tag{7}$$

where η_i is the pseudo-density of the ith density point. η^l is the lower bound of the density variables, which is set to be 0.001. $\boldsymbol{S}$ is the shape design variable matrix consisting of all the component location and orientation parameters. C is the global strain energy of the structure calculated by the nodal displacement $\boldsymbol{u}$ and the global stiffness matrix $\boldsymbol{K}$. V and V^u are the structural volume and its upper bound, respectively.

Sensitivity analysis consists in evaluating the gradient of the global strain energy with respect to the shape parameters of the components and the pseudo-density variables.

In FE method, the static equation of the structure is represented by

$$\boldsymbol{f}=\boldsymbol{K}\boldsymbol{u} \tag{8}$$

where $\boldsymbol{f}$ is the load vector. The gradient of the shape parameter is derived with a finite difference approach. Suppose s_j is one of three shape parameters x_j, y_j and θ_j with the chosen perturbed step size to be

$$\Delta s_j=10^{-5}\left|s_j^u-s_j^l\right| \tag{9}$$

where s_j^u and s_j^l are the upper and lower bounds of s_j, respectively. While the derivative of the global strain energy with respect to s_j can be approximated as

$$\frac{\partial C}{\partial s_j} \approx \frac{\Delta C}{\Delta s_j}. \tag{10}$$

Moreover, sensitivity with respect to the pseudo-density variables is derived in a standard analytical scheme. Suppose η_i dominates $l$local elements, the element stiffness matrix attached to η_i can be written as

$$\boldsymbol{K}_{ie} = \eta_i^p \boldsymbol{K}_{ie}^0 \ (e = 1, 2, \ldots, l) \tag{11}$$

where p is the penalty factor set to be 3 in this paper. $\boldsymbol{K}_{ie}^0$ is the element stiffness matrix when the element is full of solid material. Likewise, the sensitivity is also formulated by differentiating the static equation.

$$\frac{\partial \boldsymbol{K}}{\partial \eta_i}\boldsymbol{u} + \boldsymbol{K}\frac{\partial \boldsymbol{u}}{\partial \eta_i} = 0. \tag{12}$$

The derivative can be expressed as

$$\frac{\partial C}{\partial \eta_i} = \frac{1}{2}\boldsymbol{f}^T\frac{\partial \boldsymbol{u}}{\partial \eta_i} = \frac{1}{2}\boldsymbol{u}^T\boldsymbol{K}\frac{\partial \boldsymbol{u}}{\partial \eta_i} = -\frac{1}{2}\boldsymbol{u}^T\frac{\partial \boldsymbol{K}}{\partial \eta_i}\boldsymbol{u}. \tag{13}$$

Since only $\boldsymbol{K}_{ie}$ (e =1, 2, ..., l) are related to η_i, this equation can be further developed as

$$\begin{aligned}\frac{\partial C}{\partial \eta_i} &= -\frac{1}{2}\boldsymbol{u}^T\frac{\partial\left(\sum_{e=1}^{l}\boldsymbol{K}_{ie}\right)}{\partial \eta_i}\boldsymbol{u} \\ &= -\frac{p}{2}\boldsymbol{u}^T\frac{\sum_{e=1}^{l}\boldsymbol{K}_{ie}}{\eta_i}\boldsymbol{u} = -\frac{p}{\eta_i}\sum_{e=1}^{l}C_{ie}\end{aligned} \tag{14}$$

where C_{ie} is the strain energy of the eth element.

3.3 Multi-component system design examples

Some problems with solid components are tested in this section to illustrate how to optimize the CSTO design of the component packing and supporting structures of a complex multi-component system. The material properties are as follows.

Supporting structure: Elastic modulus E = 1 pa; Poisson's ratio $\nu = 0.3$;
Solid component: Elastic modulus E = 3 pa; Poisson's ratio $\nu = 0.3$;

The latter is stiffer than the supporting structure.

The first design domain as shown in Figure 12 is a plate with a size of 6 m × 12 m. The basic mesh is divided into 30 × 60 finite elements. Suppose the density points are defined at the centroids of the basic elements. The load and boundary conditions are also described in Figure 12.

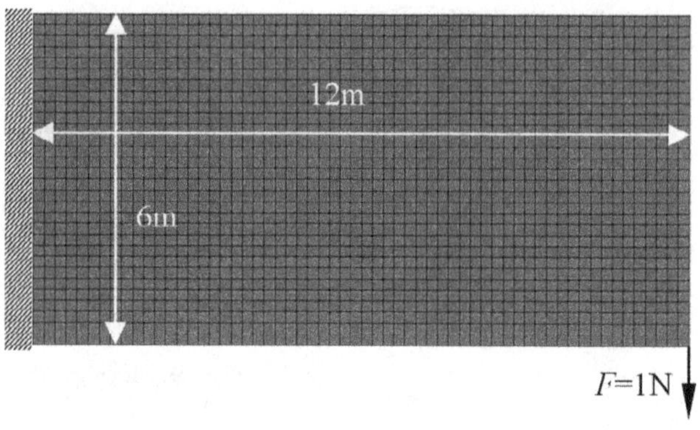

Fig. 12. Design domain and basic mesh of the plate.

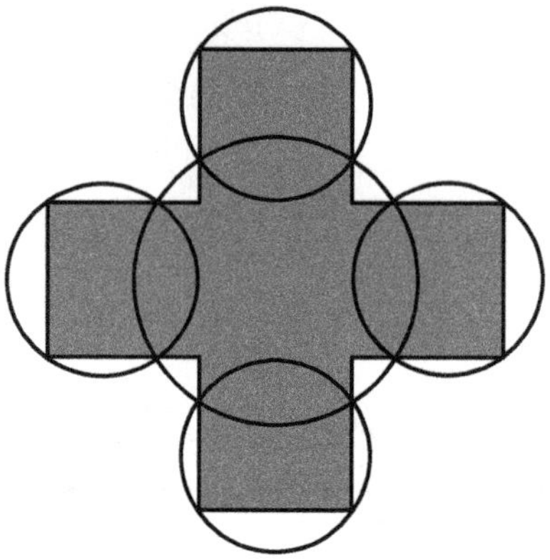

Fig. 13. Component and its circles.

As shown in Figure 13, one solid component will be approximated with 5 circles. The component is initially placed at the center of the plate. By minimizing the strain energy and constraining the volume of the supporting structures to 50% of the total, this problem is carried out with CSTO. The optimization is converged at the 32nd iteration. Several iterations and the final configuration are shown in Figure 14, where the component is located as a part of the structure.

The second design domain and its dimension are shown in Figure 15a. An L shape beam is modeled with its upper side fixed and one concentrated force is assigned at the right side. The basic mesh, its density points and the FCM boundary approximation are all presented in Figure 15b. For the concave part of the design domain, a fictive fixed triangle component is defined and approximated with 3 circles.

A solid component as shown in Figure 13 will be located in the design domain. It is approximated with 4 circles and shown in Figures 16a and 16b, respectively.

By minimizing the strain energy and constraining the volume of the supporting structures to 50% of the total volume, this problem is solved. Several iterations of the configuration patterns and the convergence of the design objective are shown in Figure 17. The component is finally located as a part of the structure after 58 iterations.

The last case is about two identical components illustrated in Figure 16 in the same design domain. Figure 18 shows the iteration histories. It can also found that both

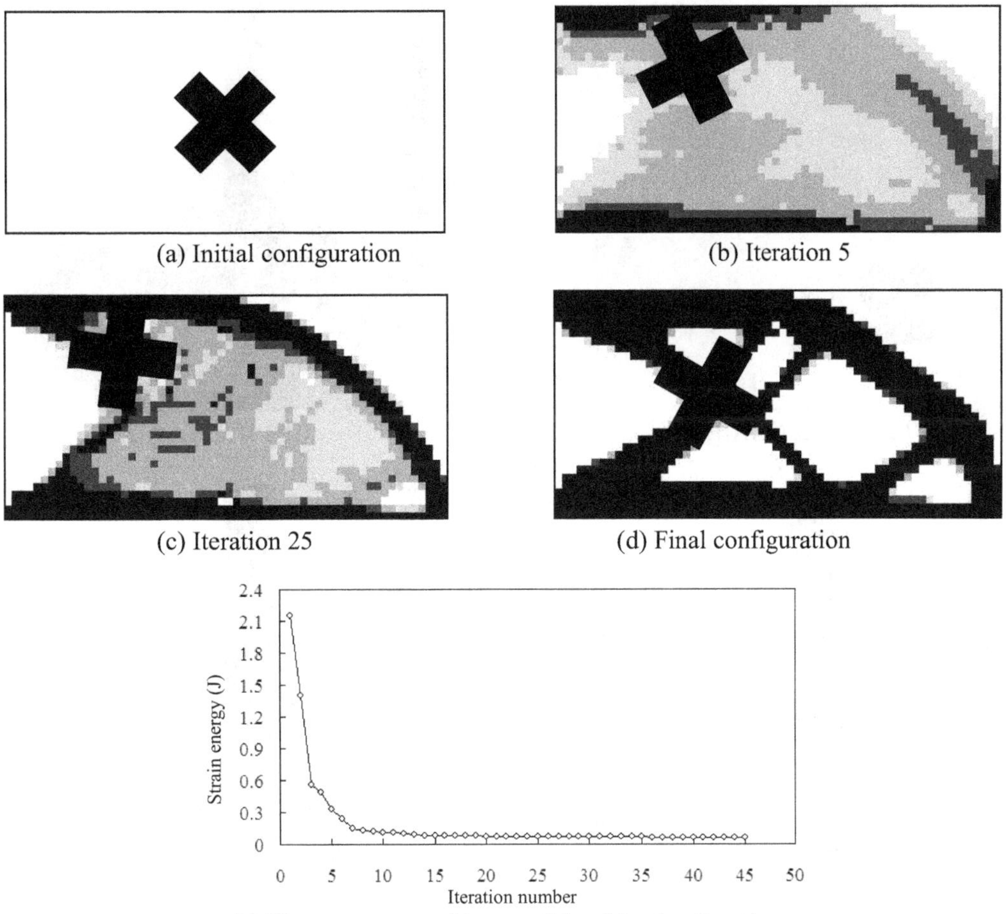

(a) Initial configuration (b) Iteration 5

(c) Iteration 25 (d) Final configuration

(e) The convergence history of the objective function

Fig. 14. Design iteration and the final configuration.

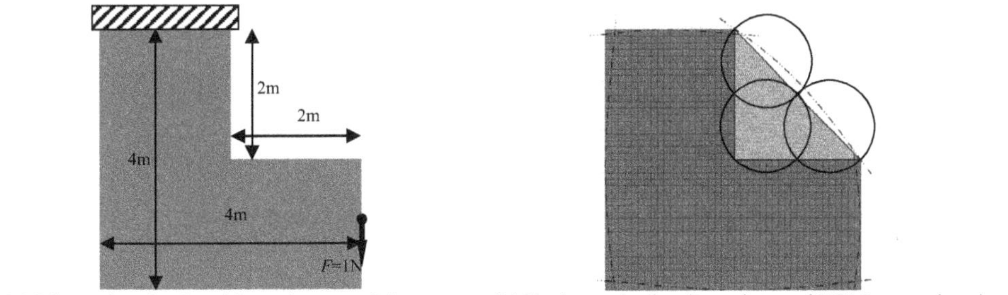

(a) Dimension, load and boundary conditions (b) Basic mesh, density points and FCM approximation

Fig. 15. The design domain.

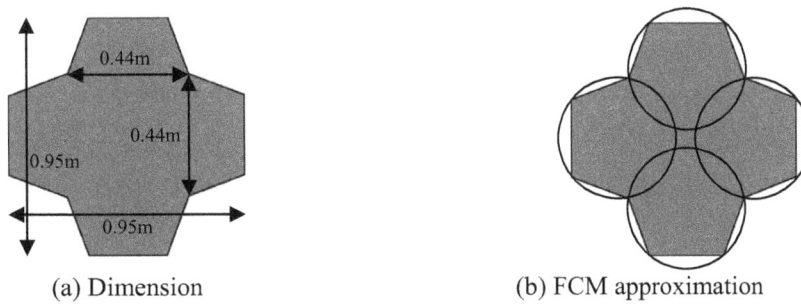

(a) Dimension (b) FCM approximation

Fig. 16. The component.

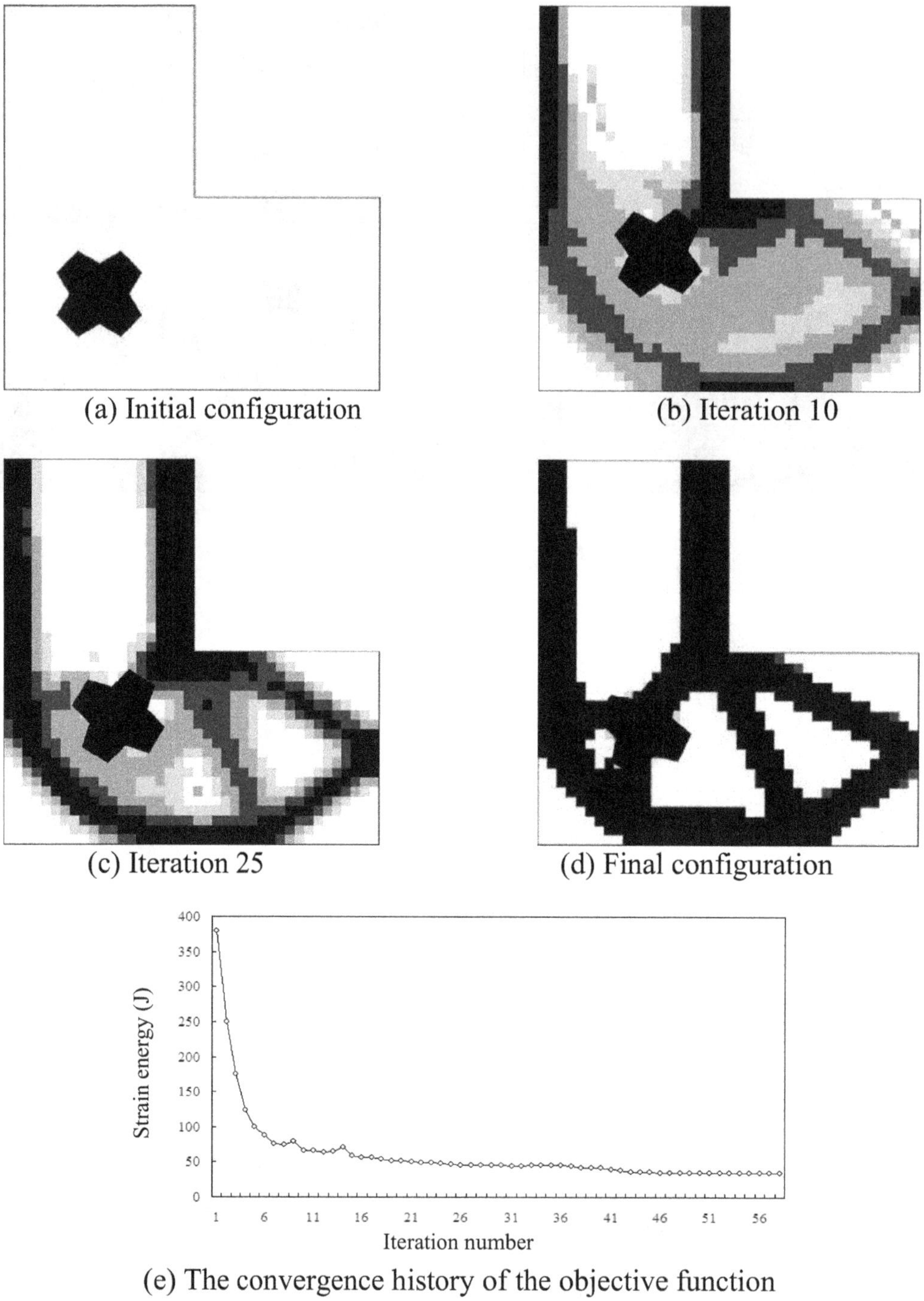

Fig. 17. The design process with one solid component.

components are now integrated as parts of the structure and demonstrate their resistance for the rigidity maximization of the system.

4 Conclusion

In this paper, components and their supporting structures are simultaneously designed with a proposed CSTO method that integrates the FCM, density point and embedded mesh techniques. The FCM aims at simplifying the complex overlap constraints and improvements are proposed for the concave case. The technique of density points is used to avoid the confliction between shape optimization and topology optimization. Meanwhile, this approach together with the embedded mesh technique can reduce the computing cost of the remeshing procedure. Two examples including concave design domain are solved. Numerical results show that CSTO is efficient to achieve reasonable design configurations.

(a) Initial configuration

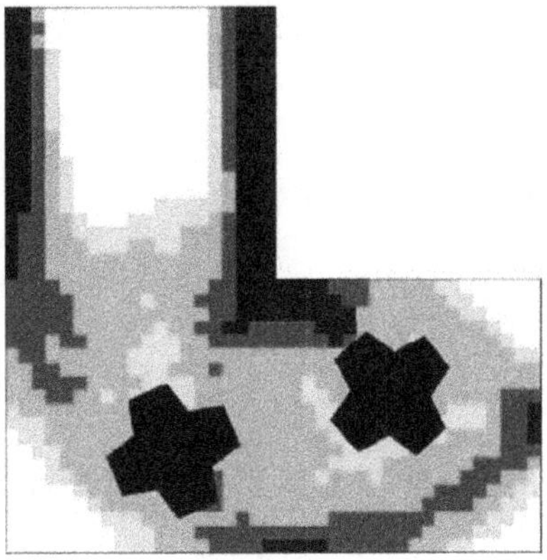

(b) Iteration 10

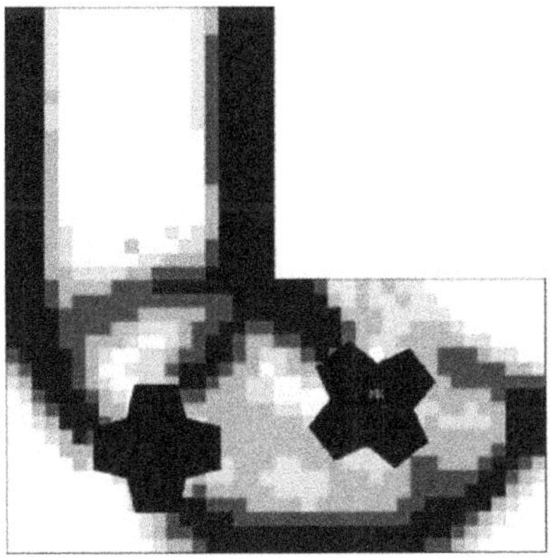

(c) Iteration 25

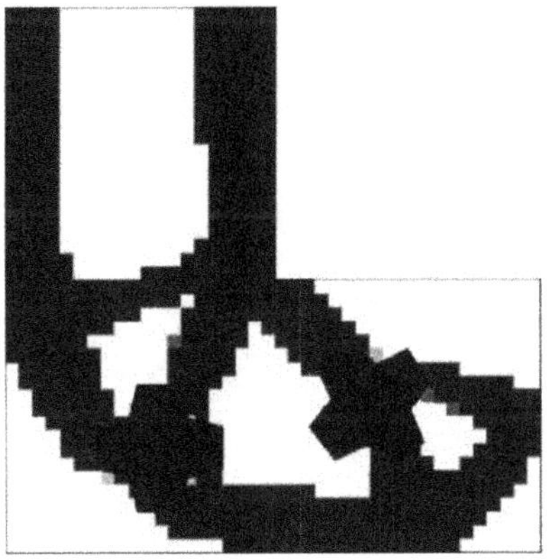

(d) Final configuration

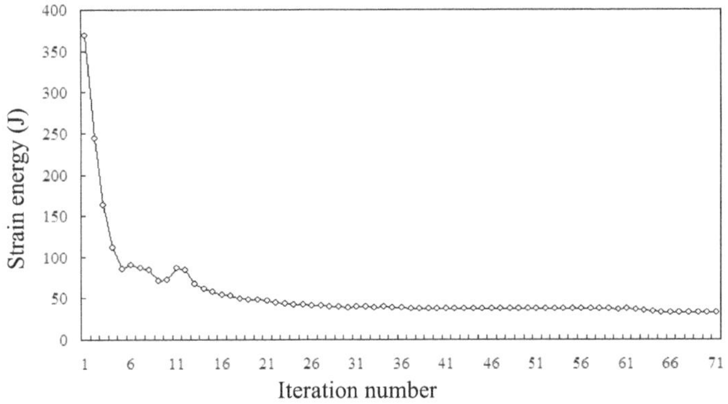

(e) The convergence history of the objective function

Fig. 18. The design process with two solid components.

Acknowledgements

This paper was published in ASMDO proceeding of First International conference on multidisciplinary optimization and Applications, April 17-20, 2007, ISBN 978-2-7598-0023-0.

This work is supported by the National Natural Science Foundation of China (10676028, 90405016), Aeronautical Science Foundation of China (2006ZA53006, 04B53080), 973 Program (2006CB601205), 863 Project (2006AA04Z122), 111 Project (B07050), Shaanxi Province Science & Technology Project (2006K05-G25) and Doctorate Foundation of Northwestern Polytechnical University (CX200508).

References

1. J. Cagan, K. Shimada, S. Yin, Computer-Aided Design **34**, 597 (2002)
2. V.Y. Blouin, Y. Miao, X. Zhou, G.M. Fadel, An assessment of configuration design methodologies, in *10th AIAA/ISSMO* (New York, 2004)
3. W.H. Zhang, J.H. Zhu, A new finite-circle family method for optimal multi-component packing design, in *WCCM VII* (Los Angeles, 2006)
4. M.P. Bendsøe, O. Sigmund, *Topology optimization: theory, method and application* (Springer, Berlin, Heidelberg New York, 2003)

5. M.P. Bendsøe, N. Kikuchi, Compt. Methods Appl. Mech. Eng. **71**, 197 (1988)
6. G. Allaire, F. Jouve, H. Maillot, Struct. Multidisciplinary Optim. **28**, 87 (2004)
7. M.P. Bendsøe, Struct. Optim. **10**, 193 (1989)
8. M. Zhou, G.I.N. Rozvany, Comput. Methods Appl. Mech. Eng. **89**, 197 (1991)
9. Y.M. Xie, G.P. Steven, *Evolutionary structural optimization* (Springer-Verlag, Berlin, 1997)
10. O. Sigmund, S. Torquato, J. Mech. Phys. Solids **45**, 1037 (1997)
11. W.H. Zhang, S.P. Sun, Internat. J. Numer. Methods Eng. **68**, 993 (2006)
12. B.C. Chen, N. Kikuchi, Finite Elements in Analysis and Design **37**, 57 (2001)
13. M. Bruyneel, P. Duysinx, Struct. Multidisciplinary Optim. **29**, 245 (2005)
14. T. Jiang, M. Chirehdast, J. Mechanical Design **119**, 40 (1997)
15. T. Buhl, Struct. Multidisciplinary Optim. **23**, 336 (2001)
16. J.H. Zhu, W.H. Zhang, Struct. Multidisciplinary Optim. **31**, 462 (2006)
17. H. Chickermane, H.C. Gea, Eng. Comput. **13**, 234 (1997)
18. Z.Y. Qian, G.K. Ananthasuresh, Mech. Based Design Struct. Mach. **32**, 165 (2004)
19. A. Remouchamps, Y. Radovcic, *Boss-Quattro Documents* (SAMTECH Inc., 2005)

15

Recent methodologies for reliability-based design optimization

G. Kharmanda[1,3,a], A. Mohsine[2,3], A. Makloufi[3] and A. El-Hami[3]

[1] Faculty of Mechanical Engineering, Aleppo University, Syrian Arab Republic
[2] INSA de Lyon, Lab. d'Automatique Industrielle, Bât. St Exupéry, 25, Avenue Jean Capelle, 69621 Villeurbanne, France
[3] INSA de Rouen, LMR, BP 08, Avenue de l'Université 76801 St Etienne du Rouvray, France

Abstract – In the field of Deterministic Design Optimization (DDO), the designer reduces the structural cost without taking into account uncertainties concerning materials, geometry and loading. This way the resulting optimum solution may represent a lower level of reliability and thus a higher risk of failure. But in the Reliability-Based Design Optimization (RBDO) model, the mean values of uncertain system variables are usually used as design variables, and the cost is optimized subject to prescribed probabilistic constraints as defined by a nonlinear mathematical programming problem. Therefore, a RBDO solution that reduces the structural weight in uncritical regions does not only provide an improved design but also a higher level of confidence in the design. In this paper, we present the advantage of the DDO and RBDO models and next some recent methodologies in order to show that the RBDO model is a practical tool for structural engineers.

Key words: Reliability-based design optimization; structural reliability; probabilistic model

Nomenclature

$\mathbf{x}$	:	Deterministic variable
$\mathbf{y}$	:	Random variable
$\mathbf{u}$	:	Normalized variable
$\mathbf{m}_i$	:	Mean of random variable
σ_i	:	Standard-deviation of random variable
β_t	:	Target reliability (safety) index
$\beta(\mathbf{x}, \mathbf{u})$	:	Reliability (safety) index in u-space
$d_\beta(\mathbf{x}, \mathbf{y})$	:	Reliability (safety) index in xy-space
$g_k(\mathbf{x}) \leqslant 0$	:	Geometrical, physical, functional constraints
$H(\mathbf{x}, \mathbf{u}) \leqslant 0$	:	Limit state in u-space
$G(\mathbf{x}, \mathbf{y}) \leqslant 0$	:	Limit state in xy-space

1 Introduction

In the last 40 years, there has been extensive research focused on structural optimization with dynamic constraint because the response of a structure to dynamic excitation depends, to a large extent, on the first few natural frequencies of the structure. Excessive vibration occurs when the frequency of the dynamic excitation is close to one of the natural frequencies of the structure. The optimum design of structures with frequency constraints is of great importance, particularly in the aeronautical and automotive industries. In designing most structures, it is often necessary to restrict the fundamental frequency or several of the lower frequencies of the structure to a prescribed range in order to avoid severe vibration. However, the designer reduces the structural cost without taking into account uncertainties concerning material properties, geometric dimension and loading. This way the resulting optimal design may represent a lower level of reliability and thus a higher risk of failure. Since structural problems are non-deterministic, it is clear that the introduction of the reliability concept plays an important role in the structural optimization field. Deterministic design optimization enhanced by reliability performances and formulated within the probabilistic framework is called Reliability-Based Design Optimization (RBDO). The RBDO problem is often formulated as a minimization of the initial structural cost under constraints imposed on the values of elemental reliability indices corresponding to various limit states (Madsen and Friis Hansen [13]; Kleiber et al. [11]). The objective of the RBDO model is to design structures which should be both economic and reliable where the solution reduces the structural weight in uncritical regions. It does not only provide an improved design but also a higher level of confidence in the design. The classical approach (Feng and Moses [3]) can be carried out in two separate spaces: the physical space and the normalized space. Since very many repeated searches are needed in the above two spaces, the computational time

[a] Corresponding author:
mgk@scs-net.org; amohsine@gmail.com;
amaklouf@insa-rouen.fr; aelhami@insa-rouen.fr

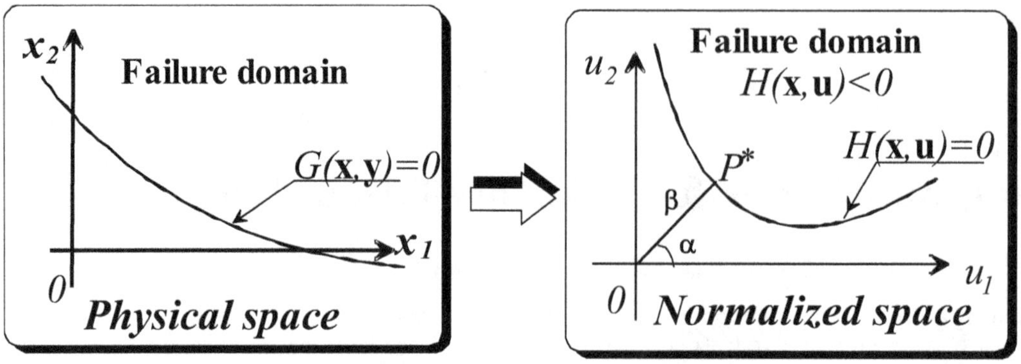

Fig. 1. The transformation between the physical space and normalized one.

for such an optimization is a big problem. To overcome these difficulties, two points of view have been considered. From **reliability** view point, RBDO involves the evaluation of probabilistic constraints, which can be executed in two different ways: either using the Reliability Index Approach (RIA), or the Performance Measure Approach (PMA) (Tu et al. [20]; Youn et al. [22]). Recently, the enhanced hybrid mean value (HMV+) method is proposed by Youn et al. [23], to improve numerical stability and efficiency in the Most Probable Point (MPP) search. The major difficulty lies in the evaluation of the probabilistic constraints, which is prohibitively expensive and even diverges with many applications. However, from **optimization** view point, an efficient method called the Hybrid Method (HM) has been elaborated by Kharmanda et al. [6] where the optimization process is carried out in a Hybrid Design Space (HDS). This method has been shown to verify the optimality conditions relative to the classical RBDO method. The advantage of the hybrid method allows us to satisfy a required reliability level for different cases (static, dynamic, ...), but the vector of variables here contains both deterministic and random variables. The hybrid RBDO problem is thus more complex than that of deterministic design. The major difficulty lies in the evaluation of the structural reliability, which is carried out by a special optimization procedure. Next, an Optimum Safety Factor (OSF) methodology has been proposed by Kharmanda et al. (2004) to simplify the optimization problem (reduction of number of variables) and aims to find at least a local optimum solution because it is based on the optimality condition. However, this method cannot be used in some dynamic cases. For example, in the freely vibrating structures, if the failure interval [*fa,fb*] is given, we cannot determine the reliability-based optimum solution using the optimum safety factor method. So Mohsine et al. [15, 16] proposed to develop the hybrid method with object of improving the optimum value of the objective function. The proposed method is called the Improved Hybrid Method (IHM). This method has been also shown to verify the optimality conditions relative to the classical RBDO method. We note that the OSF method can provide a big part of the RBDO problem analytically. So it reduces efficiently the computing time of the problem relative to the hybrid and the improved hybrid methods. In this paper, a new methodology called Safest Point method (SP) is presented as a conjoint to the OSF method in order to solve the freely vibrating structures.

2 Reliability analysis

Structural reliability analysis is a tool that assists the design engineer to take into account all possible uncertainties during the design and construction phases and the lifetime of a structure in order to estimate the probabilities of failure P_f. The vector of deterministic variables $\mathbf{x}$ to be used in the system design and optimization, the uncertainties are modeled by a vector of stochastic physical variables affecting the failure scenario. The knowledge of these variables is not, at best, more than statistical information and we admit a representation in the form of random variables. For a given design rule, the basic random variables are defined by their joint probability distribution associated with some expected parameters; the vector of random variables is denoted herein $\mathbf{Y}$ whose realizations are written $\mathbf{y}$. The safety is the state in which the structure is able to fulfill all the functioning requirements (e.g. strength and serviceability) for which it is designed. To evaluate the failure probability with respect to a chosen failure scenario, a limit state function $G(\mathbf{x}, \mathbf{y})$ is defined by the condition of good functioning of the structure. In Figure 1, the limit between the state of failure $G(\mathbf{x}, \mathbf{y}) < 0$ and the state of safety $G(\mathbf{x}, \mathbf{y}) > 0$ is known as the limit state (failure) surface $G(\mathbf{x}, \mathbf{y}) = 0$. The failure probability is then calculated by:

$$P_f = P_r \left[G(\mathbf{x}, \mathbf{y}) \leqslant 0 \right] = \int_{G(\mathbf{x},\mathbf{y}) \leqslant 0} f_{\mathbf{Y}}(\mathbf{y}) \, dy_1 ... dy_n. \tag{1}$$

This multi-dimensional integral becomes very expensive to compute when the dimension of $\mathbf{X}$ is large. For these reasons, the *First and the Second Order Reliability Methods FORM/SORM* (Ditlevsen and Madsen [2]) have been developed, these methods are based on calculation of most probable point. A convenient way to represent reliability is by a reliability index, β which can be defined as:

$$\beta = -\Phi^{-1}(P_f) \tag{2}$$

where Φ is the standard Gaussian cumulated function. In Figure 1b, the Most Probable Point (MPP) of failure can be found in a standard normalized space, $\mathbf{U}$, for a single hard or failure driven constraint $G(\mathbf{x}, \mathbf{y})$. The components of $\mathbf{U}$ are normally distributed with zero means and unit variance and are statistically independent. Any set of continuous random variables, $\mathbf{X}$, can be transformed to $\mathbf{U}$ using an one-to-one transformation i.e. $\mathbf{u} = T(\mathbf{x}, \mathbf{y})$, where $\mathbf{Y}$ are distribution parameters of $\mathbf{X}$. Many such transformations are possible, but the one typically used is called the *Rosenblatt* transformation (1976). The MPP, $\mathbf{u}^*$, lies on the failure surface or the limit state surface $G(\mathbf{x},\mathbf{y}) = H(\mathbf{x},\mathbf{u}) = 0$, and its location is the closest point on the limit state surface to the origin in U-space. The MPP can be found by solving the following constrained optimization.

$$\begin{aligned} &\beta = \min_{\mathbf{u}} \left(\mathbf{u}^T\mathbf{u} = \min \sqrt{\sum_1^m u_j^2} \right) \\ &\textit{subject to} \quad H(\mathbf{x},\mathbf{u}) \leqslant 0 \end{aligned} \tag{3}$$

where the standard space, the limit state function takes the form $H(\mathbf{x},\mathbf{u}) = G(\mathbf{x},\mathbf{y})$.

For more details, the interested reader can the different works of Hasofer and Lind [5], Liu and Der Kiureghian [12] and Ditlevsen and Madsen [2].

3 Deterministic design optimization

In Deterministic Design Optimization (DDO), the system safety may be taken into account by assigning safety factors to certain structural parameters. Using these safety factors, the optimization problem which is carried out in the physical space (Fig. 1a), consists in minimizing an objective function (cost, volume of material, ...) subject to geometrical, physical or functional constraints in the form

$$\begin{aligned} &\min_{\mathbf{x}} \ : f(\mathbf{x}) \\ &\text{subject to: } g_k(\mathbf{x}) \leqslant 0, k = 1, ..., K \end{aligned} \tag{4}$$

where $\mathbf{x}$ designates the vector of deterministic design variables. The values of the proposed safety factors principally depend on the engineering experience, but, when designing a new structure, we cannot pre-determine the real critical points, and the choice of these coefficients may therefore be wrong. Suitable geometry, material properties and loads are assumed, and an analysis is then performed to provide a detailed behavior of the structure. However, changes of the loads, variability of material properties, and uncertainties regarding the analytical models all contribute to the probability that the structure does not perform as intended. To address this concern, analysis methods have been developed to deal with the statistical nature of the input information. As structures are becoming still more complex (e.g. space shuttle engine components, space structures, advanced tactical fighters, etc.) and performance requirements are becoming more ambitious, the need for analyzing the influence of uncertainties and computation of reliability has been growing.

Over the last ten years there has been an increasing trend in analyzing structures using probabilistic information on loads, geometry, material properties, and boundary conditions. In order to evaluate the structural safety level (see problem (3)), a reliability analysis must be carried out without taking into account the safety factor from problem (4). After having followed the Deterministic Design Optimization (DDO) procedure by a reliability analysis, we can distinguish between two cases:

Case 1. High reliability level: when choosing high values of safety factors for certain parameters, the structural cost (or weight) will be significantly increased because the reliability level becomes much higher than the required level for the structure. So, the design is safe but very expensive.

Case 2. Low reliability level: when choosing small values of safety factors or bad distribution of these factors, the structural reliability level may be too low to be appropriate. For example, Grandhi and Wang [4] found that the resulting reliability index of the optimum deterministic design of a gas turbine blade is $\beta = 0.0053$ under some uncertainties. This result indicated that the reliability at the deterministic optimum is quite low and needs to be improved by probabilistic design.

For both cases, we can find that there is a strong need to integrate the reliability analysis in the optimization process in order to control the reliability level and to minimize the structural cost or weight in the non-critical regions of the structure. In the next section, we show how this can be performed efficiently. The integration of reliability analysis into engineering design optimization is termed Reliability-Based Design Optimization (RBDO). Numerical applications in Section 5, shows the advantage of the RBDO procedure relative to DDO one and the efficiency of the RBDO methods for different analysis cases (static, modal, transient).

4 Reliability-based design optimization

In the literature, many studies are reported in the field of RBDO. To the best of our knowledge, Stevenson [19] was the first who introduced reliability analysis in connection with design optimization. On the one hand, structural reliability research concentrates on probabilistic descriptions of phenomena and application to code oriented safety design. On the other hand, optimization research works toward efficient algorithms for locating optima particularly in large-scale systems using prescribed deterministic constraints. In the design of structural members, it is possible to establish the connection between optimization and reliability. This approach can be unified in the sense that all the uncertain quantities can be modeled as random variables. Hence, a lot of numerical computation is required in the space of random variables in order to evaluate the system reliability. Furthermore, the optimization process itself is executed in the space of design variables that are deterministic. Consequently, in order to search for an optimum structure, the design variables are repeatedly changed, and each set of the design variables corresponds to a new random variable space which then needs to be manipulated to evaluate the structural reliability at that point. Since too many repeated searches are needed in the above two spaces, the computational time for such an optimization is a great problem.

Feng and Moses [3, 14] attempted to unify these efforts. In this case, optimization procedures should explicitly consider safety either directly in its cost function or as one of its primary constraints. Next, the use of approximations in structural reliability and optimization in order to reduce the computer time was presented in several research papers. In the works of Chandu and Grandhi (1995) and Grandhi and Wang [4], an efficient reliability analysis based on approximating the limit state functions using two-point adaptive nonlinear approximation, has been developed to reduce the computational time. On the other hand, Madsen and Friis Hansen [13] integrated the expected failure cost in the objective function in order to compare it with a classical reliability-based optimization procedure, but this algorithm required a high number of iterations to converge. The computational cost of this combined approach is higher than that of the nested RBDO model, i.e., it requires about 50% more computational effort to converge as compared with the classical RBDO approach. Der Kiureghian and Polak [1] and Tu et al. [20] proposed several forms in order to satisfy the constraints and simplify the problem.

It is clear that efforts were directed towards the development of efficient techniques and general purpose programs to perform the reliability analysis. These programs and procedures compute the reliability index of a structure for a defined failure mode, but do not provide an optimum set of design parameters for improving the reliability of a structure for defined random information. Since the reliability index is computed iteratively, an enormous amount of computer time is involved in the whole design process. A Hybrid Method (HM) based on simultaneous solution of the reliability and the optimization problem has successfully reduced the computational time problem (Kharmanda et al. [6]). The HM allows us to satisfy a required reliability level, can be used in several cases (static, dynamic...), may provide local optimal solutions and has a big number of optimization variables (deterministic and random). An Optimum Safety Factor (OSF) method has been elaborated (Kharmanda et al. 2004) to reduce the problem scale and to improve the optimum value of the objective function. The OSF method cannot be used in special case of dynamic studies. Therefore, an Improved Hybrid Method (IHM) has been developed by Mohsine et al. [16] which can improve the optimum value of the objective function relative to HM. Recently; we developed a very simple method, called Safest Point (SP) method which is conjoint to the OSF one to solve the special dynamic cases.

4.1 Classical method (CM)

Traditionally, for the reliability-based optimization procedure we use two spaces: the physical space and the normalized space (Fig. 1). Therefore, the reliability-based optimization is performed by nesting the two following problems:

1 *Optimization problem:*

$$\begin{aligned} &\min \quad : f(\mathbf{x}) \\ &\text{subject to} \;:\; g_k(\mathbf{x}) \leqslant 0 \text{ and } \beta(\mathbf{x},\mathbf{u}) \geqslant \beta_t \end{aligned} \tag{5}$$

where $f(\mathbf{x})$ is the objective function, $g_k(\mathbf{x}) \leqslant 0$ are the associated constraints, $\beta(\mathbf{x},\mathbf{u})$ is the reliability index of the structure, and β_t is the target reliability.

2 *Reliability analysis:*
The reliability index $\beta(\mathbf{x},\mathbf{u})$ is determined by solving the minimization problem:

$$\begin{aligned} &\beta = \min \ \text{dis}(\mathbf{u}) = \sqrt{\sum_1^m u_j^2} \\ &\text{subject to}: \ H(\mathbf{x},\mathbf{u}) \leqslant 0 \end{aligned} \tag{6}$$

where dis($\mathbf{u}$) is the distance in the normalized random space, given by:

$$\text{dis} = \sqrt{\sum u_i^2} \tag{7}$$

and $H(\mathbf{x},\mathbf{u})$ is the performance function (or limit state function) in the normalized space, defined such that $H(\mathbf{x}, \mathbf{u}) < 0$ implies failure, see Figure 1b. In the physical space, the image of $H(\mathbf{x},\mathbf{u})$ is the limit state function $G(\mathbf{x},\mathbf{y})$, see Figure 1. In general, when a probabilistic approach is used instead of a conventional deterministic approach, some of the uncertain quantities should be modelled as random variables. A lot of numerical calculations are required in the space of random variables to evaluate the system reliability. Furthermore, the optimization process itself is executed in the space of design variables which are deterministic. Consequently, in order to search for an optimal structure, the design variables are changed repeatedly, and each set of design variables corresponds to a new random variable space which then needs to be manipulated to evaluate the structural reliability at that point. Since a very large number of repeated searches are needed in the above two spaces, the computational time for such an optimization is a big problem. To reduce the effects of this difficulty, a hybrid method (HM) based on simultaneous solution of the reliability and the optimization problem has been elaborated (Kharmanda et al. 2001 and [2]).

4.2 Hybrid method (HM)

The solution of the above nested problems leads to very large computational time, especially for large-scale structures. In order to improve the numerical performance, the hybrid approach consists in minimizing a new form of the

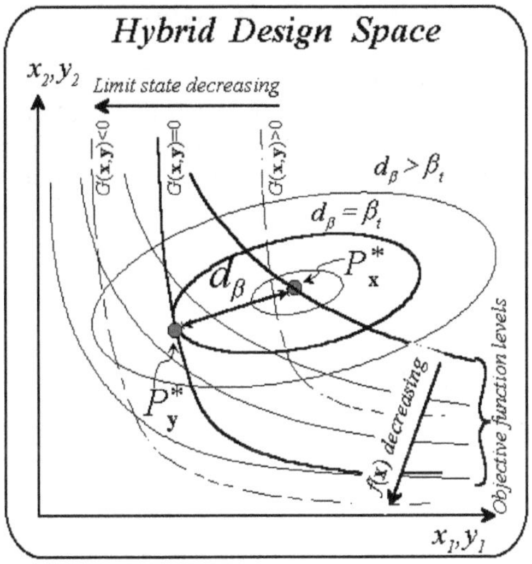

Fig. 2. Hybrid design space for normal distribution.

objective function $F(\mathbf{x},\mathbf{y})$ subject to a limit state and to deterministic as well as to reliability constraints, as:

$$\begin{array}{lll} \min\limits_{\mathbf{x},\mathbf{y}} & : F(\mathbf{x},\mathbf{y}) = f(\mathbf{x}) \cdot d_\beta(\mathbf{x},\mathbf{y}) & \\ \text{subject to} & : G(\mathbf{x},\mathbf{y}) \leqslant 0 & \\ & : g_k(\mathbf{x}) \leqslant 0 & \\ \text{and} & : d_\beta(\mathbf{x},\mathbf{y}) \geqslant \beta_t. & \end{array} \tag{8}$$

Here, $d_\beta(\mathbf{x}, \mathbf{y})$ is the distance in the hybrid space between the optimum and the design point, $d_\beta(\mathbf{x},\mathbf{y}) = \mathrm{dis}(\mathbf{u})$. The minimization of the function $F(\mathbf{x}, \mathbf{y})$ is carried out in the Hybrid Design Space (HDS) of deterministic variables $\mathbf{x}$ and random variables $\mathbf{y}$.

An example of this HDS is given in Figure 2, containing design and random variables, where the reliability levels d_β can be represented by ellipses in case of normal distribution, the objective function levels are given by solid curves and the limit state function is represented by dashed level lines except for $G(\mathbf{x},\mathbf{y}) = 0$. We can see two important points: the optimal solution P^*_x and the reliability solution P^*_y (i.e. the design point found on the curves $G(\mathbf{x},\mathbf{y}) = 0$ and $d_\beta = \beta_t$).

In the paper of Kharmanda et al. [2], we had demonstrated that the HM reduced the computational time almost 80% relative to the classical RBDO approach. Using the HM, the optimization process is carried out in the Hybrid Design Space (HDS) where all numerical information about the optimization process can be modeled. Furthermore, the classical method (CM) has weak convergence stability because it is carried out in two spaces (physical and normalized spaces).

The efficiency of the hybrid approach allows us extending for solving more complex problems. When considering the standard-deviation values as variables, we can treat a big number of optimization variables (see [15]). For an industrial application, the HM has been efficiently applied on a lorry brake system to find the reliability-based optimum design of the front body shape on dynamic behavior (see [17]).

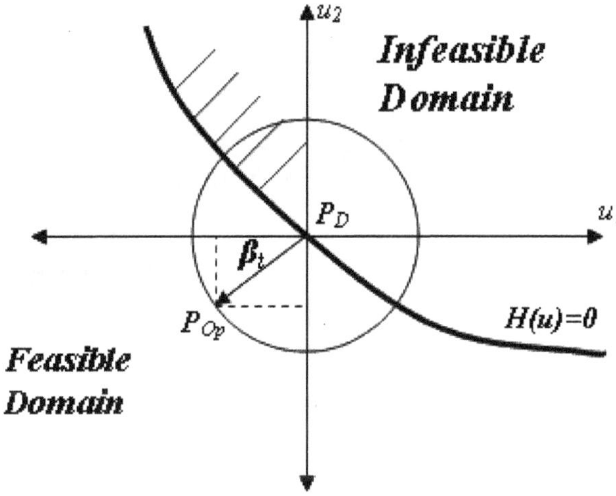

Fig. 3. Illustrated example for two-dimensional space.

4.3 Optimum safety factor method (OSF)

In fact, when using the HM, we have a complex optimization problem with many variables. Solving this problem, we get a local optimum. When changing the starting point, we may get another local optimum. This way the designer has to repeat the optimization process to get several local optima. However, to overcome these drawbacks, an optimum safety factor method has been proposed (Kharmanda et al. 2004). In general, when considering the normal distribution law, the normalized variable u_i is given by:

$$u_i = \frac{y_i - m_i}{\sigma_i}, \; i = 1, ..., n. \tag{9}$$

The standard deviation σ_i can be related to the mean value m_i by:

$$\sigma_i = \gamma_i \cdot m_i \text{ , or } \sigma_i = \gamma_i \cdot x_i, \; i = 1, ..., n. \tag{10}$$

This way we introduce the safety factors S_{f_i} corresponding to the design variables x_i. The design point can be expressed by:

$$y_i = S_{f_i} \cdot x_i \text{ , } i = 1, ..., n. \tag{11}$$

For an assumed failure scenario $G(\mathbf{y}) < 0$, the equation of the optimum safety factor for a single limit state case can be written in the following form (see the developments in Kharmanda et al. 2004):

$$S_{f_i} = 1 \pm \gamma_i \cdot \beta_t \sqrt{\frac{\left|\frac{\partial G}{\partial y_i}\right|}{\sum\limits_{i=1}^{n} \left|\frac{\partial G}{\partial y_i}\right|}} \text{ , } i = 1, ..., n. \tag{12}$$

Here, the sign $\pm$ depends on the sign of the derivative, i.e.,

$$\frac{\partial G}{\partial y_i} > 0 \Leftrightarrow S_{f_i} > 1, i = 1, ..., n \tag{13a}$$

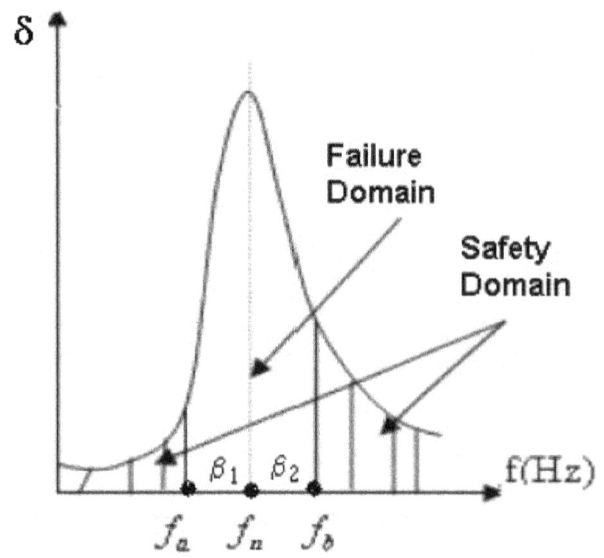

Fig. 4. The safest point at frequency *fn*.

$$\frac{\partial G}{\partial y_i} < 0 \Leftrightarrow S_{f_i} < 1, i = 1, ..., n. \tag{13b}$$

Let us consider an example of only two normalized variables u_1 and u_2 (see Fig. 3).

For an industrial application, Yang et al. [21] compared the results and efficiencies of different RBDO methods on an exhaust system. The objective was to minimize the weight of the system subject to constraints that the reliability of the resultant forces in each frequency region should be less than specified values. All in all 144 constraints were imposed, but many of them were inactive. After having tested several RBDO methods, they concluded that:

"*Kharmanda et al. (2004) also used structural safety factors, based on the sensitivity of the limit-state function, for RBDO. In addition to its simplified computational framework to completely decouple the optimization and the reliability analyses, the method has two advantages:*

1. *It incorporates the partial safety-factor concept with which most designers are familiar. And, theoretically, safety factors do not have to be tied to the individual random variables and thus the MPPs (Most Probable Points).*
2. *It produces progressively improved reliable designs in the initial steps that help designers keep track of their designs.*"

According to this experience, this method is considered as a very good active constraint strategy (for problems with many constraints). A numerical application in Section 5.2 shows the advantage of the OSF procedure relative to HM on a tri-material beam under static distributed load.

4.4 Improved hybrid method (IHM)

Using the hybrid method, we can obtain local optima and designer may then select the best optimum. In the improved hybrid method, we introduce the design point and the optimum solution in the objective function and the constraints at the design point and at the optimum solution as follows:

$$\begin{array}{ll} \min\limits_{\mathbf{x},\mathbf{y}} & : F(\mathbf{x},\mathbf{y}) = f(\mathbf{x}) \,.\, d_\beta(\mathbf{x},\mathbf{y}) .f(\mathbf{m}_y) \\ \text{subject to} & : G(\mathbf{x},\mathbf{y}) \leqslant 0 \\ & : g_k(\mathbf{x}) \leqslant 0 \\ & : g_j(\mathbf{m}_y) \leqslant 0 \\ \text{and} & : d_\beta(\mathbf{x},\mathbf{y}) \geqslant \beta_t. \end{array} \tag{14}$$

The random vector $\mathbf{x}$ has mean values $\mathbf{m}_y$ and standard-deviations $\sigma_y .f(\mathbf{m}_y)$ is the optimal objective function and $g_i(\mathbf{m}_y)$ is the constraint at which we can control the optimal configuration. The solution of this problem depends on two important points. It can be carried out simultaneously in the hybrid design space (HDS).

We consider two models: the optimal model corresponding to $\mathbf{x}$ and the design one corresponding to $\mathbf{y}$ to be first evaluated. The functions $f(\mathbf{x})$, $f(\mathbf{m}_y)$ and $d_\beta(\mathbf{x},\mathbf{y})$ are next calculated to get the value of the new objective function:

$$F(\mathbf{x},\mathbf{y}) = f(\mathbf{x}) \,.\, d_\beta(\mathbf{x},\mathbf{y}) .\, f(\mathbf{m}_y). \tag{15}$$

We estimate the different constraints to test the convergence. If the convergence is not verified, we update the vectors x and y, re-evaluate the models, all functions until the convergence (see the developments in [16]). A numerical application in Section 5.3 shows the improved optimum objective produced by the IHM relative to HM on a triangular plate structure under dynamic behavior.

4.5 Safest point method (SP)

The reliability-based optimum structure under free vibrations for a given interval of eign-frequency is found at the safest position of this interval where the safest point has the same reliability index relative to both sides of the interval. A simple method has been proposed here to meet the safest point requirements relative to a given frequency interval.

Let consider a given interval [*fa*,*fb*]. For the first shape mode, to get the reliability-based optimum solution for a given interval, we consider the equality of the reliability indices:

$$\beta_a = \beta_b \quad \text{or } \beta_1 = \beta_2 \tag{16}$$

with

$$\beta_a = \sqrt{\sum_{i=1}^{n} u_i^a} \text{ and } \beta_a = \sqrt{\sum_{i=1}^{n} u_i^b} \quad i = 1, ..., n.$$

To verify the equality (16), we propose the equality of each term. So we have:

$$u_i^a = -u_i^b, \quad i = 1, ..., n. \tag{17}$$

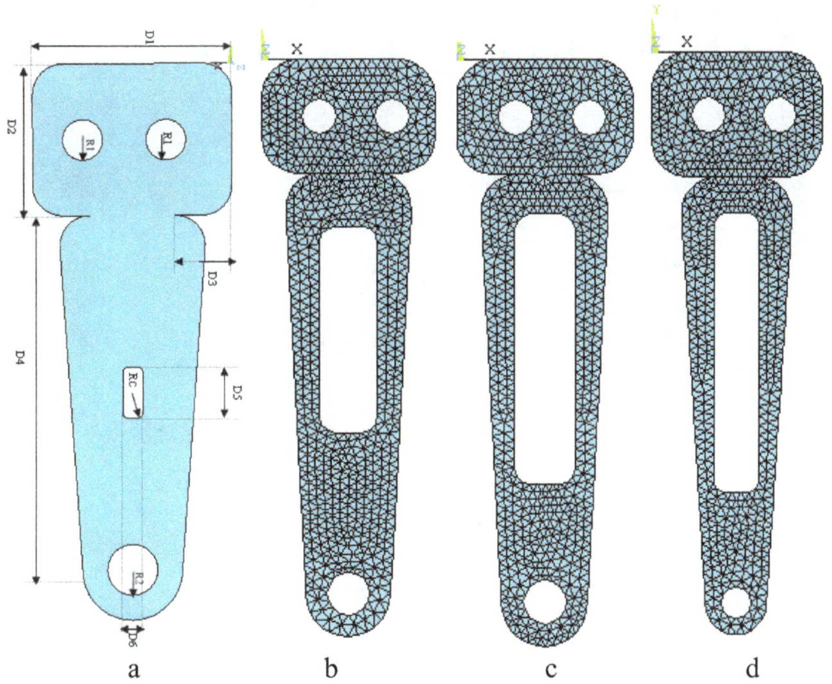

Fig. 5. (a) Geometrical models of the structure. (b), (c), (d) Different optima when safety factors 1.5, 1.2 and 1.1, respectively.

Table 1. Initial values parameter.

Parameter	E (MPa)	Density (kg/mm^3)	Poisson's ratio	$D1$ (mm)	$D2$ (mm)	$D3$ (rad)	$D4$ (rad)	$D5$ (mm)	$D6$ (mm)	Rc (rad)	$R0$ (mm)
values	2.1×10^6	78e-7	0.3	20	15	2.5	35	5	1	0.5	3

According to the normal distribution law, the normalized variable u_i is given by (17), we get:

$$\frac{y_i^a - m_i}{\sigma_i} = -\frac{y_i^b - m_i}{\sigma_i}, \quad \text{or}$$
$$\frac{y_i^a - x_i}{\sigma_i} = -\frac{y_i^b - x_i}{\sigma_i}, \; i = 1, ..., n. \tag{18}$$

To obtain equality between the reliability indices (see Eq. (16)), the mean value of variable corresponds to the structure at f_n. So the mean values of safest solution are located in the middle of the variable interval $[y_i^a, y_i^b]$ as follows:

$$m_i = x_i = \frac{y_i^a + y_i^b}{2}, i = 1, ..., n. \tag{19}$$

In a recent publication [10], we found that the safest point method is suitable for modal analysis more than the other methods that are complex to implement and to converge in this kind of study. A numerical application in Section 5.4 shows the efficiency (computing time reduction) of the SP procedure relative to HM on an aircraft wing under dynamic behavior.

5 Numerical applications

5.1 A bracket: DDO & RBDO

The objective of this example is to demonstrate the advantage of the RBDO procedure relative to the DDO one. Here, we minimize the volume of the bracket structure which is illustrated in Figure 5 subject to stress constraint and the target reliability index constraint. This structure is supported at its two upper holes of radius $R1 = 2$ mm and loaded at its lower hole of radius $R2 = 2.5$ mm by a pressure $P = 150$ MPa. The material is steel with Young's modulus $E = 210$ GPa and yield stress $\sigma_y = 235$ MPa. The target reliability index must be 3.8. Table 1 shows the different parameters of this structure.

To optimize the structure, the mean values $m_{D1}, m_{D2}, m_{D3}, m_{D5}, m_{D6}, m_{R0}$ and m_{Rc}, of the dimensions *D1, D2, D3, D5, D6, R0* and *Rc* are the design variables. The physical dimensions *D1, D2, D3, D5, D6, R0* and *Rc*, are elements of the vector of random variables and assumed to be normally distributed. The standard-deviations are given by $\sigma_{D1} = 1.5, \sigma_{D2} = 1.02, \sigma_{D3} = 0.24, \sigma_{D5} = 2.3, \sigma_{D6} = 0.13, \sigma_{R0} = 0.08$ and $\sigma_{Rc} = 0.1$ mm. During the subsequent design optimization processes, we consider all variables to be bounded by upper and lower limits.

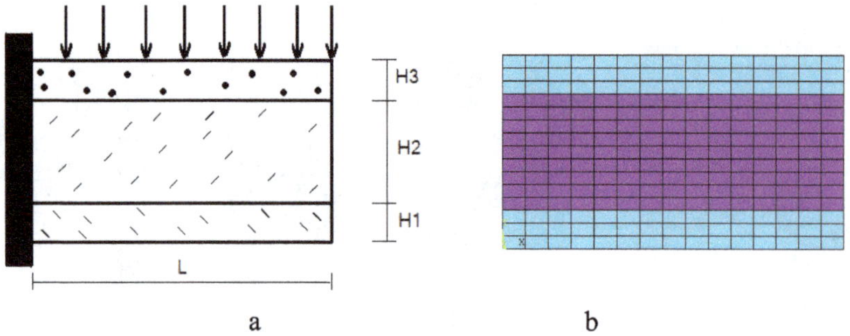

Fig. 6. Layout of tri-material cantilever beam.

5.1.1 DDO procedure

The objective is to minimize the volume subject to the design constraints and consider a safety factor that is applied to the load and based on engineering experience. The optimization problem is written as:

$$\min_{D1,...,Rc} : \text{Volume}(D1, ..., Rc)$$

$$\text{subject to} : \sigma_{\max}(D1, ..., Rc) \leqslant \sigma_y / S_f \qquad (20)$$

$$\text{and} : \sigma_{\min}(D1, ..., Rc) \geqslant -\sigma_y / S_f$$

and the corresponding reliability evaluation as:

$$\min_{u_{D1},...,u_{Rc}} : \text{dis} = \sqrt{u_{D1}^2 + ... + u_{Rc}^2}$$

$$\text{subject to} : \sigma_y - \sigma_{\max}(u_{D1}, ..., u_{Rc}, D1, ..., Rc) \leqslant 0$$

$$: -\sigma_y - \sigma_{\min}(u_{D1}, ..., u_{Rc}, D1, ..., Rc) \leqslant 0. \qquad (21)$$

In equation (20), $\sigma_{\max}$ and $\sigma_{\min}$ refer to the most critical points in terms of the maximum and minimum values of the von Mises stress, and we consider three cases for the global safety factor applied to the distributed pressure $P(S_f = 1.5, 1.2, 1.1)$ but this is not the case for equation (21).

After having optimized the structure, the resulting volumes are $V_{DDO}^{Sf=1.5} = 563.40$, $V_{DDO}^{Sf=1.2} = 502.63$ and $V_{DDO}^{Sf=1.1} = 434.75$ mm^3 and correspond to reliability indices $\beta_{DDO}^{Sf=1.5} = 5.73$, $\beta_{DDO}^{Sf=1.2} = 4.84$ and $\beta_{DDO}^{Sf=1.1} = 2.54$ respectively. This way we cannot control the required reliability level (higher or smaller than the required reliability index $\beta_t = 3.8$), however, by integration of the reliability concept into the optimization process (RBDO case), we can satisfy the reliability constraint.

5.1.2 HM procedure

It has been demonstrated that the classical method implies very high computational cost and exhibits weak convergence stability (see Kharmanda et al. [2]). So we use the Hybrid Method (HM) as a flexible RBDO method to satisfy the required reliability level (with admissible tolerances 1%). To perform the hybrid RBDO problem, we can write as follows:

$$\min_{m_{D1},...,m_{Rc},D1,...,Rc} : \text{Volume}(D1, ..., Rc) \cdot d_\beta (m_{D1}, ..., m_{Rc}, D1, ..., Rc)$$

$$\text{subject to} : \sigma_{\max}(D1, ..., Rc) \leqslant \sigma_y$$

$$: \sigma_{\min}(D1, ..., Rc) \geqslant -\sigma_y$$

$$\text{and} : d_\beta(m_{D1}, ..., m_{Rc}, D1, ..., Rc) \geqslant \beta_t. \qquad (22)$$

The main advantage of the RBDO model relative to the DDO one is that the reliability requirements can be automatically satisfied, but when using the DDO procedure, the designer has to change manually the structure parameters to satisfy the target reliability index. So we consider the hybrid RBDO as an efficient tool to control the structural reliability levels. But the DDO may lead to high or low reliability levels because it does not control the reliability.

5.2 A tri-material cantilever beam: RBDO for static analysis

The objective of the following example is to show the advantages of the Optimum Safety Factor (OSF) strategy relative to the Hybrid Method (HM) applied to a compound tri-material beam. We first show that both the OSF and hybrid methods can satisfy the required reliability levels. Next, we demonstrate that the OSF method provides the designer with the lowest structural volume for the same reliability levels relative to the HM.

The design problem under consideration pertains to a short tri-material cantilever beam of length $L = 100$ mm, height $H = 50$ mm and width $T = 20$ mm, which is loaded by a distributed pressure $q = 15$ N/mm^2. The beam structure is composed of three layers of material (Fig. 6) of different Young's moduli $E_1 = 200$ GPa, $E_2 = 100$ GPa and $E_3 = 150$ GPa, Poisson's ratios $\nu_1 = 0.3$, $\nu_2 = 0.1$ and $\nu_3 = 0.2$, and yield stresses $\sigma_1^y = 48$, $\sigma_2^y = 18$ and $\sigma_3^y = 42$ MPa. The heights of the three layers are: $H1 = 10$ mm, $H2 = 30$ mm, and $H3 = 10$ mm. To optimize the tri-material beam structure, the mean values m_{H1}, m_{H2} and m_{H3} of the heights $H1, H2$ and $H3$ are the design variables. The physical heights $H1, H2$ and $H3$ are elements of the vector of random variables. The target reliability index is taken to be: $\beta_t = 3$, and the standard-deviations are given by $\sigma_{H1} = 0.1 m_{H1}$,

Table 2. Results of DDO procedure for different values of safety factors.

Parameter	$Sf = 1.5$		$Sf = 1.2$		$Sf = 1.1$	
	Design point	Optimum solution	Design point	Optimum solution	Design point	Optimum solution
$D1$ (mm)	15.215	14.552	15.119	14.861	15.160	14.910
$D2$ (mm)	10.125	9.3327	10.021	9.3502	10.008	10.208
$D3$ (mm)	2.3730	1.8710	2.3664	1.8965	2.3775	2.4092
$D5$ (mm)	21.004	17.153	22.029	22.155	21.994	22.996
$D6$ (mm)	1.8632	2.4532	1.8476	2.5404	1.8546	1.8936
Rc (rad)	1.0688	1.4214	1.0303	1.3042	1.0118	0.84577
$R0$ (mm)	1.1603	1.6066	1.1744	1.5083	1.1752	1.3853
$Stress$ (MPa)	235.05	156.73	235.10	195.89	235	213.75
Vol (mm^3)	563.40		502.63		434.75	
β	5.73		4.84		2.54	

Table 3. Result of reliability-based design optimization.

Parameter	Design point	Optimum solution
$D1$ (mm)	15.994	15.129
$D2$ (mm)	10.515	11.857
$D3$ (mm)	2.4546	2.0451
$D5$ (mm)	21.543	21.167
$D6$ (mm)	1.9851	1.8881
Rc (rad)	0.84454	0.90157
$R0$ (mm)	1.2200	1.6739
$Stress$ (MPa)	235.07	151.75
$Volume$ (mm^3)	666.02	
β	3.8	

$\sigma_{H2} = 0.1 m_{H2}$ and $\sigma_{H3} = 0.1 m_{H3}$. During the subsequent design optimization processes, we consider all variables to be bounded by upper and lower limits.

5.2.1 HM procedure

In the HM, we minimize the product of the volume and the reliability index subject to the limit state functions and the required reliability level. The RBDO problem is written as

$$\begin{aligned} \min_{m_{H1}, m_{H2}, m_{H3}, H1, H2, H3} \quad & \text{Volume}(H1, H2, H3) \cdot d_\beta \\ & (m_{H1}, m_{H2}, m_{H3}, H1, H2, H3) \\ \text{subject to} \quad & \sigma_1^{\max}(H1, H2, H3) \leqslant \sigma_1^y \\ & \sigma_2^{\max}(H1, H2, H3) \leqslant \sigma_2^y \\ & \sigma_3^{\max}(H1, H2, H3) \leqslant \sigma_3^y \\ & d_\beta(m_{H1}, m_{H2}, m_{H3}, \\ & H1, H2, H3) \geqslant \beta_t. \end{aligned} \tag{23}$$

This optimization process is carried out in a hybrid design space. The resulting optimal values of the reliability index are found to be: $d_\beta = 3.0001 \approx \beta_t$ (i.e., 0.03% higher than the target reliability index). The resulting optimum volume is determined as: $Vol^* = 41\,782$ mm^3. The experience of the designer on finite element software plays a very important role in improving the objective function and controlling the convergence. Although the method yields results that satisfy the required reliability level within admissible tolerances, the problem is a complex optimization problem and needs a large number of iterations to converge and improve the value of objective function.

Table 4. Safety factor values.

Var	$\partial\sigma_1/\partial y_i$	$\partial\sigma_2/\partial y_i$	$\partial\sigma_3/\partial y_i$	$S_f^{\beta=3}$
$H1$	−1.052	−0.2160	−0.7318	0.8255
$H2$	−0.7452	−0.2041	−0.6119	0.84582
$H3$	−0.8432	−0.6796	−0.8271	0.81084

5.2.2 OSF procedure

We minimize the volume subject to the design constraints in order to find the coordinates of design point (or MPP). This way the optimization problem is simply written as:

$$\begin{aligned} \min_{H1, H2, H3} \quad & \text{Volume}(H1, H2, H3) \\ \text{subject to} \quad & \sigma_1(H1, H2, H3) \leqslant \sigma_1^y \\ & \sigma_2(H1, H2, H3) \leqslant \sigma_2^y \\ & \sigma_3(H1, H2, H3) \leqslant \sigma_3^y. \end{aligned} \tag{24}$$

The design point is found to correspond to the maximum von Mises stresses $\sigma_1^{\max} = 47.335$ MPa, $\sigma_2^{\max} = 17.177$ MPa and $\sigma_3^{\max} = 41.999$ MPa, that are almost equivalent to the yield stresses σ_1^y, σ_2^y and σ_3^y.

Next, we compute the optimum safety factors using (12) for normal distributions. In this example, the number of the deterministic variables is equal to that of the random ones. During the optimization process, we obtain the sensitivity values of the limit state with respect to all variables. So there is no need for additional computational cost. Table 1 shows the results leading to the values of the safety factors, namely the sensitivity results for the different limit state functions. The optimum solution can be calculated by reevaluating of the model. Table 5 presents the different results of the HM and OSF procedure. In order to demonstrate the efficiency of the

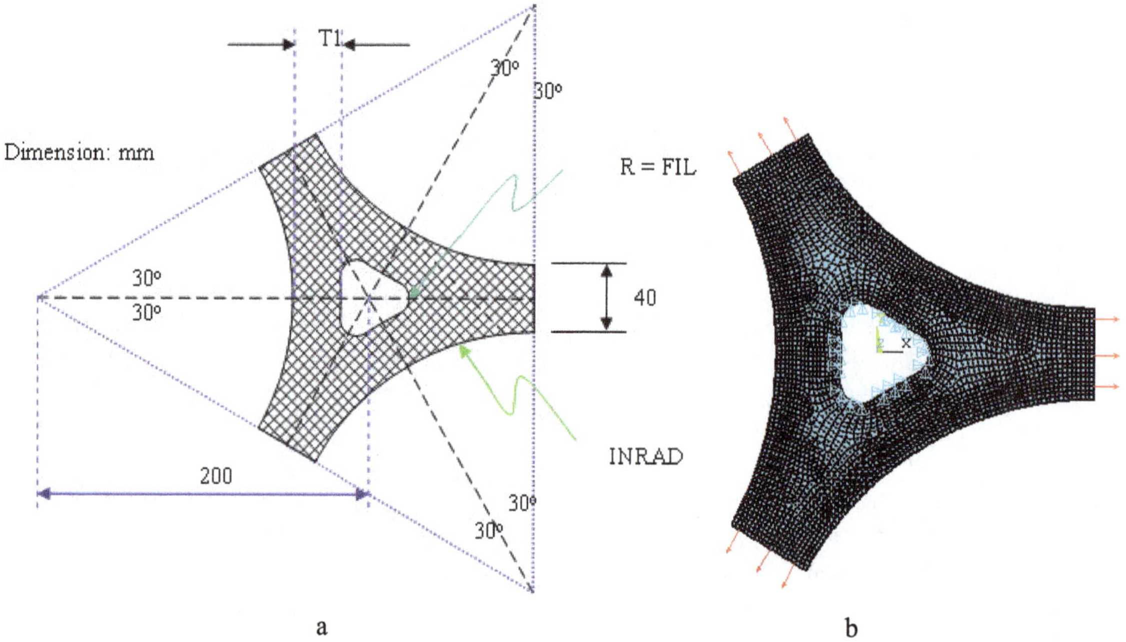

Fig. 7. (a) Geometry model and (b) Boudary conditions of triangular plate structure.

Table 5. Results of the OSF and HM.

Var	*HM*	*OSF*
m_{H1}	8.3992	9.3974
m_{H2}	24.753	21.851
m_{H3}	8.6298	9.1176
$\sigma_1^{\max}$	33.096	34.347
$\sigma_2^{\max}$	12.059	12.134
$\sigma_3^{\max}$	29.718	30.915
$H1$	7.4942	7.7576
$H2$	18.726	18.482
$H3$	7.4368	7.3930
σ_1^y	47.488	47.335
σ_2^y	17.075	17.177
σ_3^y	41.997	41.999
β	3.0001	3.0000
Vol	41 782	40 366

OSF method relative to the HM, we can see that the resulting design obtained by the OSF method is better than that obtained by the HM where the objective is to provide the best compromise between cost and safety. The OSF methodology satisfies the required reliability level $\beta_t = 3$ and gives a smaller structural volume than the HM for the reliability level. In order to improve the resulting structure by the HM, the designer can obtain several local optima and then select the best solution. The resulting optimum volume obtained by OSF ($V_{OSF} = 40\,366$ mm^3) is smaller than the resulting volume determined by the HM by 3.39%.

5.3 Triangular plate: RBDO for transient analysis

A triangular plate structure being illustrated in Figure 7a, is submitted to pressure 200 Mpa. The Young's modulus is: 207 GPa and Poisson's ratio is: 0.3. The thickness of this plate is: $R0 = 10$ mm and $T1 = 30$ mm. The radius of fillet is: $FIL = 10$ rad. The yield stress is: $\sigma_y = 235$ Mpa.

The objective of this study is to show the efficiency and the robustness of the IHM relative to the HM. Here, we can regroup $T1, R0$ and FIL in a random vector $\mathbf{y}$ but to optimize the design, the means m_{T1}, m_{R0} and m_{FIL} are regrouped in a deterministic vector $\mathbf{x}$, and their fix standard-deviation equals to $0.1m_x$ according the normal distribution. The optimization problem is to find the optimum value of the structural volume subject the maximum stress (transient response).

5.3.1 HM procedure

We seek to minimize the hybrid form of the objective function subject to the stress constraint and the reliability one as follows:

$$\begin{aligned}
&\min_{T1,FIL,R0,\,m_{T1},m_{FIL},m_{R0}} && : \text{Volume}(T1,FIL,R0).\ d_\beta\ (T1,FIL,R0,\ m_{T1},m_{FIL},m_{R0})\\
&\text{subject to} && : \sigma_{\max}(T1,FIL,R0\,) - \sigma_y \leqslant 0\\
&\text{and} && : d_\beta(T1,FIL,R0\,,m_{T1},\ m_{FIL},m_{R0}) \geqslant \beta_t. \qquad (25)
\end{aligned}$$

5.3.2 IHM procedure

We seek to minimize the improved hybrid form of the objective function subject to the stress constraint and the reliability one. This problem can be expressed as:

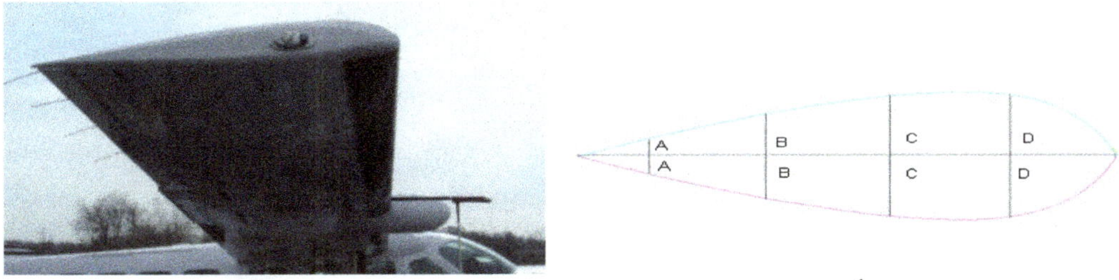

a b

Fig. 8. Aircraft wing.

Table 6. Result of HM and IHM.

Parameter	HM	IHM
$T1$	24.985	24.058
FIL	8.5833	9.1013
$R0$	7.3251	9.8216
σ_y	234.92	235.04
m_{T1}	29.678	26.092
m_{FIL}	10.600	9.1062
m_{R0}	7.6991	6.0869
σ	204.51366	216.420899
Volume	105874.1	78250.5
β	3.8096	3.8

$$\begin{aligned}
&\min_{T1,FIL,R0,m_{T1},m_{FIL},m_{R0}} && : \mathrm{Vol}\,(T1, FIL, R0, m_{T1}, m_{FIL}, m_{R0}).\ d_\beta(T1, FIL, R0, m_{T1}, m_{FIL}, m_{R0}).\ Vol(m_{T1}, m_{FIL}, m_{R0})\\
&\text{subject to} && : \sigma_{\max}(T1, FIL, R0) - \sigma_y \leqslant 0\\
&\text{and} && : d_\beta(T1, FIL, R0, m_{T1}, m_{FIL}, m_{R0}) \geqslant \beta_t \qquad (26)
\end{aligned}$$

Table 6 shows the results of both HM and IHM. Both two methods satisfy the required reliability level $\beta_t \approx 3.8$. However, the optimal volume obtained by the IHM is less than that obtained by the HM. This way the volume value reduction is almost 26% that leads to economic structures. This study shows the efficiency and the robustness of the improved hybrid method.

5.4 Aircraft wing: RBDO for modal analysis

The wing is uniform along its length with cross sectional area as illustrated in Figure 8a. It is firmly attached to the body of the airplane at one end. The chord of the airfoil has dimensions and orientation as shown in Figure 8b. The wing is made of low density **polyethylene** with a Young's modulus of 38e3 psi, Poisson's ration of 0.3, and a density of 8.3E-5 lbf-sec2/in4. Assume the side of the wing connected to the plane is completely fixed in all degrees of freedom. The wing is solid and material properties are constant and isotropic.

The objective is to find the eigen-frequency for a given interval [16,18], that is located on the safest position of this interval. So $fa = 16$ Hz, $fb = 18$ Hz and $fn = ?$ Hz, where fn must verify the equality of reliability indices: $\beta_a = \beta_b$. We can deal with three models: The first structure must be optimized subject to the first frequency value of the given fa, the second one must be optimized at the end frequency value of the interval fb, and the third structure must be optimized subject to a frequency value fn that verifies the equality of reliability indices relative to both sides of the given interval (see Fig. 8) [10].

5.4.1 HM procedure

We minimize the composite form of the objective function subject to the different frequencies constraint and the reliability one as follows:

$$\begin{aligned}
&\min_{A_1,\ldots,m_A,\ldots,A_2,\ldots} && : \mathrm{Vol}_n(m_A, m_B, m_C, m_D).\ d_{\beta 1}(A_1, B_1, C_1, D_1, m_A, m_B, m_C, m_D).\ d_{\beta 2}(A_2, B_2, C_2, D_2, m_A, m_B, m_C, m_D)\\
&\text{subject to} && : f^1_{\max}(A_1, B_1, C_1, D_1) - f_a \leqslant 0,\\
& && f^2_{\max}(A_2, B_2, C_2, D_2) - f_a \leqslant 0\\
&\text{and} && : d_{\beta 1}(A_1, B_1, C_1, D_1, m_A, m_B, m_C, m_D) - d_{\beta 2}(A_2, B_2, C_2, D_2, m_A, m_B, m_C, m_D) \leqslant 0. \qquad (27)
\end{aligned}$$

5.4.2 SP procedure

We have two simple optimization problems:

- The first is to minimize the objective function of the first model subject to the frequency fa constraint as follows:

$$\begin{aligned}
&\min_{A_1,\ldots} && : \mathrm{Vol}_a\,(A_1, B_1, C_1, D_1)\\
&\text{subject to} && : f^1_{\max}(A_1, B_1, C_1, D_1) - f_a \leqslant 0. \qquad (28a)
\end{aligned}$$

- The second is to minimize the objective function of the second model subject to the frequency fb constraint as follows:

$$\begin{aligned}
&\min_{A_2,\ldots} && : \mathrm{Vol}_b\,(A_2, B_2, C_2, D_2,)\\
&\text{subject to} && : f^2_{\max}(A_2, B_2, C_2, D_2) - f_a \leqslant 0 \qquad (28b)
\end{aligned}$$

Table 7. Results for the aircraft wing.

	Variables	Initial	Optimum design with SP	Optimum design with HM
FN	*A*	0.13295	0.123005	0.13526
	B	0.24112	0.228385	0.26225
	C	0.30834	0.299635	0.29445
	D	0.26316	0.2266	0.20797
FA	*A*1	0.09295	0.11300	0.13526
	*B*1	0.16112	0.21557	0.20527
	*C*1	0.22834	0.29071	0.28851
	*D*1	0.18316	0.18443	0.21867
FB	*A*2	0.17295	0.13301	0.13999
	*B*2	0.32112	0.2412	0.25565
	*C*2	0.38834	0.30856	0.30846
	*D*2	0.34316	0.2689	0.26600
	FA	14.056	16.000	16.025
	FB	21.744	17.913	17.960
	FN	17.800	16.85	16.94
	$DIF = \beta 1\text{-}\beta 2$	$8.88E-16$	$-1.33E-15$	$0.60E-03$
	volume	6.17144217	5.72973783	5.84788645
	Time(S)	–	35.21	252.6

and next we compute the coordinates of the third model which corresponds to fn according to equation (19). Table 7 shows the results of the SP method and presents the reliability-based optimum point for a given interval [16,18]. The value of fn is 16.85 Hz, presents the equality of reliability indices. The SP method reduces the computing time by 86% relative to the hybrid method. So the advantage of the SP method is simple to be implemented on the machine and to define the eigen-frequency of a given interval. It can be also a conjoint of the OSF method.

6 Conclusion

A RBDO solution that reduces the structural weight in uncritical regions both provides an improved design and a higher level of confidence in the design. The classical RBDO approach can be carried out in two separate spaces: the physical space and the normalized space. Since very many repeated searches are needed in the above two spaces, the computational time for such an optimization is a big problem. The structural engineers do not consider the RBDO as a practical tool for design optimization. Fortunately, an efficient method called the Hybrid Method (HM) has been elaborated where the optimization process is carried out in a Hybrid Design Space (HDS). The different applications present the flexible uses of the HM that allows us to satisfy a required reliability level for different cases (static, dynamic...). However, the vector of variables here contains both deterministic and random variables. The RBDO problem by HM is thus more complex than that of deterministic design. The major difficulty lies in the evaluation of the structural reliability, which is carried out by a special optimization procedure. Next, an OSF (Optimum Safety Factor) methodology has been proposed to simplify the optimization problem (reduction of number of variables) and aims to find at least a local optimum solution because it is based on the optimality condition, which is shown in the second example. However, the OSF method cannot be used in some dynamic cases like freely vibrating structures. For example, if the failure interval [*fa*,*fb*] is given, we cannot determine the reliability-based optimum solution using the optimum safety factor method. Furthermore, we cannot guarantee even a local optimum when using HM. So there is strong motivation to improve the optimum value of the HM. An Improved Hybrid Method (IHM) has been proposed with object of improving the optimum value of the objective function. The third example shows the power of the IHM to provide the designer with a better optimum value of the objective function relative to the HM. The procedure of IHM is still complex to be implemented on the machine and the OSF method can provide a big part of the RBDO problem analytically (high reduction of the computing time). So since the OSF procedure cannot be used for some special dynamic cases, we here propose a new methodology called Safest Point method (SP) as a conjoint to the OSF method in order to solve the freely vibrating structures as it shown in the last example.

References

1. A. Der Kiureghian, E. Polak (1998), Reliability-based optimal structural design: A Decoupled Approach, in: Andrzej S. Nowak (ed.), *Reliability and Optimization of Structural Systems* 197–205
2. O. Ditlevsen, H. Madsen (1996), *Structural Reliability Methods*. John Wiley & Sons
3. Y.S. Feng, F. Moses (1986), A method of structural optimization based on structural system reliability. J. Struct. Mech. **14**, 437–453
4. R.V. Grandhi, L. Wang (1998), Reliability-based structural optimization using improved two-point adaptive

nonlinear approximations. Finite Elements in Analysis and Design **29**, 35–48
5. A.M. Hasofer, N.C. Lind (1974), *An exact and invariant first order reliability format.* J. Eng. Mech. **100**, 111–121
6. G. Kharmanda, A. Mohamed, M. Lemaire (2002), *Efficient reliability-based design optimization using hybrid space with application to finite element analysis.* Structural and Multidisciplinary Optimization **24**, 233–245
7. G. Kharmanda, A. Mohsine, A. El-Hami (2003), Efficient Reliability-Based Design Optimization for Dynamic Structures, In: *Proc. Fifth World Congress of Structural and Multidisciplinary Optimization*, WCSMO-5, eds. C. Cinquini, M. Rovati, P. Venini, R. Nascimbene, May 19–23 (2003), Lido di Jesolo-Venice, Italy. University of Pavia, Italy, 6 pp. (2003)
8. G. Kharmanda, A. El-Hami, N. Olhoff (2004a) Global Reliability–Based Design Optimization, in: *Frontiers on Global Optimization*, C.A. Floudas, ed., 255 (20), Kluwer Academic Publishers
9. G. Kharmanda, N. Olhoff, A. El-Hami (2004b), Optimum safety factor approach for reliability-based design optimization with extension to multiple limit state case. Structural and Multidisciplinary Optimization, **26**
10. G. Kharmanda, A. Altonji, A. Elhami (2006), Safest point method for reliability-based design optimization of freely vibrating structures, CIFMA01-IFCAM01, Aleppo, Syria 02–04 May
11. M. Kleiber, A. Siemaszko, R. Stocki (1999), Interactive stability-oriented reliability-based design optimization. Computer Methods in Applied Mechanics and Engineering **168**, 243–253
12. P.-L. Liu, A. Der Kiureghian (1991), Optimization algorithms for structural reliability. Structural Safety **9**, 161–177
13. H.O. Madsen, P. Friis Hansen (1991), Comparison of some algorithms for reliability-based structural optimization and sensitivity analysis, in: Brebbia, C.A. and Orszag, S.A. (eds.): *Reliability and Optimization of Structural Systems*, pp. 443–451 (Springer-Verlag, Germany)
14. F. Moses (1977), Structural system reliability and optimization. Comput. Struct. **7**, 283–290
15. A. Mohsine, G. Kharmanda, A. El-Hami (2005), Reliability-based design optimization study using normal and lognormal distributions with applications to dynamic structures, in: *The Fifth International Conference on Structural Safety and Reliability ICOSSAR05*, June 19–22 (2005), Rome (Italy)
16. A. Mohsine, G. Kharmanda, A. El-Hami (2006), Improved hybrid method as a robust tool for reliability-based design optimization. Struct. Multidiscipl. Optim. **32**, 203–213 (2006)
17. A. Mohsine (2006), Contribution à l'optimisation fiabiliste en dynamique des structures mécaniques, Ph.D. thesis, INSA de Rouen, France (French version)
18. E. Rosenblatt (1976), Optimum Design for Infrequent Disturbances. J. Struct. Div. ASCE **102**, ST9, 1807–1825
19. J.D. Stevenson (1967), Reliability analysis and optimum design of structural systems with applications to rigid frames. Division of Solid Mechanics and Structures, 14, Case Western Reserve University, Cleveland, Ohio
20. J. Tu, K.K. Choi, Y.H. Park (1999), A new study on reliability-based design optimization. Journal of Mechanical Design, ASME **121**, 557–564
21. R.J. Yang, C. Chuang, L. Gu, G. Li (2005), Experience with approximate reliability-based optimization methods II: an exhaust system problem. Struct. Multidiscipl. Optim. **29**, 488–497
22. B.D. Youn, K.K. Choi, Y.H. Park (2003), Hybrid analysis method for reliability-based design optimization. Journal of Mechanical Design **125**, 221–232
23. B.D. Youn, K.K. Choi, L. Du (2005), Adaptive probability analysis using an enhanced hybrid mean value method. Struct. Multidiscipl. Optim. **29**, 134–148

16

Modern fracture mechanics for structural optimization with incomplete information

N.V. Banichuk[a], S.Yu. Ivanova and E.V. Makeev

Institute for Problems in Mechanics, Russian Academy of Scienses, Moscow, Russia

Abstract – Questions described in this paper are concerned with the shape optimization of brittle or quasi-brittle (axisymmetric) elastic shells. These questions take into consideration the possibilities of crack arising and damage accumulation in the process of application of cyclic load to the shell structure. Initial structural defects, arising cracks and damage accumulation are characterized by incomplete information concerning initial crack sizes, crack position and its orientation. In this context we develop the statements of the optimization problems based on guaranteed approach for the considered problems with incomplete information. For many realistic cases it is reasonable to use variants of the mini-max optimization, named as optimization for "the worst case scenario". Considered in this paper the structural optimization problems consist in finding of the shape and thickness distribution of axisymmetric quasi-brittle elastic shells with arising cracks in such a way that the cost functional (volume or weight of the shell material) reaches the minimum, while satisfying some constraints on the stress intensity factor and geometrical constraints. In the case of cycling loadings, we consider the number of loading cycles before fracture as the main constraint.

Key words: Shells; cracks; structural design; shape optimization; incomplete information.

Introduction

In the theory of optimal structural design there are several principal directions. One of these directions is devoted to conventional problems of structural optimization under strength constraints without taking into account initial damages and the possibilities of cracks arising and their growth. However, because the elements of many important structures are usually brittle or quasi-brittle, they are prone to cracking under very low applied stresses. Cracking not only reduces structural stiffness, but can lead to other undesirable effects such as delamination and global fracture. A few studies have been devoted to this class of problems of brittle and quasi-brittle bodies optimization on the basis of modern fracture mechanics criterion. Thomsen et al. [1], Thomsen and Karihaloo [2], Wang and Karihaloo [3] investigated some problems of crack appearance in the considered bodies. Crack appearance was also taken into account by Papila and Haftka [4] and Vitali et al. [5]. In these investigations all parameters of crack appearing have been supposed as given. The problem of the shape design of complex cracked shells under the optimization criteria of the weight of the structure and the strength constraint on a critical stress intensity factor has been studied by Abdi et al. [6]. In this paper the optimal cracked structure is that with minimum weight and with a stress intensity factor which does not reach the critical stress intensity factor.

In the most cases the number of cracks, their position, orientation, size and modes (opening cracks, shear cracks, mixed cracks) are unknown beforehand and so the optimal structural design (especially for brittle and quasi-brittle bodies) intrinsically contains uncertainty factors or randomness. In these cases it is difficult or impossible in principle to obtain complete information concerning the problem parameters and to formulate the structural optimization problem as the classical pure deterministic problem. In such cases to apply fracture mechanics representation in the design process and to formulate optimal design problem we can use min-max guaranteed approaches, probabilistic approaches and some mixed probabilistic-guaranteed approaches, developed early in the theory of optimal control and in the theory of differential games.

Banichuk [7,8] formulated general structural optimization problem under fracture mechanics constraints adopting arbitrary crack-appearance using min-max guaranteed approach. Some probabilistic approaches for optimal design of structures and structural elements and mixed probabilistic-guaranteed approaches have been developed by Yu et al. [9], Banichuk et al. [10, 11]. In what follows [12] investigated macrofailure criterion and performed the optimization of homogeneous structures and composite structures with edge delamination.

An important damage scenario is realized when a through – the thickness cracks are propagated under cyclic loading. Corresponding questions have been

[a] Corresponding author: banichuk@ipmnet.ru

investigated in the context of optimal design of beams and beam-like structures by Banichuk et al. [13] and Serra [14]. In this context we can mention also the paper by Lyubimov and Makarenko [15] devoted to the optimization problem of a reinforced plate with a crack.

Basic relations and the problem of the optimal structural design

In the paper we consider the elastic shell that has the form of surface of revolution. The position of the meridian plane is defined by the angle θ, measured from datum meridian plane and the position of a parallel circle is defined by the angle φ made by the normal to the surface and the axis of rotation (see Fig. 1) or by the coordinate x, measured along the axis of rotation, $0 \leqslant x \leqslant L$, L – given value. The meridian plane and the plane perpendicular to the meridian are the planes of principal curvatures. The geometry of the shell is defined when a shape of the middle surface is given. We restrict our consideration to an axially symmetric shape of a middle surface (a profile of each cross-section of a shell is a circle) and will use a distance $r(x)$ from the axis of rotation to a point of the middle surface as the variable describing the shape of the middle surface. These variables $r(x)$ and the thickness of the shell $h(x)$ will be considered as the design variables. The geometrical relations between meridian radius of curvature $r_\varphi(x)$, circumferential radius of curvature $r_\theta(x)$ and the radius $r(x)$ are given by the following expressions

$$r_\varphi = -\left(1+\left(\frac{dr}{dx}\right)^2\right)^{3/2}\left(\frac{d^2r}{dx^2}\right)^{-1},$$

$$r_\theta = r\sqrt{1+\left(\frac{dr}{dx}\right)^2} \qquad (1)$$

$$r = r_\theta \sin\varphi, \quad \frac{dr}{d\varphi} = r_\varphi \cos\varphi, \quad \frac{dx}{d\varphi} = r_\varphi \sin\varphi. \qquad (2)$$

It is assumed that a through crack can be arisen in the shell during its manufacturing or exploitation and it is taken into account that the material of the shell is quasi-brittle. The arisen crack is supposed to be rectilinear and its length is very small with respect to characterized geometric sizes of the shell without the restriction on the location of the crack in the shell, its orientation and its initial length $l_i \leqslant l_{cr}$. The value l_{cr} determines the moment when the global fracture can be realized.

The parameters of the crack characterizing its size, location, shape and orientation (in the case of crack appearance) can not be specified rigorously. Taking this into account we specify only the set Λ which contains all possible vectors

$$\omega = \{l, x_c, \alpha\} \qquad (3)$$

where x_c – coordinate of the crack center, l – length of the crack, α – angle setting the crack inclination with respect to the meridian. The second coordinate θ_c of the crack

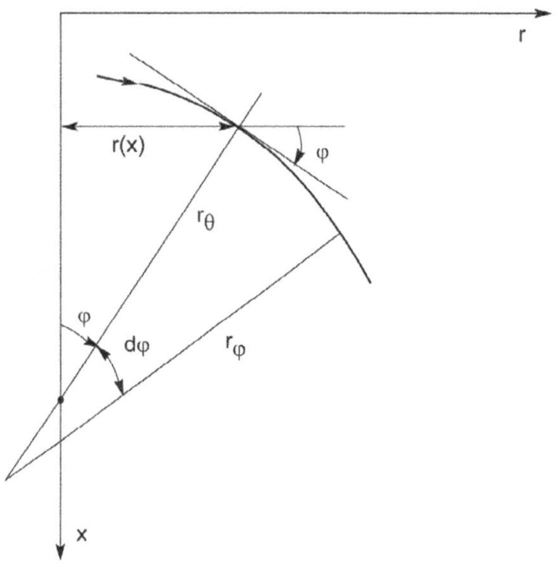

Fig. 1. Geometric terminology of the axisymmetric shell.

midpoint is nonessential and omitted because we consider axisymmetric problems and admit all possible locations of the crack in parallel direction ($0 \leqslant \theta_c \leqslant 2\pi$). If $\alpha = 0$, the crack is oriented in the meridian direction (axial crack) and if $\alpha = \pi/2$, the crack peripherally oriented in parallel direction. We shall use the following condition

$$\omega \in \Lambda. \qquad (4)$$

Accepted assumptions and available additional data concerning the most dangerous parts of the shell suggest that the set Λ to be considered as given.

The safety condition as it is well known from quasi-brittle fracture mechanics [16,17], is written as

$$K_1 \leqslant K_{1C} \qquad (5)$$

and is approximated by the modified inequality constraint [18,19]

$$K_1 \leqslant K_{1\varepsilon}, K_{1\varepsilon} = K_{1C} - \varepsilon, \varepsilon \geqslant 0 \qquad (6)$$

where K_1 is the stress intensity factor occurring in asymptotic representations of the stresses nearby the crack tip (in the opening mode) and K_{1C} – given quasi-brittle strength constant (toughness of the material). Here we introduced small positive parameter ε and reduced meaning $K_{1\varepsilon}$ of the limiting value of the stress intensity factor K_1. This approximation is used for convenience of application of the shape optimization procedure.

After application of cyclic loading the initial crack goes ahead, and its length l monotonically increases ($l_i \leqslant l \leqslant l_{cr}$). The process of the fatigue crack growth under cyclic loading can be adequately characterized in the following form [16, 17]

$$\frac{dl}{dn} = C\left(\Delta K_1\right)^m \tag{7}$$

$$l_i \leqslant l \leqslant l_{cr}, \quad 0 \leqslant n \leqslant n_{cr}. \tag{8}$$

Here C and $m(2 \leqslant m \leqslant 4)$ are some material constants, K_1 – is the stress intensity factor for opening cracks and the increment ΔK_1 is given by

$$\Delta K_1 = \left(K_1\right)_{max} - \left(K_1\right)_{min} = \left(K_1\right)_{p=p_{max}} - \left(K_1\right)_{p=p_{min}}. \tag{9}$$

Here $\left(K_1\right)_{max}$ and $\left(K_1\right)_{min}$ are the maximum and minimum values of the stress intensity factor K_1 in any given cycle, respectively.

The ordinary differential equation (7) defines the quasistatical process of crack growth and determines the dependence of the crack length l on the number of cycles n. This equation is valid up to the moment, when

$$n = n_{cr}, \quad l = l_{cr} \tag{10}$$

and the unstable crack growth (catastrophic fracture of the shell) is attained. To find l_{cr} we will use the fracture criterion

$$\left(K_1\right)_{l=l_{cr}, p=p_{max}} = K_{1\varepsilon}. \tag{11}$$

In what follows we will assume that not only the initial crack but also all temporary cracks ($l_i \leqslant l \leqslant l_{cr}$) are rectilinear and that the crack length l is bigger than h and is small with respect characterized size r_m of the shell.

In the considered design process the longevity constraint can be taken as $n_{cr} \geqslant n_*$, where n_* is a given minimum value of the number of cycles. Taking into account incompleteness of the information concerning the possible location of initial cracks we can rewrite the longevity constraint in the following manner

$$\min_{\omega \in \Lambda} n_{cr} \geqslant n_*. \tag{12}$$

Considered optimization problem consists of finding the shape $r = r(x)$ and the thickness distribution $h(x)$, $(0 \leqslant x \leqslant L)$ of the shell such that the volume of the shell material

$$J = 2\pi \int_0^L rh\left(1 + \left(\frac{dr}{dx}\right)^2\right)^{1/2} dx \tag{13}$$

is minimized, while satisfying the longevity constraint (12) and additional geometrical constraint

$$r(x) \geqslant r_g(x), \quad 0 \leqslant x \leqslant l \tag{14}$$

where $r_g(x) \geqslant 0$ – given function. This restriction do not change the common sense of the problem but permits to avoid possible singularities.

The shape of the shell will be called optimal if for any shell with a smaller weight it is possible to select a vector of unknown parameters ω belonging to the admissible set Λ such that some assigned constraints have been violated.

Taking into account that the longevity constraint (12) is reduced to the system of inequalities for stress components [20, 21] we shall consider (non-dimensional) problem of minimization of the functional (13) under the following constraints

$$\psi_1 = \frac{\gamma}{hr}\sqrt{1 + \alpha\left(\frac{dr}{dx}\right)^2} \int_x^1 rh\sqrt{1 + \alpha\left(\frac{dr}{dx}\right)^2} dx - \sigma_1 \leqslant 0 \tag{15}$$

$$\psi_2 = \frac{\beta}{h}\left(rh\frac{dr}{dx} - \frac{\frac{d^2r}{dx^2}}{\sqrt{1 + \alpha\left(\frac{dr}{dx}\right)^2}} \times \int_x^1 rh\sqrt{1 + \alpha\left(\frac{dr}{dx}\right)^2}\right) - \sigma_2 \leqslant 0 \tag{16}$$

$$\psi_3 = -\frac{\beta}{h}\left(rh\frac{dr}{dx} - \frac{\frac{d^2r}{dx^2}}{\sqrt{1 + \alpha\left(\frac{dr}{dx}\right)^2}} \times \int_x^1 rh\sqrt{1 + \alpha\left(\frac{dr}{dx}\right)^2}\right) - \sigma_1 \leqslant 0 \tag{17}$$

$$r_g \leqslant r(x) \leqslant 1, \quad 0 \leqslant x \leqslant 1 \tag{18}$$

$$h_{min} \leqslant h(x) \leqslant 1, \quad 0 \leqslant x \leqslant 1 \tag{19}$$

$$\psi_4 = V_0 - \pi \int_0^1 r^2 dx \leqslant 0. \tag{20}$$

Here $\alpha, \beta, \gamma, \sigma_1, \sigma_2, V_0, h_{min}$ are given dimensionless parameters of the problem.

Note that the problem of brittle and quasi-brittle shape optimization formulated in this paragraph is investigated for the class of axisymmetric shells that is important as from theoretical as from practical points of view. Besides this class of optimization problems is characterised by the possibility to arrive to one-dimensional formulation of the optimization problem.

Computations of the optimal shapes by genetic algorithm

To find the solution of the optimization problem (13), (15)–(20) for the axisymmetric shell under the action of force of gravity and the reaction force we shall use the method of penalty functions in combination with genetic algorithm. To this purpose we introduce the augmented functional

$$J^a = J + \sum_{i=1}^{4} \mu_i J_i \tag{21}$$

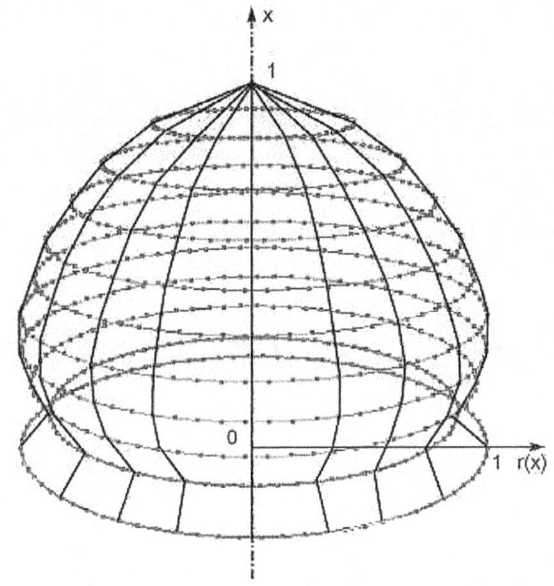

Fig. 2. A typical optimal shape of the axisymmetric shell.

where μ_i – arbitrary positive constants and J_i – penalty functionals defined as

$$J_i = \int_0^1 \Psi_i dxi, i = 1,2,3 \quad \Psi_i = \begin{cases} 0, \text{ if } \psi_i \leqslant 0 \\ \psi_i, \text{ if } \psi_i > 0 \end{cases} \quad (22)$$

$$J_4 = \begin{cases} 0, \text{ if } \psi_4 \leqslant 0 \\ \psi_4, \text{ if } \psi_4 > 0. \end{cases} \quad (23)$$

Thus the original problem of optimal design is reduced to finding of design variables $r(x)$, $h(x)$ satisfying inequalities (18), (19) and minimizing the augmented functional (20).

Optimal shape and thickness distributions of axisymmetric shells have been found numerically (one of the optimal projects is shown in Fig. 2) with application of the genetic algorithm [22].

It is supposed that the interval [0, 1] of the variable x is divided by the points $x_i, i = 1, 2, \ldots, n$ into $n-1$ subintervals of equal length Δx. Populations under consideration consist of N individuals. Each j-individual of the population is described by two sets of values $r(j,i)$, $h(j,i)$ representing the meanings of design variables at the nodes. The "best" individual, i.e. the sets $r(j,i)$, $h(j,i)$ minimizing augmented functional, is sought with application of genetic algorithm. The first step of the algorithm consists in initialization of population that is performed by means of assignment of random values taken from $[r_g(i),1]$ to each element $r(j,i)$ and from $[h_{min},1]$ to each element $h(j,i)$. The values $r(j,i)$ at the nodes with numbers $i = 1$ amd $i = n$ are fixed. For created individuals ($j = 1, ..., N$) of initial population we complete augmented functional $J^a(j)$ and find the individual having the minimal value of the functional. Using initial data on the next steps of the algorithm it is possible to determine new population consisting of N individuals and to perform in what follows successive minimization of the functional J^a. At the second step of the algorithm we perform successive selection of $N/2$ individual pairs – "parents" to obtain $N/2$ pairs of individuals – "children", which constitute new population. Selection of the first parent ("a") is performed by the following manner. Some natural number N^T is choused and then N^T individuals are selected by the random way. From this set of individuals we preserve and use only one individual having the minimal value of augmented functional J^a. Similarly we find the second parent ("b") and put together the first pair of individuals. All together we compose $N/2$ of such pairs. The third step of the algorithm consists in obtaining of two children from each parent pair of individuals. To this purpose we take some constant value from the interval [0, 1], that is called as crossover probability p_{co}. Then for each parent pair it is generated random number p_r from interval [0, 1] and random natural number m_r from [1, ..., n]. If it is appeared that $p_r \leqslant p_{co}$ then the values of design variables of children at the nodes $i = 1, 2, \ldots, m_r$ are copied from their parents "a" and "b", but the meanings of these values at the nodes $i = m_r+1, \ldots, n$ are obtained with the help of crossover. The latter means that for child "a" we copy the values in the corresponding nodes of the parent "b" and vice versa. Successive sorting of all parent pairs and performing of described operations lead to obtaining of N individuals – children, that compose new population. The fourth step of the algorithm consists in mutation of the obtained new population. This step is necessary not to stick at the local minimum of the considered functional. To realize the mutation procedure we take some small enough (~0.005) parameter p_m (probability of mutation). Then for all nodes of each individual of the population we generate the random number p_r from the interval [0, 1]. If $p_r \leqslant p_m$ then the value of design variable at this node is replaced by the arbitrary value, satisfying given constraints. Mutation procedure is not performed for the values $r(j,i)$ at the nodes with numbers i=1 amd $i = n$. For obtained new population we compute the functionals $J^a(j)$ and select the best individual. Then we go to the second step of applied algorithm. Note, that if the best child from new population is appeared to be worse then the best parent from previous population then we replace it by this parent. Thus we arrive the monotony in the process of finding of global minimum. Finish of the algorithm work is when the number of populations is large enough for convergence of the functional to be minimized.

With application of described algorithm we determined the optimal shapes of axisymmetric shell for the following values of problem parameters: $n = 11$, $N = 10$, $N^T = 4$, $p_{CO} = 0.5$, $p_m = 0.05$ $\alpha = 1$, $\beta = 0.1$, $\gamma = 0.1$, $h_{min} = 0.1$, $r_g(x) = 0$, $\sigma_1 = \sigma_2 = 1$, $\mu_i = 1$ ($i = 1, \ldots, 4$). Calculations were completed after generations of 10 000 populations. Optimal shapes of shells for boundary conditions $r(0) = 1$, $r(1) = 0$ and given volumes $V_0 = 2$, $V_0 = 1.5$, $V_0 = 1$ are represented by the curves 1, 2, 3 in Figure 3. Dependence of the functional J^a on the number of population of genetic algorithm (number of iteration) characterizing the speed of convergence

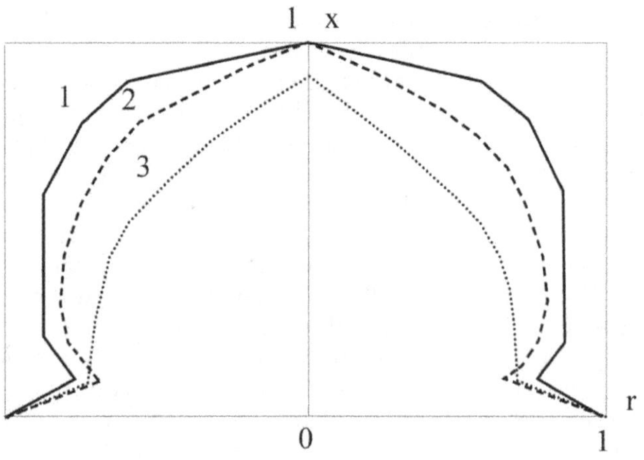

Fig. 3. Optimal shapes distributions for different values of volume of the shell.

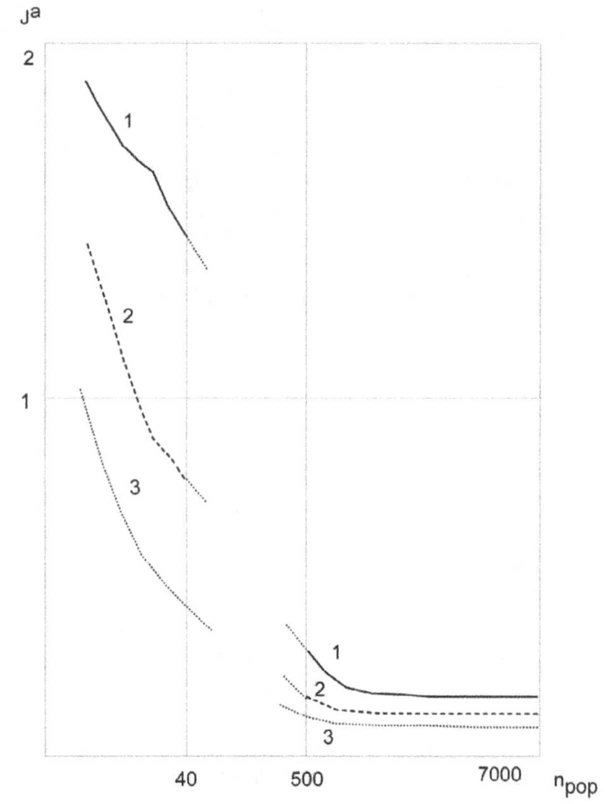

Fig. 4. Dependence of the functional J^a on the number of population of genetic algorithm.

is shown in Figure 4. Optimal shapes of shells for $V_0 = 2$ are shown in Figure 5 by curves 1 (for $r(0) = 1$) and 2 (for $r(0) = 0.9$). In all considered cases the optimal distributions of shell thickness have been also found. In Figure 6 the curves 1 and 2 correspond to optimal thickness distributions for $r(0) = 1$, $r(1) = 0$ and two values of shell volume $V_0 = 2$ and $V_0 = 1.5$, respectively. We can see the increasing of thickness along the boundary circle

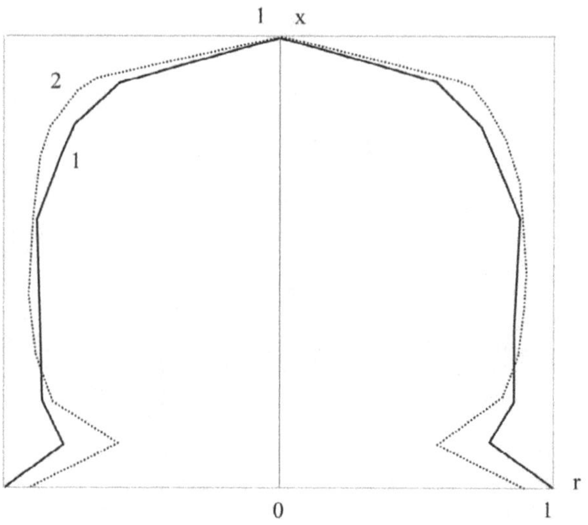

Fig. 5. Optimal shapes distributions for $V_0 = 2$ and different values of $r(0)$.

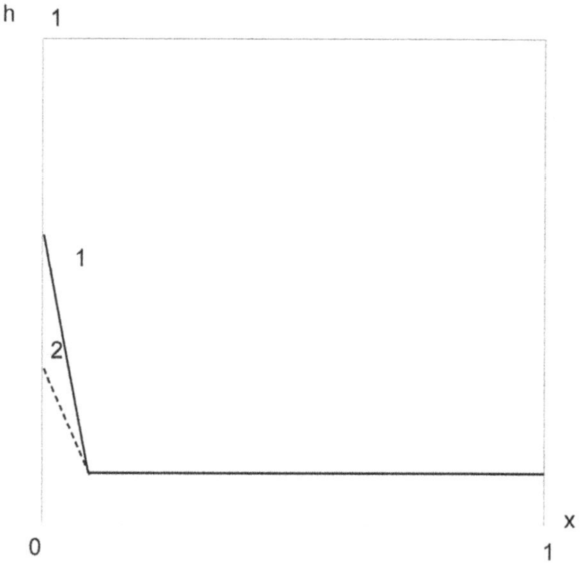

Fig. 6. Optimal thickness distributions for $V_0 = 2$ and $V_0 = 1.5$.

with radius $r_0 = r(0)$. This is the characteristic feature of all considered optimal projects.

Conclusions

In this paper the problem of optimal shell design based on fracture mechanics of brittle and quasi-brittle bodies was considered. Some possibilities of effective application of minimax (guaranteed) approach and "worst case scenario" were demonstrated. The original optimization problem with uncertainties in damage characteristics (crack position and orientation) was reformulated using the transformation of the longevity constraint into the direct constraints imposes on the stresses and was effectively solved with the help of application of the penalty function method and genetic algorithm. Performed investigations are related to designing of axisymmetric shells loaded by axisymmetric forces. However, the optimization method described in this paper can also be applied to problems of optimal design of shells under more general geometrical and loading conditions in the frame of guaranteed approach.

Acknowledgements

The paper is performed under support of RFBR (grants № 05-08-18094-a, № 07-08-08010-3, № 08-08-00025-a), RAS Program № 14 ("Accumulation of damages, fracture and ..."), Program of Support of Leading Scientific Schools (grant № 169.2008.1).

References

1. N.B. Thomsen, J. Wang, B.L. Karihaloo, Structural Optimization **8**, 9 (1994)
2. N.B. Thomsen, B.L. Karihaloo, J. Am. Ceram. Soc. **78**, 3 (1995)
3. J. Wang, B.L. Karihaloo, Composite Structures **32**, 453 (1995)
4. M. Papila, R.T. Haftka, Structural Optimization **25**, 327 (2003)
5. R. Vitali, R.T. Haftka, B.V. Sankar, Struct. Multidisc. Optim. **23**, 347 (2002)
6. R. Abdi, M. Touratier, P. Convert, Optimal design for minimum weight in cracked pressure vessel of a turboshaft, in: *Communications in Numerical Methods in Engineering* (1996), pp. 271–280
7. N.V. Banichuk, Analete Stiintifice, Universitatii. OVIDIUS, Constanta **5**, 13 (1996)
8. N.V. Banichuk, Mechanics of Structures and Machines **26**, 365 (1998)
9. X. Yu, K.K. Choi, K.X. Chang, Structural Optimization **14**, 81 (1997)
10. N.V. Banichuk, F. Ragnedda, M. Serra, Meccanica **34**, 29 (1999)
11. N.V. Banichuk, F. Ragnedda, M. Serra, Mechanics Based Design of Structures and Machines **31**, 459 (2003)
12. A. Borovkov, V. Palmov, N. Banichuk, E. Stein, V. Saurin, F. Barthold, Yu. Misnik, Int. J. Comput. Civil Struct. Eng. **1**, 91 (2000)
13. N.V. Banichuk, F.J. Barthold, M. Serra, Meccanica **40**, 135 (2005)
14. M. Serra, Struct. Multidisc. Optim. **19**, 159 (2000)
15. A.K. Lyubimov, F.V. Makarenko, Probability approach to the optimization problem of a reinforced plate with a crack, *in Applied Problems of Strength and Plasticity* (KMK Scientific Press, Moscow, 1996), pp. 120–131
16. K. Hellan, *Introduction to Fracture Mechanics* (New York, Mc-Graw-Hill, 1984)
17. M.F. Kanninen, C.H. Popelar, *Advanced Fracture Mechanics* (Oxford University Press, New York, 1985)
18. N.S. Streletskii, Izv. AN SSSR, OTN **1** (1947)
19. V.V. Bolotin, *Statistical Methods in Structural Mechanics* (Holden-Day, inc., San-Francisco, 1969)
20. N.V. Banichuk, S.Yu. Ivanova, E.V. Makeev, A.V. Sinitsin, Mechanics Based Design of Structures and Machines **33**, 253 (2005)
21. N.V. Banichuk, P. Neittaanmäki, Mechanics Based Design of Structures and Machines **35**, 76 (2007)
22. D.E. Goldberg, *Genetic Algorithm in Search, Optimization and Machine Learning* (Westley Publ. Comp., inc., M.A. 1989), p. 154

17

Reliability based robust design optimization of steel structures

Nikos D. Lagaros[a], Vagelis Plevris and Manolis Papadrakakis

Institute of Structural Analysis & Seismic Research, National Technical University of Athens, 9, Iroon Polytechniou Str., Zografou Campus, GR-15780 Athens, Greece

Abstract – In this work the uncertainty of a structural system is taken into account in the framework of a structural Reliability based Robust Design Optimization (RRDO) formulation where probabilistic constraints are incorporated into the robust design optimization formulation. A robust design optimization problem is formulated as a multi-criteria optimization problem. The Pareto front representing the solution of the RRDO problem is composed by designs with a state of robustness, since their performance is the least sensitive to the variability of the uncertain variables. The cross section dimensions together with other structural parameters, such as the modulus of elasticity, the yield stress and the applied loading, are considered as random variables. For the solution of the RRDO problem, the non-dominant Cascade Evolutionary Algorithm is employed combined with a weighted Tchebycheff metric.

Key words: Probabilistic constraints; robust design optimization; cascade evolutionary algorithms; Monte Carlo simulation

1 Introduction

Improving the performance of a structural system in terms of constructional cost and structural response is the main objective during its development. Improvements can be achieved either by simply using design rules based on the experience or on an automated way using optimization methods. Strictly speaking, optimal means that no better solution exists under certain constraints. Considering the complexity of a structural optimization problem it is obvious that finding the global optimum solution is a very difficult task. In the real world, given the uncertainty or scatter of the structural parameters, the significance of such an optimum solution, where the uncertainty has not been taken into account, would be limited. Although in a computing environment nearly perfect structural models can be simulated, real world structures always have imperfections or deviations from the nominal state defined by the design codes. An optimum solution obtained computationally is never materialized in an absolute way and as a result a near optimal solution is always implemented in practice. A deterministic based formulation of a structural optimization problem ignores scatter of any kind of the structural parameters. It is possible to achieve a feasible optimum according to the deterministic formulation, yet once this solution is implemented in a real physical system, its optimal performance may vanish due to of the parameters scatter which is unavoidable. Consequently, the performance of the "implemented" design may be far worse than expected. In order to account for the randomness of some parameters that affect the response of the structure, a different formulation of the optimization problem has to be used based on stochastic analysis.

In recent years, probabilistic based formulations of the optimization problem have been developed to account for uncertainty and randomness through stochastic simulation and probabilistic analysis. The development, over the last two decades [1], on the stochastic analysis methods has stimulated the interest for probabilistic structural optimization problems. There are two distinguished design formulations to account for the probabilistic systems response: Robust Design Optimization (RDO) [2–4] and Reliability-Based Design Optimization (RBDO) [5–8]. RDO formulations primarily seek to minimize the influence of stochastic variations on the mean design. On the other hand, the main goal of RBDO formulations is to design for safety with respect to extreme events. In this work a Reliability based Robust Design Optimization (RRDO) formulation is examined, where probabilistic constraints are incorporated into the RDO formulation.

The main goal of this work is to account for the influence of probabilistic constraints in the framework of a structural RDO problem, by comparing the RRDO formulation with the RDO one. The objective functions considered for both formulations are the weight and the variance of the response of the structure represented by a characteristic node displacement. In the case of the RDO

[a] Corresponding author:
{nlagaros, vplevris, mpapadra}@central.ntua.gr

formulation each design is checked whether it satisfies the provisions of the European design codes for steel structures [9]. On the other hand, additional probabilistic constraints are taken into account in the case of the RRDO formulation, where the code provisions are checked if they are satisfied with a prescribed probability of violation. The uncertainty of loads, material properties, and cross section dimensions is taken into consideration using the Monte Carlo Simulation (MCS) method combined with the Latin hypercube sampling. The solution of the multi-criteria optimization problem is performed with the non-dominant Cascade Evolutionary Algorithm combined with a weighted Tchebycheff metric.

2 Deterministic based structural optimization

2.1 Single objective optimization problem

In a single objective Deterministic Based sizing Optimization (DBO) problem the aim is often to minimize the weight of the structure subject to certain deterministic behavioural constraints usually on stresses and displacements. A discrete DBO problem can be formulated in the following form

$$\begin{aligned} \min \quad & f(\mathbf{s}) \\ \text{subjectto} \quad & g_j(\mathbf{s}) \leqslant 0 \quad j = 1, ..., k \\ & s_i \in R^d, \quad i = 1, ..., n \end{aligned} \tag{1}$$

where $f(\mathbf{s})$ is the objective function, $\mathbf{s}$ is a vector of the design variables, which can take values only from a discrete given set R^d, while $g_j(\mathbf{s})$ are the deterministic constraints. Most frequently the deterministic constraints refer to the member stresses and nodal displacements or the interstorey drifts in building structures.

2.2 Multi-objective optimization problem

In practical applications of sizing optimization problems, the weight itself rarely gives a representative measure of the performance of the structure. In fact, several conflicting and usually incommensurable criteria usually exist in real-life design problems that have to be dealt with simultaneously. This situation forces the designer to look for a good compromise among the conflicting requirements. Problems of this kind are defined as multi-objective optimization problems. The consideration of such problems in their present form originated towards the end of the 19th century when Pareto [10] presented the optimality concept in economic problems with several conflicting criteria.

Criteria and conflict in multi-objective structural optimization

An engineer looking for the optimum design of a structure is faced with the question of selecting the most suitable criteria for measuring the economy, strength, serviceability or any other factor that affects the performance of the structure. Any quantity that has a direct influence on the structural performance can be considered as a criterion. On the other hand, quantities that must satisfy imposed requirements are not considered as criteria but they can be treated as constraints.

One important property in the multi-criteria formulation is the conflict that may exist among the criteria. Only those quantities that are competing with each other should be treated as independent criteria whereas the others can be combined into a single criterion representing the whole group. Two functions f_i and f_j are called locally collinear with no conflict at point $\mathbf{s}$ if there is $c > 0$ such that $\nabla f_i\ (\mathbf{s}) = c\nabla f_j\ (\mathbf{s})$. Otherwise, the functions are called locally conflicting at point $\mathbf{s}$. According to this definition any two criteria are locally conflicting at a point of the design space if their maximum improvement is achieved at different directions. On the other hand, two functions f_i and f_j are called globally conflicting in the feasible region F of the design space when the two optimization problems $\min\{f_i\ (\mathbf{s}),\ \mathbf{s} \in F\}$ and $\min\{f_j\ (\mathbf{s}),\ \mathbf{s} \in F\}$ have different optimal solutions.

Formulation of a multi-objective optimization problem

The selection of the design variables, criteria and constraints represents undoubtedly the most important part in the formulation of an optimization problem. In general, the mathematical formulation of a multi-objective problem with n design variables, m objective functions and k constraint functions and can be defined as follows

$$\begin{aligned} \min_{\mathbf{s}\in \mathrm{F}} \quad & [f_1(\mathbf{s}), f_2(s), \ldots, f_m(s)]^T \\ \text{subject to} \quad & g_j(\mathbf{s}) \leqslant 0 \quad j = 1, ..., k \\ & s_i \in R^d, \quad i = 1, ..., n \end{aligned} \tag{2}$$

where $\mathbf{s}$ is a design variable vector and F is the feasible set in the design space R^n which is defined as the set of design variables that satisfy the constraint functions $g(\mathbf{s})$

$$F = \{\mathbf{s} \in R^n | g_j(\mathbf{s}) \leqslant 0 \quad j = 1, ..., k\} \tag{3}$$

If the m objective functions are globally conflicting, there exists no unique point which represents the optimum for all m criteria. Thus the common optimality condition used in single-objective optimization must be replaced by a new concept, the so called Pareto optimum. A design vector $\mathbf{s}* \in \mathrm{F}$ is called Pareto optimum for the problem of equation (2) if and only if there is no other design vector $\mathbf{s} \in \mathrm{F}$ such that

$$\begin{aligned} & f_i(\mathbf{s}) \leqslant f_i(\mathbf{s}^*) \quad \text{for} \quad i = 1, ..., m \\ \text{with} \quad & f_i(\mathbf{s}) < f_i(\mathbf{s}^*) \quad \text{for at least one objective } i \end{aligned} \tag{4}$$

The geometric locus of the Pareto optimum solutions is called Pareto front curve and represents the solution of the optimization problem with multiple objectives. A typical Pareto front curve is depicted in Figure 1 for two conflicting objective functions $f(x)$ and $g(x)$ to be minimized.

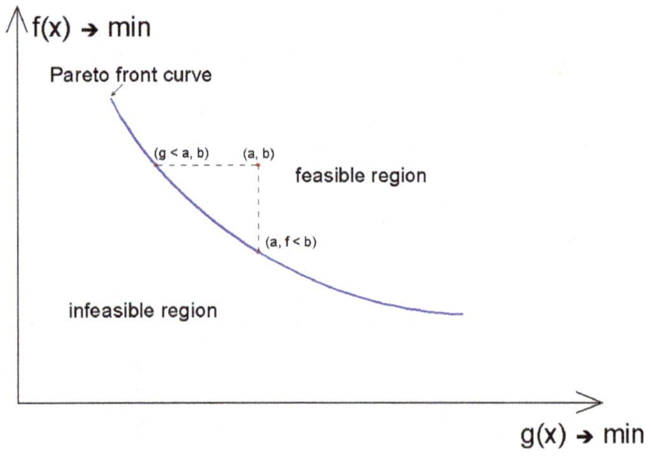

Fig. 1. Typical Pareto front curve.

Solving the multi-objective optimization problem

A number of techniques have been developed to deal with the multi-objective optimization problem [11, 12]. The algorithm employed in this work belongs to the hybrid methods, where an evolutionary algorithm is combined with a scalarizing function. In general, when scalarizing functions are used, local Pareto optimal solutions are obtained. Global Pareto optimality can be guaranteed only when the objective functions and the feasible region are both convex or quasi-convex and convex, respectively. For non-convex cases, such as the majority of structural multi-objective optimization problems, a global single objective optimizer is required.

Single objective optimizer

The optimization methods for single objective problems can be classified in two general categories: deterministic and probabilistic. Mathematical Programming (MP) methods, such as Sequential Quadratic Programming (SQP) belong to the first category. The main advantage of such optimizers is that they capture the right path to the nearest optimum quickly, but they cannot guarantee that the global optimum has been achieved. Consequently, in real world structural optimization applications, MP algorithms encounter great difficulties in dealing with multiple local minima, large and non-convex search spaces and several (possibly conflicting) constraints that have to be satisfied [13].

On the other hand, probabilistic optimization techniques, such as Evolutionary Algorithms (EA), are robust and present a better global behaviour than the MP methods when dealing with complex optimization problems [14, 15]. EA are not vulnerable to being trapped in local optima and therefore can be considered as reliable in approaching the global optimum for non-convex constrained optimization problems. Due to their random search, EA proceed toward the optimal solution with slower rate than gradient-based optimizers and most times need a greater number of analysis steps. However, these analyses are less time consuming than the corresponding computations in MP algorithms, since EA do not require gradient calculations. Thus, an EA-based optimization procedure is implemented in the present work for the solution of the single objective optimization part.

Augmented weighted Tchebycheff problem

The weighted Tchebycheff metric that was used by the authors in a previous work [16] is stated as follows

$$\min_{s \in F} \max_{i=1,\dots,m} \left[w_i \frac{|f_i(\mathbf{s}) - z_i^*|}{f_i(\mathbf{s})} + \rho \sum_{i=1}^{m} \frac{|f_i(\mathbf{s}) - z_i^*|}{f_i(\mathbf{s})} \right] \quad (5)$$

where ρ is a sufficiently small positive scalar (in this work $\rho = 0.1$) while z_i^* is the utopian objective function value. The weight parameters w_i are random numbers, uniformly distributed between 0 and 1. These weight parameters have to fulfil the requirement

$$\sum_{i=1}^{m} w_i = 1 \quad (6)$$

and if not, they are updated according to the following expression:

$$w_i = \begin{cases} w_i + \frac{1-\sum_i^m w_i}{\sum_i^m w_i} w_i, & \text{if } \sum_i^m w_i \neq 1 \\ w_i, & \text{if } \sum_i^m w_i = 1 \end{cases} \quad (7)$$

Cascade optimization

It is generally accepted that there is still no unique optimization algorithm capable of handling all existing optimization problems effectively. Cascade optimization has been proposed as a remedy to this observation [15,17,18]. According to this optimization strategy, a multi-stage procedure is applied, in which various optimizers are implemented successively. In the first stage of the cascade procedure the initial optimizer starts from a user-specified design known as 'cold-start'. The intermediate optimal solutions, called 'hot-starts', are used to initiate the next cascade stages. Each optimization stage of the cascade procedure starts from the (possibly perturbed) optimum design achieved in the previous stage. Thus, each cascade stage initiates from a hot-start and produces a new hot-start for the next stage, coupling the autonomous computations of successive optimization stages.

The optimization algorithm implemented in each stage of the cascade process (called cascade step) may or may not be the same. Cascade optimization has been implemented using different deterministic optimizers [17], as well as using both deterministic and probabilistic optimizers [15, 18]. The main advantage in combining different optimizers in a successive manner has to do with maximizing the exploitation of the advantages offered by various optimization algorithms and alleviating the influence of the corresponding disadvantages on the optimum design achieved. The selection of the optimizers to be included in a cascade procedure, their exact sequence and the number of cascade stages performed can be determined via a trial-and-error process.

Non-dominant CEA multi-objective search using the Tchebycheff metric

In the present work the idea of Cascade Evolutionary Algorithms (CEA) is implemented for solving multi-objective structural optimization problems. The resulting cascade evolutionary procedure consists of a number of optimization stages, each of which employs the same EA optimizer. In order to differentiate the search paths followed by the same optimization algorithm during the cascade stages, the initial conditions of individual optimization runs are suitably controlled by using at each stage: (a) a different initial design (each stage initiates from the solution of the previous stage) and (b) a different seed for the random number generator of the EA process.

The multi-objective optimization procedure used in this study is based on the combination of the CEA procedure and the Tchebycheff metric employing the non-dominant search. The multi-objective optimization procedure initiates with a set of parent vectors and a set of weighting coefficients for each independent run, representing a single objective optimization problem of the series. The independent runs can be implemented in parallel and in every global generation non-dominant search is applied. A global generation is defined as the state where a local generation (a generation in one of the independent runs) is completed in every independent run. The non-dominant search can be applied either in every global generation or locally for a number of local generations before it is applied globally. In this work the first option is used where each independent run corresponds to a CEA procedure using the Tchebycheff metric. According to this procedure in every global generation a local Pareto front is produced which evolves towards the global one. The basic idea of the procedure is to create an evolutionary process of the Pareto front curve.

The optimization algorithm used in this study is denoted as: non-dominant $CEATm(\mu{+}\lambda)_{nrun,csteps}$ where μ, λ are the number of the parent and offspring vectors respectively, *nrun* is the number of independent CEA procedures and *csteps* is the number of cascade steps employed. A detailed description of the CEATm method can be found in a previous work of the authors [16]. The basic steps inside an independent run of the multi-objective algorithm, as adopted in this study, are the following:

Independent run i, i=1,. . . ,nrun

Generate the parameters $w_{i,j}$ $j = 1,\ldots,m$ of the Tchebycheff metric. Check if the requirement of equation (6) is fulfilled, if not change the weight parameters using equation (7).

CEATm loop

1. *Initial generation:*
   - 1a. *Generate* $\mathbf{s}_k$ ($k = 1,\ldots,\mu$) vectors
   - 1b. *Analysis step*
   - 1c. *Evaluation of the Tchebycheff metric, equation (5)*
   - 1d. *Constraint check:* if satisfied $k = k+1$ else $k = k$. Go to *step 1a*
2. *Global non-dominant search:* Check If global generation is accomplished. If yes, then non-dominant search is performed, else wait until global generation is accomplished.
3. *New generation:*
   - 3a. *Generate* $\mathbf{s}_\ell$ ($\ell = 1,\ldots,\lambda$) vectors
   - 3b. *Analysis step*
   - 3c. *Evaluation of the Tchebycheff metric, equation (5)*
   - 3d. *Constraint check:* if satisfied $\ell = \ell{+}1$ else $\ell = \ell$. Go to *step 3a*
4. *Selection step:* selection of the next generation parents according to $(\mu + \lambda)$ or (μ,λ) scheme
5. *Global non-dominant search:* Check If global generation is accomplished. If yes, then non-dominant search is performed, else wait until global generation is accomplished.
6. *Convergence check:* If satisfied stop, else go to *step 3*

End of CEATm loop

End of Independent run i

3 Reliability based robust design optimization

In a reliability based robust design structural sizing optimization problem an additional objective function is considered which is related to the influence of the random nature of some structural parameters on the performance of the structure. In the present study the aim is to minimize both the weight and the variance of the response of the structure. In a RRDO problem the constraint functions are also varied due to variations of the random structural parameters. The mathematical formulation of the RRDO problem implemented in this study is as follows

$$\begin{aligned} \min \quad & \Phi(\mathbf{s},\mathbf{x}) \\ \text{subject to} \quad & g_j(\mathbf{s}) \leqslant 0 \quad j = 1,\ldots,k \\ & p_{v,j}(\mathbf{s},\mathbf{x}) \leqslant p_{all} \quad j = 1,\ldots,k \\ & s_i \in R^d, \quad i = 1,\ldots,n \end{aligned} \tag{8}$$

where $\Phi(\mathbf{s})$ is the multi-objective function which is defined by equation (5), $\mathbf{s}$ is the vector of the design variables, which can take values only from the given discrete set R^d, $g_j(\mathbf{s})$ are the deterministic constraints while $p_{v,j}$ is the probability of violation of the j-th constraint bounded by an upper allowable probability equal to p_{all}, while $\mathbf{x}$ is the vector of the random variables. Probabilistic constraints define the feasible region of the design space by restricting the probability that a deterministic constraint is violated within the allowable probability of violation. The probabilistic constraints that are employed in this study enforce the condition that the probabilities of violation of the structure are smaller than a certain value.

Figure 2 depicts the difference between the DBO and the RDO optimum for a single variable problem. In the

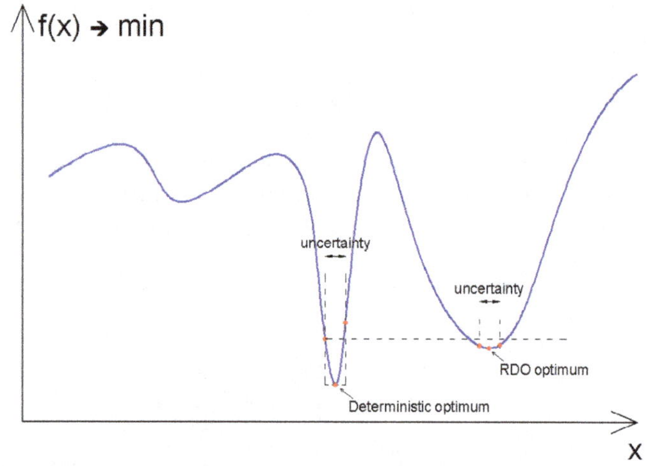

Fig. 2. The concept of robust design optimization.

case of DBO formulation the minimization of the objective function $f(x)$ is the only criterion, on the other hand according to the RDO formulation the minimization of the objective function $f(x)$ in conjunction with its variance is considered. It can be seen that the DBO formulation leads to a design having the minimum objective function value ignoring its variance while the RDO formulation leads to a compromise optimum solution corresponding a worse objective function value compared to the one of the DBO solution, but it is significantly better with respect to the variance.

In this study three types of constraints are imposed: (i) stress, (ii) compressive force (for buckling) and (iii) displacement constraints. The stress constraint can be written as follows

$$\sigma_{\max} \leqslant \sigma_a \qquad \sigma_a = \frac{\sigma_y}{1.10} \tag{9}$$

where $\sigma_{\max}$ is the maximum axial stress in each element group for all loading cases, σ_a is the allowable axial stress according to Eurocode 3 [9] and σ_y is the yield stress. For the members under compression an additional constraint is used

$$|P_{c,\max}| \leqslant P_{cc} \qquad P_{cc} = \frac{P_e}{1.05} \qquad P_e = \frac{\pi^2 EI}{L_{eff}^2} \tag{10}$$

where $P_{c,\max}$ is the maximum axial compressive force for all loading cases, P_e is the critical Euler buckling force in compression, taken as the first buckling mode of a pin-connected member, and L_{eff} is the effective length. The effective length is taken equal to the actual length of each element. Similarly, the displacement constraints can be written as

$$|d| \leqslant d_a \tag{11}$$

where d_a is the allowable value of the displacement at a certain node or the maximum nodal displacement.

4 Monte Carlo simulation

In stochastic analysis of structures the MCS method is particularly applicable when an analytical solution is not attainable. This is mainly the case in problems of complex nature with a large number of random variables (random structural parameters), where all other stochastic analysis methods are not applicable. Despite the fact that the mathematical formulation of the MCS is simple and the method has the capability of handling practically every possible case regardless of its complexity, this approach has not received an overwhelming acceptance due to the excessive computational effort that it requires. Despite the improvements achieved on the efficiency of the computational methods for treating reliability analysis problems, they still require disproportional computational effort for practical reliability problems. This is the reason why relatively few successful numerical investigations are known in the field. The Latin Hypercube Sampling (LHS) method was introduced by MacKay et al. [19] in an effort to reduce the required computational cost of purely random sampling methodologies. LHS is generally recognized as one of the most efficient size reduction techniques. In the current study the MCS with Latin hypercube samples has been employed for the calculation of the probability of violation of the behavioural constraints and the variance of the response of the structure which are required in the framework of an RDO problem.

In structural stochastic analysis problems, where the probability of violation of some behavioural constraints is to be calculated, MCS can be stated as follows: expressing the limit state function as $G(\mathbf{x}) < 0$, where $\mathbf{x} = [x_1, x_2, \ldots, x_M]^{\mathrm{T}}$ is a vector of the random structural parameters, the probability of violation of the behavioural constraints can be written as

$$p_{viol} = \int_{G(x)\geqslant 0} f_x(\mathbf{x})d\mathbf{x} \tag{12}$$

where $f_x(\mathbf{x})$ denotes the joint probability of violation for all random structural parameters. Since MCS is based on the theory of large numbers (N_∞) an unbiased estimator of the probability of violation is given by

$$p_{viol} = \frac{1}{N_\infty}\sum_{j=1}^{N_\infty} I(\mathbf{x}_j) \tag{13}$$

where $\mathbf{x}_j$ is the j-th vector of the random structural parameters, and $I(\mathbf{x}_j)$ is an indicator for successful and unsuccessful simulations defined as

$$I(x_j) = \begin{cases} 1 & \text{if} \quad G(\mathbf{x}_j) \geqslant 0 \\ 0 & \text{if} \quad G(\mathbf{x}_j) < 0 \end{cases} \tag{14}$$

In order to estimate p_{viol} an adequate number of N independent random samples is produced using a specific, uniform probability density function of the vector $\mathbf{x}_j$. The value of the violation function is computed for each random sample $\mathbf{x}_j$ and the Monte Carlo estimation of p_{viol}

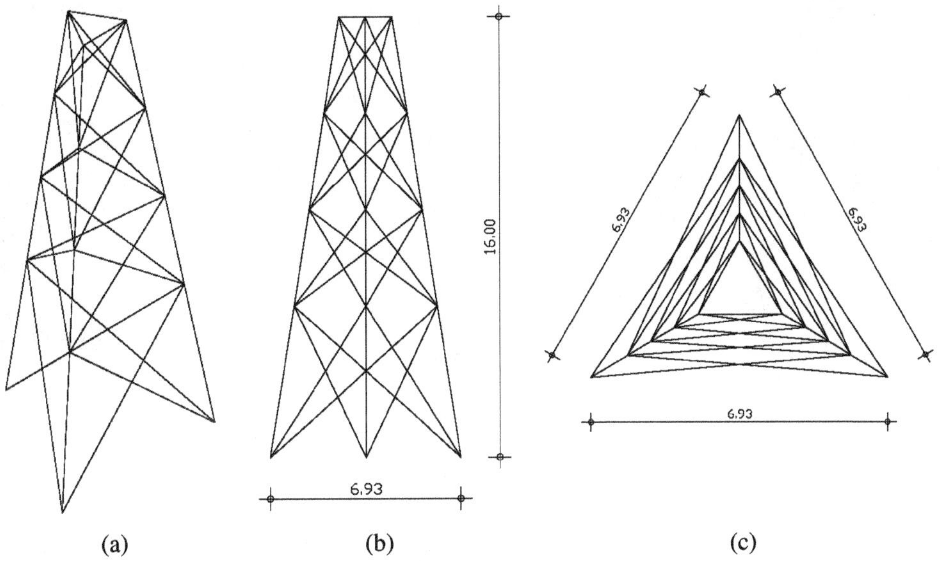

Fig. 3. 39-bar truss: (a) 3D view, (b) Side view, (c) Top view.

Table 1. 39-bar truss example: characteristics of the random variables.

Random variable		Probability Density Function	Mean value μ	CoV
E (kN/m^2)	Young's Modulus	Normal	2.10E+08	7.14%
σ_y**(kN/m^2)**	Allowable stress	Normal	3.55E+05	10.00%
F (kN)	Horizontal loading	Normal	8	37.50%
D	CHS Diameter	Normal	d_i *	2%
t	CHS Thickness	Normal	t_i *	2%

* Taken from the Circular Hollow Section (CHS) table of the Eurocode, for every design

is given in terms of sample mean by

$$p_{viol} \cong \frac{N_H}{N} \qquad (15)$$

where N_H is the number of successful simulations and N the total number of simulations.

5 Numerical results

For the purposes of this study two 3D steel truss structures have been considered. For both test examples, two objective functions have been taken into account, the weight of the structure and the standard deviation of a characteristic node displacement representing the response of the structure, subject to constraints on stresses, element buckling and displacements imposed by the European design codes [9]. The design variables considered are the dimensions of the members of the structure, taken from the Circular Hollow Section (CHS) table of the Eurocode. The random variables related to the cross sections are the external diameter D and the thickness t of the circular hollow section. Apart from the cross-sectional dimensions of the structural members, the material properties (modulus of elasticity E and yield stress σ_y) and the lateral loads have been considered as random variables.

5.1 Three dimensional 39-bar truss

The first test example considered is the 39-bar truss structure shown in Figure 3. The height of the structure is 16 m, while its basis is an equilateral triangle of side 6.93 m. The model consists of 15 nodes and 39 elements divided into 4 design variables. A vertical load $V = 2$ kN is applied to all nodes, while a probabilistic horizontal load F of mean value 8 kN is applied to the top nodes at the x-direction. The type of probability density function, the mean value, and the variance of the random variables are shown in Table 1.

In order to examine the performance of the MCS combined with LHS a parametric study is performed that monitors the influence of the number of LHS simulations with respect to the accuracy in the calculation of the statistical quantities. For this reason, the probability of violation of the displacement constraint and the standard deviation of the characteristic node displacement are calculated with respect to the number of LHS simulations for the case of a randomly selected design vector. The results of this study can be seen in Figure 4. For this test example 3,000 LHS simulations have been considered enough in order to calculate with sufficient accuracy the statistical quantities under consideration.

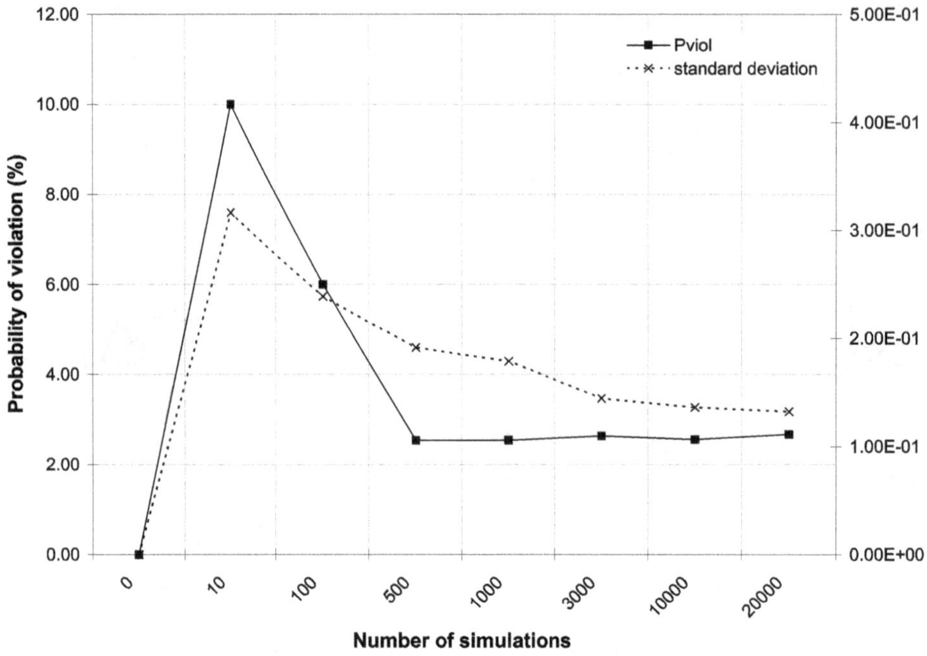

Design D×t	Weight (kN)	P_{viol} (%)	St. Dev. (m)
114.3×6.3, 88.9×8, 101.6×8, 114.3×6.3	107.91	2.40	0.14

Fig. 4. 39-bar truss: Verification.

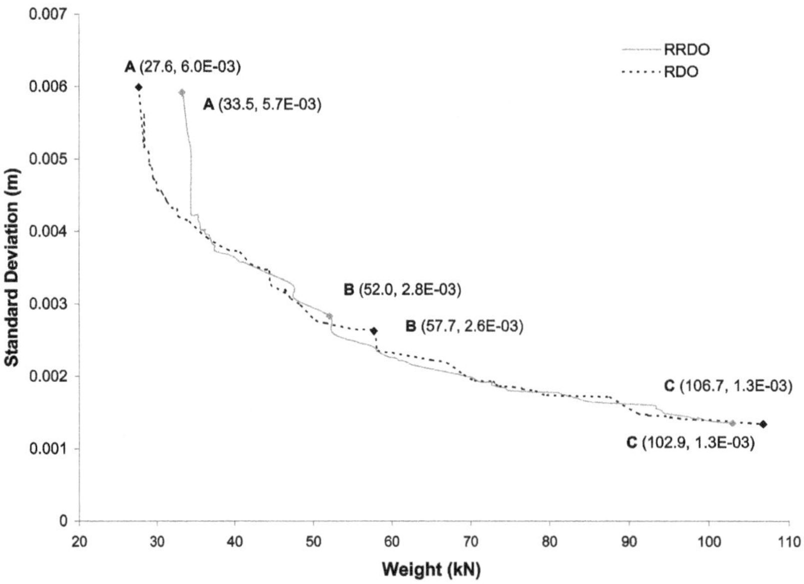

Fig. 5. 39-bar truss: Comparison of the Pareto front curves.

For the solution of the multi-objective optimization problem in question the non-dominant $CEATm(\mu + \lambda)_{nrun,csteps}$ optimization scheme was employed where $\mu = \lambda = 5$, $nrun = 10$ and $csteps = 3$. These values were found in a previous study by the authors [16] to be appropriate for providing good quality Pareto front curves. Two different formulations of the RDO problem have been considered in this study: (i) RRDO with probabilistic constraints and (ii) the RDO formulation. In the case that probabilistic constraints have been taken into account, a probability of violation equal to 2% has been considered. The resultant Pareto front curves for the two RDO formulations are depicted in Figure 5, with the structural weight on the horizontal axis and the standard deviation of the characteristic node displacement on the vertical axis.

It can be seen from Figure 5 that the influence of considering the probabilistic constraints is significant when the weight of the structure is the dominant criterion (designs A). The two Pareto front curves almost coincide

Table 2. 39-bar truss example: characteristic optimal solutions.

	RDO			RRDO		
	DBO (A)	(B)	(C)	DBO (A)	(B)	(C)
(D×t)*	(D×t)*	(D×t)*	(D×t)*	(D×t)*	(D×t)*	(D×t)*
Sec_1	139.7 × 4	168.3 × 6.3	193.7 × 12.5	139.7 × 5	139.7 × 10	193.7 × 12.5
Sec_2	139.7 × 5	219.1 × 7.1	193.7 × 12.5	168.3 × 4.5	114.3 × 10	193.7 × 12.5
Sec_3	101 × 6.3	168.3 × 8	193.7 × 12.5	101.6 × 10	139.7 × 8	193.7 × 12.5
Sec_4	139.7 × 4	193.7 × 6.3	193.7 × 12.5	139.7 × 4	101.6 × 10	273 × 7.1
Weight (kN)	27.6	57.7	106.7	33.5	52.0	102.9
Variance (m)	6.04E-03	2.62E-03	1.34E-03	5.71E-03	2.83E-03	1.35E-03
P_{viol} (%)	4.2E-00	3.9E-01	9.0E-03	1.9E-00	5.1E-01	9.0E-03

* Taken from the Circular Hollow Section (CHS) table of the Eurocode.

when the importance of the second criterion (standard deviation of the response) increases. In Table 2 three designs are compared which have been selected from the two Pareto front curves. Designs B_{RRDO} versus B_{RDO} and C_{RRDO} versus C_{RDO} are similar, with respect to both weight and standard deviation of the response, leading to similar probabilities of violation. On the other hand designs A_{RRDO} and A_{RDO} differ by 17.5% with respect to the weight and by 6.0% with respect to the standard deviation of the characteristic node displacement. Moreover the probability of violation of the constraints, in the case of the A_{RDO} design, is equal to 4.2% while it becomes equal to 1.9% for the A_{RRDO} design case.

5.2 Three dimensional truss tower

The second test example considered is the 3D truss tower shown in Figure 6. The height of the truss tower is 128 m, while its basis is a rectangle of side 17.07 m. The model consists of 324 nodes and 1254 elements divided into 12 groups that play the role of the design variables. The applied loading consists of: (i) self weight (dead load), (ii) live loads and (iii) wind actions according to the Eurocode [20]. The type of probability density function, the mean value, and the variance of the random variables are shown in Table 3.

A similar to the previous test example parametric study, is also performed here, in order to examine the applicability of the MCS with LHS. The probability of violation of the displacement constraint and the standard deviation of the characteristic node displacement are calculated with respect to the number of LHS simulations for the case of a randomly selected design vector. The results of this test example are depicted in Figure 7. For this test example 3,000 LHS simulations have also been considered enough in order to calculate with sufficient accuracy the statistical quantities.

For the solution of the multi-objective optimization problem in question the non-dominant $CEATm(\mu + \lambda)_{nrun,csteps}$ optimization scheme was employed where $\mu = \lambda = 5$, $nrun = 10$ and $csteps = 3$. The resultant Pareto front curves for the two RDO formulations are depicted in Figure 8. As it can be seen from Figure 8 the trend on the influence of considering the probabilistic constraints is similar to the one of the first test example, i.e. it is significant near the area where the weight of the structure is the dominant criterion. When the importance of the standard deviation of the response increases, the two Pareto front curves almost coincide. In Table 4 three designs are compared which have been selected from the two Pareto front curves. Designs B_{RRDO} versus B_{RDO} and C_{RRDO} versus C_{RDO} are similar with respect to both weight and standard deviation of the response while they

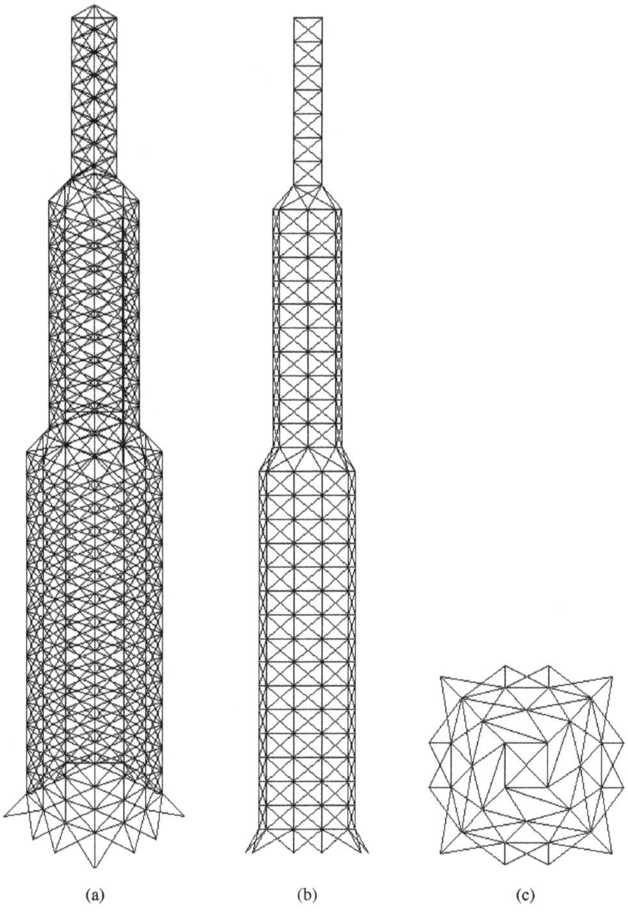

Fig. 6. 3D truss tower: (a) 3D view, (b) side view, (c) top view.

Table 3. 3D truss tower example: characteristics of the random variables.

Random variable		Probability Density Function	Mean value μ	CoV
E (kN/m^2)	Young's Modulus	Normal	2.10E+08	7.14%
σ_y**(kN/m^2)**	Allowable stress	Normal	3.55E+05	10.00%
F (kN/m^2)	Wind loading	Normal	F_μ	40.00%
D	CHS Diameter	Normal	d_i *	2%
t	CHS Thickness	Normal	t_i *	2%

* Taken from the Circular Hollow Section (CHS) table of the Eurocode, for every design.

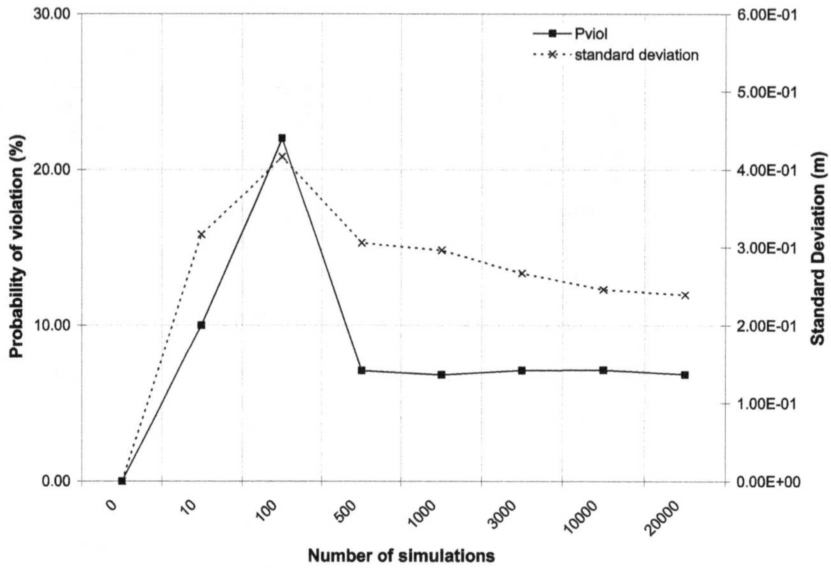

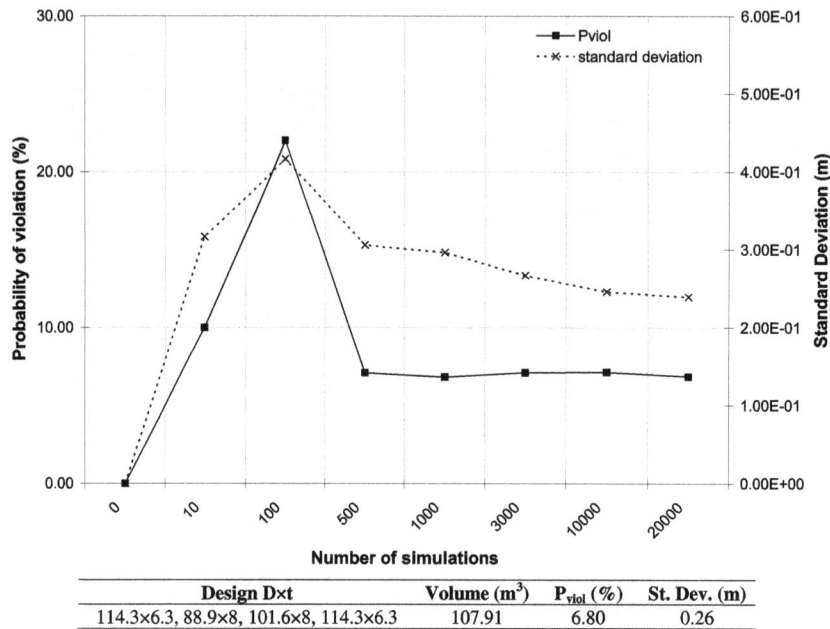

Design Dxt	Volume (m^3)	P_{viol} (%)	St. Dev. (m)
114.3x6.3, 88.9x8, 101.6x8, 114.3x6.3	107.91	6.80	0.26

Fig. 7. 3D truss tower: verification.

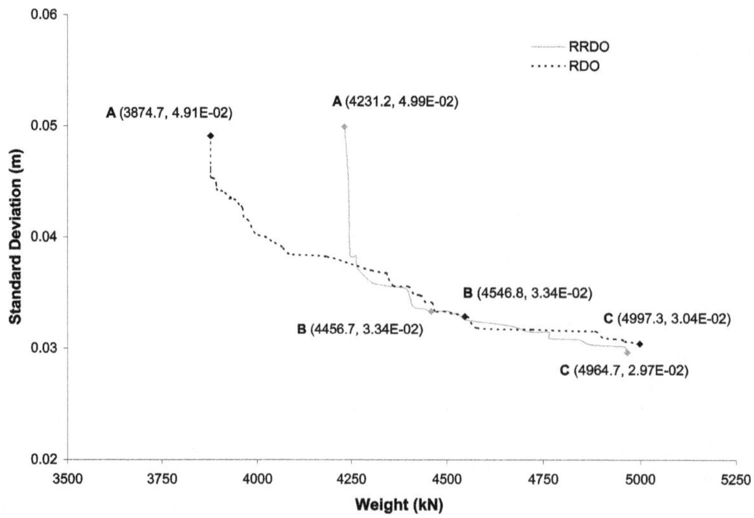

Fig. 8. 3D truss tower: comparison of the Pareto front curves.

Table 4. 3D truss tower example: characteristic optimal solutions.

	RDO			RRDO		
	DBO (A)	(B)	(C)	DBO (A)	(B)	(C)
(D×t)*	(D×t)*	(D×t)*	(D×t)*	(D×t)*	(D×t)*	(D×t)*
Sec_1	355.6 × 8.0	219.1 × 20.0	406.4 × 10.0	323.9 × 8.0	323.9 × 8.0	323.9 × 12.5
Sec_3	355.6 × 10.0	355.6 × 10.0	406.4 × 8.8	355.6 × 10.0	355.6 × 10.0	323.9 × 12.5
Sec_4	323.9 × 10.0	244.5 × 16.0	323.9 × 12.5	355.6 × 10.0	219.1 × 20.0	406.4 × 8.8
Sec_5	406.4 × 8.8	406.4 × 8.8	406.4 × 8.8	406.4 × 8.8	323.9 × 12.5	406.4 × 8.8
Sec_6	355.6 × 8.0	355.6 × 8.0	355.6 × 10.0	355.6 × 8.0	355.6 × 8.0	273.0 × 12.5
Sec_7	323.9 × 8.0	273.0 × 8.0	323.9 × 10.0	323.9 × 8.0	323.9 × 8.0	219.1 × 16.0
Sec_8	323.9 × 10.0	323.9 × 12.5	244.5 × 20.0	355.6 × 10.0	355.6 × 10.0	273.0 × 16.0
Sec_9	323.9 × 8.0	323.9 × 8.0	355.6 × 8.0	323.9 × 8.0	323.9 × 8.0	323.9 × 8.0
Sec_{10}	219.1 × 7.1	323.9 × 8.0	273.0 × 12.5	355.6 × 8.0	355.6 × 8.0	219.1 × 20.0
Sec_{11}	323.9 × 10.0	244.5 × 20.0	244.5 × 20.0	406.4 × 8.8	244.5 × 20.0	244.5 × 20.0
Sec_{12}	323.9 × 8.0	323.9 × 12.5	273.0 × 16.0	323.9 × 10.0	273.0 × 12.5	273.0 × 16.0
Weight (kN)	3874.7	4546.8	4997.3	4231.2	4456.7	4964.7
Variance (m)	4.91E-02	3.34E-02	3.04E-02	4.99E-02	3.34E-02	2.97E-02
P_{viol} (%)	3.8E-00	2.8E-01	8.5E-03	1.8E-00	2.9E-01	8.0E-03

* Taken from the Circular Hollow Section (CHS) table of the Eurocode.

have similar probabilities of violation. On the other hand designs A_{RRDO} and A_{RDO} differ by 10.0% with respect to the weight and by 2.0% with respect to the standard deviation of the characteristic node displacement. Moreover the probability of violation of the constraints, in the case of the A_{RDO} design, is equal to 3.8% while it becomes equal to 1.8% for the A_{RRDO} design.

6 Concluding remarks

In most cases optimum design of structures is based on deterministic parameters and is focused on the satisfaction of the deterministically defined code provisions. The deterministic optimum is not always a "safe" design, since many random factors might affect the design, the manufacturing and the performance of a structure during its lifetime. In order to find the "real" optimum the designer has to take into account all necessary random parameters. For this purpose two separate formulations have been proposed in the past: the reliability based optimization and the robust design optimization. In the present work the reliability based robust design optimization formulation is examined where probabilistic constraints are incorporated into the robust design optimization formulation.

The Pareto front curves obtained for the RRDO and RDO formulations appear to be quite distinct when the weight objective function is predominant. In particular for the same standard deviation the weight of the RRDO optimum solution is larger compared to the optimum solution achieved through the RDO formulation. The probability of violation of the constraints of the RRDO solution is up to 60% lower than the one of violation of the RDO solution. Furthermore it was observed that considering the standard deviation of the response as an additional objective function forces the solution of the RDO formulation to coincide with to of the RRDO formulation.

Acknowledgements. This work has been funded by the project PENED 2003. The project is part of the Operational Programme "Competitiveness" (measure 8.3) of the 3rd Community Support Programme and is co-funded, 75% of public expenditure through EC - European Social Fund, 25% of public expenditure through the Greek Ministry of Development - General Secretariat of Research and Technology and through private sector.

References

1. G.I. Schueller, *Computational stochastic mechanics – recent advances*, Comp. & Struct. **79**, 2225–2234 (2001)
2. K.-H. Lee, G.-J. Park, *Robust optimization considering tolerances of design variables.* Comput. Struct. **79**, 77–86 (2001)
3. A. Messac, A. Ismail-Yahaya, *Multiobjective robust design using physical programming.* Struct. Multidiscip. O. **23**, 357–371 (2002)
4. Hans-Georg Beyer, Bernhard Sendhoff, *Robust optimization – A comprehensive survey.* Comput. Method. Appl. M. **196**, 3190–3218 (2007)
5. M. Papadrakakis, N.D. Lagaros, *Reliability-based structural optimization using neural networks and Monte Carlo simulation.* Comput. Method. Appl. M. **191**, 3491–3507 (2002)
6. X. Qu, R.T. Haftka, S. Venkataraman, *Deterministic and Reliability-Based Optimization of Composite Laminates for Cryogenic Environments.* AIAA Journal **41**, 2029–2036 (2003)
7. M. Allen, K. Maute, *Reliability-based shape optimization of structures undergoing fluid-structure interaction phenomena.* Comput. Method. Appl. M. **194** (30-33), 3472–3495 (2005)

8. J.S. Kong, D.M. Frangopol, *Life-cycle reliability-based maintenance cost optimization of deteriorating structures with emphasis on bridges*. Journal of Structural Engineering - ASCE **129**, 818–828 (2003)
9. Eurocode 3, Design of steel structures, Part1.1: General rules for buildings, CEN, ENV 1993-1-1/1992
10. V. Pareto, Cours d'économique politique, Vol. 1&2, Rouge, Lausanne (1896)
11. R.T. Marler, J.S. Arora, *Survey of Multi-objective Optimization Methods for Engineering*. Structural and Multidisciplinary Optimization **26**, 369–395 (2004)
12. C.A. Coello Coello, *An Updated Survey of GA-Based Multi-objective Optimization Techniques*. ACM Computing Surveys **32**, 109–143 (2000)
13. J. Nocedal, S.J. Wright, *Numerical Optimization*, Springer, New York (1999)
14. N.D. Lagaros, M. Papadrakakis, G. Kokossalakis, *Structural optimization using evolutionary algorithms*. Comput. Struct. **80**, 571–587 (2002)
15. M. Papadrakakis, Y. Tsompanakis, N.D. Lagaros, *Structural shape optimization using Evolution Strategies*. Eng. Optimiz. **31**, 515–540 (1999)
16. N.D. Lagaros, V. Plevris, M. Papadrakakis, *Multi-objective design optimization using cascade evolutionary computations*. Comput. Method. Appl. M. **194**, 3496–3515 (2005)
17. S.N. Patnaik, R.M. Coroneos, J.D. Guptill, D.A. Hopkins, *Comparative evaluation of different optimization algorithms for structural design applications*. Int. J. Numer. Meth. Eng. **39**, 1761–1774 (1996)
18. D.C. Charmpis, N.D. Lagaros, M. Papadrakakis, *Multi-database exploration of large design spaces in the framework of cascade evolutionary structural sizing optimization*. Comput. Method. Appl. M. **194** (30-33) 3315–3330 (2005)
19. M.D. McKay, R.J. Beckman, W.J. Conover, *A comparison of three methods for selecting values of input variables in the analysis of output from a computer code*, Technometrics **21**, 239–245 (1979)
20. Eurocode 1 : Basis of design and actions on structures – Part 2-4: Actions on Structures – Wind actions, CEN, ENV 1991-2-4, May 1995

18

Reliability of the connections used in IGBT modules, in aeronautical environment

A. Zéanh[1,2,3,4,a], O. Dalverny[2], M. Karama[2], E. Woirgard[3], S. Azzopardi[3], A. Bouzourene[1], J. Casutt[1] and M. Mermet-Guyennet[4]

[1] THALES Avionics Electrical Systems, 41 boulevard de la République, BP 53, 78401 Chatou, France
[2] ENI de Tarbes - LGP, 47, avenue d'Azereix, BP 1629, 65016 Tarbes Cedex, France
[3] Université Bordeaux - Laboratoire IMS, 351 cours de la Libération, 33405 Talence Cedex, France
[4] PEARL, Alstom Transport Tarbes, rue du Docteur Guinier, BP4, 65600 Séméac, France

Abstract – In this paper, two IGBT modules assembling technologies with double side cooling capabilities and high level of integration were proposed for aeronautic applications after a state of the art and failures analysis. These technologies are compared using design of experiment based on non-linear finite element analysis with various materials, with respect to their potential failures under thermal and power loading profiles. The configurations optimizing the lifetime and reliability level were pointed out by loading profile and failure mode. Recommendations were then done in order to choose the optimal configuration of assembly for each application. Finally, these design rules were followed by the study of parts dimensions effects on the design outputs in order to help dimensioning the IGBT modules.

Key words: Power electronic packaging; constitutive laws; failures criteria; FEM; DoE.

1 Introduction

Within the framework of the electric plane programs which policy aims at replacing hydraulic actuators by electric ones for weight reduction and reliability improvement, the aircraft industry is facing a higher demand of electric power, fact which involves an increasing use of IGBT modules in power converters. Although such modules have been well studied and known in railway and the automotive domains, they will be subjected to stresses and operational cycles specific to the aeronautical environment. It was predicted that these modules will be used in harsh environment as in the engine nacelle, near the actuators they supply. Consequently, this requires manufacturers to answer some questions about their lifetime and reliability issues.

Many works studied solders lifetime [1] and evaluated modules reliability by probabilistic approaches [2], but the question of an optimal design for a specific application was usually studied by packaging techniques overviews.

The objective of this work is to propose specific technologies of IGBT modules, materials choice and mechanical dimensioning rules, for aeronautic applications.

To tackle this problem, the followed steps consist in analyzing the failures observed in railway and automotive environments, (wire bonding lift-off, solders crack, substrates delamination...), in order to propose solutions concerning connection techniques and materials more appropriate to aeronautical operational environment. Two sandwich assemblies with ceramic substrates and metallizations, brazed on two base plates for double side cooling capabilities were proposed, with respect to aeronautic criteria.

These technologies were then compared with different materials chosen according their availability and in compliance with regulations about the use of hazardous substances [1].

In aeronautic applications, IGBT modules can be considered as operating under a combination of two loading profiles: a power loading at chips levels when switched at high frequency to ensure good input and output current waveforms, and a thermal loading due to environmental temperature variation.

For both profiles, Design of Experiments method was used basing on non-linear Finite Element Modelling results regarding the chips junction temperatures (thermal impedances), the chips and ceramic substrates brittle fracture (maximal principal stresses) and the critical solders joints fatigue (Inelastic Strain Energy Densities – ISED). A 2^3 factorial design built according to the Yates algorithm helped reducing the number of simulations while getting the maximum information on how materials and their interactions affect the module design outputs. The most significant factors with their effects were pointed out for each yield response, and then recommendations were proposed.

[a] Corresponding author: adrien.zeanh@enit.fr

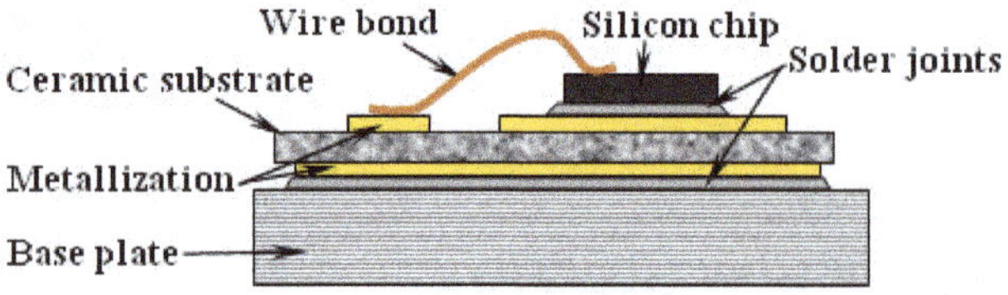

Fig. 1. Structure of a wire bonding IGBT module.

Then, an analysis was performed on module geometrical parameters (bumps sizes and layers thicknesses) to determine how they affect the yield responses. This second analysis brought additional information about the effect of these parameters design rules for each specific application.

2 IGBT modules technologies overview

For high power electronic applications, such as in power transmission, industrial drives, and locomotive traction control systems, many IGBT modules packaging approaches were proposed by industrials and research laboratories. The most common of these technologies are presented hereafter, with their main advantages and drawbacks.

2.1 Wire bonding connection module

This approach is the most used in IGBT module packaging. The chips are brazed on a metallized substrate and base plate, the electrical connections being achieved by wire bonds (Fig. 1). These wires bonds are main cause of failure under power cycling [3] as depicted in Figure 2. Large solder joints cracks (Fig. 3a) are main failures mode under thermal cycling [4]. As shown in Figure 3b, the delamination starts at solder joint corners, causing thermal impedance degradation and then chip excessive heat [5]. The substrates and silicon chips also exhibits high risk of brittle fracture. Other concerns with the use of this technology in aeronautical environment are about the worst electromagnetic compatibility due to high parasitic inductances, the poor thermal management and low integration level. Some other approaches were proposed.

2.2 Pressure pack module

Sandwich assembly with chips pressed between two attachments flanges with molybdenum interleaving part to accommodate the coefficients of thermal expansions (Fig. 5). This technology exhibits a good reliability because of the absence of solder joints, a good thermal management but needs special isolation and cooling systems, leading excessive weight and cost.

2.3 Layer connection technologies

This approach was first proposed in 2001 by General Electric [6], the chips are mounted on metallized substrate with top connections obtained by alternating conducing and dielectric layers (Fig. 6). Other variants of this technology are Planar Power Polymer Packaging (General Electric – 2004) and Embedded Power (CPES – 2004) [7]. Despite of the good electrical and thermal performances with high integration level, these approaches exhibit high parasitic capacitances and require complex and delicate manufacturing processes.

2.4 Bump array contact modules

This technology is based on chips brazed on metallized substrate, the connections being realized by solder bump arrays (Fig. 7). Many variants of this technology were proposed. The most known of this approach are Dimple Array Interconnexion (CPES – 2004) [6], Flip Chip on Flex (CPES, General Electric – 2004) [6] and Power Bump Connection (Eupec, PEARL – 2004). This approach presents a better integration level and thermal management compared to wire bonding technology, but requires complex and delicate manufacturing processes. Questions about lifetime of connection solders during thermal cycling also need to be answered.

2.5 Direct solder contact module

This approach first investigate in 2000 by INPG – LEG, CEA – LETI and Alstom Technology – PERT [8], aims at improving the thermal impedance and reliability by balancing the module structure [9]. The chips are sandwiched by two metallized substrates with connections by direct solder (Fig. 8). Despite of the good electrical and thermal performances with high integration level [8, 10], this technology needs designers to answer questions about dielectric withstand, because of the low distance between metallizations.

3 Choice of IGBT modules configurations

Selection criteria were defined on the basis of thermal and thermomechanical efficiency, (chips connection possibilities, integration level, thermal management, electromagnetic compatibility, weight, volume, materials safe operating areas and toxicities, cost, processing...). Two sandwich technologies with Power Bump (PB) (Fig. 9) and Direct Solder (DS) (Fig. 10) connection technique were retained. Both assemblies allow double side cooling with good integration levels.

3.1 Potential failures modes

From mechanical point of view, two main failures are susceptible to occur in these technologies: solder joints thermomechanical fatigue and brittle materials (ceramics and

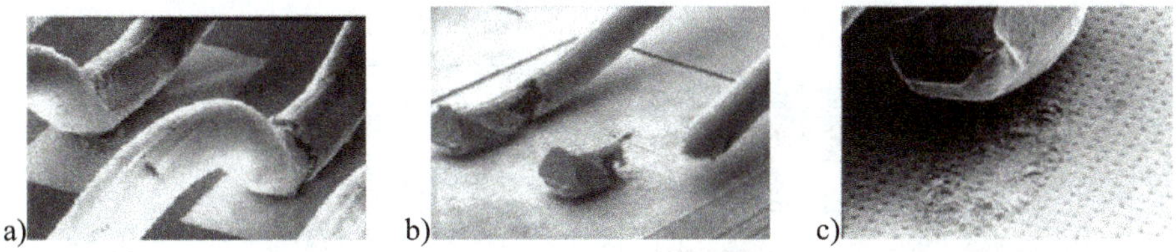

Fig. 2. Rupture (a), corrosion (b), and lift-off (c) of wire bonds [3].

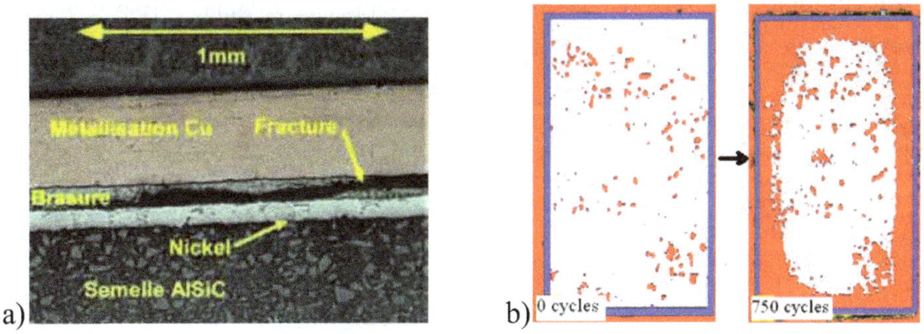

Fig. 3. Base plates solder fatigue fracture (a) [4], and delamination progression (b) [5].

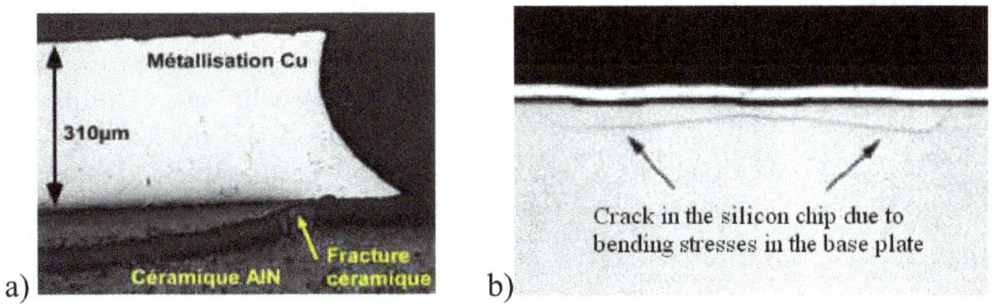

Fig. 4. Fracture of ceramic substrates (a) [4] and silicon chip (b) [3].

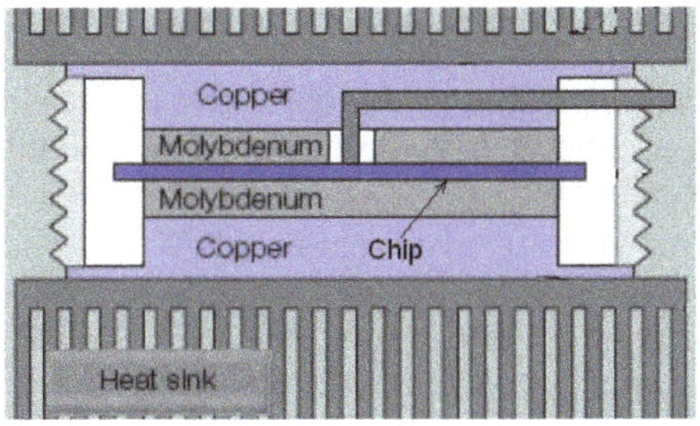

Fig. 5. Pressure Pack assembling stack.

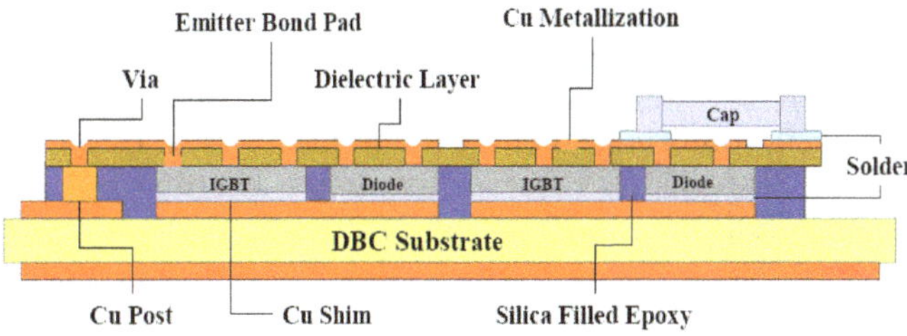

Fig. 6. Power Overlay Technology (General Electric – 2001).

silicon chips) fracture. Beside these failures, the electronic components (IGBT and Diodes) failures by excessive heat due to poor thermal management were also taken into account.

3.2 Materials

The two basic assemblies will be compared with different materials chosen according to their availability, their compliance with aeronautic criteria and with the directives RoHS and WEEE on the use of hazardous substances in electrical and electronics equipments [1, 11].

Aluminium nitride (AlN) and silicon nitride (Si_3N_4) ceramic substrates were retained with copper and aluminium as metallizations. The assembling process of these packaging requires the use of two solders with high and low melting point. According to their good mechanical strength, melting points and wetting properties, the eutectic $Sn_{96.5}Ag_{3.5}$ were retained for all the connections, except the chips bottom (collectors or anodes) solders which are realized with $Pb_{92.5}Sn_5Ag_{2.5}$ solder. Two metal matrix composites (Al-SiC(63%) and Cu-C(40) [12]) are considered as base plates.

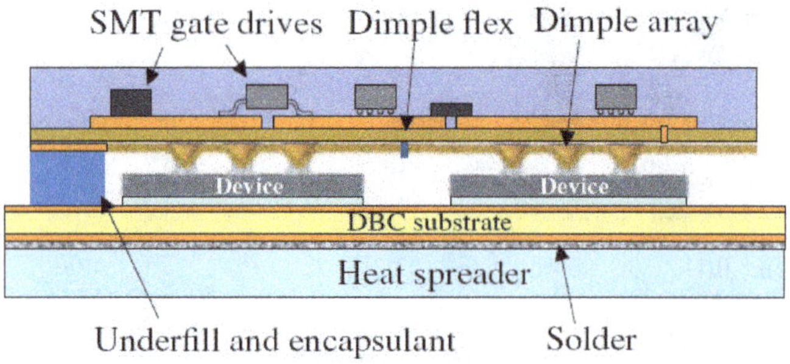

Fig. 7. Dimple array interconnexion (CPES – 2004).

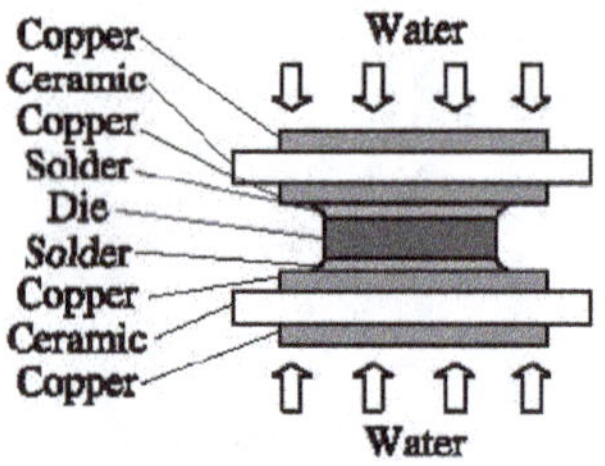

Fig. 8. Direct solder contact module [8].

Table 1. Base plates constitutive laws parameters [4,12].

	Al-SiC(63%)	Cu-C(40)
λ (W/(K m))	175	300// 160⊥
C (J/(kg °C))	741	420
ρ (kg/m^3)	4000	6100
CTE (10^{-6}/K)	7.9	8.5
E (GPa)	192	75
ν	0.24	0.3

4 Materials thermal and mechanical behaviour

Thermal and mechanical properties were gathered from literature researches, for all the materials regarding their operating conditions, melting point, yield stress, Ultimate Tensile Strength (UTS), Coefficients of Thermal Expansion (CTE), etc.

4.1 Ceramics and metallizations

The base plates thicknesses lead stresses to remain in their elastic domains. Linear elastic law was used to model their mechanical behaviour. The corresponding material law parameters are presented in Table 1.

As suggested by Dupont in [4], elastoplastic law with linear kinematic hardening was used to model metallization mechanical behaviour. Table 2 shows the corresponding material parameters.

4.2 Silicon chips and ceramic substrates

Regarding the observed failures at their level, and their traditional operating conditions, these materials will be modelled with elastic linear law, with brittle fracture.

Table 2. Metallization constitutive laws parameters [4].

	Aluminium	Copper
λ (W/(K m))	220	398
C (J/(kg °C))	880	380
ρ (kg/m^3)	2700	8850
CTE (10^{-6}/K)	24	17.3
E (GPa)	70.6	128
ν	0.34	0.36
Yield Stress (MPa)	17.8	98.7
Tangent modulus (MPa)	350	1000

Table 3. Chips and ceramics constitutive laws parameters [4].

	Silicon chip	AlN ceramic	Si_3N_4 ceramic
λ (W/(K·m))	146	190	60
C (J/(kg·°C))	750	750	800
ρ (kg/m^3)	2330	3300	3290
CTE (10^{-6}/K)	2.5	4.5	3.3
E (GPa)	130	344	310
ν	0.22	0.25	0.27
UTS (MPa)	200	400	800

Many models based on weakest link theory were proposed to describe the ceramics rupture [13]. These approaches need the material parameters to be identified, but a more simple way is to consider the Rankine's (maximal principal stress) criterion which is widely sufficient within a comparison purpose. The corresponding material parameters are given in Table 3.

4.3 Solders joints

The solders, operating at temperatures above the third of their melting points, were described using Anand's unified viscoplastic model [14].

This law was not available in ABAQUS^{TM} Software, it was integrated according to the steps suggested by [15], and implemented via the user interface UMAT. A FORTRAN subroutine, calculating the stress increment and material Jacobian matrix, knowing the strain field, was written. Simulations and verifications with various solders revealed that there are good agreements between the user subroutine and bibliography results presented in [16,17]. Table 4 lists hereafter the two solders constitutive laws parameters.

Concerning the solders fatigue, inelastic dissipation is believed to better capture the accumulated damage. Many authors proposed cyclic Inelastic Strain Energy Density (ISED) based models for solders joints lifetime prediction [1,18]. These models show that the number of loading cycles before solder failure is a monotonic decreasing function of the ISED dissipated per cycle. The analysis of the various assembling configurations is for this reason done hereafter according to the ISED dissipated in solder joints per loading cycle.

A. Z´

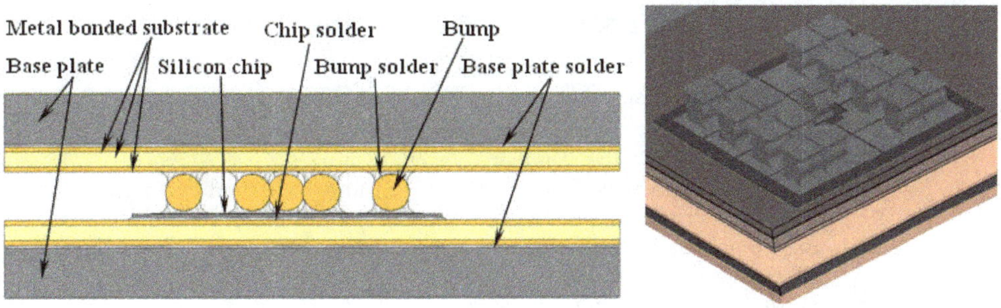

Fig. 9. Power Bump (PB) assembly (Infineon 1200 V – 150 A chip).

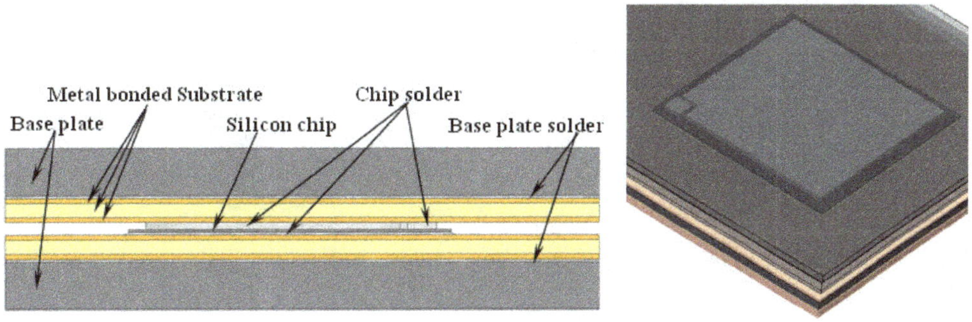

Fig. 10. Direct Solder (DS) assembly (ABB 1200 V – 150 A chip).

Table 4. Solders constitutive laws parameters [16, 17].

	$Pb_{29.5}Sn_5Ag_{2.5}$	$Sn_{96.5}Ag_{3.5}$
λ (W/(K m))	35	33
C (J/(kg °C))	129	200
ρ (kg/m^3)	11300	7360
CTE (10^{-6}/K)	29	30.2
E (MPa)	24028−28·T(°C)	47200−191·T(°C)
ν	0.44	0.4
	Anand's Parameters	
s_0 (MPa)	33.07	7.72
Q/R (K)	11024	14100
A (s^{-1})	105000	1630000
ξ	7	1.61
m	0.241	0.13
h_0 (MPa)	1432	58700
$\hat{s}$ (MPa)	41.63	11.99
n	0.002	0.017
a	1.3	2.09

5 Numerical design of experiment

With respect to the potential failure modes and loading profiles, nine design responses were defined and considered for this analysis: thermal impedances for chips junction temperatures, maximal principal stresses for chips and ceramic substrates brittle fracture, and the inelastic strain energy densities for critical solders joints fatigue, for both loadings profiles. Table 5 lists the responses considered, with their given labels.

The nine plans was built following the Yates algorithm [19], 3 factors listed in Table 6, with 2 levels each other, were considered with their first order interactions. The responses calculations were performed with the nonlinear finite element models presented below.

5.1 Finite element modelling

Two parameterised finite element models (Fig. 11) generated using python scripts under ABAQUS^{TM} were used. The geometries of the elementary modules are based on standard thicknesses used in automotive and railway domains: base plates (3 mm), ceramic substrates (635 μm) metallizations (300 μm), base plate solder joints (100 μm) bumps cylinders (ϕ 1.4 mm × 1.5 mm) and direct connection solder joints (200 μm).

Three-dimensional solid linear brick and tetrahedron elements were used for meshing the geometry, the interfaces being supposed perfect.

As suggested by Guédon-Gracia et al. in [20], the assembling process and storage were simulated for the various configurations in order to compute residual stresses across the whole assemblies before cycling. The two loading profiles considered are presented below.

5.1.1 Power cycling

To model the power cycling, 250 W heat dissipation was generated in the whole volume of the IGBT chip within relative short cycles as shown in Figure 12. An overall heat transfer coefficient corresponding to water cooling, at 70 °C reference temperature was applied on the two external sides of the base plates as boundary conditions.

Table 5. DoE responses definition.

Responses	
$Y1$	Thermal impedance
$Y2$	Base plate solder cyclic ISED under thermal cycling
$Y3$	Chip solder cyclic ISED under thermal cycling
$Y4$	Substrate stress ratio (max. stress/UTS) under thermal cycling
$Y5$	Chip stress ratio (max. stress/UTS) under thermal cycling
$Y6$	Base plate solder cyclic ISED under power cycling
$Y7$	Chip solder cyclic ISED under power cycling
$Y8$	Substrate stress ratio (max. stress/UTS) under power cycling
$Y9$	Chip stress ratio (max. stress/UTS) under power cycling

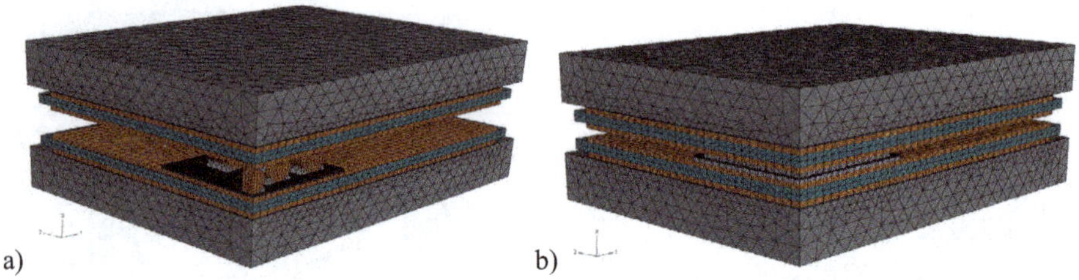

Fig. 11. Elementary power bump (a) and direct solder (b) modules finite elements models.

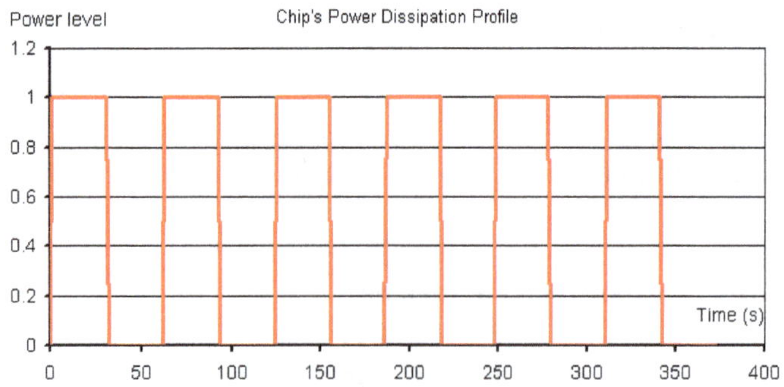

Fig. 12. Power cycling profile.

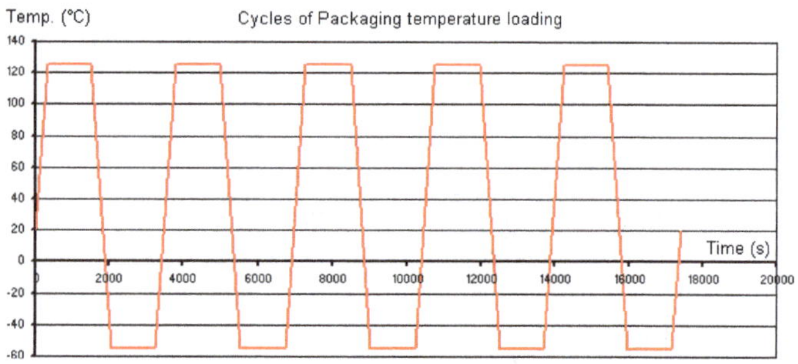

Fig. 13. Thermal cycling profile.

Table 6. DoE factors and levels.

	Factors	Levels	
		-1	1
A	Connection	PB	DS
B	Metallization	Copper	Aluminium
C	Ceramic Substrate	Si_3N_4	AlN

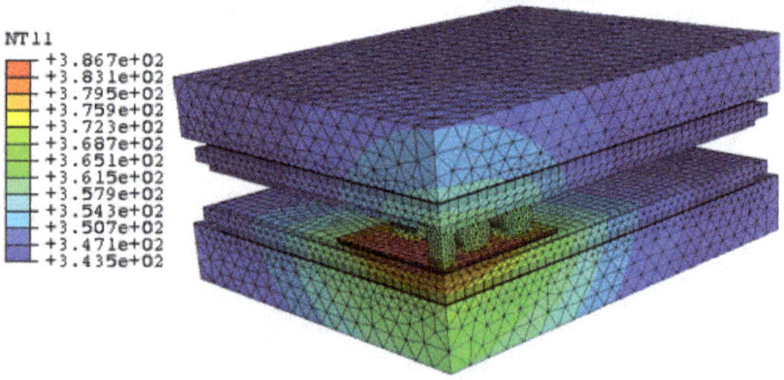

Fig. 14. Temperature distribution in solder bump assembly (K).

5.1.2 Thermal cycling

The International Standard of Atmosphere (ISA) gives the aircraft external temperature profile during a flight cycle. To represent these loadings, an accelerated thermal profile defined according to the Military Standard Handbook 883F was considered.

As depicted in Figure 13, the profile starts at 25 °C, the ramp rate was 20 °C/min and the dwell time at −55 °C and 125 °C was 20 min.

5.2 Simulation results

The heat flux balance across assemblies and the thermal impedances were evaluated from the power cycling simulations (Fig. 14). The thermal impedances ($Y1$) were calculated from the chip maximal temperatures, measured after the steady state is reached. For all the configurations, the design responses $Y4$, $Y5$, $Y8$ and $Y9$ were computed by the ratio of maximal principal stresses in the ceramic substrates (Fig. 15) and chips (Figs. 16 and 17), over their mechanical strengths.

The maximal principal stress in the ceramic substrate is localized at the periphery of the bonded metallization.

A. Z´

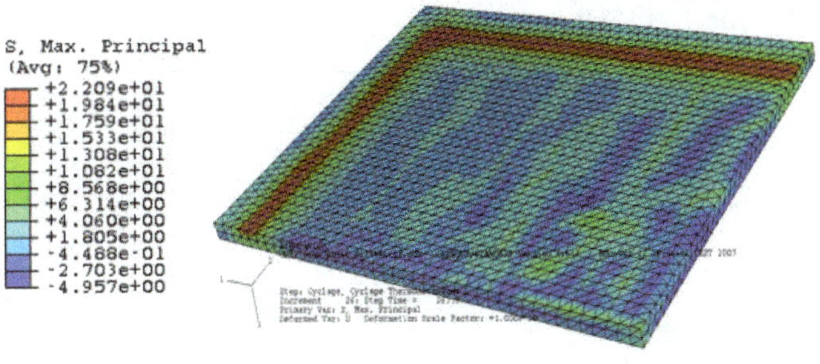

Fig. 15. Maximal principal stress distribution in ceramic substrate (MPa).

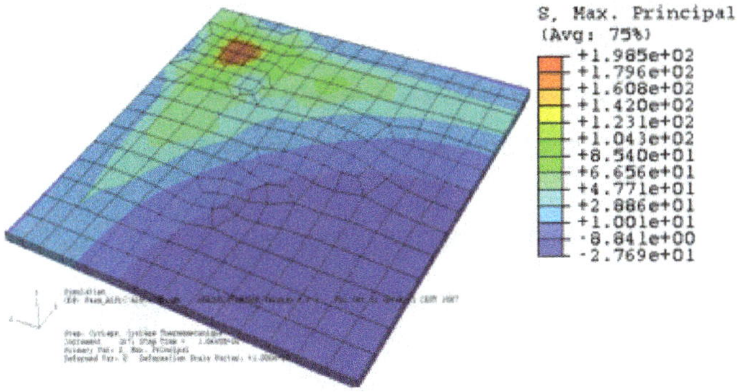

Fig. 16. Maximal principal stress distribution in silicon chip with DS connection (MPa).

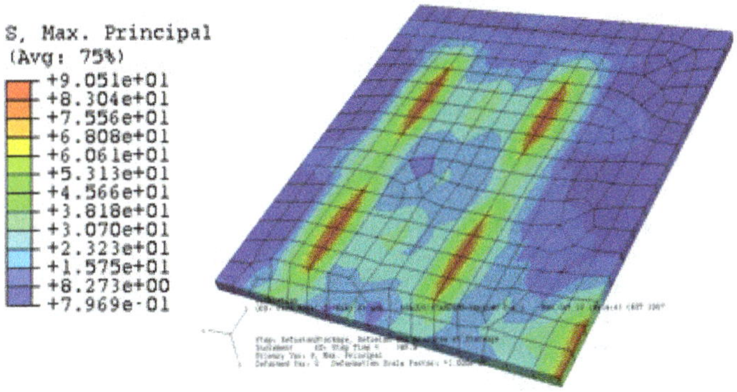

Fig. 17. Maximal principal stress distribution in silicon chip with PB connection (MPa).

In the chips, the maximal principal stress were localized around bump contacts for solder bump connection (Fig. 16) and just under the gate for direct solder connection (Fig. 17).

In order to evaluate $Y2$, $Y3$, $Y6$ and $Y7$, inelastic strain energy densities were computed over the solder joints (Figs. 18–20). Many works [5] showed that the crack in large area solders propagates from solder joint corner (Fig. 3b). The ISED distribution in base plate solder (Fig. 18) is consistent with this result, and then, for more accuracy in this analysis, $Y2$ and $Y6$ were evaluated in the corner elements instead of the whole solder joint.

Regarding connection solders, the ISED were found to be maximal in the bump solder in solder bump assemblies (Fig. 19) and in the gate solder for direct solder assemblies

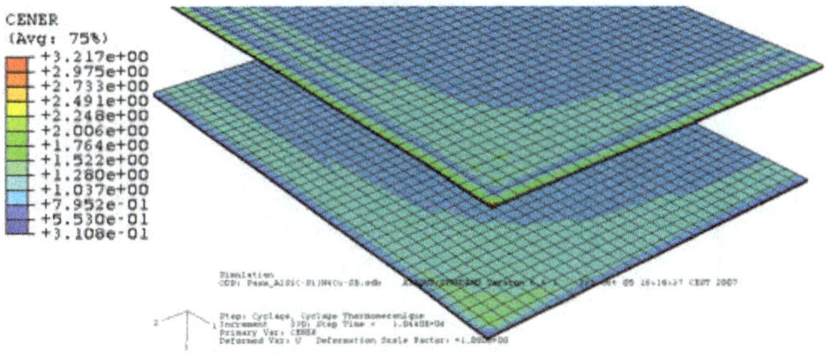

Fig. 18. ISED distribution in base plate solder after thermal cycling (mJ/mm^3).

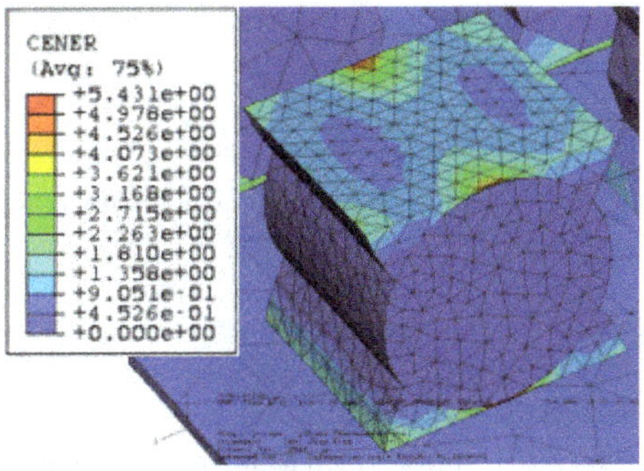

Fig. 19. ISED distribution in bumps solder after thermal cycling (mJ/mm^3).

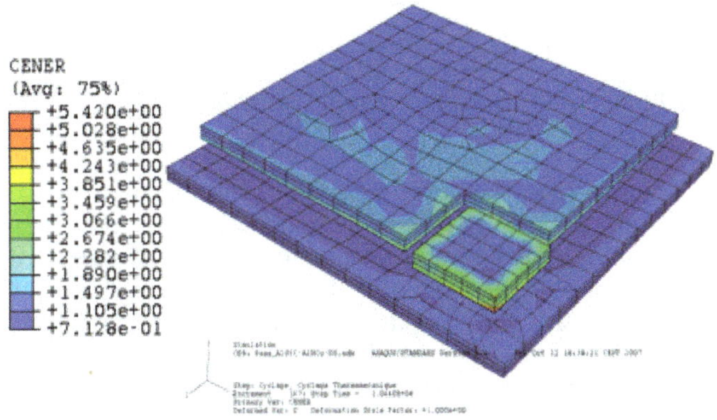

Fig. 20. ISED distribution in direct connection solder after thermal cycling (mJ/mm^3).

(Fig. 20). Due to the relative small size of these solders, $Y3$ and $Y7$ were evaluated in their whole volume.

As observed in Figure 21, the inelastic strain energy density accumulation per cycle, quickly reaches a saturated value within three loading cycles. This observation allows reducing the computation time, by calculating $Y2$, $Y3$, $Y6$ and $Y7$ within the third cycle, while having good accuracy on the results.

5.3 DoE results and analysis

Basing on simulation results, the effects of the factors were evaluated for the nine responses. A student test with a risk of 5% helped identifying the most significant factors for the nine models. All the 9 responses models built

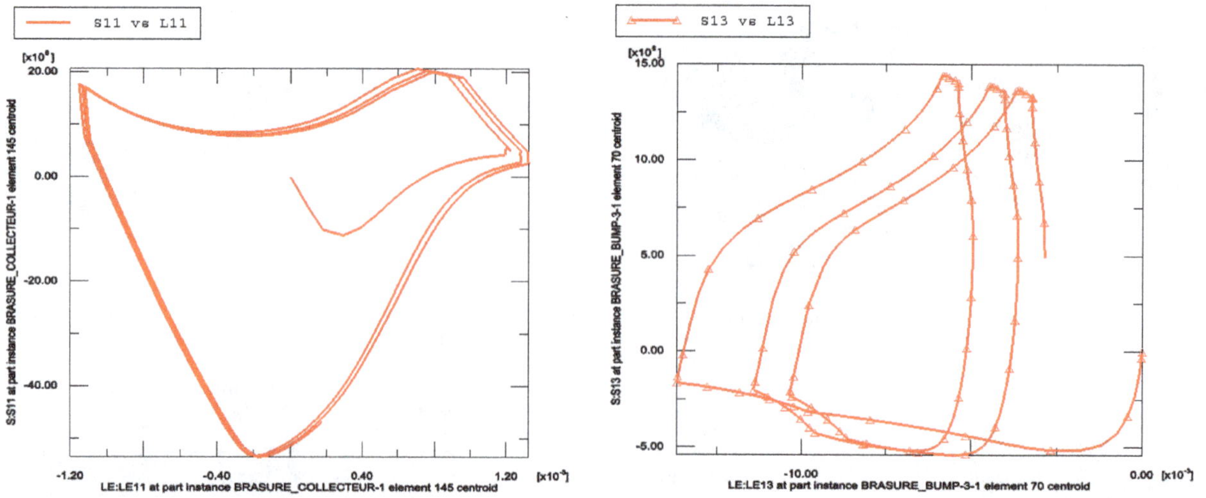

Fig. 21. Normal and shear stress versus strain, during thermal cycling without residual stresses computation.

with the identified factors passed successfully a Fischer-Snedecor test with a 5% risk. Table 7 sums these results, with the significant effects highlighted in bold.

The assembling configurations minimising the design responses, and therefore maximising the modules lifetimes and reliabilities were then identified with respect to the loading profiles.

5.3.1 Thermal impedance and fluxes balance

It appears that the three factors are significant, without their interactions. Direct solder connection with AlN substrate and copper metallization should be preferred, but considering the intrinsic contribution of the metallization within the thermal impedance model (4.6%), aluminium could be an excellent alternative. The fluxes should be more balanced across the assemblies in order to minimize thermal stress concentrations. Regarding this characteristic, direct solders assemblies are superior with 47% of thermal flux balance through bottom and 53% through top cooling faces, compared to solder bump ones which lead to a partition of 26–74%. Other simulation with the two base plates showed that Cu-C is superiors than Al-SiC with about 6.3% of thermal impedance improvement.

5.3.2 Base plates solder fatigue

The only significant factors are the ceramics and metallizations. The use of AlN substrate with aluminium metallization reduces significantly the inelastic strain energy density for both loading profiles. Regarding base plates, Cu-C are superiors than Al-SiC in power cycling, due to their better thermal performances, but less interesting in thermal cycling because of higher coefficient of thermal expansion.

5.3.3 Chip connections solder fatigue

Only aluminium metallizations are recommended for thermal cycling. Considering power cycling loading, direct solder connection, AlN substrate with copper metallization should be recommended.

5.3.4 Ceramic fracture

The stress ratio in the ceramic only depends on the metallization and the ceramic substrate for both loading profiles. Aluminium and Si_3N_4 ceramic minimises ceramic cracking failures risk.

5.3.5 Chip fracture

During thermal cycling, the recommendations are direct solder connection, AlN substrate with aluminium metallization. None of the factors considered have not significant effects on chips stresses during power cycling.

5.4 Discussion

Some contradiction appears in the above analysis when trying to take into account the recommendations for all the responses and loading profiles together. To solve these contradictions, the materials choices could be done with respect to the most critical design outputs and the most preponderant profile, knowing that IGBT modules really operate under a combination of thermal and power loadings. The significant effects presented in Table 7 could then help doing the appropriate design.

The direct solder connection assembly with aluminium bonded AlN appears as the most consensual configuration.

A good solution could be the use of AlN substrate with copper bonded at the chip side and aluminium bonded at base plate side, with suitable thicknesses in order to avoid substrates fracture by bimetallic effect.

Table 7. DoE effects and significance test results.

Responses	I	A	B	C	AB	AC	BC
$Y1$	0.152	−0.031	0.007	−0.013	−0.002	0.003	−0.001
$Y2$	471749	1927	−14631	−16313	−2083	−342	2487
$Y3$	319290	27350	−51018	−13670	4212	−278	−2644
$Y4$	0.138	0.005	−0.046	0.040	0.008	0.001	−0.018
$Y5$	0.621	−0.054	−0.006	−0.046	−0.047	−0.006	0.005
$Y6$	31656	−713	−29262	−7259	821	861	6989
$Y7$	15983	−7660	2198	−4235	−855	1775	−418
$Y8$	0.109	0.001	−0.029	0.032	−0.004	−0.001	−0.008
$Y9$	0.399	−0.035	−0.037	−0.040	−0.078	−0.034	−0.012

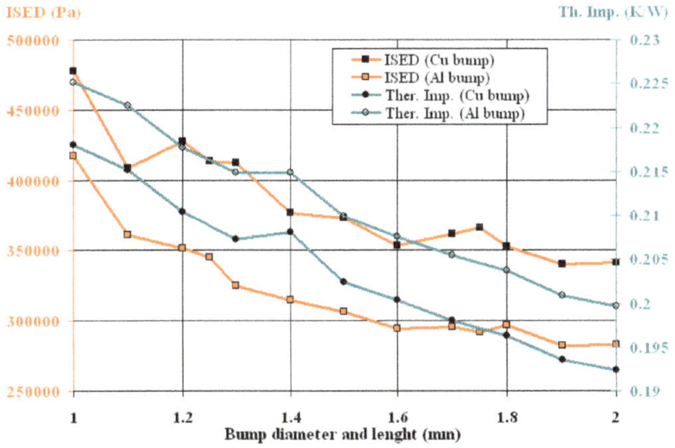

Fig. 22. Evolution of bump solders ISED and module thermal impedance with bump size and material.

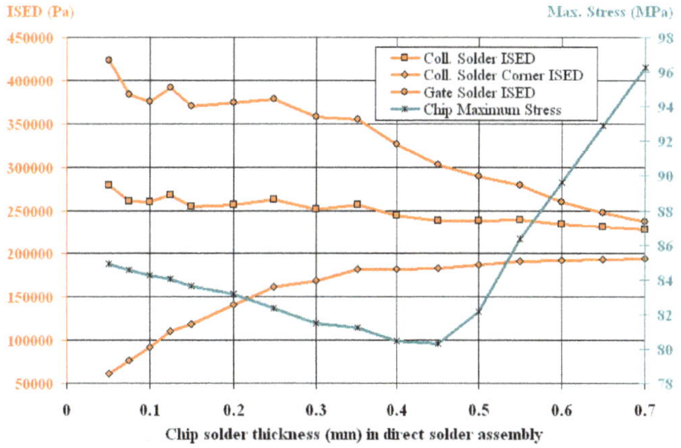

Fig. 23. Evolution of direct connection solders ISED and chip maximal principal stress with chip solder joint thickness.

6 Influence of geometric parameters on module lifetime

The objective of this study is to investigate how the thicknesses of base plate solder, metallizations, ceramic substrates, direct connection solder and bumps dimensions affect the module lifetime. For this purpose, different parameterized models with four IGBT chips, AlN substrates and Al-SiC base plates were generated under ABAQUS^{TM} using python scripts. The results are presented here after.

6.1 Bump nature and sizes

The design parameters chosen for sizing the bumps were the module thermal impedance and the bump solders fatigue. Two materials (cooper and aluminium) were considered with sizes (diameter and length) varying from 1 to 2 mm. Figure 22 presents the evolution of two design parameters for the two kinds of bump. It appears that the module thermal impedance and bumps solder cyclic ISED decrease with the bump size. Moreover, the use of aluminium bump instead of copper reduces bump solders cyclic ISED of about 18% while increasing the thermal impedance of about 3.6%.

6.2 Direct connection solder thickness

The design parameters chosen for sizing the solder thickness were the module thermal impedance, the chip maximal stress and the solder joints fatigue. The simulations were done with thicknesses varying from 50 to 700 μm. As presented in Figure 23, the inelastic strain energy density seems to be constant in the whole chip solder joint, while increasing in the corners. The gate solder joint which is the most critical, decreases with the solder thickness. The chip maximal principal stress presents a minimum for a solder joint thickness around 0.45 mm, but the module thermal impedance increases with a rate of about 0.132 K/(W mm).

6.3 Base plate solder thickness

The design parameters chosen for sizing the base plate solder thickness were the module thermal impedance and the base plate solders fatigue. The simulations were done with thicknesses varying from 20 to 300 μm. Figure 24 presents the evolution of the two design parameters considered. The thermal impedance increases linearly with a rate of 0.122 K/(W mm), while the cyclic ISED seems to be constant when calculated in the whole solder joint, but quickly decreases at the corners until 0.1 mm before decreasing linearly with a rate of about 187 000 Pa/mm.

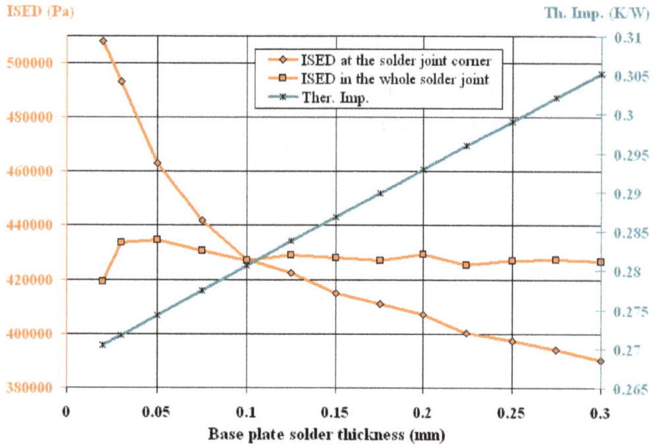

Fig. 24. Evolution of base plate solder ISED and module thermal impedance with base plate solder thickness.

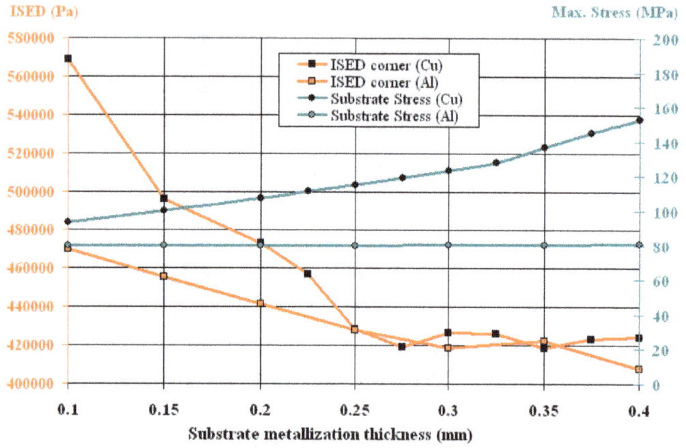

Fig. 25. Evolution of base plate solder corner ISED and substrate maximal principal stress with metallization nature and thickness.

6.4 Metallization nature and thickness

The design parameters chosen for sizing the metallization thickness were the module thermal impedance, the substrate maximal stress and the base plate solders fatigue. Two materials (cooper and aluminium) were considered, the simulations being done with thicknesses varying from 100 to 500 μm. Figure 25 presents the evolution of cyclic ISED at the corner of base plate solder joint. The maximal principal stress in the substrate seems to be constant with aluminium metallization, but increases linearly with copper metallization with a rate of about 150 MPa/mm. Regarding the module thermal impedance, it evolves linearly, with a rate of 0.004 K/(W mm) for aluminium and −0.037 K/(W mm) for copper metallization.

6.5 Substrate thickness

The design parameters chosen for sizing the substrate thickness were the module thermal impedance, the substrate maximal stress and the base plate solders fatigue. The simulations were done with thicknesses varying from 200 to 1400 μm. The maximal principal stress in the ceramic decreases with the substrate thickness (Fig. 26),

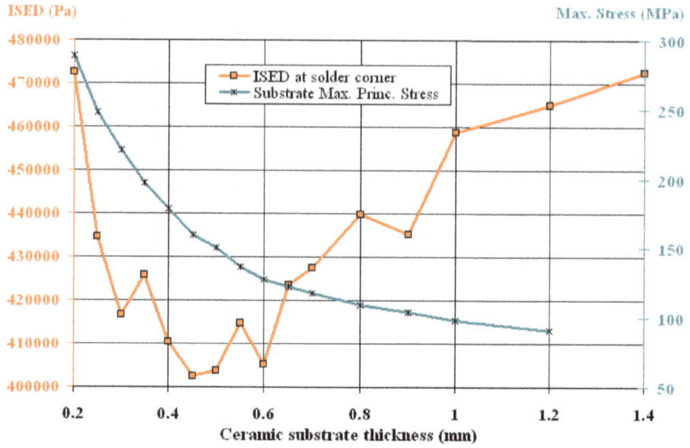

Fig. 26. Evolution of base plate solder corner ISED substrate maximal principal stress with substrate thickness.

while the thermal impedance increases linearly with a rate of about 0.013 K/(W mm). Regarding the base plate solder corner cyclic ISED, the simulations done here with a 300 μm copper metallized AlN substrate showed an optimum of substrate thickness between 0.4 and 0.6 mm. Additional simulations could be necessary for each specific application in order to determine the ceramic optimal thickness, with respect to the metallization material and thickness.

7 Conclusion

The objective of this study was to propose IGBT modules technologies based on materials available today, and propose design rules to optimise the lifetime for aeronautic applications. FEM based DoE was done with respect to the main failure modes under thermal and power cycling loading profiles.

This work showed that there was not optimal configuration of assembly (connection techniques, materials and dimensions) for all the applications, the designer should then take into account the most representative profile with the most critical design parameters to define the module. To do this, the results presented in this paper could be very helpful. It is obvious that complementary simulations could be necessary for each specific application in order to verify the interactions between all the parameters.

Five configurations are being manufactured for thermal and power cycling tests, in order to verify the predicted thermomechanical performances and failures criteria.

References

1. A. Guédon-Gracia, *Contribution à la conception thermo-mécanique optimisée d'assemblages sans plomb*, Ph.D. Thesis of Université Bordeaux 1, France (2006)
2. A. Micol, M. Karama, O. Dalverny, C. Martin, M. Mermet-Guyennet, *Reliability Design of Power Modules Using Probabilistic Approaches*, Proc. of the 8th Conf. on

Computational Structures Technology (12–15 September, 2006)
3. M. Ciappa. Microelectronics Reliability **42**, 653–667 (2002)
4. L. Dupont, *Contribution à l'étude de la durée de vie des assemblages de puissance dans des environnements haute température et avec des cycles thermiques de grande amplitude*, Ph.D. Thesis of ENS Cachan - France (June 2006)
5. T. Lhommeau, C. Martin, M. Karama, R. Meuret, M. Mermet-Guyennet, *Base-plate solder reliability study of IGBT modules for aeronautical Application, Proc. of EPE* (Aalborg, 2007)
6. F.C. Lee, J.D. van Wyk, D. Boroyevich, G.Q. Lu, Z. Liang, P. Barbosa. IEEE Circuits and Systems Magazine **2**, 4–23 (2002)
7. S. Dieckerhoff, *Thermal performance and advanced power electronics packages*, ECPE Seminar (Nürnberg, November 09, 2006)
8. C. Gillot, C. Schaeffer, C. Massit, L. Meysenc. IEEE Transactions on Components and Packaging Technologies **24** (2001)
9. J.G. Bai, J.N. Calata, G.Q. Lu, *Comparative Thermal and Thermomechanical Analyses of Solder-Bump and Direct-Solder Bonded Power Device Packages Having Double-Sided Cooling Capability, IEEE 0-7803-8269* (February 2004)
10. C.M. Johnson, C. Buttay, S.J. Rashidt, F. Udreat, G.A.J. Amaratungat, P. Ireland, R.K. Malhan, *Compact Double-Side Liquid-Impingement-Cooled Integrated Power Electronic Module*, Proceedings of the 19th International Symposium on Power Semiconductor Devices & ICs (Jeju, Korea, May 27–30, 2007)
11. European parliament directives concerning the use of hazardous substances http://ec.europa/environment/index_en.htm
12. J.F. Silvain, J. Soccart, *Copper matrix composite materials used for thermal management of high power electronic devices*, Imaps France (La Rochelle, 2006)
13. L.J.M.G. Dortmans, G. de With. J. Eur. Ceramic Soc. **6**, 369 (1990)
14. G.Z. Wang, Z.N. Cheng, K. Beker, J. Wilde. Journal of Electronic Packaging **123** (2001)
15. A.M. Lush, G. Weber, L. Anand. Int. J. Plasticity **5**, 521 (1989)
16. M. Pei, J. Qu, *Constitutive Modeling of Lead-Free Solders*, School of Mechanical Engineering, Georgia Institute of Technology, 0-7803-9085 – IEEE (July 2005)
17. J. Wilde, K. Becker, M. Thoben, W. Blum, T. Jupitz, G. Wang, Z.N. Cheng. IEEE Transactions on Advanced Packaging **23** (2000)
18. R. Darveaux, *Effect of Simulation Methodology on Solder Joint Crack Growth Correlation*, 50th Electronic Components and Technology Conference (2000)
19. M. Pillet, *Les plans d'expériences par la méthode Taguchi* (Les Éditions d'Organisation, Paris, 1997)
20. A. Guédon-Gracia, P. Roux, E. Woirgard, C. Zardini. Microelectronics Reliability **44**, 1331 (2004)

19

Optimum design of SRC composite beams

Shan-Suo Zheng[1,a], Lei Zeng[1], Wei-Hong Zhang[2], Jie Zheng[1], Lei Li[1] and Bin Wang[1]

[1] School of Civil Engineering, Xi'an University of Architecture and Technology, Xi'an 710055, P.R. China
[2] Northwestern Polytechnical University, Xi'an 710049, P.R. China

Abstract – Based on experimental study of bond-slip behaviours between steel shape and concrete, an optimal design scheme of single objective and discrete variables is proposed to design steel reinforced concrete (SRC) frame beams. In the optimum scheme, design variables include the layout dimensions of SRC frame structure, structural member sections, strength of concrete and steel, dimensions of steel shapes. The objective function is cost of the entire materials applied to construct SRC frame structure. The constraint conditions are main requirements stated in Chinese code for design of SRC structures, basic design rules, reasonable calculating theories and indispensable constructions, as well as some mature and consistent conclusions confirmed by experimental studies on calculating methods of SRC structures based on the bond-slip theory between steel shape and concrete. These may be a reference in analyzing and designing SRC frame structures and provide a practical method satisfying civil engineering practice and requirements stated in the codes.

Key words: SRC beam; bond-slip theory; optimum design; mixing penalty function method

1 Introduction

Steel reinforced concrete composite structure, SRC structure for short, is a main kind of steel and concrete composite structure that steel shapes are embedded in the reinforced concrete [1]. SRC Structure has been applied in many practical engineering, such as high-rise buildings, heavy-duty factory buildings, long-span bridges, submarine tunnels, drilling platforms, etc.

SRC structural members have a relatively small sectional dimension and a high bearing capacity. SRC structures obtain a better integral stiffness, and possess an excellent seismic behaviour. The fine bond behaviour between steel shape and concrete is the base of their cooperation. The cooperation is the reflection of the advantages of SRC structures. The bond behaviour directly affects the bearing capacity, failure modalities, crack, deformation and calculating methods etc [2].

Most of the researches on SRC structures at home and abroad focused on approximate calculation of member's strength, stiffness etc, but few of them take the bond-slip behaviour between steel shape and concrete into account in analyzing member's ultimate bearing capacity. Nowadays, test and research on SRC structures indicate that bond-slip phenomenon exists between steel shape and concrete, and this bond-slip behaviour has a significant impact on mechanical behaviour, failure form and calculation hypothesis of structural members, especially after arriving 80% of ultimate bearing capacity, the influence is more apparent.

Optimum design converts structural analysis from the traditional and passive method to an active optimum design method, which solves optimal problems under certain preset conditions by applying computer and greatly improves economic benefit and design efficiency [3,4]. Optimum design of SRC structures is a design method that acquires maximum economic value at minimal cost. Few researchers study on optimum design of SRC structures at home and abroad, so a perfect optimization design system of the structures has not been formed. Based on experimental study of bond-slip behaviours between steel shape and concrete, an optimal design scheme of single objective and discrete variables is proposed to design SRC frame beams in this paper.

2 Calculation of SRC frame beams based on bond-slip theory

2.1 Theoretical frame

In SRC structural members, there exists bonding action between the two different materials of steel and concrete, which enables stress to transfer effectively each other on the interface of steel and concrete and furnishes working stress needed to fulfill loading capacity of structures in steel and concrete [5]. Bond stress is a kind of shear force in macroscopical view. Experiments indicate that

[a] Corresponding author: zhengshansuo@263.net

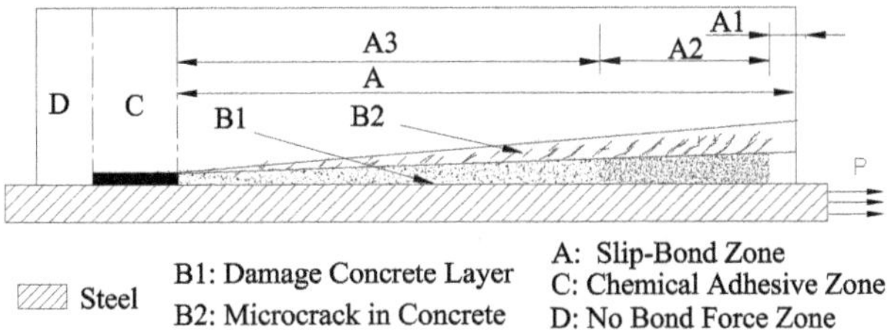

Fig. 1. Distribution of interfacial bond stress.

the distribution of bond stress on the interface of steel and concrete is illustrated as Figure 1.

Based on theory of mechanics of materials, the distribution of bond stress along steel surface is given as follows

$$\tau(x) = \frac{A_s}{u}\frac{d\sigma_x}{dx} = \frac{A_s}{u}E_s\frac{d\varepsilon_x}{dx} = \frac{-k_1 A_s \varepsilon_{\max}}{u}E_s e^{-k_1 x} \quad (1)$$

where E_s is elastic modulus of steel; $d\sigma_x/dx$, $d\varepsilon_x/dx$ are steel cross-section normal stress and normal strain increment along anchorage length respectively; σ_x, A_s and u, for full section, are average stress, area and perimeter of steel section respectively, for flange or web, are average stress, area, perimeter of flange or web section respectively.

The distribution of slip between the interface of steel and concrete is given as follows

$$s(x) = s_{\max}e^{-k_2 x} \quad (2)$$

where $s(x)$ is the slip quantity at different part of cross-section along anchorage length; $s_{\max}$ is the maximum slip quantity along anchorage length; k_2 is characteristic exponent of slip distribution.

2.2 Failure criteria of bond-slip

According to the difference of use requirements to the SRC structures, the failure criterion is different. In the special working conditions, the composite action of encased steel and concrete must be fully required, and even the local bond slippage at the loading end is forbidden [6]. This state is defined as local bond-slip failure. General SRC structures should meet the requirements of the normal use state and ultimate bearing capacity state, which requires that the bond-slip failure along the whole embedment should not take place. The state is defined as complete bond-slip failure.

Criterion I to the failure of local bond-slip

$$T_1 = F_1(\tau, \sigma_z)/P_0 - 1 \quad (3)$$

Criterion II to the failure of complete bond-slip

$$T_2 = F_2(\tau, \sigma_z)/P_u - 1 \quad (4)$$

where P_0 is local bond-bearing capacity at the loading end; P_u is ultimate bond-bearing capacity along the

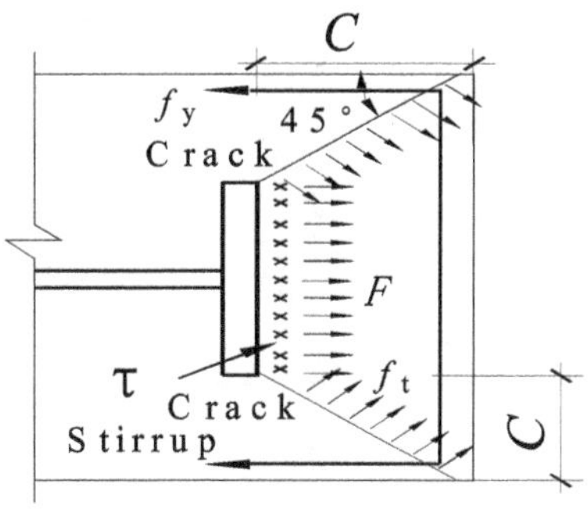

Fig. 2. Computation sketches for Q_u.

embedment; $F_1(\tau, \sigma_z)$ and $F_2(\tau, \sigma_z)$ are resultant forces of τ and σ_z.

$$P_0 = C_a l_0 \tau_0', \quad \tau_0' = 0.364 f_t + 0.1991 l_a/h_a + 1.5209\rho_{sv} + 0.4998 C_a/h_a - 0.3027 \quad (5)$$

$$P_u = 4 b_f l_a \tau_u, \quad \tau_u = 0.004 f_t + 0.0874\rho_{sv} + 0.8624 C_a/h_a - 0.0454 l_a/h_a + 0.9107 \quad (6)$$

where f_t is axial tensile strength; l_a is embedment length; ρ_{sv} is traverse stirrup ratio; C_a and h_a are respectively perimeter and height of traverse section of steel shape; l_0 is the diffusion length of chemical bond force which is approximately equal to the height of traverse section of steel shape; τ_0' is the average bond stress on the whole steel shape surface within the diffusion length l_0; τ_u is the average bond stress on the steel flange surface within the embedment length l_a, which neglects the bond action between web and concrete.

Providing the complete bond-slip failure is known with respect to Criterion II, the failure model of bond-slip can be obtained uniquely with Criterion III to traverse crack in concrete, which is given as follows

$$T_3 = G(\sigma_r, \sigma_\theta)/Q_u - 1 \quad (7)$$

where $G(\sigma_r, \sigma_\theta)$ is resultant force of σ_r and σ_θ; Q_u is ultimate load-bearing capacity of traverse crack- resistance, which is greatly influenced by the confinement of concrete encasement and stirrups, and it can be deduced exactly when concrete cover thickness and stirrups ratio are relatively large.

There is an assumption: the tension stress in the concrete cover reaches f_t and the tension stress in stirrup reaches f_y after the full crack of transverse concrete. The computation sketches are shown in Figure 2. Then

$$Q_u = 2Cf_t + 2A_s f_y/s \quad (8)$$

where C is the concrete cover thickness, mm; f_y is the yield strength of stirrup, N/mm^2; s is the space of stirrups, mm; A_s is the cross-sectional area of stirrups, mm^2. The failure models in the SRC structure are shown in Table 1.

2.3 Constitutive relationship of bond-slip

According to the present experimental data [7], it is generally accepted that the bond stress is commonly evaluated

Table 1. Failure criteria of bond-slip.

Failure criteria	Results of bond-slip failure		
$T_1 < 0$	Steel shape cooperates with concrete entirely		
$T_1 \geqslant 0$	Local bond-slip failure		
$T_2 \geqslant 0$	Complete bond-slip failure	$T_3 \geqslant 0$	Splitting failure of concrete
		$T_3 < 0$	Embedment failure of steel shape
$T_2 < 0$	Yield failure of steel shape		

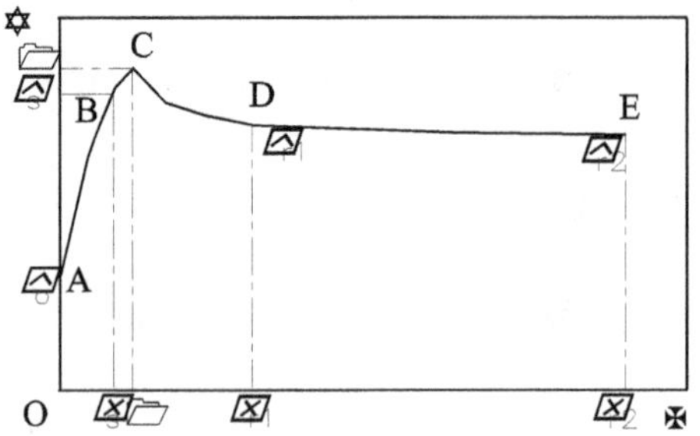

Fig. 3. Standard $\tau - S$ curve.

as the average bond stress, which is the load transferred between the flange of steel shape and concrete encasement, divided by the total surface area of the flange within concrete, in which the bond stress on surface of the web is neglected. Based on the statistic analyses of the force and slip relationship, a typical bond-slippage ($\tau - S$) curve is obtained, as is shown in Figure 3. The $\tau - S$ curve includes ascending stage and descending stage. The ascending stage includes no slip (OA), linearly ascending (AB), and nonlinearly ascending (BC) parts. The descending stage includes sharply descending and smooth parts. The typical bond-slippage curve can be standardized as follows: $x = S/S_u$ and $y = \tau/\tau_u$, where S_u is the slippage at the loading end in the ultimate bond-bearing capacity state. The characteristics of the standard curve are as follows:

a. Starting point: $x = 0, y = y_0 > 0$. Specimens have virtually no slip for loads up to 40% P_u, and local chemical adhesive force between encased steel and concrete at the loading end equals the applied load. The composite action of encased steel and concrete is fully developed.
b. Ascending stage: $0 \leqslant x \leqslant 1$, $\frac{d^2y}{dx^2} > 0$, $\frac{dy}{dx} \geqslant 0$, $x = 1, y = 1$. When the relative slippage occurs, the applied load equals the friction and mechanical interlock forces within the slip zone and chemical adhesive force within the bond zone. Slip increases linearly with increasing of load before the control point (P_s, S_s), where P_s is about 70% P_u. After P_s, the standard curve tends to ascend nonlinearly.
c. Descending stage: $1 \leqslant x$, $\frac{d^2y}{dx^2} < 0$, $\frac{dy}{dx} < 0$, $y_{x\to\infty} = y_r$. After P_u, the increase of the friction and interlock forces is smaller than the fall of adhesive force. After the load descends to 70–80% P_u, cracks in concrete are steady. The part of the curve is nearly horizontal.

The ascending stage and descending stage of the standard $\tau - S$ curve can be fitted with mathematical expressions respectively

$$x \leqslant 1 \quad y = ax^2 + bx + c \quad x \geqslant 1 \quad y = \frac{x}{mx + n}. \tag{9}$$

The parameters in the equation (9) can also be specified with the control points of $(0, y_0)$, (x_s, y_s), $(1, 1)$ and (x_r, y_r) acquired in the test, as follows

$$a = \frac{y_0 + x_s - y_s - y_0 x_s}{x_s - x_s^2}, \quad b = \frac{y_0 + x_s^2 - y_s - y_0 x_s^2}{x_s - x_s^2},$$

$$c = y_0, \quad m = \frac{x_r - x_r y_r}{y_r - x_r y_r}, \quad n = \frac{x_r - y_r}{x_r y_r - y_r}. \tag{10}$$

The values of control points in the standard $\tau - S$ curve can directly be obtained by test or from the regressive formulas of control points. The $\tau - S$ curves obtained in the test and ones fitted with mathematical expressions agree well with each other.

With the positional difference of the embedment, the $\tau - S$ curves are different [8,9]. The $\tau - S$ curves in the different location can be obtained in test. The characteristics of the $\tau - S$ curves are as follows: With the increase of the embedment length, the threshold bond strength, the maximum bond stress and local slippage decrease. Near the loading end, the $\tau - S$ curves include ascending stage and descending stage; near the free end, the curves only include a descending stage. The residual bond stress and slippage are in steady state for all the $\tau - S$ curves. The $\tau - S$ curves of the loading end are similar to that of different location. It is assumed that there are constant ratios between the values of control points in the $\tau - S$ curves, then the below simplified equations can be obtained

$$\frac{\tau_0(x)}{\bar{\tau}_0} = \frac{\tau_s(x)}{\bar{\tau}_s} = \frac{\tau_u(x)}{\bar{\tau}_u} = A(x), \quad \frac{\tau_r(x)}{\bar{\tau}_r} = 1,$$
$$\frac{S_s(x)}{\bar{S}_s} = \frac{S_u(x)}{\bar{S}_u} = B(x), \quad \frac{S_r(x)}{\bar{S}_r} = 1. \tag{11}$$

All the $\tau - S$ curves of the different embedment length can be specified by the standard $\tau - S$ curve and position functions of $A(x)$ and $B(x)$, which can be deduced from the experimental results. With respect to the separation between the steel and concrete at the loading end, $\tau(0)$ is zero.

$$A(x) = -384.52x^6 + 1256.1x^5 - 1604.6x^4 + 1011x^3 - 322.04x^2 + 44.276x$$
$$B(x) = 671x^2 - 1.6373x + 1.0 \tag{12}$$

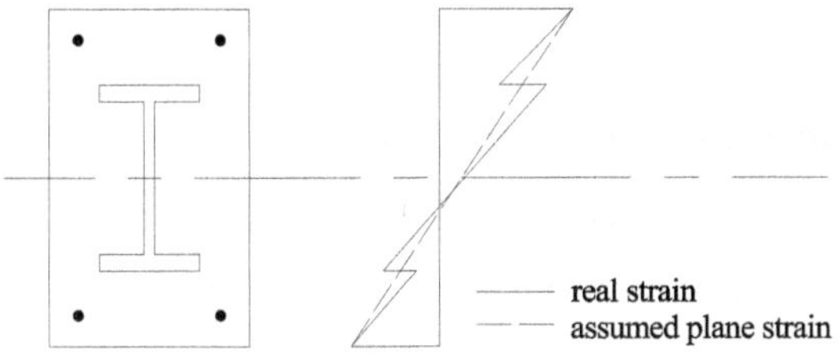

Fig. 4. Section strain diagram of SRC.

where $x = X/l_a$, X is the distance from the anchorage point to the loading end. $A(x)$ and $B(x)$ do not indicates the characteristics of the other three parameters, and the researcher will follow up them particularly.

Due to obvious slipping zones on the interface of steel and concrete, the bearing capacity of solid-web steel reinforced concrete beams is lower and is reduced to some extent. As a result of the impact of bond-slip, SRC beam's section strain is clearly not in accordance with the assumption of plane section, a strain mutant occurs at the top and bottom steel flange, as shown in Figure 4. To assure that the depth of compression zone and the limit of bearing capacity remain unchanged, a modified plane section will be used to replace the multi-line cross-section strain. While calculating bearing capacity of solid-web steel reinforced concrete beams, the following assumptions were made to take into account the influence of bond-slip cracks.

a. Cross-section strain should conform to the assumption of modified plane section.
b. When damaged, the ultimate compression strain of concrete at edge of compression zone is 0.003.
c. At ultimate state, compressive stress can be described as rectangle distribution, where $f_{cm} = f_c$, depth of compression zone is $0.8x_0$, x_0 is actual depth of compression zone.
d. At ultimate state, the concrete can not bear any pulling tension in tensile area.

3 Optimal calculating model

3.1 Design variables

The design scheme of SCR frame beams may be determined by length of beam (l), strengths of concrete, steel, longitudinal reinforcement and stirrup (f_c, f_a, f_y, f_{yv}), dimensions of cross-section (depth h, width b), cross-section dimensions of steel (bottom flange width b_{af}, bottom flange thickness t_{af}, top flange width b'_{af}, top flange thickness t'_{af}, web depth h_w, web thickness t_w), the distance from bottom (top) flange to tension (compression) zone edge, which is just the thickness of covering layer (a_a, a'_a), diameter and quantity of longitudinal reinforcement at bottom (d_s, n_s), diameter and quantity of longitudinal reinforcement at top (d'_s, n'_s), the distance from tensile (compression) reinforcement to compression (tension) zone edge (a_s, a'_s), diameter and range interval of stirrup (d_{sv}, s_{sv}), length coverage and range interval of added stirrup (l_{sv}, s_1), anchorage length of steel (l_a).

3.2 Objective function

Objective function of SCR frame beams is total cost per unit length, as construction and labor costs for different design change not large and can be considered as an invariable value, so they are not considered in the objective function. The cost of engineering materials is composed of four parts: concrete cost, steel cost, longitudinal reinforcement cost, and stirrup cost. The expression of objective function is given as follows

$$\begin{aligned}\mathrm{Cost}\,(X) = \mathrm{CostC}\,(X) + \mathrm{CostA}\,(X) \\ + \mathrm{CostS}\,(X) + \mathrm{CostSV}\,(X)\end{aligned} \tag{13}$$

where $\mathrm{CostC}\,(X)$, $\mathrm{CostA}\,(X)$, $\mathrm{CostS}\,(X)$, $\mathrm{CostSV}\,(X)$ are concrete cost, steel cost, longitudinal reinforcement cost, and stirrup cost respectively.

3.3 Constraint conditions

3.3.1 Bearing capacity requirements

According to the computational theory of SRC frame beams based on the bond-slip theory mentioned above, bearing capacity constraint conditions were given as follows [10]

a. Bending carrying capacity requirements

$$\begin{aligned}M \leqslant \frac{1}{\gamma_{\mathrm{RE}}} [f_c bx\,(h_0 - x/2) + f'_y A'_s\,(h_0 - a'_s) \\ + f'_a A'_{af}\,(h_0 - a'_a) + M_{aw}]\end{aligned} \tag{14}$$

$$f_c bx + f'_y A'_s + f'_a A'_{af} - f_y A_s - f_a A_{af} + N_{aw} = 0. \tag{15}$$

b. Shear carrying capacity requirements

$$V \leqslant \frac{1}{\gamma_{\mathrm{RE}}} \left[0.06 f_c bh_0 + 0.8 f_{yv} \frac{A_{sv}}{s} h_0 + 0.58 f_a t_w h_w\right] \tag{16}$$

$$V_b \leqslant 0.36 f_c bh_0 \tag{17}$$

$$\frac{f_a t_w h_w}{f_c bh_0} \geqslant 0.10. \tag{18}$$

3.3.2 Bearing capacity requirements

a. Requirements on height, width and ratio of height and width of beam

$$b \geqslant 300\ \mathrm{mm}, \quad h > 300\ \mathrm{mm}, \quad h \leqslant 4b. \tag{19}$$

b. Dimension requirements of steel and panel.
Requirements on thickness of covering layer:

$$b_{af} \leqslant b-200, b'_{af} \leqslant b-200 \quad (20)$$

$$b'_{af} \leqslant 2b/3,\ \ b_{af}>0,\ \ b'_{af}>0,\ \ h_w>0. \quad (21)$$

Requirements on steel welding and constructional details:

$$t_{af} \geqslant 6\text{ mm},\quad t'_{af} \geqslant 6\text{ mm},\quad t_w \geqslant 6\text{ mm}. \quad (22)$$

c. Constructional detail requirements on longitudinal reinforcement.
Requirements on diameter:

$$16\text{ mm} \leqslant d_s \leqslant 50\text{ mm}; 16\text{ mm} \leqslant d'_s \leqslant 50\text{ mm} \quad (23)$$

Requirements on clear spacing:

$$\frac{h-50-n_s d_s}{n_s-1} \geqslant \mathrm{MAX}\left(1.5d_s, 30\right) \quad (24)$$

$$\frac{h-50-n'_s d'_s}{n'_s-1} \geqslant \mathrm{MAX}\left(1.5d'_s, 30\right). \quad (25)$$

Requirements on number of longitudinal reinforcements:

$$n_s \geqslant 2,\quad n'_s \geqslant 2. \quad (26)$$

Requirements on longitudinal reinforcement ratio:

$$\frac{n_s p d_s^2}{4b\left(h-a_s\right)} > 0.003. \quad (27)$$

d. Constructional detail requirements on stirrup.

Requirements on diameter:

$$d_{sv} \geqslant 6+2\min\left[1, \mathrm{int}\left(\frac{h}{800}\right)\right] \quad (28)$$

$$d_{sv} \geqslant 0.25d'_s \max\left\{\min\left[1, \mathrm{int}\left(\frac{n'_s}{3.0}\right)\right], \min\left[1, \mathrm{int}\left(\frac{d'_s}{18.0}\right)\right]\right\}. \quad (29)$$

Requirements on range interval of stirrups:

$$S_{sv} \leqslant \frac{15d'_s}{\left\{\max\left[\min\left(1, \frac{n'_s}{6}\right), \min\left(1, \frac{d'_s}{18.0}\right)\right]\right\}^3} - 5d'_s \min\left[1, \mathrm{int}\left(\frac{n'_s}{6}\right)\right]\min\left[1, \mathrm{int}\left(\frac{d'_s}{20}\right)\right] \quad (30)$$

$$S_{sv} \leqslant \min\left[1, \mathrm{int}\left(\frac{0.999\cdot V}{0.07f_c b h_0}\right)\right] \times 50\left\{4+\min\left[1, \mathrm{int}\left(\frac{h}{525}\right)\right]+\min\left[1, \mathrm{int}\left(\frac{h}{825}\right)\right]\right\} + \mathrm{int}\left[\frac{\min\left(V, 0.07f_c b h_0\right)+1}{V+1}\right]300. \quad (31)$$

$$S_{sv} > 100. \quad (32)$$

Requirements on the stirrup ratio:

$$A_{sv}/bS_{sv} \geqslant 0.24f_t/f_{yv}. \quad (33)$$

e. Requirements on thickness of covering layer

$$a_a - 0.5t_{af} \geqslant 100\text{ mm} \quad (34)$$

3.3.3 Deformation restrictions

SRC beams satisfying the above conditions can generally meet the deformation requirements of design code, so crack and deflection restrictions will not be regarded as constrained condition, which will work as a final check for the design [11].

3.3.4 Constraint conditions of discrete variables

All of the 15 variables of SRC frame beam optimization design are discrete variables except top and bottom steel flange width, web depth and steel covering depth, and discrete variables are defined by specification of materials and sectional dimensions commonly used in engineering.

4 Mixing penalty function method of discrete variables

The Optimum design of SRC composite beams can be solved by using sequential unconstrained minimization technique (SUMT for short), which is an indirect method translating constraint optimization into an unconfined one. The following is its primary principles [12, 13].

The mathematic model of constraint optimization is written as

$$\begin{aligned} \min\ & \boldsymbol{f}(\boldsymbol{X}) \quad \boldsymbol{X} \in \boldsymbol{R}^n \\ s.t.\ & \boldsymbol{g}_i(\boldsymbol{X}) \leqslant 0\,(i=1,2\ldots,p) \\ & \boldsymbol{h}_j(\boldsymbol{X}) = 0\,(j=1,2\ldots,p) \end{aligned} \quad (35)$$

where $\boldsymbol{X} = [x_1, x_2, \ldots, x_n]^T$ is decision-making vector; $\boldsymbol{f}(\boldsymbol{x})$ is objective function vector; $\boldsymbol{g}_i(\boldsymbol{X}) \leqslant 0$, $\boldsymbol{h}_j(\boldsymbol{X}) = 0$ is constraint functions.

Adding $\boldsymbol{g}(\boldsymbol{X})$ and $\boldsymbol{h}(\boldsymbol{X})$ to $\boldsymbol{f}(\boldsymbol{x})$, the original optimization question is translated into an equivalent unconfined one shown as

$$\begin{aligned} \min\ & \boldsymbol{F}(\boldsymbol{X}, r_k, t_k) \quad \boldsymbol{X} \in \boldsymbol{R}^n \\ & \boldsymbol{X} = [x_1 x_2 \ldots x_n]^T \\ & k = 0, 1, 2, \ldots \end{aligned} \quad (36)$$

$\boldsymbol{F}(\boldsymbol{X}, r_k, t_k)$ is an artificial objective function, named as penalty function, and it is expressed as

$$\min\ \boldsymbol{F}(\boldsymbol{X}, r_k, t_k) = \boldsymbol{f}(\boldsymbol{X}) + r_k \sum_{i=1}^{p} \boldsymbol{G}\left[\boldsymbol{g}_i(\boldsymbol{X})\right] + t_k \sum_{j=1}^{q} \boldsymbol{H}\left[\boldsymbol{h}_j(\boldsymbol{X})\right] \quad (37)$$

where $\boldsymbol{G}[\boldsymbol{g}_i(\boldsymbol{X})]$, $\boldsymbol{H}[\boldsymbol{h}_j(\boldsymbol{X})]$ are fonctionelles of $\boldsymbol{g}_i(\boldsymbol{X})$ and $\boldsymbol{h}_j(\boldsymbol{X})$ respectively, a group of inequality constraint conditions and equality constraint conditions in regard to the original optimization question; r_k and t_k are called penalty factors or penalty parameters, which are adjusted according to increase of k; $\sum_{i=1}^{p}\boldsymbol{G}[\boldsymbol{g}_i(\boldsymbol{X})]$, $\sum_{j=1}^{q}\boldsymbol{H}[\boldsymbol{h}_j(\boldsymbol{X})]$ are called penalty items, and they are non-negative.

It is visible that value of $\boldsymbol{F}(\boldsymbol{X}, r_k, t_k)$ is usually larger than value of the original objective function $\boldsymbol{f}(\boldsymbol{x})$. In order to astringe penalty function $\boldsymbol{F}(\boldsymbol{X}, r_k, t_k)$ to the constraint optimum solution $\boldsymbol{x}^*$ of original question, the penalty must own the following character

$$\left.\begin{aligned} \lim_{k\to\infty} r_k \sum_{i=1}^{p} \boldsymbol{G}[\boldsymbol{g}_i(\boldsymbol{X})] = 0 \\ \lim_{k\to\infty} t_k \sum_{j=1}^{q} \boldsymbol{H}[\boldsymbol{h}_j(\boldsymbol{X})] = 0 \end{aligned}\right\} \tag{38}$$

which means penalty effect on penalty function will disappear gradually along with continuous adjustment of penalty factors, i.e.

$$\lim_{k\to\infty} |\boldsymbol{F}(X, r_k, t_k) - \boldsymbol{f}(\boldsymbol{X})| = 0. \tag{39}$$

If objective function and constraint function are both continuous and differentiable, it is necessary to satisfy the following equation in gaining extreme point of penalty function, which is K-T condition for constraint extreme point.

$$\begin{aligned} \nabla \boldsymbol{F}(\boldsymbol{X}, r_k, t_k) = \nabla \boldsymbol{f}(\boldsymbol{X}) + r_k \sum_{i=1}^{p} \nabla \boldsymbol{G}[\boldsymbol{g}_i(\boldsymbol{X})] \\ + t_k \sum_{j=1}^{q} \nabla \boldsymbol{H}[\boldsymbol{h}_j(\boldsymbol{X})] = 0. \end{aligned} \tag{40}$$

Optimum question with both equality and inequality constraints can be solved by combining inner point method and outer point method, which is mixing penalty function method.

When constraint conditions are $\boldsymbol{g}_i(\boldsymbol{X}) \leqslant 0$ and $\boldsymbol{h}_j(\boldsymbol{X}) = 0$, the general expression of penalty function is as follows:

$$\begin{aligned} \boldsymbol{F}(\boldsymbol{X}, r_k) = \boldsymbol{f}(\boldsymbol{X}) - r_k \sum_{i\in I_1} \frac{1}{g_i(\boldsymbol{X})} \\ + t_k \sum_{i\in I_2} \{\max[0, g_i(\boldsymbol{X})]\}^2 + t_k \sum_{j=1}^{q} [\boldsymbol{h}_j(\boldsymbol{X})]^2 \end{aligned} \tag{41}$$

$$\left\{\begin{aligned} I_1 = \{i \,|\boldsymbol{g}_i(\boldsymbol{X}^{(0)}) < 0 \quad (i = 1, 2, \ldots, p)\} \\ I_2 = \{i \,|\boldsymbol{g}_i(\boldsymbol{X}^{(0)}) \geqslant 0 \quad (i = 1, 2, \ldots, p)\} \end{aligned}\right. \tag{42}$$

where r_k is a decreasing sequence of positive real number; t_k is an increasing sequence of positive real number, $\boldsymbol{X}^{(0)}$ is initial point; I_1, I_2 are two constraint sets.

In the optimization of practical structural engineering, parts even all of the design variables are often discrete variables, which can only take special and discrete values. This means will add several equation constraint conditions to mathematical model, so it can be solved by using mixing penalty function method.

On the assumption that the number of discrete variables is l in design variables, and the rest are continuous variables, in which discrete variables are given as $\boldsymbol{X}^d = [x_1, x_2, \ldots, x_l]^T$, then penalty function can be expressed as

$$\begin{aligned} \boldsymbol{F}(\boldsymbol{X}, r_k, t_k, s_k) = \boldsymbol{f}(\boldsymbol{x}) - r_k \boldsymbol{G}_1[\boldsymbol{g}_i(\boldsymbol{X})] \\ + t_k \boldsymbol{H}[\boldsymbol{h}_j(\boldsymbol{X})] + s_k \boldsymbol{D}(x_u) \end{aligned} \tag{43}$$

where r_k and t_k are the same as formula (42). On the right side of equation, the first item is original objective function; the second is punitive item of interior point method considering the constraint condition of $\boldsymbol{g}_i(\boldsymbol{X}) \leqslant 0$; the third is punitive item of outside point method considering the constraint condition of $\boldsymbol{h}_j(\boldsymbol{X}) \leqslant 0$; the fourth is punitive item to assure specified discrete value for the design variables, s_k is penalty factor, which is an increasing sequence of positive real number.

The items in penalty function are as follows

$$\boldsymbol{G}_1[\boldsymbol{g}_i(\boldsymbol{X})] = \sum_{i\in I_1} \frac{1}{\boldsymbol{g}_i(\boldsymbol{X})} \tag{44}$$

$$\boldsymbol{H}[\boldsymbol{h}_j(\boldsymbol{X})] = \sum_{j=1}^{q} [\boldsymbol{h}_j(\boldsymbol{X})]^2 \tag{45}$$

$$D(x_u) = \sum_{u=1}^{l} \left[\prod_{v=1}^{m_u} \left(\frac{x_u - z_{uv}}{\overline{x}_u - z_{uv}}\right)\right]^2 \tag{46}$$

$$\boldsymbol{D}_u(x_u) = \left[\prod_{v=1}^{m_u} (x_u - z_{uv})\right]^2 \tag{47}$$

where $z_{uv}(v = 1 \sim m_u)$ is discrete value of variablex_u; m_u is the discrete value number of variable x_u; $\overline{x}_u$ is the average value of z_{uv} and $z_{u,v+1}$, i.e. $\overline{x}_u = \frac{1}{2}(z_{uv} + z_{u,v+1})$.

5 An example of calculation

A SRC frame beam bears uniform load and its shear span ratio equals zero. The maximum internal force at beam end section is given as $M_A = -500$ kNm and $V_A = 250$ KN; the maximum bending moment in the middle section of the span equals 300 kNm. The calculation diagram is shown in Figure 5. The problem is to choose beam section dimensions, reinforcement consumption and steel consumption to make total cost lowest.

The known conditions are as followings: concrete strength grade is C30; steel grade is Q235; HPB235 is used for longitudinal reinforcement and stirrup; the price of concrete is 500 yuan per cubic meter; the price of steel is 3700 yuan per ton, the price of longitudinal reinforcement

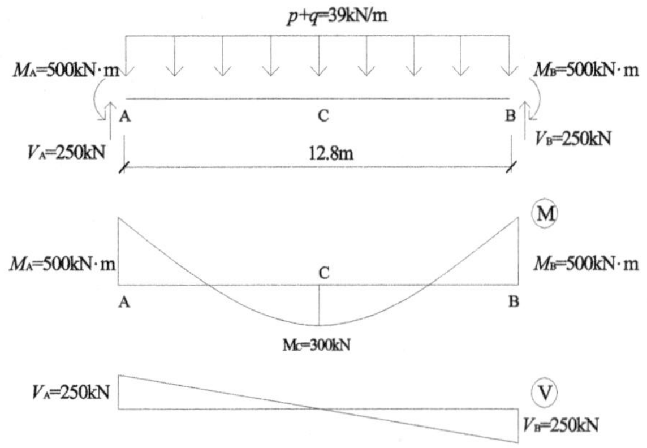

Fig. 5. Simplified computing sketch of SRC beam.

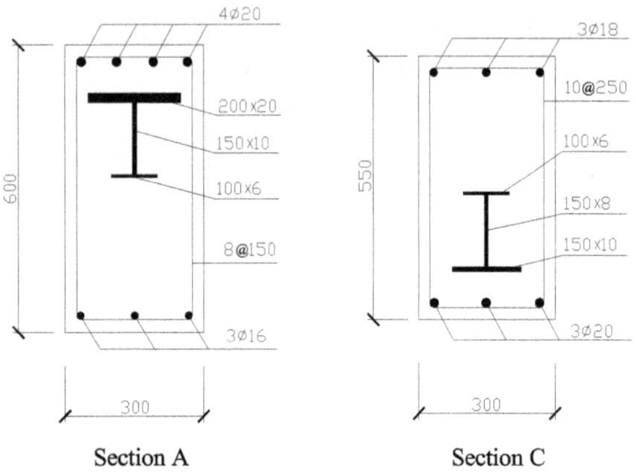

Fig. 6. Optimized section of SRC beam.

is 3000 yuan per ton, the price of stirrup is 2700 yuan per ton.

The program first checks up initial values. If the values do not satisfy all constraint conditions, the preset initial values will be considered ineffective, and then the initial values satisfied will be generated stochastically. This example is carried out two computations. First time, the initial values cannot satisfy all constraint conditions, so the program produces automatically new initial values, and second time the initial values satisfy all constraint conditions, then the procedure make the optimized computation directly.

Optimum results indicated that the cost of beam end section converges to 318.3 Yuan from the initial 552.4 Yuan, and the cost of middle section of the span converges to 226.4 Yuan from the initial 348.6 Yuan. The final sections are shown in Figure 6.

6 Conclusion

This paper uses a discrete variable mixing penalty function optimization algorithm to design SRC frame beams, which is one of mathematical programming methods. The method has extremely strict logic and is suitable for the SRC frame design.

As some of the parameters impact the convergence process in optimum scheme, for different initial condition the convergence process varies obviously, and sometimes the process will be longtime. To improve versatility and stability of the method, it is needed to do more systematic analysis research on the entire optimum process and its parameters.

The authors would like to thank National Natural Science Foundation of China and the Educational Office of Shaan'xi Province in China for their support throughout this research.

References

1. JSJ138-2001, *Technical specification for steel reinforced concrete composite structures*, China Architecture and Building Press, Beijing (2002)
2. Zhao Hongtie, *Steel and concrete composite structure*, Science Press, Beijing (2001)
3. S. Pezeshk, C.V. Camp, D. Chen, Design of nonlinear framed structures using genetic optimization. J. Struct. Engrg. **126**, 382–388 (2000)
4. H. Huang, R.W. Xia, Two-level multipoint constraint approximation concept for structural optimization. Structural Optimization **9**, 38–45 (1995)
5. J.O. Bryson, R.G. Mathey, Surface condition effect on bond strength of steel beams in concrete. J. ACI **59**, 397–406 (1962)
6. M. Keuser, G. Mehlhorn, Finite element models for bond problems. J. Struct. Engrg. **113**, 2160–2173 (1987)
7. S.S. Zheng, G.Z. Deng, Y. Yang, M.H. Yu, Experimental study on bond-slip performance between steel shape and concrete in SRC structures. Engineering Mechanics **20**, 63–69 (2003)
8. C.W. Roeder, R. Chmielowski, C.B. Brown, Shear connector requirements for embedment steel sections. J. Struct. Engrg. **125**, 142–151 (1999)
9. R. Furlong, Binding and Bonding to Composite Columns, *Composite and mixed construction*. ASCE, 227–253, 330–336 (1984)
10. Zhou Qijing, Jiang Weishan, Pan Taihua, *Manual for design and construction of composite structures*, China Architecture and Building Press, Beijing (1991)
11. N.M. Hawkins, Strength of concrete encased steel beams. Civil Engineering Transaction of Institution of Australia Engineer **E15** (1/2), 39–46 (1973)
12. Charles V. Camp, Shahram Pezeshk, Håkan Hansson, Flexural design of reinforced concrete frames using a genetic algorithm. J. Struct. Engrg. **129**, 105–115 (2003)
13. C.K. Choi, N.T. Cheng, Integrated genetic algorithms for optimization of space structures. J. Aerosp. Eng. **6**, 3015–328 (1993)

20

Experimental study of the drying and modelling of the humidity migration in a clay matrix

K.-E. Atcholi[1,a], E. Padayodi[1], J. Vantomme[2], K. Kadja[1] and D. Perreux[3]

[1] Laboratoire d'Études et de Recherches sur les Matériaux, les Procédés et les Surfaces (LERMPS) Équipe de Recherche sur les Agro-Matériaux et la Santé Environnementale (ERAMSE), Département de Génie Mécanique et Conception, Université de Technologie de Belfort-Montbéliard, 90010 Belfort Cedex, France

[2] Civil Engineering Department, École Royale Militaire de Bruxelles, Belgique

[3] Laboratoire de Mécanique Appliquée, R. Chaléat (LMARC), Université de Franche-Comté, 25030 Besançon, France

Abstract – Available almost everywhere, therefore economic and even so durable, clays have stimulated the creativity of men who have imagined since millennia and under all latitudes, multiple processes to develop pottery (faïence enameled, porcelain, ceramics) and construction materials. These materials compete with modern materials in civil engineering for important construction works such as bridges, roads, tunnels, monuments etc. In state-of-the-art technologies, clay ceramics are used as anti-thermal coatings: turbine vanes, etc... The weakness of our scientific knowledge limits applications of these natural materials. This study on tropical clays, has for objective on one hand to develop new processing of construction materials for lodging needs that are more and more increasing in the developing countries, and the other hand to develop new refractory materials, anti-thermal coatings or ceramic prostheses. This study is part of a research program and academic cooperation between the laboratories LMARC-Besançon/France, LERMPS-Belfort/France, Civil Eng. Dept./ERM-Bruxelles/Belgique, URMA-Lomé/Togo (K.-E. Atcholi, *Rapport d'activités de Recherches*, Université de Lomé, Février 1997). In this paper, we present an experimental modelling of a clay matrix drying in order to better understand and resolve material deformations and cracking. The complex behaviour of the clay during drying, leads us to propose two complementary approaches: an experimental approach that allows to establish the drying kinetic and a modelling of moisture transfer in a clay matrix in order to optimise the drying. A diffusive model based on Fick's laws allowed to highlight the internal stresses causing matrixes damage.

Key words: Clay; dough; drying; building materials; diffusion laws

1 Introduction

In the process of clay materials manufacture, drying is a critical operation that determines the resulting products quality. So, this operation requires particular cautions taking in account the complex micro and macro transformations occurring inside the clay material, and that generate material damaging such deformations and cracks.

The clay matrix moisture evaporation is correlatively related to the matrix dimensions variations by shrinking and to the matrix densification. These related phenomena are the principle causes of the material damage during drying.

Understanding the mechanisms of a clay matrix shrinking, densification and evaporation will make it possible to overcome to these damages or to reduce damage risks during drying.

Due to the complexity of these phenomena occurring in a clay matrix during drying, we propose in this study two complementary approaches: an experimental approach that allowed the quantification of the phenomena at the macroscopic scale and a theoretical approach that is about the modelling of the moisture transfer in a clay flat matrix during drying using a diffusive model.

2 Experimental approach of a clay matrix drying

This experimental approach deals with the matrix water content variation during drying and the matrix sizes variation due to the shrinkage.

[a] Corresponding author:
kokou-esso.atcholi@utbm.fr, katcholi@hotmail.com

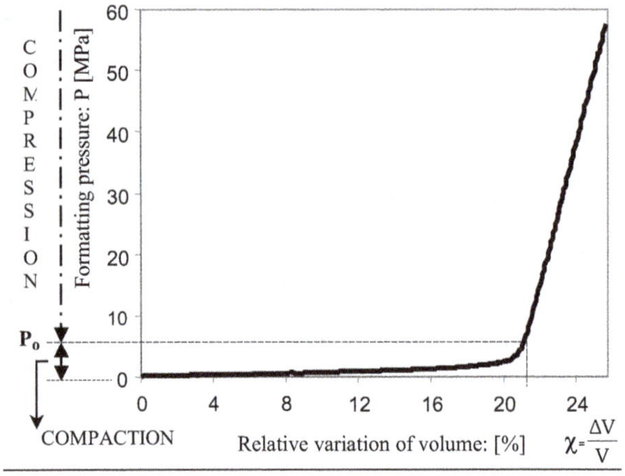

Fig. 1. Compaction curve of the clay paste: variation of the relative volume of the matrix in function of pressure.

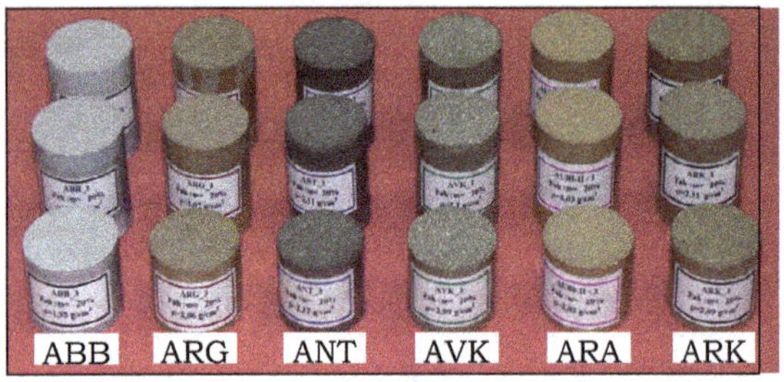

Fig. 2. Cylindrical test-tubes made of clay matrix. Six clay varieties: white clay of Bangéli (ABB); red clay of Guérin-Kouka (ARG); black clay of Togblékopé (ANT); green clay of Kouvé (AVK); red clay of Albi (ARA) and red clay of Kouvé (ARK).

2.1 Elaboration of clay matrix test-tubes

A sample of clay is dried, grinded and sieved by means of a sieve of 0.5 mm of aperture size. Clay powders are moistened to a suitable shaping water content (Padayodi [2]) of 18%. The obtained pastes are hermetically sealed and kept in a wet place during 8 weeks.

The cylindrical test-tubes are shaped by static pressing on a conventional machine of traction and compression. They are compressed to present a wet matrix density of about 2 g/cm^3 (Collard [3], Ho-Yick-Cheong [4]), a density advised in brickyard and tilery.

Figure 1 shows the test-tubes compaction curve giving the volume variation according to the pressure. This curve shows that the matrix undergoes a maximum compaction for a pressure of 8 MPa (Ho-Yick-Cheong [4]). At this pressure, the clay test-tubes or matrixes present a density of about 1.94 g/cm^3 to 2.1 g/cm^3 according to the clay variety. These values meet the "Industrial Development Center" requirements (ACP-CEE [5]). This center advise a density between 1.87 g/cm^{3}and 2.2 g/cm^3.

Figure 2 is an illustration of test-tubes of six clay varieties elaborated according to the above protocol.

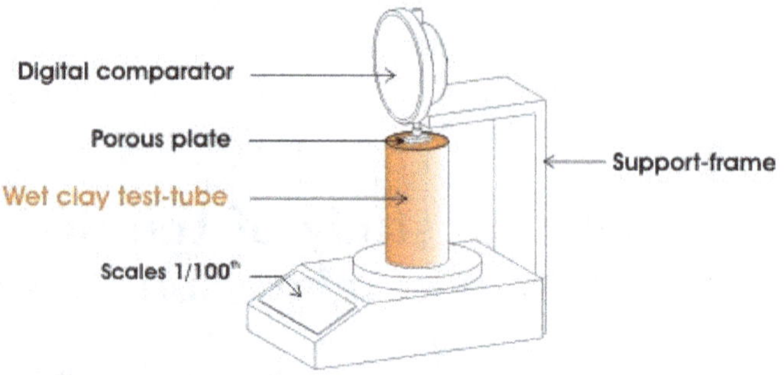

Fig. 3. Device for measuring the test-tube shrinkage and water content variation during drying.

2.2 Clay matrixes drying: water content variation

2.2.1 Experimental devices and protocol

The experimental device (Fig. 3) consists of an accurate scales on which lies the test-tube during drying and of a comparator that measures the test-tube shrinkage. A computer records the test-tube moisture evaporation according to the time and the test-tube size variation due to the shrinkage. The drying temperature is maintained between 20 °C and 22 °C and the relative moisture of the ambient air is maintained at 80%.

2.2.2 Results and discussions

From the measured mass $m(t)$, one establishes for every clay the variation of the average water content $\varpi(t)$ of the matrix during drying. $\varpi(t)$ is given by the expression (Ho-Yick-Cheong [4], Magnan [6], Puffeney [7], Suri [8]):

$$\varpi(t) = \frac{m(t) - m_0}{m_0} \tag{1}$$

where m_o is the dried sample mass. The derivative $d\varpi/dt$ gives is the drying velocity.

Figure 4 gives the curves of the average water content variation of different clay matrixes.

Table 1 gives the drying velocity at the beginning of drying and the balance water content ϖ_{eq} at the end of drying. The balance water content ϖ_{eq} is the residual water content of the matrix at the end of drying.

The experimental curves show that the clay matrix drying is carried out in two stages:

- A linear evaporation at the drying beginning. The moisture evaporation is intensive in this stage. The average values of the maximum drying velocity lie between –10 and –5 mg/h.g according to the clay variety. The negative sign of the drying velocity expresses the loss of moisture by evaporation.

  According to Figure 4 and Table 1, the drying velocity is more important in the case of the ANT clay and lower for the ARG clay. This variation is related to their affinity with the moisture and the matrix porosity.

Table 1. Drying velocity at the beginning of drying and the balance water content at the end of drying.

	ABB	ANT	ARG	AVK	ARK	ARA
Drying velocity (mg /g *of dried matter* /hour)	– 5.5	– 10	– 4.5	– 6.5	– 5	– 5.8
Balance water content ϖ_{eq} (%)	3	4	5	9	4	3

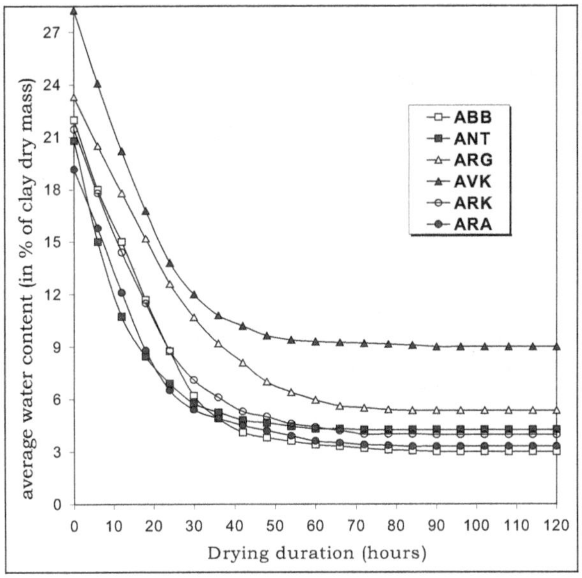

Fig. 4. Variation of the average water content of clay matrixes during drying.

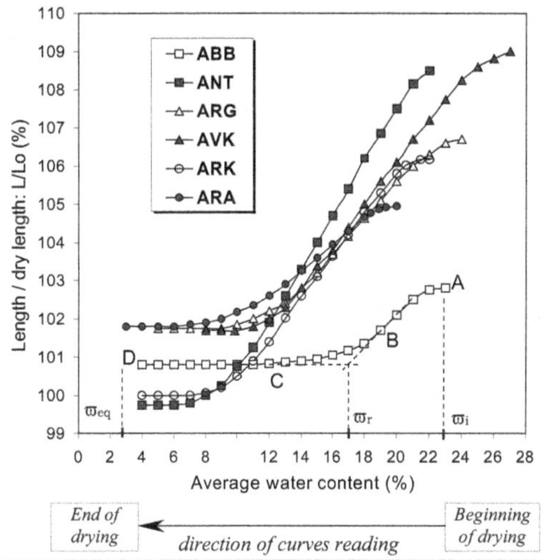

	ABB	ANT	ARG	AVK	ARK	ARA
Shrinkage limit ϖ_r (%):	17	8	12	13	9	10

Fig. 5. Representative shrinking curves: Bigot's curves, shrinkage limits.

- A second stage in which the drying velocity decreases gradually to zero. When the velocity is about zero, the matrix is in hygroscopic balance with the drying air and its water content does not vary any more. This water content is the balance water content ϖ_{eq}. ϖ_{eq} depends on the hygroscopic characteristics of the drying air and its affinity with a considered clay. Table 1 shows that the AVK clay presents a more important balance water content, so the more affinity with drying air.

The curves of Figure 4 highlight this affinity with ambient air. Although all the clay pastes are moistened to a water content of 18%, the curves indicate the initial water contents $\varpi(0)$ (or ϖ_i) different from 18% and variable according to clays. The drying curve of the AVK clay whose balance water content is about 9% gives for example an initial water content of 28% (Fig. 4). This value corresponds approximately to the sum of 18% and 9%.

2.3 Experimental study of the drying Shrinkage: the representative shrinking curves or Bigot's curves

2.3.1 Bigot's curves measurement

The matrix dimensions variation according to the average water content $L(\varpi)/Lo$ is called Bigot's curve (Collard [3], Magnan [6], Puffeney [7], Alviset [9], Sigg [10]). *Lo* is the length of the dried matrix. This curves expressed the relation between the evaporation and the shrinkage, the shrinkage being due to the moisture evaporation.

2.3.2 Results and discussions

Figure 5 gives Bigot's curves of the studied clays and their shrinkage limits.

Bigot's curves allow to determine the shrinking limit ϖ_r of these clays. This limit is the value of the water content bellow which the moisture evaporation does no more generate the shrinkage. It corresponds to the abscissa of the intercept point of the linear part of the curve and the tangent of the horizontal part, as illustrated in Figure 5.

The Bigot's curves are interpreted by the shrinkage mechanism occurring in the matrix during drying. This mechanism is illustrated in Figure 6.

Referring to Figures 5 and 6, drying decreases the moisture of the matrix from the shaping water content or initial water content ϖ_i to the hygroscopic balance water content ϖ_{eq}. The matrix undergoes various transformations:

- from ϖ_i to ϖ_r (zone AB of Figs. 5 and 6(a)), the matrix dimensions decrease linearly. The moisture evaporation moves solid particles closer together and causes the variation of the matrix dimensions (Fig. 6(a)). The

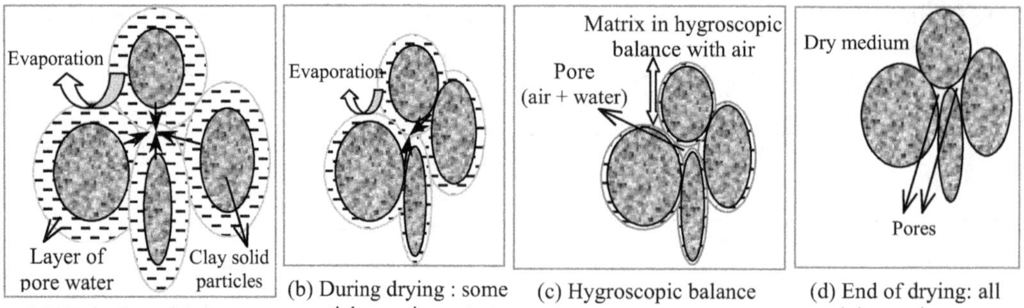

(a) Beginning of drying : evaporation brings particles closer together ($\varpi > \varpi_r$)

(b) During drying : some particles are in contact (ϖ is close to ϖ_r)

(c) Hygroscopic balance ($\varpi = \varpi_{eq}$)

(d) End of drying: all particles are in contact ($\varpi = 0$)

Fig. 6. Schematic representation of the shrinking mechanism in a clay matrix during drying.

matrix volume variation is proportional to the volume of the evaporated water: shrinkage is linear;

- around ϖ_r, (zone BC, Fig. 5 and 6(b)), some solid particles are already in contact and the variation of matrix dimensions decreases. The slope of the curve decreases as the number of particles in contact increases. The matrix volume variation is no more proportional to the evaporated water volume;
- from ϖ_r to ϖ_{eq} (zone CD of Fig. 5 and 6(c), 6(d)), the shrinkage is almost finished. Almost all of the solid particles are in contact and the water evaporation does not any more bring clay particles together. The dimensions of the matrix will remain constant until its complete dehydration (Fig. 6(d)).

3 Modelling of the moisture transfer in a clay matrix

During the clay matrix drying, the water evaporation is carried out by a mechanism of transfer due to the water content gradient. In a dissymmetric drying, the water transfer generates internal stresses in the matrix that causes the matrix deformation. The modelling of the water transfer in a clay matrix will make it possible to better understand that mechanism and to optimise clay matrixes drying.

3.1 Moisture transfer by diffusion: diffusive model for the clay matrix drying

There are several modes of water transport through a porous structure. There are transfers by diffusion, by capillarity, by pressure gradient, by gravity (Collard [3], Charpin and Rasneur [15]). But according to many authors among whom Sherwood [16], Lewis [17], the transfer by diffusion is the most important mode of water transfer in a porous structure. A diffusive model based on Fick's laws is representative of the water transfer in a clay matrix during drying.

Let $S(x_1, x_2, x_3)$ be a point located at the centre of a volume element of the matrix and of which elementary volume is $dx_1.dx_2.dx_3$; let $\omega(x_1, x_2, x_3, t)$ be the

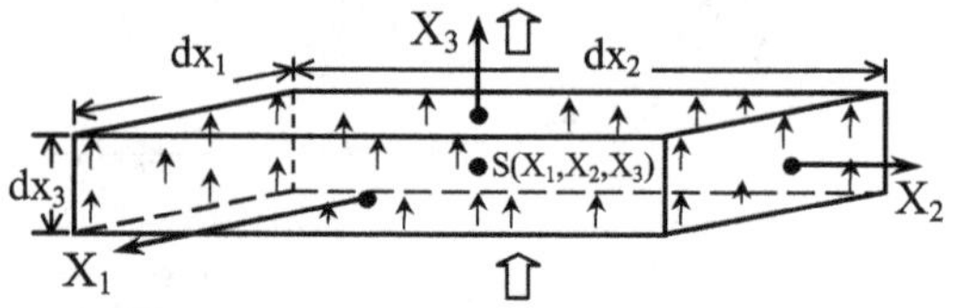

Fig. 7. Water diffusion in a volume element $dx_1.dx_2.dx_3$.

water content at this point at a given time t (Fig. 7). Let us consider the flow of water diffusing along the axis x_3 (Fig. 7).

The equations of Fick's diffusion laws (Collard [3], Crank [18]) are expressed by:

$$\begin{cases} \Phi_1 = -\left[D_{11}\frac{\partial\omega}{\partial x_1} + D_{12}\frac{\partial\omega}{\partial x_2} + D_{13}\frac{\partial\omega}{\partial x_3}\right] \\ \Phi_2 = -\left[D_{21}\frac{\partial\omega}{\partial x_1} + D_{22}\frac{\partial\omega}{\partial x_2} + D_{23}\frac{\partial\omega}{\partial x_3}\right] \\ \Phi_3 = -\left[D_{31}\frac{\partial\omega}{\partial x_1} + D_{32}\frac{\partial\omega}{\partial x_2} + D_{33}\frac{\partial\omega}{\partial x_3}\right] \end{cases} \tag{2}$$

$$\frac{\partial\omega}{\partial t} = \frac{\partial}{\partial x_1}\left(D_1\frac{\partial\omega}{\partial x_1}\right) + \frac{\partial}{\partial x_2}\left(D_2\frac{\partial\omega}{\partial x_2}\right) + \frac{\partial}{\partial x_3}\left(D_3\frac{\partial\omega}{\partial x_3}\right). \tag{3}$$

Φ_i is the flux density according to the direction x_i, D_{ij} the diffusion coefficient (m^2/s), $\omega(x_1, x_2, x_3, t)$ the water content at a point $S(x_1; x_2; x_3)$ of the matrix at a time t.

In an isotropic medium where the coefficients D_{ij} are identical in every direction, and by considering them constant, and in the case of a flat shape matrix where the flux of water diffusion is unidirectional and carried out according to the axis x_3, the above equations are simplified (Crank [18], William and Callister [19], Shelby [20], Asheland [21], Evans and Keey [22]):

$$\Phi = -D\frac{\partial\omega}{\partial x_3} \quad \text{and} \quad \frac{\partial\omega}{\partial t} = D\frac{\partial^2\omega}{\partial x_3^2}. \tag{4}$$

The solutions of these equations depend on the initial conditions and the boundary conditions, as showed by Crank [18]. These conditions are illustrated in Figure 8.

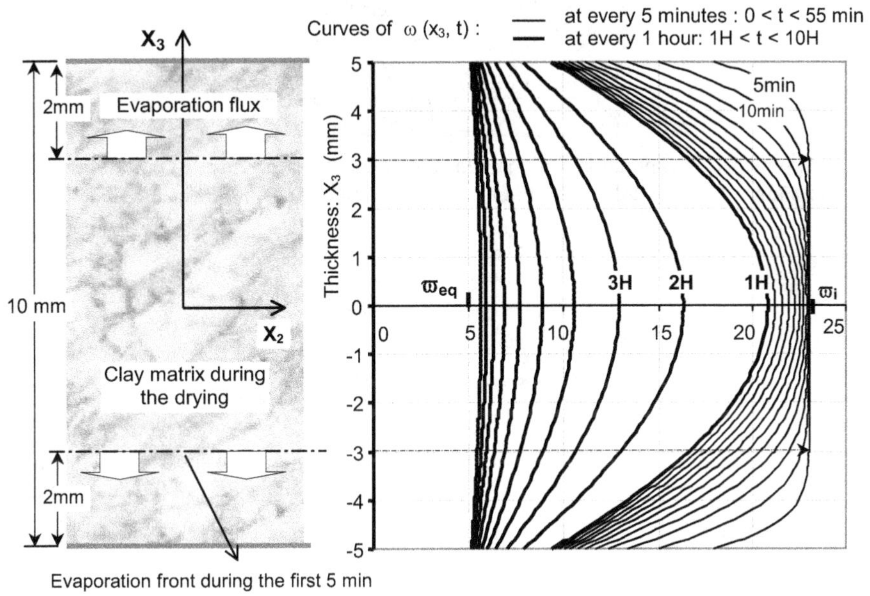

Fig. 9. Simulation of a flat clay matrix drying (case of the red clay of Guérin-Kouka): evolution of the water content at every 5 min and at every 1 h.

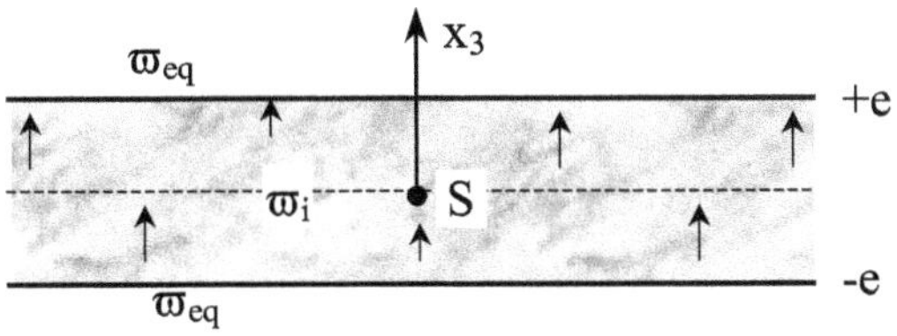

Fig. 8. Initial and boundary conditions.

The initial conditions and the boundary conditions are then expressed as follows:

$$\omega(x_3,t)=\varpi_i \quad \text{for} \quad -e<x_3<+e \quad \text{and} \quad t\leqslant 0 \qquad (5)$$

$$\omega(x_3,t)=\varpi_{eq} \quad \text{for} \quad x_3=\pm e \quad \text{and} \quad t>0. \qquad (6)$$

The solution giving the value of the water content at each point of the clay flat matrix at every time during drying is expressed as follows (Collard [3], Polubarinova-Kochina [12], Crank [18]):

$$\frac{\omega(x_3,t)-\varpi_{eq}}{\varpi_i-\varpi_{eq}}=\frac{4}{\pi}\sum_{k=0}^{\infty}\frac{(-1)^k}{(2k+1)}\text{Cos}\left(\frac{(2k+1).\pi}{2e}x_3\right) \times \text{Exp}\left(-\frac{(2k+1)^2\pi^2}{4e^2}Dt\right) \qquad (7)$$

$$\forall x_3\in[-e,+e] \quad \text{and} \quad \forall t\geqslant 0.$$

From the above expression of $\omega(x_3,t)$, the average water content ϖ(t) of the whole matrix can be deducted as:

$$\varpi(t)=\frac{1}{2e}\int_{-e}^{e}\omega(x_3,t)dx_3 \quad \text{or} \qquad (8)$$

$$\frac{\varpi(t)-\varpi_{eq}}{\varpi}_i-\varpi_{eq}=\frac{8}{\pi^2}\sum_{k=0}^{\infty}\frac{1}{(2k+1)^2}\text{Exp}\left(-\frac{(2k+1)^2\pi^2}{4e^2}Dt\right)$$

$$\forall t\geqslant 0. \qquad (9)$$

3.2 Modelling results and discussion

The diffusion coefficient has been determined experimentally and its value during a transitory water diffusion regime differs slightly from that of a permanent regime:

- during the transitory regime: $D=10^{-9}$ m^2/s;
- during the permanent regime: $D=10^{-10}$ m^2/s.

These values are very closed to those given by the literature (Collard [3], Evans and Keey [22]). The simulation is carried out on a flat matrix of 10 mm of thickness.

The initial water content ϖ_i and the balance water content ϖ_{eq} are given by Table 1 and Figure 4.

In the case of the red clay of Guérin-Kouka (ARG): $\varpi_i=23\%$; $\varpi_{eq}=5\%$ and $e=5$ mm.

The computer calculation is reduced to the first 52 terms of the sum of the above equations or $0\leqslant k\leqslant 51$.

Figure 9 gives the evolution of the water content at every 5 min during the first 55 min of drying and at every 1 h during the first 10 h.

Figure 9 shows that the evaporation front progresses from faces towards the middle of the matrix thickness. At the end of the first 5 min, the evaporation front is at 2 mm from faces, the matrix interior still having the initial water content ϖ_i. Evaporation front reaches the middle of the matrix thickness after a half an hour of drying. The water exchange balance will be established only after 30 to 40 h of drying.

At this point the balance water content ϖ_{eq} is about 5%.

The unequal water content $\omega(x_3,t)$ (Fig. 9) in the matrix during drying presents two consequences:

- evaporation generates the matrix shrinkage that is increasingly important from faces towards the matrix thickness middle. This generates internal tensions and stresses correlatively more important from faces

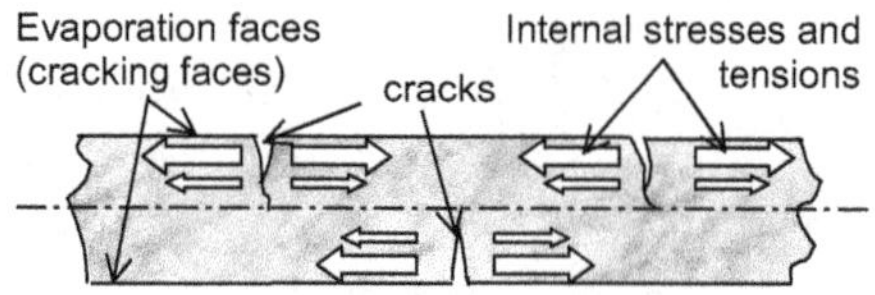

(a) Symmetrical evaporation: internal tensions and stresses are more important from faces towards the middle of the flat matrix thick: appearance of cracks on the faces

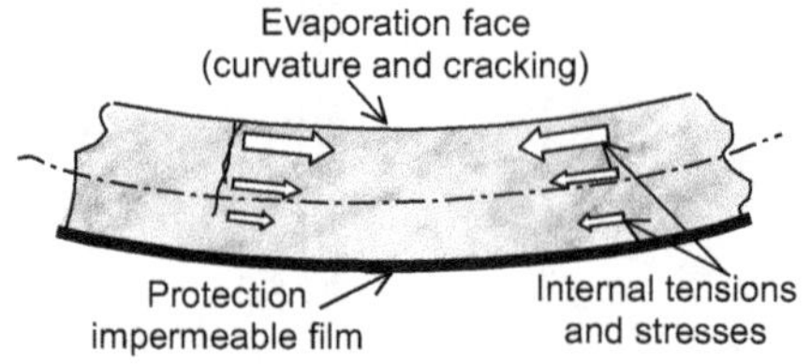

(b) Dissymmetrical evaporation: dissymmetrical distribution of internal stresses in a flat matrix of which a face is covered with an impermeable film: curvature on the side of the evaporation face

Fig. 10. Internal stresses distribution in a flat matrix thickness during a symmetrical drying (a) and a dissymmetrical drying (b).

towards the matrix thickness middle (Puffeney [7], Jouenne [13], Bottin [14]). When these internal stresses and tensions are important, they generate the matrix cracking, the cracks occurring on the flat faces as illustrated in Figure 10(a) (Collard [3], Jacquemin [23], Paul [24]). To reduce the internal stresses and tensions, the evaporation velocity must be controlled and lowered;

- when the two faces of a flat matrix are not exposed in an identical manner to the drying air, the dissymmetrical resulting evaporation generates a dissymmetrical shrinkage and correlatively an unequal distribution of tensions and stresses on the two faces. This causes the curvature of the flat. The concavity appears on the face which is most exposed to the air because of its more important shrinkage and its more important internal stresses and tensions.

This is observed when drying is performed on a clay flat matrix of which one of its faces is covered with an impermeable film in order to reduce the evaporation on that face (Fig. 10(b)) (Collard [3], Alviset [9]).

The simulation results can be validate by superimposing the experimental curve of the average water content and the calculated average water content $\varpi(t)$ (Fig. 11).

The calculated $\varpi(t)$ is reduced to the first 52 terms of the sum of the equation of $\varpi(t)$:

$$\varpi(t) = 5 + \frac{144}{\pi^2}\sum_{k=0}^{50}\frac{1}{(2k+1)^2}\mathrm{Exp}\left(-10^{-6}\pi^2(2k+1)^2 t\right) \quad \forall t \geqslant 0. \quad (10)$$

Referring to Figure 11, the maximum error resulting from the simulation is about 10%. In spite of this high value of the error, the experimental curve and the simulated curve are globally close and superimposed. This coincidence allows to validate the Fick's diffusive model applied to the clay matrixes drying.

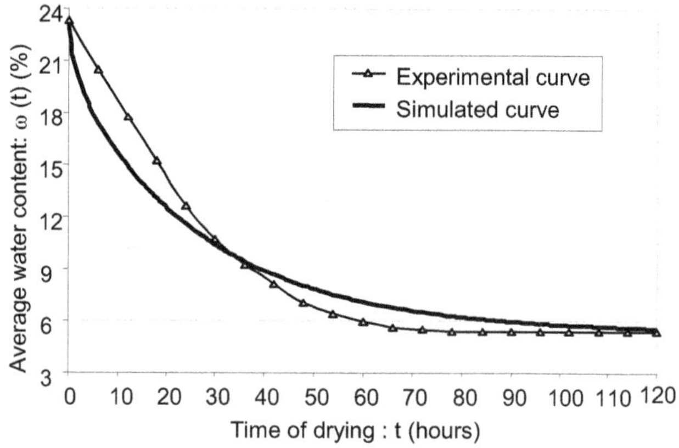

Fig. 11. Superimposition of the experimental curve and the simulated curve giving the average water content in the clay matrix during drying.

4 Conclusion

The validation of the simulation model allows to apply the experimental results and the modelling results to the drying of the various studied clay varieties.

A clay matrix drying comprises a critical zone corresponding to a zone where the water content of the matrix is still above its shrinkage limit and a zone without risks corresponding to a zone where the water content is below the shrinkage limit.

During the critical zone, the risks of the matrix cracking and deformation are more important when the evaporation velocity is important and when the evaporation is dissymmetrical.

To avoid cracking and deformations, drying must be controlled and performed with precautions. These precautions are summarized in Table 2. Table 2 gives some advised evaporation velocities required according to the clay variety. The evaporation must not exceed these velocity values in order to prevent matrixes from cracking

Table 2. Drying parameters and controlling advisable for the six clay varieties matrixes drying.

Critical drying zone		Shrinkage limit	Drying without risk
- Symmetrical drying - Evaporation velocity controlling			– Drying in a dry atmosphere
Clay variety	Advisable drying velocity (mg *of water /g of dried matter /hour)*	ϖ_r (%)	– Increasing of the drying velocity without risk
ABB	-6	17	
ANT	-10	8	
ARG	-5	12	
AVK	-7	13	
ARK	-5	9	
ARA	-6	10	

and deforming. For this purpose, drying will be performed in a wet atmosphere or in a quasi hermetic medium for example (see Tab. 2).

During the zone without risk o below the shrinkage limit, drying can be continued in a dry atmosphere and the evaporation velocity can be increased without risk.

References

1. K.-E. Atcholi, *Rapport d'activités de Recherches*, Université de Lomé, Février 1997
2. E., Padayodi *Contribution à l'élaboration des matériaux de construction à base de produits naturels : caractérisation thermomécanique et physico-chimique des argiles et des fibres naturelles cellulosiques*, Thèse UFR Sciences et Techniques, Université de Franche-Comté (2001)
3. J.-M. Collard, *Étude des transferts d'humidité et des déformations pendant le séchage d'une plaque d'argile*, Thèse de l'Université de Poitiers (1989)
4. R. Ho-Yick-Cheong, *La brique de terre crue comprimée et stabilisée au ciment: caractéristiques et propriétés physico-mécaniques*, Thèse de l'Université Joseph-Fourier-Grenoble I (1989)
5. CDI Centre pour le Développement Industriel (A.C.P-C.E.E), *Blocs de terre comprimée: choix du matériel de production*, 1ère édition, Sept. 1988
6. J.-P. Magnan, *Description, identification et classification des sols*, Technique de l'Ingénieur, Vol. C2I, N° C208. Ed. Istral BL (1996)
7. J.-P. Puffeney, *Système à base de connaissances d'aide à la décision pour la conduite d'une tuilerie automatisée*, Thèse UFR Sciences et Techniques, Université de Franche-Comté (1997)
8. C. Suri, *Étude du couplage des phénomènes d'absorption et d'endommagement dans un composite verre-epoxy*, Thèse UFR Sciences et Tech., Univer. de Franche-Comté (1995)
9. L. Alviset, *Matériaux de terre cuite*, Technique de l'Ingénieur, Vol. C1, N° C905. Ed. Istral BL (1996)
10. J. Sigg, *Les produits de terre cuite*, Paris, Septima (1991)
11. Publication du Centre Technique des Tuiles et Briques, *Facteurs influençant le séchage*, 30 pages
12. P.Ya Polubarinova-Kochina, *Theory of ground water movement* (Gostekhizdat, Moscou), Princeton Univ. Press, N.J. (1962)
13. C.A. Jouenne, *Traité de Céramiques et Matériaux Minéraux*, Editions Septima. Paris. 1980, pp. 368–500
14. C. Bottin, *La fabrication artisanale de tuiles romanes*, Edition CRATerre (1988)
15. Charpin, Rasneur, Mesure des surfaces spécifiques, Technique de l'Ingénieur, Analyse et caractérisation, N° P1050 (1987).
16. T.K. Sherwood, *The drying of solid*, Tome I and II **21**, 12–16 and 976–980 (1929)
17. R.W. Lewis, Drying induced stresses in porous bodies. Int. J. Numer. Meth. Eng. **11**, 1175–1184 (1977)
18. J. Crank, *The Mathematics of diffusion*, Clarendon Press, Oxford (1975)
19. D. William, Jr. Callister, *Materials Science and Engineering*, Dep. of Metal. Eng. of The University of Utah. 4th edn., 90–100 (1996)
20. J.A. Shelby, *Introduction to glass science and technology*, New York State College of Ceramics and Alfred Univ., 158–160 (1997)
21. Donald R. Askeland, *The Science and Engineering of Materials*, 3th ed., 429–430 (1994)
22. A.A. Evans, R.B. Keey, *The water diffusion coefficient of an shrinking clay on drying.* Chem. Eng. J., 126–135 (1975)
23. F. Jacquemin, *Modélisation et simulation des contraintes internes dans les structures tubulaires composites épaisses*, Thèse de l'Université Blaise Pascal-Clermont II (2000)
24. D. Paul, *Contraintes Hygrothermiques dans les matériaux composites stratifiés : Modélisation et mise en évidence expérimentale*, Thèse de l'université Jean Monnet de Saint-Etienne (1996)

21

A Monte-Carlo method used to study the fragment impact effect on the industrial facilities

Q.B. Nguyen[1,2,a], A. Mebarki[1], R. Ami Saada[1], F. Mercier[2] and M. Reimeringer[2]

[1] Université Paris-Est, Laboratoire de Modélisation et Simulation Multi-échelle, MSME FRE 3160 CNRS, 5 boulevard Descartes, 77454 Marne la Vallée, France
[2] Institut National de l'Environnement Industriel et des Risques (INERIS), 10 boulevard Lahitolle, 18000 Bourges, France

Abstract – The generated fragments due to the explosion of vessel under pressure may be threats to the surrounding equipments at the impact. The fragment characteristics are number, shape, mass, departure angles, departure velocity. These characteristics are considered as uncertain variables. The authors propose the probability distribution for each variable. A fragment trajectory model concerning its shape is then presented in order to study the impact problem. The numerical simulations using the Monte-Carlo method allow calculating the probability of impact, P_{imp} between fragments and surrounding equipments.

Key words: Explosion; vessel; fragment; source term; trajectory; impact; probability.

1 Introduction

The industrial site shelters cylindrical vessels under pressure of gas or liquids. This is a closed system. The overpressure or the fragment aggression is the cause of the vessel explosions. The fragment series can be generated after an explosion. Nevertheless, all the generated fragments aren't entirely sent in to space. Certain fragments remain around the origin, others land up far from that. This paper considers only the sent fragments which characteristics are number, shape, mass, departure angles, and departure velocity (denominated "source terms" and considered as uncertain variables) (see Sect. 2). The fragments can impact the surrounding equipments. This possibility corresponds to the probability of impact, P_{imp} [1]. Next the authors write the fragment trajectory equations in 3D in order to study the impact problem between a fragment and an equipment (see Sect. 3). The Monte-Carlo simulation allows to determine the impact probability, P_{imp} (according to the source terms probability distributions). Thus the numerical simulations using data collected from the accident in Mexico City [2] have been performed. This study allows evaluating the fragment impact range and the arriving fragment energies at the impact on the surrounding equipments.

[a] Corresponding author:
Quoc-Bao.Nguyen@univ-paris-est.fr

Nomenclature

n	fragment number,
v	fragment velocity,
m	fragment mass,
φ	vertical angle,
θ	horizontal angle,
D	drag force,
L	lift force,
C_D	drag coefficient,
C_L	lift coefficient,
S_D	frontal surface,
S_L	projected surface to the ground,
ρ_{air}	air density,
P_{imp}	probability of impact.

2 Threat

2.1 Fragment number, n

According to an experimental study on cylindrical vessels [3–6], a crack initiates at circumferential welding and detaches one of end-caps, then propagates itself axially to unfold the ring of cylindrical vessel. On the other hand, the crack can also initiate at the middle of vessel because of the manufacture defects. It unfolds the ring, then propagates itself to the edges of the vessel, at last, following the circumference, detaches one or two end-caps. In two cases, the ring is either unfolded and remains at the origin, or it generates one or two fragments [7, 8]. This explains why the quantity of plate is smaller than the one of end-cap and oblong end-cap (see Sect. 2.2). The fragment number

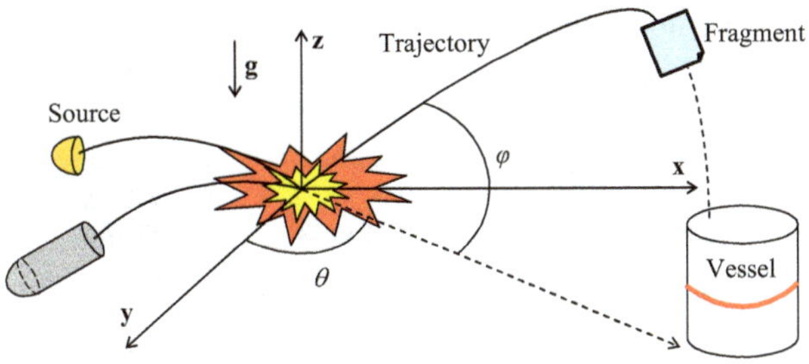

Fig. 1. Explosion generates fragment series.

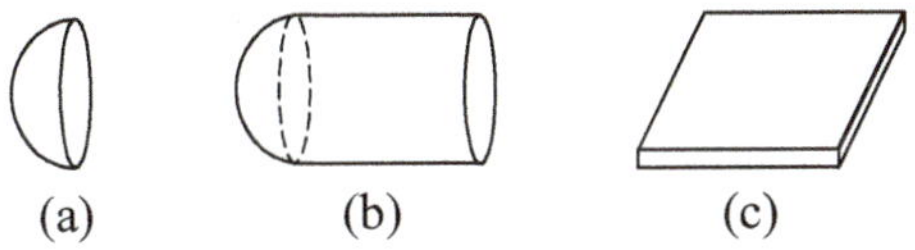

Fig. 2. Fragment shapes (a) end-cap; (b) oblong end-cap; (c) plate.

varies from one to three (rarely four). Simultaneously basing on the report of Holden [2], the authors determine the probability distribution for this variable while using the principle of maximum entropy [9]. It allows applying a discreet distribution for the fragment number that varies from one to four.

2.2 Fragment shape

As described previously, the fragment types are the end-cap, the end-cap attached to completed or partial ring (the oblong end-cap) and the flattened piece of ring (the plate). Moreover, according to the accident report of Holden [2], the generated fragments due to an explosion are end-caps (44 pieces), plates (57 pieces) or oblong end-caps (86 pieces) (Fig. 2). In first approach, the end-caps represent 25% of generated fragments, 30% for the plates and 45% for the oblong end-caps.

2.3 Fragment mass, m

The crack can initiate at the circumferential welding, or at the middle of vessel (see Sect. 2.1). That allows applying a constant probability distribution for the oblong end-cap length and for the plate dimensions. When the shape, the size, as well as the constituent material of fragment are known, its mass is then determined.

2.4 Vertical departure angle, φ

With the hypothesis that the crack can initiate anywhere on the vessel, the vertical departure angle is situated in the interval $[-90°, 90°]$. Basing on this hypothesis, the vertical departure angle is supposed follow a constant distribution in the interval $[-90°, 90°]$.

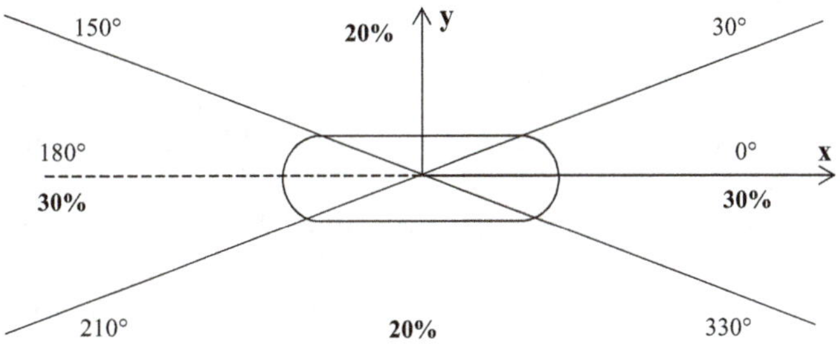

Fig. 3. Fragment distribution.

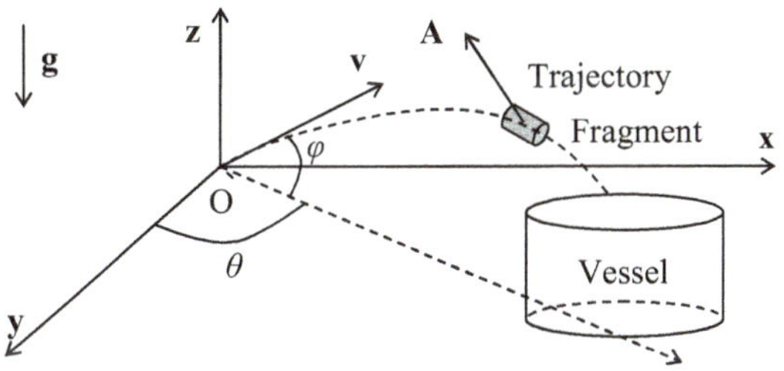

Fig. 4. Fragment trajectory.

2.5 Horizontal departure angle, θ

Basing on the accident analysis [2, 7], a fragment distribution in relation to the principal vessel axis is carried out: 24 on 42 fragments are projected in an angular sector of 60° around the principal axis and the remainder is projected in the perpendicular direction. An approximate distribution is considered for the horizontal departure angle: 20% of fragments in the interval [30°, 150°]; 30% of fragments in [150°, 210°]; 20% of fragments in [210°, 330°] and 30% of fragments in [330°, 30°] (Fig. 3).

Moreover a big number of end-caps and oblong end-caps is situated in the angular sector of the longitudinal vessel axis and of the perpendicular axis for the plate fragments. Nevertheless, some end-caps and oblong end-caps are found on the perpendicular direction and on the vessel axis for the plates. A possible interpretation of this phenomenon is the rotation of the end-caps, oblong end-caps and plates when they were cut [7].

2.6 Departure velocity, v

The departure velocity is determined while knowing its kinetic energy, E_c and its mass, m. E_c is deducted from the total energy, E with a multiplying factor situated in the interval [0.2, 0.5] (0.2 is the advised value). This energy serves to expansion, to vessel rupture, to fragment kinematics, etc. and it is calculated by the equations of Brode, Baker and Baum [10]. On the other hand, a probabilistic study based on experiments of Baum [3,4] allows applying a log-normal probability distribution for the departure velocity, v.

3 Impact

3.1 Fragment trajectory

During the flight, a fragment is subjected to the following forces:

- the weight $\overrightarrow{\mathbf{G}} = m\overrightarrow{\mathbf{g}}$, when $\overrightarrow{\mathbf{g}}$ is the acceleration of gravity vector;
- the aerodynamic force, $\overrightarrow{\mathbf{A}}$ is decomposed into two parts: the drag and the lift,
  - the drag, $\overrightarrow{\mathbf{D}}$ is a combination of friction forces due to the air flow. It is parallel and opposed to the fragment trajectory
$$D = \frac{1}{2}\rho_{air}v^2 C_D S_D; \qquad (1)$$
  - the lift, $\overrightarrow{\mathbf{L}}$ is created by the aspiration in the depression zone. It is perpendicular to the fragment movement
$$L = \frac{1}{2}\rho_{air}v^2 C_L S_L. \qquad (2)$$

In order to determine the impact point between a fragment and an equipment, these forces are projected on three principal axis $\overrightarrow{\mathbf{x}}$, $\overrightarrow{\mathbf{y}}$, $\overrightarrow{\mathbf{z}}$ including the horizontal and vertical angles (θ and φ, respectively):

$$\begin{cases} -k_D\cos\varphi - (-1^q\, k_L\sin\varphi)\cdot\left(\dot{x}^2+\dot{y}^2+\dot{z}^2\right) - \dfrac{\ddot{x}}{\sin(\theta)} = 0 \\ -k_D\cos\varphi - (-1^q\, k_L\sin\varphi)\cdot\left(\dot{x}^2+\dot{y}^2+\dot{z}^2\right) - \dfrac{\ddot{y}}{\cos(\theta)} = 0 \\ (-(-1)^q\, k_D\sin\varphi + k_L\cos\varphi)\cdot\left(\dot{x}^2+\dot{y}^2+\dot{z}^2\right) - \ddot{z} - g = 0 \end{cases} \qquad (3)$$

$$k_D = \frac{1}{2}\frac{\rho_{air}C_D S_D}{m}; \quad k_L = \frac{1}{2}\frac{\rho_{air}C_L S_L}{m} \qquad (4)$$

with $q = 1$ for the descending; $q = 2$ for the ascending; x, y, z, are the fragment center coordinates; m, the fragment mass; C_D, the drag coefficient; S_D, the frontal surface; C_L, the lift coefficient; S_L, the projected surface to the ground; ρ_{air}, the air density.

The lift and drag coefficients (C_L and C_D) are obtained experimentally [11]. In the first approach, the functions of the angle between the velocity and the fragment axis are created for each coefficient.

3.2 Interaction

The fragment shapes are the end-cap, the oblong end-cap and the plate. The equipment shapes are cylindrical, spherical or cube. Previously, the impact problem between two ellipsoids ($\mathbf{E}_1$, $\mathbf{E}_2$) is analyzed:

$$(\mathbf{E}_1) \quad \mathbf{X}^{\mathrm{T}}\mathbf{A}\mathbf{X} = 0 \qquad (5)$$

$$(\mathbf{E}_2) \quad \mathbf{X}^{\mathrm{T}}\mathbf{B}\mathbf{X} = 0 \qquad (6)$$

with $\mathbf{A}$, $\mathbf{B}$, their characteristic matrices; $f(\lambda) = \det(\lambda\mathbf{A} + \mathbf{B}) = 0$, their characteristic equation; and $\mathbf{X} = (x, y, z, 1)^{\mathrm{T}}$.

Table 1. Data for the numerical simulations.

Cylindrical vessel (A)		Spherical vessel (B)	
	$r_x = 2$ m		$r_x = 2$ m
Radius	$r_y = 8$ m	Radius	$r_y = 8$ m
	$r_z = 8$ m		$r_y = 8$ m
Length	6 m	Thickness	0.05 m
Thickness	0.07 m	Center	$0 \times 20 \times 10$ m

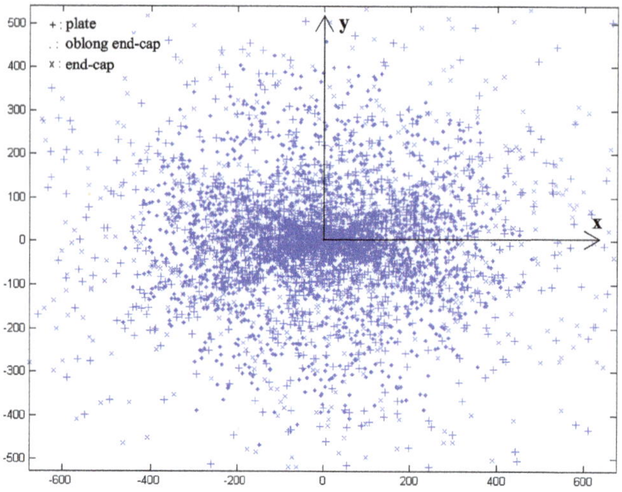

Fig. 5. Fragment distribution.

According to the results justified by Wang et al. [12]:

- the equation $f(\lambda) = 0$ has always at least of two negative roots;
- two ellipsoids ($\mathbf{E}_1$, $\mathbf{E}_2$) are separated by a plan if $f(\lambda) = 0$ has two different positive roots;
- two ellipsoids ($\mathbf{E}_1$, $\mathbf{E}_2$) intercept themselves if $f(\lambda) = 0$ has a double positive root, λ_0.

In case of tangency, the intersection point, $\mathbf{X}$ is determined by solving equation (7):

$$\left(\lambda_0\mathbf{I} + \mathbf{A}^{-1}\mathbf{B}\right)\mathbf{X} = 0 \qquad (7)$$

with $\mathbf{I}$, the matrix identity.

In a second time, the development of this model allows to obtain the impact condition between fragment and equipment.

4 Numerical simulations

Using the Monte-Carlo method, the numerical simulations are realized while considering two cylindrical vessels **A** and **B** with volumes equals to: 100 m^3 and 2100 m^3, respectively. The distance between them is about 20 m. The data used to carry out these simulations are given in Table 1 and the results are presented in Figures 5–7.

One can notice (Fig. 5) that after explosion of vessel **A** (placed at the origin) the generated fragments are sent into a circular zone having a diameter equal to 1 km approximately. The majority of fragments is distributed in the angular sector of 60° (±30° in comparison with the

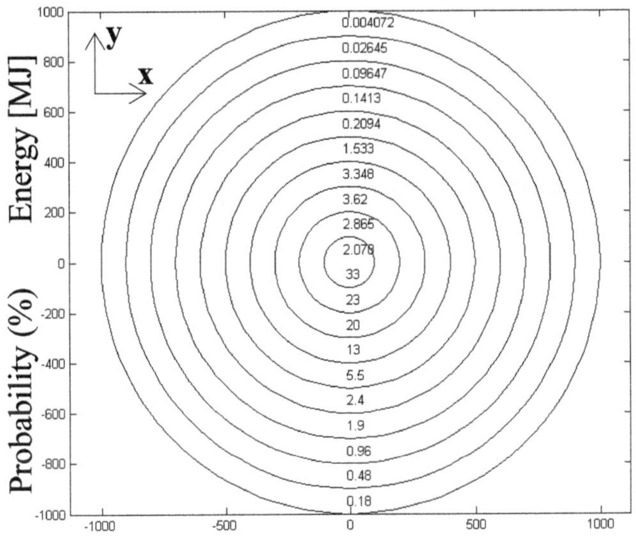

Fig. 6. Arriving fragment probability and their energy.

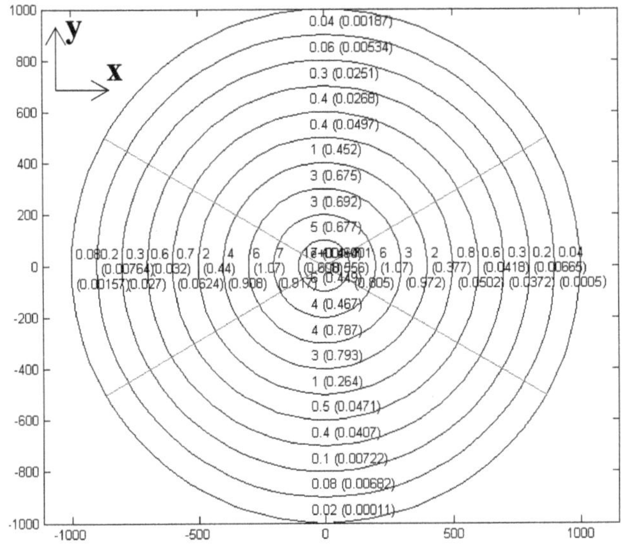

Fig. 7. Arriving fragment probability and their energy per sector. (The number between brackets represent energies in [MJ], the others correspond to the arriving projectile probabilities.)

axis $\overrightarrow{\mathbf{X}}$ of vessel **A**). The probability of the arriving fragment energies are reported in Figures 6, 7. One can notice that the last two quantities decrease according to the distance from origin. The arriving fragment velocity average is 150.19 m/s. The arriving fragment energy average is 35.06 MJ, the energy lost during movement is 6.50 MJ. The impact probability is $P_{imp} = 0.038\%$.

5 Conclusions

The authors analyze the generated fragments reaches after a cylindrical vessel explosion. The fragment characteristics are considered as uncertain variables. A fragment movement model considering its shape and its size is then developed. This allows evaluating the impact probability, P_{imp}, between a fragment and an equipment in the industrial domain.

This research work has received the support of Institut National de l'Environnement Industriel et des Risques (INERIS), France.

References

1. A. Mébarki, R. Ami Saada, Q.B. Nguyen, F. Mercier, M. Reimeringer, Mechanical study integrated probabilistic approaches used to analyze the industrial risks, *International Conference on Computational & Experimental Engineering and Sciences* (Miami, USA, 03–08 January 2007)
2. P.L. Holden, *Assessment of Missile Hazards: Review of Incident Experience Relevant to Major Hazard Plant* (Safety and Reliability Directorate, Health & Safety Directorate, 1988)
3. M.R. Baum. J. Loss Prev. Process Ind. **8**, 149 (1995)
4. M.R. Baum. J. Loss Prev. Process Ind. **12**, 137 (1999)
5. M.R. Baum. J. Loss Prev. Process Ind. **14**, 199 (2001)
6. Ineris, *Les éclatements de réservoirs, Phénoménologie et modélisation des effets* (N° Ineris-DRA-2004-46055, 2004)
7. Ineris, *Calculs des effets mécaniques d'un BLEVE de citerne ferroviaire* (N° Ineris-DRA-72293, 2005)
8. Q.B. Nguyen, A. Mébarki, F. Mercier, R. Ami Saada, M. Reimeringer, in *Proceedings of the Eighth International Conference on Computational Structures Technology*, edited by B.H.V. Topping, G. Montero, R. Montenegro (Civil-Comp Press, Stirlingshire, U K, 2006)
9. C. Soize, *Probabilités et modélisation des incertitudes* (2004)
10. U. Hauptmanns. J. Loss Prev. Process Ind. **14**, 395 (2001)
11. P. Liu, X. Deng. J. Aéronautique et Spatial du Canada **49**, 31 (2003)
12. W. Wang, J. Wang, M. Kim. Computer Aided Geometric Design **18**, 531 (2001)

22

An adjacency representation for structural topology optimization using genetic algorithm

B. Sid, M. Domaszewski[a] and F. Peyraut

M3M Laboratory, University of Technology of Belfort-Montbeliard, 90010 Belfort Cedex, France

Abstract – A new approach for continuum structural topology optimization using genetic algorithms is presented in this paper. The proposed approach is based on a representation by adjacency where the principle is founded on the concept of connectivity of finite elements, considered as cells. This principle is expressed by an adjacency matrix similar to that used in the graph theory. The encoding of the structure solutions uses this matrix by transforming it into a binary string. The research of optimal solution, i.e. the optimal material distribution, is interpreted in this approach by the determination of the connectivity of elements (cells). Using density variable, the approach has some common points with the homogenization techniques. The proposed approach is tested with simple benchmark applications.

Key words: Topology optimization; genetic algorithm; adjacency representation

1 Introduction

Genetic algorithms have proved a good efficiency to resolve complex optimization problems [1]. However, this efficiency is strongly dependent on a good choice of the algorithm parameters. The representation of solutions is the most important parameter because it must give a good formulation of the problem. Additionally, the choice of appropriate representation requires domain knowledge in order to make the search more efficient.

In topology optimization using genetic algorithm, the bit-array representation is usually used. Sandgren et al. [2] are among the first researchers to develop a genetic algorithm based approach for continuum structural topology optimization. In their work, the design domain is discretized into small elements, where each element contains material or void and thus no intermediate densities are allowed. This is a typical bit-array representation approach, in which a bit-array is used to define the design variables and can be directly mapped into the design domain discretized by a fixed regular mesh, where each of the small elements contains either material or void. Hence, the original '0-1' optimization problem was attacked directly by using a bit-array representation and a genetic algorithm. The work of Sandgren and his co-workers, using bit-array representation, has been extended by Jakiela and his co-workers [3–5], by Schoenauer and his co-workers [6, 7, 9], by Fanjoy and Crossley [8, 10], and, more recently, by Wang and Tai [11]. Although all these extensions can well prevent checkerboard patterns by exploiting a connectivity restriction, the other numerical instabilities in structural topology optimization such as mesh dependency and one-node connections still exist. More important, the issue of design connectivity analysis, which affects significantly the computational results, has not been completely solved.

The bit-array representation is founded in the intuitive concept of absence or presence of material. But structural topology optimization can be also defined as the determination of the structure elements nature and connectivity. The approach that we present in this paper is founded on this definition. We consider each finite element as a cell and then we define a representation based on the connectivity of these cells. This connectivity is expressed by a binary adjacency matrix. The encoding operation consists on transforming the matrix in one-dimensional binary string. The proposed representation seems more adequate to the topology optimization problem than the bit-array representation. Moreover, the one-dimensional string code is well matched to a genetic algorithm optimization process.

The research of optimal solution, i.e. the optimal material distribution, is interpreted in this approach by the determination of the connectivity of elements (cells). The mutation plays the most important role to accomplish this task. The operation of mutation is inspired partially from cellular automata method [12, 13]. We assign to every cell an artificial density variable whose value is altered by mutation operation. The value of the artificial density is then used for the search of the element connections associated with the mutant cell. Using density variable, the approach has some similarity to the SIMP

[a] Corresponding author: matthieu.domaszewski@utbm.fr

techniques [14]. The proposed approach is validated with simple benchmark applications.

2 Structural optimization using GAs

Using an evolutionary, survival of the fittest optimization mechanism, genetic algorithms [1] allows designs in a population to compete against one another to serve as a parent designs. Parents then pair and mate, swapping portions of their 'chromosome' to create a generation of child designs of hopefully higher performance. After undergoing infrequent, random mutation, the child generation replaces the original generation, and the process then iterates until an optimal design is reached.

3 Problem formulation

This work is limited to linear elasticity topology optimization problems with a single objective function. The design domain is a continuous 2-D solid composed of elastic, homogeneous, isotropic material. Topology optimization consists in choosing a redistribution of material in each element according to the objective and constraint functions.

The principle of the genetic algorithms is a search of the fittest individual (global optimum solution) in the population. If the optimization problem is to maximize the objective function then the fitness function for evaluation of each individual is the same as the objective function. For the minimization problem, the fitness function is defined as the reciprocal objective function. The optimal solution should satisfy some constraints of inequality type. Finally, the optimization problem is stated in the following form

$$\begin{aligned} &\text{Maximize: } Fitness\,(x) = \frac{1}{f(x)} \\ &\text{Subject to: } g_i(x) \geqslant 0 \quad i = 1, K, n \end{aligned} \tag{1}$$

where, x is the solution vector, $f(x)$ is the objective function to be minimized, $g_i(x)$ is the i-th inequality constraint function, n is the number of inequality constraint functions.

Although genetic algorithms were initially developed to solve unconstrained optimization problems, during the last decade several methods have been proposed for handling constrained optimization problems as well [15]. The methods based on the use of penalty functions are employed in majority of cases for treating constraint optimization problems with GA. In this study, an efficient constraint handling method proposed by Deb [16], witch is also based on penalty function approach, is further used. Using the Deb penalty method the fitness function of constrained problem (1) is transformed into the following form

$$\text{Max: } Fitness(x) = \begin{cases} \frac{1}{f(x)} & \text{if } x \in \Omega^F \\ \frac{1}{f_{\max} \times (1 + \sum_{i=1}^{n} \max[0, -g_i(x)])} & \text{otherwise} \end{cases} \tag{2}$$

where, Ω^F is the feasible region of design domain, $f_{\max}$ is the objective function value of the worst feasible solution in the population.

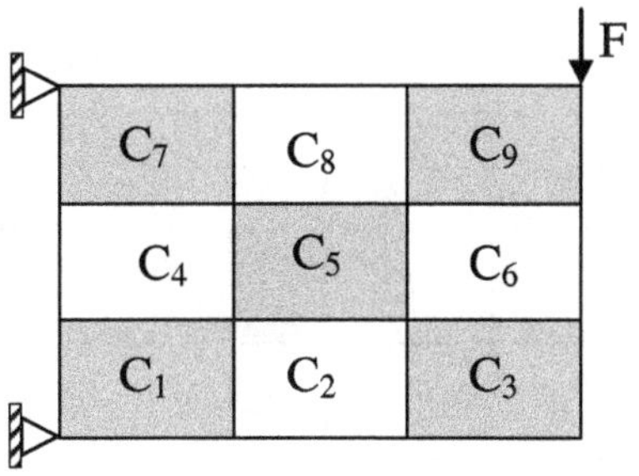

Fig. 1. Principle of adjacency representation.

4 Adjacency representation approach

4.1 Principle

The principle of this new representation is based on the adjacency or neighborhood's concept. Each finite element is considered as a cell (Fig. 1). By considering the gray finite elements as cells, the white finite elements which connect these cells between them are considered as connection elements. The remove of a connection element involves the elimination of the material which occupied the finite element geometrical domain. On the other hand, the search of optimal material distribution can be done by adding or removing element connections. Practically, the finite element considered as cell in the present generation will be considered as connection in the next generation, and vice versa for an element connection type. Thus, each finite element alternates, as cell or connection type, from one generation to another. This alternation makes possible to explore the solution research space. That can not be possible if elements remain fixed in a particular role.

A particular case concerns the element cells which are on boundary condition surfaces. For these elements, we add virtual cells which represent the boundary condition surfaces (Fig. 2). The cells C_1, C_7 and C_9 have the virtual neighborhood cells C_{1f}, C_{7f} and C_{9f} respectively. The virtual cells are considered empty spaces without physical or mechanical properties.

4.2 Chromosome representation

The cell elements neighborhood relative to Figure 2 can be expressed by a binary adjacency matrix as follows.

The structure representation is only the gray part of the adjacency matrix (Fig. 3) which explains the element

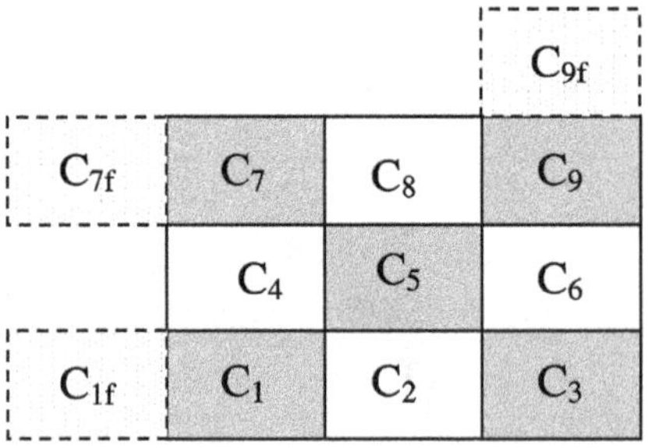

Fig. 2. Virtual cells on boundary condition surfaces.

	C_1	C_3	C_5	C_7	C_9
C_2	0	1	1	-	-
C_4	1	-	1	1	-
C_6	-	1	1	-	1
C_8	-	-	1	0	1
C_{1f}	1	-	-	-	-
C_{7f}	-	-	-	1	
C_{9f}	-	-	-	-	1

Fig. 3. Representation through adjacency matrix.

connectivity. This part, indicated by the gray color, is that which we use to build the chromosome, i.e. encoding operation.

4.3 Chromosome encoding

The encoding operation consists to transform the adjacency matrix, given in Figure 3, into one-dimensional binary string in order to facilitate genetic operations. The column, corresponding to an element cell in the adjacency matrix, represents the element cell associated gene and the element connection associated gene is corresponding to a line in the adjacency matrix. The encoding operation consists to place these genes one beside other in order to forming one-dimensional string. The resulting chromosome is given in Figure 4.

Unlike in the encoding based on the 'bit-array' representation, where each finite element is represented by one bit, in the adjacency approach each element is represented by a gene. Each gene associated to an element cell is composed by some bits where each bit corresponds to an element connection. Practically, each finite element is a cell surrounded by connection elements. Thus, cell elements and connection elements can divide some bits between them.

4.4 Evaluation

The evaluation needs structural analysis by finite element method. However, in the present approach we use the physical properties relaxation technique as it used by SIMP method (Bendsøe et al. [14]). The elastic modulus of each element, E_{ie}, is modeled as a function of the relative density, ρ_{ie}, using the power law. This can be expressed as

$$E(ie) = \rho_{ie}^{p} E_0 \qquad (p > 1) \tag{3}$$

where E_0 is the elastic modulus of the solid material and p is a coefficient used to penalize intermediate relative density values and drive the design to black and white structure. The relative density ρ_{ie} of each cell is expressed as

$$\rho(ie) = \frac{\sum_{j=1}^{Nbits(ie)} bit_j(ie)}{Nbits(ie)} \tag{4}$$

where $Nbits(ie)$ is the length of the gene associated to the ie element cell.

Expression (4) gives the density of the element cell ie according to the average of the associated gene bits values. Knowing that these values are binary, the density value is equal 0 if all the bits values are zeros or equal 1 if all the bits values are equal 1. Practically, in order to avoid the singularity of the stiffness matrix we affect a low value of about 10^{-4} to the empty element cells. Then the global stiffness matrix is calculated as follows

$$K_G = \sum_{ie=1}^{nelt} \rho_{ie}^{p} k(ie) \tag{5}$$

where K_G is the global stiffness matrix, $k(ie)$ is the elementary matrix associated to ie element and $nelt$ is the total number of finite elements.

4.5 Mutation

The mutation operation consists to alter one or some genetic information of an individual arbitrary chosen. Once the chromosome chosen, the mutation consists to modify the information of some genes. Practically, in our approach we use the cell densities as variables. The number of the element connections is proportional to the density of the associated element cell (expression (4)). Thus, the modification of the density of the mutate cell leads to the modification of the configuration and the number of the associated element connections. The proposed technique is based on the adaptation of each cell with its neighborhood (Tatting et al. [12], Tovar et al. [13]).

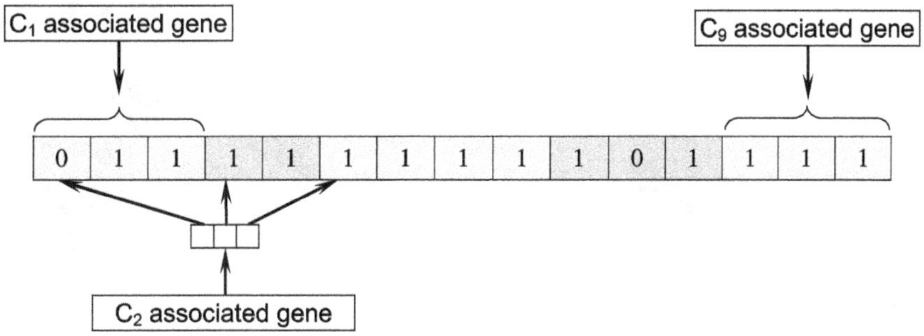

Fig. 4. Chromosome storage.

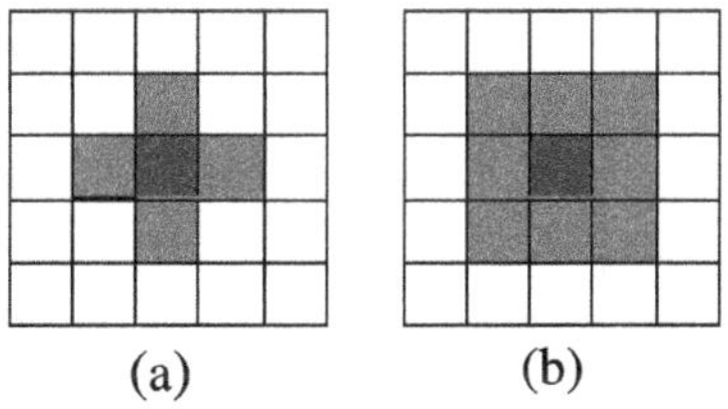

Fig. 5. Cell neighborhoods (a) Von Neumann (b) Moor.

The principle consists in interpolating the density of the mutate cell according to the adjacent cells layout. The most commonly used are the von Neumann layout that includes four neighboring cells ($N = 4$) (Fig. 5a) and the Moore layout that includes eight neighboring cells ($N = 8$) (Fig. 5b).

The average cell density is then altered according to the following expression

$$x_{ie} = \hat{x}_{ie} + \Delta x_{ie} \tag{6}$$

$$\text{where} \quad \hat{x}_{ie} = \frac{\rho_{ie} + \sum\limits_{je=1}^{Nc} \rho_{je}}{Nc + 1} \tag{7}$$

$$\text{and} \quad \Delta x_{ie} = rand \cdot (-1)^{round(rand)} \tag{8}$$

where x_{ie} is density variable, $\hat{x}_{ie}$ is the ie element density computed using the expression 7 and Δx_{ie} is real value randomly generated using expression (8).

The new density value done by the expression (6) should be always positive and less than one.

$$x_{ie} = \begin{cases} 0 & \text{if } x_{ie} < 0, \\ 1 & \text{if } x_{ie} > 1, \\ x_{ie} & \text{if } 0 < x_{ie} < 1. \end{cases} \tag{9}$$

Consequently, if the density value decreases that leads to material elimination and, on the contrary, it leads to material addition if the density value increases.

The corrected density (expression (9)) is then used to find the new element connection number:

$$Nbr = round(x_{ie} \times Nbit) \tag{10}$$

where *Nbit* is the mutate element gene length corresponding to the number of associated connection elements.

4.6 Selection method

Selection operator is applied to the current population to create an intermediate one. The so-called "Stochastic Universal Sampling" method is used in this paper. The probability of selection of each individual is proportional to the performance calculated in function of its rank in the population. According to a linear version of this method proposed by Baker [17] a selection pressure encloses to two is applied in genetic algorithm. The artificial performance of an individual of rank r_i is calculated in the following way

$$Fit_{Rk}(r_i) = -\frac{2 \times (r_i - N)}{N - 1} \tag{11}$$

where, $Fit_{Rk}(r_i)$ is the artificial fitness function value of the i-th individual, N is the population size, r_i is the rank of the i-th individual (the rank number one designates the best individual).

5 Numerical implementation

The algorithm of the method developed in this paper has been implemented in MATLAB programming language [18]. The computer code includes also a finite element solver for isotropic linear elasticity with plane stress elements. Figure 6 shows the flow chart of the proposed approach.

The employed strategy involves also the transfer of best individual of each population into the next generation without transformations. This individual replaces the best of the current generation if this last one is worst. If the best individual remainders the same during fifty successive generations, the optimization process is halted. This is the first stopping criteria we are using in our approach. The second one, often used, is based on a prefixed maximum number of generations.

5.1 Application

The proposed approach is tested using standard benchmarks of structural topology optimization problems. One of these benchmarks involves a rectangular structure supported at the lower corners, with a downward force of

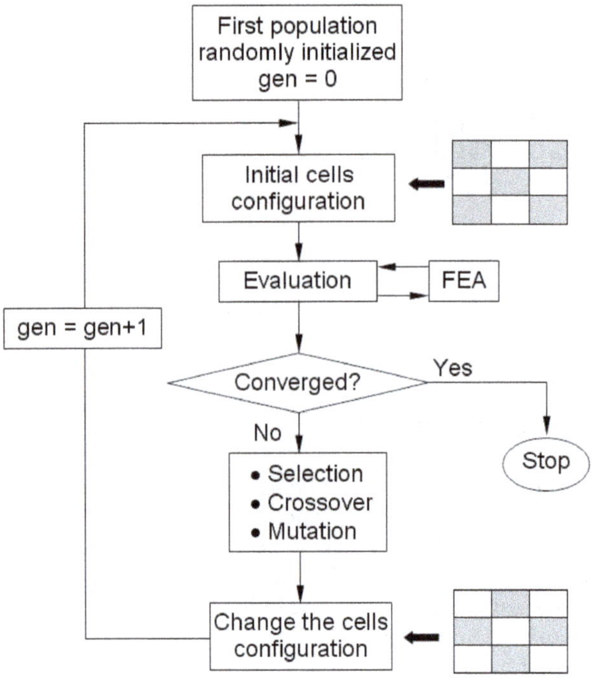

Fig. 6. Flow chart of the proposed approach.

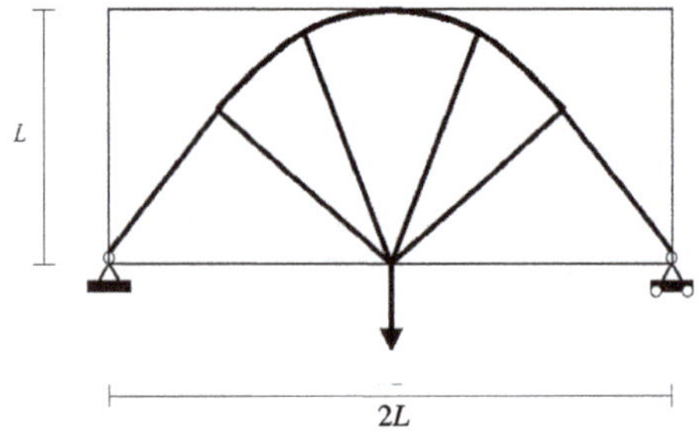

Fig. 7. Ideal solution of the Michell truss.

magnitude 4×10^3 N applied at the centre of the bottom edge (see Fig. 6). The design domain dimensions are 1×0.5 m^2. The material is elastic isotropic and he have the following physical properties: $E = 100$ GPa and $\nu = 0.3$.

This problem roughly corresponds to the Mitchell truss, a classical topology optimization problem. The ideal solution is shown in Figure 7.

The GA algorithm is applied on a population composed of 80 individuals randomly generated at the beginning of the algorithm. The domain conception is meshed on a 36×18 cells (finite elements) mesh modeling a half of the domain by using symmetry. The optimization problem consists to minimize de compliance subject to 50% initial structure volume limitation. The minimum compliance optimal topology design problem can be expressed as

$$\text{Minimize:}\ F^T U \tag{12a}$$

$$\text{Subject to:}\ V(x) - V_{\text{lim}} \leqslant 0 \tag{12b}$$

where F is the applied force, U is the displacement array, V is the volume and V_{lim} the imposed volume limitation.

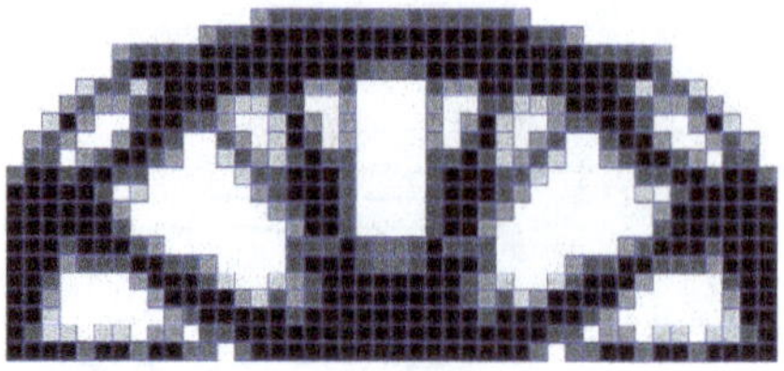

Fig. 8. Optimal solution by GA.

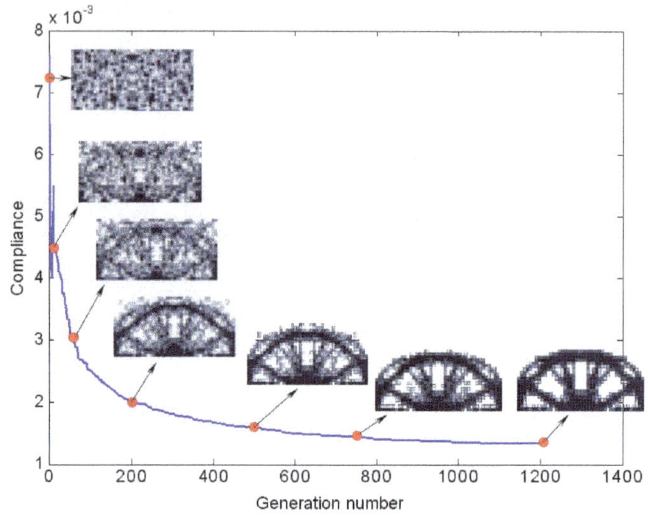

Fig. 9. Best solution and its objective value histories.

The optimal topology solution reached after 1201 generations is shown in Figure 8. The optimal solution topology tends to the ideal solution.

Figure 8 shows the best solution and its objective function history at each generation. The solution converges to the optimal topology by searching of the material distribution in each finite element. As it shown in Figure 9, the float elements are eliminated automatically without need to using the connectivity analysis technique [5].

6 Conclusions

This paper presents a topology optimization approach for 2-D structures using genetic algorithm and a new representation by adjacency. The proposed approach uses the finite element connectivity to encoding the structure solutions. The element connectivity is coded by binary adjacency matrix. By using the connectivity principle, our approach seems more compatible with the topology optimization problem than the Bit-array representation based approach.

The approach shears some concepts with the Cellular Automata Method, and uses, also the SIMP density penalization technique. Indeed, the material interpolation technique it used to analyze the structural solution using the finite element method. The density variable it also used by the mutation operator to search to the optimal distribution material. It is an efficient technique which plays an important role to eliminate checkerboards and floating elements. In other words, the technique consists in automatically correcting the possible connectivity problems.

Finally, using an appropriate chromosome representation, GA overcome the use of connectivity repair techniques which are expensive and often leads to chromosome degeneracy.

References

1. D.E. Goldberg, *Genetic Algorithms in Search, Optimization and Machine Learning*, Addison-Wesley (1989)
2. E. Sandergen, E.D. Jensen, J. Welton, *Topological design of structural components using genetic optimization methods*, in Sensitivity Analysis and optimization with Numerical Methods. AMD, 1990, 115, Proceeding of Winter Annual Meeting of the American Society of Mechanical Engineers, Dallas, TX, 31–43.
3. C. Chapman, K. Saitou, M. Jakiela, *Genetic Algorithms as an approach to configuration and topology design.* ASME Journal of Mechanical Design **116**, 1005–1012 (1994)
4. C. Chapman, M. Jakiela, *Genetic algorithm-based structural topology design with compliance and topology simplification considerations.* ASME Journal of Mechanical Design **118**, 89–98 (1996)
5. M. Jakiela, C. Chapman, J. Duda, A. Adewuya, K. Saitou, *Continuum Structural topology design with Genetic Algorithms.* Comput. Method. Appl. M. **186**, 339–356 (2000)
6. M. Schoenauer, *Shape representation for evolutionary optimization and identification in structural mechanisms.* In Genetic Algorithms in Engineering and Computer Science, G. Winter, J. Periaux, M. Galain and P. Cuesta, Eds. Chichester John Wiley (1995) p. 443–464.
7. C. Kane, M. Schoenauer, *Topological Optimum Design using Genetic Algorithms.* Control and Cybernetics **25**, 1059–1088 (1996)
8. D.W. Fanjoy, W.A. Crossley, *Topology design of planar Cross-sections with a genetic algorithm: Part 1- Overcoming the obstacles.* Engineering Optimization **34**, 33–48 (2002)
9. H. Hamda, F. Jouve, E. Lutton, M. Schoenauer, M. Sebag, *Représentation non-structurée en optimisation topologique de forme par Algorithmes Evolutionnaires.* ESAIM Proceedings actes du 32e Congrès d'analyse Numérique CNUM, 1–3 (Septembre 2000)
10. D.W. Fanjoy, W.A. Crossley, *Topology design of planar Cross-sections with a Genetic Algorithm: Part 2- Bending, Torsion and combined loading applications.* Engineering Optimization **34**, 49–64 (2002)
11. S.Y. Wang, K. Tai, *Structural topology optimization using genetic algorithms with bit-array representation.* Computer Methods in Applied Mechanics and Engineering **194**, 3749–3770 (2005)
12. B. Tatting, Z. Gürdal, *Cellular automata for design of tow-dimensional continuum structures.* In Proceeding of the 8th AIAA/USAF/NASA/ISSMO Symposium of Multidisciplinary Analysis and Optimization, pages 2000–4832, Long Beach, CA, 2000, AIAA Paper.
13. A. Tovar, N. Patel, A.K. Kaushik, G.A. Letona, J.E. Renaud, *Hybrid cellular automata: a biologically-inspired structural optimization technique.* 10th AIAA/ISSMO Multidisciplinary Analysis and Optimization Conference, 30 August – 1 September 2004, Albany, New York.
14. M.P. Bendsøe, O. Sigmund, *Topology Optimization: Theory, Methods and Applications*, Springer-Verlag, New York (2003)
15. C.A.C. Coello, *Theoretical and numerical constraint-handling techniques used with Evolutionary Algorithms: A survey of the State of the Art.* Computer Methods in Applied Mechanics and Engineering **191**, 1245–1287 (2002)
16. K. Deb, *An efficient constraint handling method for genetic algorithms.* Computer Methods in Applied Mechanics and Engineering **186**, 311–338 (2000)
17. J.E. Baker, *Reducing bias and inefficiency in the selection algorithms.* in Proceeding of the Second International Conference on Genetic Algorithms. and their Applications, New Jersey, USA, p. 14–21 (1987)
18. D.J. Higham, N.J. Higham, MATLAB guide. Society for Industrial and Applied Mathematics, 2nd edition (Philadelphia 2005)

A comparative study of failure criteria applied to composite materials

E.H. Irhirane[1], J. Echaabi[1,a], M. Hattabi[1], M. Aboussaleh[2] and A. Saouab[3]

[1] Équipe de Recherche Appliquée sur les Polymères, Département de Génie Mécanique, ENSEM, Université Hassan II Aïn Chok, BP 8118, Oasis, Casablanca, Morocco
[2] Laboratoire de Mécanique, Département de Génie mécanique, ENSAM, Université Moulay Ismail, Présidence, Marjane 2, BP 298 Mekness, Morocco
[3] Laboratoire d'Ondes et Milieux Complexes, FRE 3102 CNRS, 53 rue Prony, BP 540, 76058 Le Havre Cedex, France

Abstract – This article presents an analytical and numerical investigation of the failure loads, successive failures and failure modes of laminated beams. Two simulations were developed to model various composites behaviours under transverse static loading. Progressive failure analysis has been carried out in linear and elastic range. For the two simulations, the first order shear deformation theory with shear correction factor and the finite element method have been used respectively. The main objective of this paper is to evaluate the overall damage and successive failure for different laminates obtained by failure criteria and then to conduct a comparative study of the commonly used failure criteria. Various failure criteria have been studied to predict the load, when the weakest lamina fails under flexural bending test. After the failure of the weakest ply, the stiffness is reduced to account for fibre or matrix failures. The failure criteria are tested with various stiffness reduction models. Comparisons are made between the failure loads, successive failures, failure modes, macroscopic curves and the total behaviour curves obtained by the various failure criteria used.

Key words: First ply failure; failure load; successive failur; failure criteria; stiffness reduction.

1 Introduction

Currently, composite materials are largely used in a majority of the industrial sectors in a current way. However, a substantial effort is still needed for their optimal use. Dimensioning tools must be effective, reliable and precise to improve design optimisation. An alternative to facilitate the analysis of failure behaviour of composite materials consists in using failure criteria. These criteria are largely used in the commercials finite elements codes for the dimensioning of the composites materials. However, the problem of the choice of the appropriate criterion to predict failure has not received enough interest. Indeed, a great number of formalisms exist [1–3] and new or simply modified ones are still published [4–6]. The choice among these formalisms is difficult because of their diversity and because they must be validated by biaxial tests which are both expensive and difficult to realize [3–7].

In this work, we propose to study three types of formalisms representative of the great families of failure criteria; 1. non interactive failure criteria such as the maximum stress and the maximum strain; 2. interactive failure criteria such as Tsai Hill and the well now Tsai Wu criteria; 3. finally, failure criteria which take into account physical considerations such as Hashin and Hart-Smith criteria.

In this study, we use the laminates theory with transverse shearing taking account of the coefficients of shearing and the finite element method. An analytical and a numerical simulation were developed to model the progression of damage. In general, after the first failure and for some configurations, the laminate can still support efforts which depend on its nature and the type of the stress in action. Several authors have developed approaches to relate the state of damage to the principal characteristics of the material [8, 9]. Thereafter, the total discount and the limited discount methods for the reduction of the coefficients of the matrix of rigidity are used. For certain configurations, the three points bending test makes it possible to highlight how a material is degraded gradually until the final failure [10, 11]. For this effect, we carry out the simulation of three points bending test on standardized specimens [12]. The failure loads, the successive failures, the failure modes, the progression of the damage, the effects of the geometrical parameters of the specimens and of the stacking sequences were studied.

[a] Corresponding author: j.echaabi@ensem-uh2c.ac.ma

2 Theoretical failure analysis

2.1 Constitutives equations

The formalism used in this work is based on the laminates law with transverse shearing and take account of the coefficients of correction. The constitutive equation is written in the following form:

$$\begin{bmatrix} N_{ij} \\ M_{ij} \\ Q_y \\ Q_x \end{bmatrix} = \begin{bmatrix} A_{ij} & B_{ij} & 0 & 0 \\ B_{ij} & D_{ij} & 0 & 0 \\ 0 & 0 & F_{44} & F_{45} \\ 0 & 0 & F_{45} & F_{55} \end{bmatrix} \begin{bmatrix} \varepsilon^0_{ij} \\ \kappa_{ij} \\ \gamma^0_{yz} \\ \gamma^0_{xz} \end{bmatrix}. \quad (1)$$

N_{ij}, Q_x and Q_y are the resulting forces and M_{ij} are the resulting moments. A_{ij}, B_{ij}, D_{ij} $(i,j = 1, 2, 6)$ and F_{ij} $(i,j = 4,5)$ are the matrices of rigidity of the laminate, with:

$$\begin{cases} A_{ij} = \sum\limits_{k=1}^{n} \left(Q'_{ij}\right)_k (h_k - h_{k-1}); \\ B_{ij} = \frac{1}{2} \sum\limits_{k=1}^{n} \left(Q'_{ij}\right)_k (h_k^2 - h_{k-1}^2); \\ D_{ij} = \frac{1}{3} \sum\limits_{k=1}^{n} (Q'_{ij})_k (h_k^3 - h_{k-1}^3); \\ F_{ij} = \sum\limits_{k=1}^{n} K_i K_j (C'_{ij})_k (h_k - h_{k-1}) \end{cases} \quad (2)$$

where Q'_{ij} and C'_{ij} are the expressions for the reduced stiffness apart from its principal axes, h_k and h_{k-1} are, respectively, the distances of the higher and lower interfaces of the layer k from the center of the laminate (Fig. 1) and K_i is the correction coefficient of shearing.

2.2 Failure criteria

The failure of composite materials can be treated from the microscopic or macroscopic point of view. The analysis of composites at the microscopic level is difficult and complex to model. The macroscopic behaviour modelling begins from the data of load displacement of a specimen subjected to a test. The majority of the failure criteria are phenomenological, in the sense that they are not deduced from a micromechanical analysis. Consequently, they can be accepted or rejected only by comparison with experimental data. The failure criteria applied to composite materials are difficults to formulate because these later have a structural and material complexity. A simple criterion cannot model with precision the failure of composites. It is thus necessary to increase his complexity but at the same time it becomes more difficult to implement. In practice, each criterion is used for particular experimental results. Until date, there is no general and systematic approach to describe failure of composite materials.

The validity of the failure criteria is discussed in general according to various aspects: linear and nonlinear effects, failure modes and their interaction, the effect of shearing and the type of the laminate etc. More researches are necessary in particular on the failure modes and their

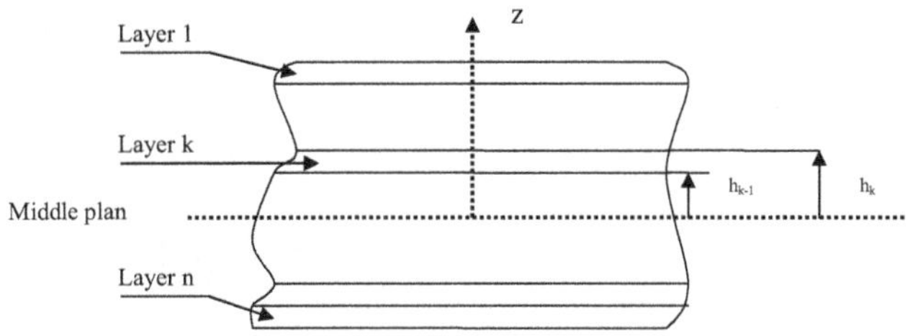

Fig. 1. Laminate geometry.

interaction. One of the objectives of our work consists in validating the reliability and the precision of the most answered failure criteria;

1. Non interactive criteria such as the maximum stresses and the maximum strains ones whose mechanisms of longitudinal, of transverse or of shearing occur independently. The failure stresses and the failure modes are predicted. But only the linear behaviour is described. These criteria over-estimate the failure stresses in the corners of the rectangle (failure envelope).
2. Interactive criteria such as Tsai Hill and Tsai Wu ones: the failure stresses are predicted but the failure modes cannot be predicted. The problem in the application of Tsai Wu criterion is the determination of the coefficients of interaction. Several methods were proposed to evaluate these coefficients. In our case, the following equation:
$$F_{ij} = -\frac{1}{2}\sqrt{F_{ii}F_{jj}} \quad (3)$$
is used to calculate the coefficients of interaction. These criteria cannot correctly describe the complexity of the failure of composite materials. However, they can be used in practice, when their precision is satisfactory and that no other method is valid.
3. A number of criticisms [3] were formulated by Hashin [13] and recently by Hart-Smith [14] about the limitations and the problems encountered when using the interactive failure criteria. Hashin has developed for transversally isotropic unidirectional laminates a quadratic polynomial criterion which takes into account physical considerations. Four distinct failure modes are considered separately. Two important items were discussed by this criterion, the incorporation of the failure modes by proposing for each mode an equation and the possibility of interaction of the stress. Until now, the matrix failure modes remains difficult to model.

The method of Hart-Smith is based on the criterion of the maximum shear stress in the space of the strains. The originality of the work of Hart-Smith is not in the choice of the criterion of maximum shear stress, which was used previously nor in its criticism of the traditional criteria which had been announced by some authors, but rather in the incorporation of the physical considerations. The criterion is derived for strong, stiff fibres embedded in a soft matrix and its validity is restricted to fibre dominated failure. To account for other failure modes, Hart-Smith

Table 1. Mechanical characteristics of test samples.

X_t11	X_C1	$X_t22 = X_t33$	$X_c22 = X_c33$	$X12 = X13$
2.647	1.723	0.0514	0.222	0.0861

Table 2. Specimen characteristics and specifications.

Laminates Type	Layup (degrees)	# of Plies	Length L (mm)	Span l (mm)	Width b (mm)	Thickness h (mm)	Ratio l/h
A	$[90_8/0_8]_s$	32	200	152.5	25	4.736	32
B	$[0_8/90_8]_s$	32	200	152.5	25	4.736	32
C	$[0/90]_{8s}$	32	200	152.5	25	4.485	34
D	$[45/0/-45]_{5s}$	30	200	152.5	25	4.236	36
E	$[45/-45/90/0]_{3s}$	24	150	115.0	10	3.600	32
F	$[45/-45/90/0]_{3s}$	24	115	57.0	10	3.600	16

Table 3. Successive failures results of laminates (A, B, C) with the total discount method.

Criterion tested	Sample A$[90_8/0_8]_s$	Sample B$[0_8/90_8]_s$	Sample C$[0/90]_{8s}$
Tsai Hill	1-2-3-4-5-6-7-8	9-10-11-32-12-13-31-14	2-4-6-8-10-1-32-30
Tsai Wu	1-2-3-4-5-6-7-8	9-10-11-12-32-13-1-31	2-4-6-8-10-12-1-31
Maximum strain	1-2-3-4-5-6-7-8	9-10-11-32-12-13-31-14	2-4-6-8-10-32-12-30
Maximum stress	1-2-3-4-5-6-7-8	9-10-11-32-12-13-31-14	2-4-6-8-10-32-12-1
Hashin	1-2-3-4-5-6-7-8	9-10-11-32-12-13-31-14	2-4-6-8-10-32-12-1
Harth Smith	1-2-3-4-5-6-7-8	1-32-31-2-3-30-29-4	32-1-31-2-30-3-29-4

xxx is the same successive failures predicted by the failure criteria.

truncated the failure envelopes. A comparison with the experimental results is still necessary to conclude on the reliability and the general information of this approach.

3 Analytical results and discussion

The present study uses a graphite epoxy AS1/3052 specimens. The failure stresses parameters are presented in Table 1. Six rectangular specimens are used to study the influence of the stacking sequence and the effects of the geometrical characteristics of specimens on the successive failure. Table 2 illustrates the geometrical characteristics and the stacking sequences of the test-specimens. The dimensions are those recommended by ASTM specifications. In the first part of this work, an analytical study of successive failure and failure modes is presented. The theory of laminates with transverse shearing taking into account the coefficients of shearing correction is used. A finite element method is used in the second part.

The objective is to evaluate the results obtained by the failure criteria used in the various specimens and to elaborate a comparative study between these criteria.

The evaluation of the stress and the strain fields makes it possible to determine the first time when failure is reached; the failure criterion is satisfied. The program elaborated detects the failure position and the corresponding ply. Then the corresponding coefficients in the matrix of stiffness of the layer are then modified. A new analysis is made with a modified stiffness matrix [ABCD]. The analytical results obtained by the six failure criteria used are represented in the form of tables. The latter clearly shows the failure succession, the failure loads, displacements and the position of each failure. The study is made on the first eight successive failures with two models of stiffness reduction: 1. total discount method; 2. limited discount method.

1. In the total discount method, the stiffness and strength of a failed ply are reduced to zero, although the ply is still physically present. The analysis of the results obtained with this approach enables us to make the following conclusions.

For all the specimens, the results of successive failures are presented in Tables 3 and 4. For type A specimens, all the failure criteria give the same predictions (successive failures, load and displacement at failure). All the failures observed in the 32 plies show that, before the failure occurred in the 0° ply, no difference between the predictions of the five criteria is observed. A light divergence was noted on the level of the breaking loads of the 0° ply (Tab. 5). After this macroscopic failure, a weak difference was noticed between the various predictions.

For type B specimens, the failure succession is identical for the maximum strain, maximum stress, Tsai Hill and Hashin criteria. A small difference between their breaking loads was observed. The first three failures obtained by Tsai Wu are identical to those predicted by the others criteria. Hart-Smith always predicts alternate successive failures. However, the others criteria predict almost the same macroscopic breaking loads (Tab. 5). A disparity was noted between the predictions of the five

Table 4. Successive failures results of laminates (D, E, F) with the total discount method.

Criterion tested	Sample D$[45/0/-45]_{5s}$	Sample E$[45/-45/90/0]_{3s}$	Sample F$[45/-45/90/0]_{3s}$
Tsai Hill	29-26-1-2-23-4-25	3-1-7-2-21-5-6-22	3-1-7-2-21-5-6-22
Tsai Wu	29-26-2-23-1-5-20-4	3-1-7-2-5-21-6-4	3-1-7-2-5-21-6-4
Maximum strain	29-1-30-3-26-4-6-28	3-7-21-1-2-5-22-17	3-7-21-1-2-5-22-17
Maximum stress	29-1-28-3-27-4-25-6	3-7-24-21-1-2-23-5	3-7-24-21-1-2-23-5
Hashin	29-1-26-2-4-3-23-5	3-24-7-23-1-2-5-20	3-24-7-23-1-2-5-20
Harth Smith	29-2-26-5-8-23-20-11	3-22-4-21-7-18-8-17	3-22-4-21-7-18-8-17

Table 5. Failure loads for samples A and B.

	Sample A$[90_8/0_8]_s$			Sample B$[0_8/90_8]_s$		
	FPF load	M.FPF load Total dicount	M.FPF load Marix unchanged	FPF load	M.FPF load Total dicount	M.FPF load Marix unchanged
Tsai Hill	278 (1)*	1102 (9)	—	2506 (9)	3826 (32)	—
Tsai Wu	278 (1)	1165 (24)	—	2454 (9)	4764 (32)	—
Maximum strain	278 (1)	1165 (24)	1911 (24)	2414 (9)	3833 (32)	3887 (32)
Maximum stress	278 (1)	1160 (24)	1894 (24)	2510 (9)	3819 (32)	3866 (32)
Hashin	278 (1)	1160 (24)	1894 (24)	2510 (9)	3819 (32)	3866 (32)
Harth Smith	278 (1)	1220 (24)	— 3031	(1)	3031 (1)	—

FPF load is the first ply failure load expressed in N,
M.FPF load is the macroscopic first ply failure load expressed in N,
()* Failure location, given by ply number,
Note: ply number starts from bottom to top.

Table 6. Failure loads for samples C and D.

	Sample C$[0/90]_{8s}$			Sample D$[45/0/-45]_{5s}$		
	FPF load	M.FPF load Total dicount	M.FPF load Marix unchanged	FPF load	M.FPF load Total dicount	M.FPF load Marix unchanged
Tsai Hill	753 (2)	2147 (1)	—	1095 (29)	1282 (2)	—
Tsai Wu	749 (2)	2639 (1)	—	861 (29)	1382 (2)	—
Maximum strain	746 (2)	2191 (32)	2323 (32)	759 (29)	968 (2)	2563 (2)
Maximum stress	753 (2)	2177 (32)	2304 (32)	963 (29)	1355 (2)	2591 (2)
Hashin	753 (2)	2177 (32)	2304 (32)	1459 (29)	1424 (2)	2591 (2)
Harth Smith	1812 (32)	1812 (32)	—	1034 (29)	1034 (29)	—

failure criteria after the failure of the ply 0°. The maximum stress and Hashin criteria practically lead to identical results.

For type C specimens, a good correlation for the successive failure is observed with the maximum stress, maximum strain and Hashin criteria, but with different breaking loads (Tab. 6).

For type D specimens, it appears that there is a notable difference between the predictions of the six failure criteria. For the laminates of the type D (mixture of 0° and 45°), It is impossible to distinguish between the results obtained for the six failure criteria. This type of laminates tends to have a uniform distribution of rigidity along the thickness compared to types A and B laminates. The criterion of Hashin predicts breaking loads larger than those obtained by the maximum stress criterion (Tab. 6).

For type E and F specimens, all the failure criteria predict the first failure in the same ply (ply 3) with a small differences between the values of the breaking loads (Tab. 7). Confrontation between the successive failures obtained by the six criteria shows that there is a great disparity between them. As a conclusion of this part, a very small variation in the results of the specimen A, B and C was observed. The maximum stress and Hashin criteria lead practically to identical predictions for the first three specimens.

For the specimens D, E and F, it is difficult to distinguish between the predictions obtained by the six failure criteria.

2. The limited discount approach considers the stiffness reduction dependent on the failure mode in action. Only failure criteria allowing the prediction of the failure modes can be used. For that, only the maximum stress, the maximum strain and the Hashin criteria are used. In this method of stiffness reduction, the failure succession, the breaking loads and the ply where failure occurred are predicted but also the failure modes. If the rupture occurs

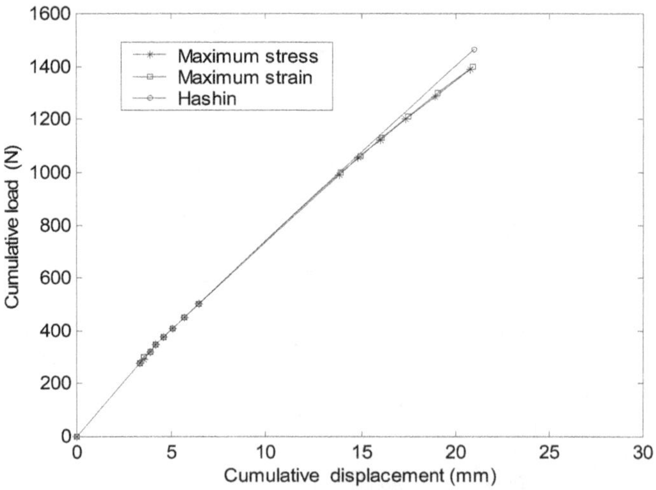

Fig. 2. Cumulative macroscopic curves of the test sample A with the limited discount method (matrix stiffness reduction).

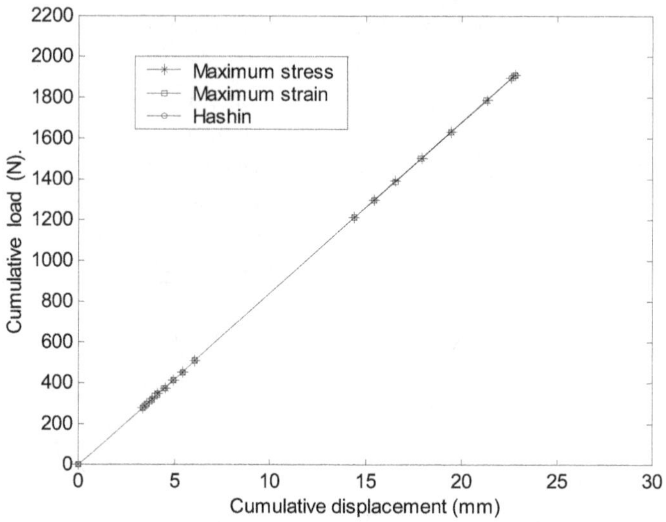

Fig. 3. Cumulative macroscopic curves of the test sample A with the limited discount method (matrix stiffness unchanged).

by fibre, the total discount approach is applied and if the rupture occurs by matrix cracking, two cases were treated; 1. in the first case, all the components of the matrix of rigidity of the resin are reduced to zero; 2. in the second one, the matrix of rigidity remains unchanged. The analysis of the results obtained for the two cases of reduction of the matrix of rigidity, makes it possible to draw the following remarks:

A good correlation between the successive failure and the failure modes is obtained for the specimens A, B and C.

The breaking loads of the specimen A are practically identical for the three failure criteria;

The breaking loads of the specimen B and C are of the same value for the maximum stress and for Hashin criteria and are slightly larger than those predicted by the maximum stress criterion.

It is impossible to distinguish between the successive failures obtained with the specimens D, E and F.

According to position of failure, one notes that there is a uniform distribution of rigidity along the thickness for the specimens D, E and F. For the same specimens, the Hashin and the maximum stress criteria predict the same macroscopic breaking load in the second case of stiffness reduction.

The macroscopic breaking loads (failure of 0° ply) obtained for the second case of stiffness reduction are slightly larger than those predicted with the first case. The approach of the limited discount of rigidity does not have a great influence on the successive failures, but it increase the breaking loads.

The total behaviour can be regarded as parameter of comparison between the failure criteria. The total or cumulative curve is the superposition of the curves of various loadings after each failure. Figures 2 to 5 present the cumulative curves obtained from the various models of stiffness reduction presented earlier. The representation is limited to specimens A and E. Each one represents a

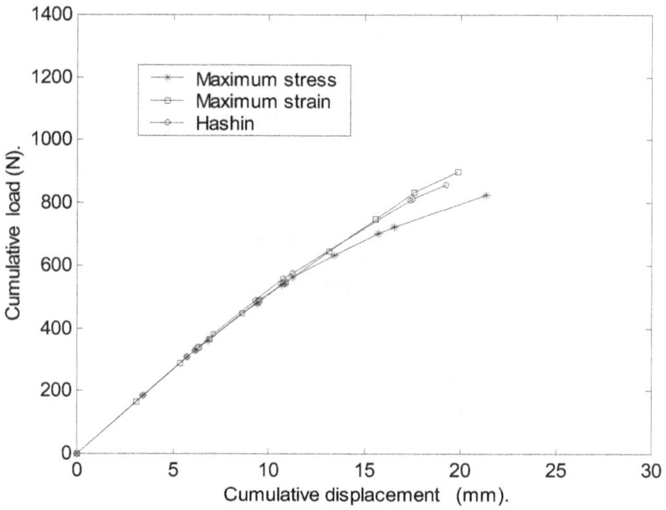

Fig. 4. Cumulative macroscopic curves of the test sample E with the limited discount method (matrix stiffness reduction).

family of the studied specimens. These curves show the maximum loads at the final failure of the laminate. This cumulative charge is the sum of the increments of the forces necessary to produce a failure in each layer. It appears that the cumulative curves obtained with the second case of stiffness reduction are almost confused. What shows that the failure criteria used predict the same total behaviour and consequently the same flexural rigidity.

4 Numerical results and discussion

A data-processing program based on the finite element method was elaborated within our laboratory to simulate the deflection of the specimens. The element used is a solid with eight nodes. The classical formulation in terms

Table 7. Failure loads for samples E and F.

	Sample E$[45/-45/90/0]_{3s}$			Sample F$[45/-45/90/0]_{3s}$		
	FPF load	M.FPF load Total dicount	M.FPF load Marix unchanged	FPF load	M.FPF load Total dicount	M.FPF load Marix unchanged
Tsai Hill	183 (3)	368 (21)	—	371 (3)	744 (21)	—
Tsai Wu	172 (3)	302 (21)	—	348 (3)	609 (21)	1012 (21)
Maximum strain	165 (3)	263 (21)	501(21)	333 (3)	531 (21)	1012 (21)
Maximum stress	185 (3)	330 (21)	615 (21)	374 (3)	666 (21)	1241 (21)
Hashin	185 (3)	407 (21)	615 (21)	374 (3)	823 (21)	1241 (21)
Harth Smith	206 (3)	345 (4)	—	834 (3)	714 (4)	—

Table 8. Successive failures results obtained by numercical analysis.

Sample	Tsai Wu	Maximum stress	Maximum strain
A$[90_8/0_8]_s$	32-31-30-29-28-27-26-25	32-31-30-29-28-27-26-25	32-31-30-29-28-27-26-25
B$[0_8/90_8]_s$	24-23-22-21-20-19-18-17	24-23-22-21-20-19-18-17	24-32-23-31-22-30-21-29
C$[0/90]_{8s}$	31-29-27-25-23-21-2-19	31-29-27-25-23-2-21-4	31-29-27-25-2-23-4-21
D$[45/0/-45]_{5s}$	30-1-3-2-28-27-25-4	30-28-27-25-1-3-2-24	1-3-2-4-5-6-7-30
E$[45/-45/90/0]_{3s}$	24-23-22-20-19-18-16-15	22-24-23-20-18-19-16-3	22-24-23-18-20-3-1-19
F$[45/-45/90\text{-}0]_{3s}$	24-23-22-20-19-18-16-3	22-24-23-20-18-19-3-16	24-23-22-20-19-16-18-3

Table 9. First ply failure loads obtained by numercical analysis.

Sample	Tsai Wu	Maximum stress	Maximum strain	Error = Maxi-Mini/Maxi
A$[90_8/0_8]_s$	227	278	274	18%
B$[0_8/90_8]_s$	1956	2236	2553	23%
C$[0/90]_{8s}$	566	894	724	36%
D$[45/0/-45]_{5s}$	543	1440	809	62%
E$[45/-45/90/0]_{3s}$	106	187	199	46%
F$[45/-45/90\text{-}0]_{3s}$	234	414	409	43%

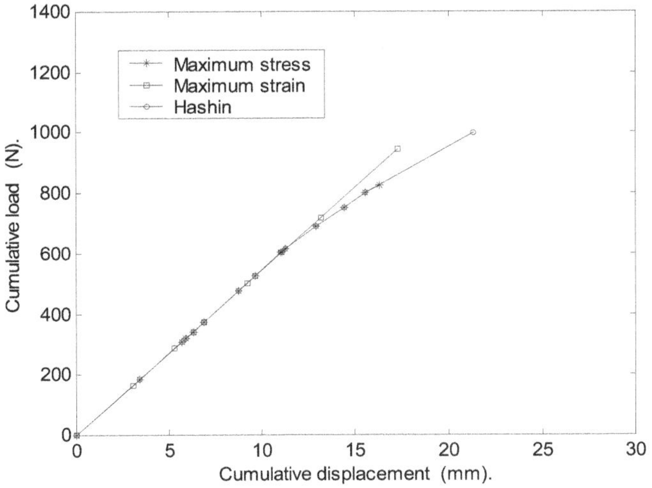

Fig. 5. Cumulative macroscopic curves of the test sample E with the limited discount method (matrix stiffness unchanged).

of the deformation is used. Information necessary for the calculation is the number of groups to calculate the stiffness matrix and each group consists of a set of layers. The convergence is studied while decreasing the number of ply by group until even two elements by ply. The results mentioned in the article are gotten for one element by ply through the thickness of the specimen. A grid where the density was increased in the central part of the specimen was adopted. The evaluation of the stress and the strain fields makes it possible to determine the minimum loading for which a failure is reached in an element. When failure is reached in the most constrained element, its rigidity will be reduced. The results are saved at each stage and then make it possible to follow the evolution of the failure behaviour. To predict the failure with this computer code, we were interested in the most used failure criteria in the commercial software: the Tsai-Wu 3d criterion which, in general, under estimate the breaking load and the maximum stress and the maximum strain criteria. The results obtained by this analysis were compared with those of the analytical method.

The results for successive failures and first ply failure load obtained with the three criteria used are respectively presented in Tables 8 and 9. For specimen A, the failure criteria used give practically the same successive ruptures. An attentive look of all the results, shows that there are small divergences between the successive failure when the damage is important. The plies where failure takes place obtained by this analysis are not in agreement with those obtained by the analytical approach. Some difference was noted on the level of the breaking loads. The Tsai Wu criterion under-estimates slightly the load of the first rupture. For the specimen B, the succession of failure is identical between Tsai Wu and the maximum stress criteria. For these last two criteria, the ply 90°, which is in the bottom of the specimen, are broken first. On the other hand, the criterion of the maximum strain predicts a failure succession different from that of the others criteria. Indeed, the failure alternate between the plies 90° and 0° of the bottom of the specimen. The maximum strain criterion predicts the maximum load of the first rupture and the criterion of Tsai Wu the minimal load. For the specimen C, the maximum strain criterion predicts a failure succession different from that predicted by the two others criteria. The first four ruptures obtained by the three criteria are practically identical. The criterion of the maximum stress predicts the maximum loads whereas Tsai Wu criterion predicts the minimal forces. For the specimen D, the successive failure obtained by the three failure criteria are largely different. According to the positions of the ruptures, we notice that this laminate has a uniform distribution of rigidity along the thickness.

For the specimen E and F, all the failure criteria predict the same successive failures but in different orders. For the two l/h ratios, the area of failure initiation is the same. Indeed, the failure starts in bottom of the specimen and is propagated to the top. The maximum strain predicts the maximum loads for the two specimens.

5 Conclusion

For the symmetrical cross ply laminates (0° and 90° plies), all the failure criteria used predict almost identical successive failures and a small difference was observed between the breaking loads. In the approach of the limited discount, without reduction of the rigidity of the matrix, the macroscopic breaking loads obtained with the majority of the failure criteria are identical for angle ply laminates formed with 45°, −45° and 0° plies. These laminates have an uniform distribution of rigidity along the thickness. Consequently, it is difficult to distinguish between the predictions obtained with the failure criteria. In the majority of the cases, Hashin criterion predicts the largest breaking loads and the Tsai Wu one the smallest forces. The whole of the results obtained with all the approximations suggested, highlights the difficulty to predict the failure loads and the successive failures. The results depend on the criterion used.

For all the failure criteria used, the eight first successive failures occur in the same plies but the order is different. It is noticed also that the difference in the results obtained by the various criteria depends crucially on the stacking sequences. This difference is the smallest for cross ply laminates containing the 0° ply in the centre of the specimen. For angle ply laminates containing 45° plies, the criteria used give largely different results. Then the choice of the criterion has a substantial influence on the results of the dimensionnning. This study shows that the dimensioning of the composites by failure criteria remains an open problem. Experimental, analytical and numerical studies are still necessary to highlight the suitable criteria for each type of laminates; stacking sequences and materials.

References

1. M.N. Nahas. Technology and Research **8**, 138 (1986)
2. W.W. Feng. J. Compos. Mater. **25**, 88 (1991)
3. J. Echaabi, F. Trochu, R. Gauvin. Polymer Composite **17**, 786 (1986)
4. Y. Hsien-Liang. J. Reinf. Plast. Compos. **22**, 517 (2003)
5. S.J. DeTeresa, G.J. Larsen. J. Compos. Mater. **37**, 1769 (2003)
6. C.G. Davila, P.P. Camanho, C.A. Rose. J. Compos. Mater. **39**, 323 (2005)
7. J. Echaabi, F. Trochu. J. Compos. Mater. **30**, 1088 (1996)
8. R. Talreja. J. Compos. Mater. **19** (1985)
9. A. EL Mahi, J.M. Berthelot, J.M. Brillaud. Comps. Struct. **30**, 123 (1995)
10. P. Pal, C. Ray. J. Reinf. Plast. Compos. **21**, 1505 (2002)
11. B.G. Prusty, S.K. Satsangi, C. Ray. J. Reinf. Plast. Compos. **20**, 671 (2001)
12. ASTM test. *American Standards of Testing and Materials* (1992), p. 790.
13. Z. Hashin. J. Appl. Mech. **47**, 329 (1980)
14. L.J. Hart-Smith. Composites Science and Technology **58**, 1151 (1998)

24

Incorporating industrial constraints for multiobjective optimization of composite laminates using a GA

D.H. Bassir[1,3,a], F.X. Irisarri[2], J.F. Maire[2] and N. Carrere[2]

[1] Institut FEMTO-ST, UMR 6174, 24 rue de l'épitaphe, 25000 Besançon, France
[2] ONERA, DMSC 29 avenue de la division Leclerc, 92322 Chatillon Cedex, France
[3] Currently visiting Aerospace Structures, Faculty of Aerospace Engineering, TU Delft, The Netherlands

Abstract – In this article, complexities related to the multicriteria (multiobjective) optimization of laminated composite structures subjected to technological constraints will be presented. So, various technological constraints will be presented and a strategy of handling each constraint (in order to use the multi-objective optimization tools based on genetic algorithms) will be also introduced.

1 Introduction

An advantage of using fibre-reinforced composites over conventional materials is that they can easily reduce the mass of structures by adapting the stiffness to the requirements of many practical applications especially for aeronautic and aerospace structures. Berthelot, and Tsai and Hahn [1,2] reported on composites behaviour and the advantage of using such kind of materials. Also, many handbooks such as the Military Handbook [3] and the Aircraft Crash Survival Design Guide [4] are summerazing the theoretical and practical benchmark examples that help the designer to design basic structures with composite material. In these industrial fields the main goal is how to obtain the highest structural performances with substantial savings in terms of weight and stiffness, since composite materials possess high values of strength to weight and elasticity to weight ratios with respect to conventional materials as steel or aluminium alloys.

To improve the mechanical performances without increasing the mass of these structures, it is essential to apply an optimization process to define the optimal conception values. In general, design optimization of laminated composite structures involves not only one objective function but several objective functions [5, 6, 20] that are in permanent conflict. This kind of problems can be handled by a multicriteria approach, leading to designs which are balanced from an overall viewpoint. There exist two approaches: we can use either a direct method [7–10] or a posteriori method based on evolutionary strategies [11, 20]. Multicriteria optimization of composite structures has been discussed widely in the literature by various researchers, Adali et al. [6] presented multicriteria optimisation of laminated plates for maximum pre-buckling, buckling and post-buckling strength, Kere et al. [9] reported on multicriteria optimization for strength design of composite laminates, while Adali et al. [10] used the multicriteria approach to design laminated cylindrical shells for maximum pressure and buckling load.

When we use the approach based on genetic algorithm (GA) to solve multicriteria optimization, the constraints are often hard to handle. Some authors such as Deb [11] have studied these complexities and gave an efficient and quick method to handle them and to obtain well-spread solutions through the Pareto front for nonlinear benchmark functions. In the next sections, we will focus on the complexities related to the optimization of laminated composite structures subjected to technological constraints that arise in the aeronautical fields and the strategy for handling each constraint in the NSGA-II program [11].

2 Multicriteria optimization

Multicriteria optimization problem is stated as follows: find the vector of design variables $\mathbf{x} = [x_1, x_2, x_3, ., x_n]^T$ which minimizes the vector of objective functions $F(x)$

$$\text{Min } \mathbf{F}(\mathbf{x}) = \text{Min}[f_1(x), f_2(x). \ldots .f_k(x)]. \qquad (1)$$

Subject to linear or nonlinear constraints:

$$g_j(x) \geqslant 0 \quad j = 1, 2, \ldots, m.$$

The feasible domain (Fig. 1) defined by the constraints will be denoted by Ω. f_i: $\Omega \rightarrow R$, $i = 1, 2, ..., k$ are called criteria or objective functions and they represent the design objectives by which the performance of the laminate is measured in the case of composite material. A vector $\mathbf{x}^* \in \Omega$ is called Pareto-optimal solution if there is no vector $\mathbf{x} \in \Omega$ which would decrease some criteria without

[a] Corresponding author: hbassir@univ-fcomte.fr

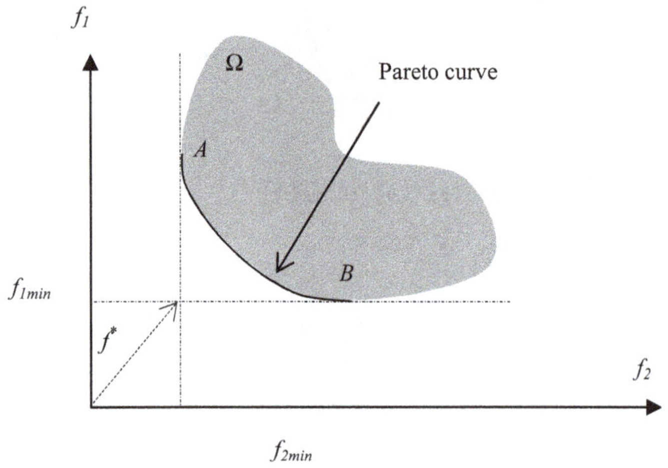

Fig. 1. Pareto curve (Min (f_1 and f_2)).

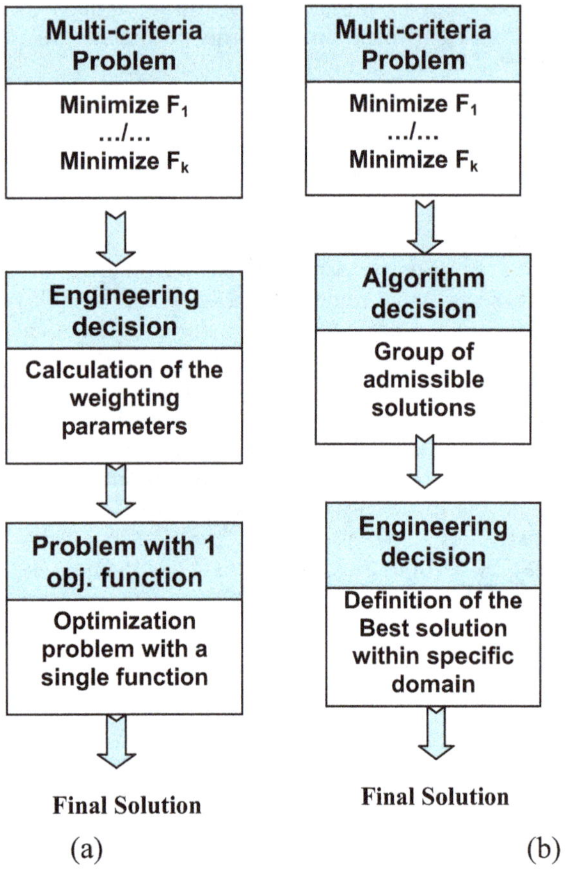

Fig. 2. Direct (a) and posteriori (b) methods for multicriteria optimisation.

causing a simultaneous increase in at least one criteria function. Usually several Pareto-optimal solutions exist.

The optimal solutions from the Pareto domain can be reached by two main strategies: either by a direct approach or by a posteriori approach (Fig. 2).

The direct method is based on the transformation of the initial problem into a single optimization problem. This transformation can be done by using for example the weighting method which reduces a vector-optimization problem to a scalar optimization problem where, for instance, the scalar objective function $f(x)$ can be defined as the weighted sum of the individual objective functions. The Pareto-optimal solutions obtained by this method depends on the choice of the weighting factors α_i that will generate a unique Pareto optimal solution of the original problem. This approach needs to define appropriate weighting factors that will guide the convergence of the optimal design process [7]. However, this definition depends on the decision-maker and the shape of the Pareto domain (continuity, convexity and the number of limitations to handle). If the weighting factors are not well chosen, this approach can converge in local zone of the Pareto domain especially when the optimal solution is very sensitive to the weighting factors. This conclusion was described and demonstrated for some academic examples by different authors such as Das [12]. Other researchers have defined good strategies to calculate these weighting factors using the sensitivities evaluation of the objective functions [13] which turns out to be problematic in the case of the non differential functions. Moreover, when the final solution is reached, there is no guaranty that this solution is the best one for the decision-maker as he needs to run the process many times before making his decision.

The posteriori method consists in starting with a group of initial solutions that we spray uniformly to have a global idea about the Pareto front and to make the final decision easier to take for the decision-maker. In the case of composite materials, GAs are considered among the most efficient algorithms to solve the optimization problem. The theoretical foundations of GAs were first introduced by Holland and then extended by Goldberg [14,15]. This method uses the evolutionary survival-of-the fittest optimization mechanism. The principle of GAs is to simulate the evolution of one population of individuals to which we apply different production operators (*selection, crossover and mutation*). Many researchers have been inspired by the principles of the GA to use them with some modifications to solve their optimization problems either with one or more objective functions with some basic constraints [16, 17].

There are two main issues in using GAs. Firstly, the comparison between two solutions can not be achieved easily in the selection process. Secondly, the constraints handling is difficult to represent as an inequality or equality equation. To overcome these difficulties, many researchers like Deb for instance, proposed approaches based on their previous investigations on GAs [18,19]. Deb has introduced in his current algorithm NSGA-II the idea of non-dominated sorting to speed up the convergence and increase the performance of his previous algorithm NSGA. To choose between two solutions i and j to continue the iterative process of the GA a non-dominate sorting approach in the selection process is used as follows:

- between two admissible solutions i and j, choose the one with the best objective function;
- if solution i is admissible and j is not, choose solution i;

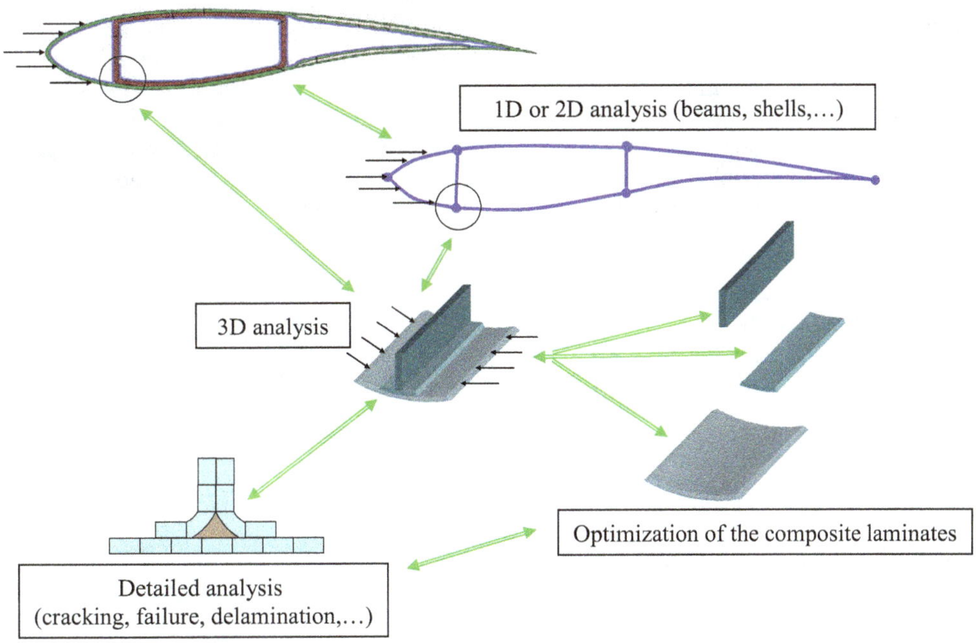

Fig. 3. Multilevel strategy for structural optimisation in the aeronautical industry.

- if the solutions i and j are not admissible, choose the one with the minimum violation of the constraints.

Next sections will be devoted to the particularities and the origin of the constraints to better understand them and find the right way to handle them as constraints in NSGA-II. Procedure as a pseudo code of these constraints will also be presented.

To handle the constraints in the GA program we define the following composite fitness function for any solution $\boldsymbol{x}$ such as: $F(\boldsymbol{x}) = f(\boldsymbol{x})$ if x is feasible, otherwise $F(\boldsymbol{x}) = f_{max} + CV(\boldsymbol{x})$. f_{max} is the objective function value of the worst feasible solution in the population and $CV(\boldsymbol{x})$ is the overall normalized constraint violation of the solution $\boldsymbol{x}$. Thus, there is no need to have any penalty parameter for handling the constraints as usually used in the common approaches. Constraints are normalized to avoid scaling problems and are equal to one in the case of feasible solution.

3 Constraints handling for laminated composite structures

In the aeronautical industry, structural optimization process is often based on the multilevel strategy that can be built as a pyramid (Fig. 3).

The global optimization process starts by the optimization of the basic components. Once the validation of each component is established, the optimization process is applied to the upper level until the final design of the whole structure is reached. Nevertheless an iterative scheme is often required because of load redistribution problems. In the following, we will focus on the industrial constraints for the detailed design of the laminated components of the structure.

In the design of laminated composite structures, the major objective is to find a laminate lay-up configuration that satisfies several requirements or conception rules. In practice, the design starts by the choice of the number of plies. The stacking sequence design is considered in a second time. From experience of manufacturing composite laminates, some rules appeared for choosing the stacking sequences. Nowadays, these rules constitute a group of constraints that increase the difficulty for laminated composite optimization. Among these rules:

1. laminate must be symmetric with respect to its mid-plane;
2. laminate must be balanced regarding to main direction of the loads;
3. maximum number of plies in one group;
4. maximal disorientation between two consecutive layers;
5. the stacking sequence must be homogeneous;
6. minimum amount of plies in each direction;
7. use of integer orientation.

3.1 Symmetric stacking

The symmetry about the mid-plane eliminates the membrane/plate couplings. In terms of manufacturing, this rule of conception cancels the twisting of the panels during their elaboration. The modelling of this rule consist in taking in the account for a vector $\boldsymbol{x}$ $(\theta_1, \theta_2, \ldots \theta_N)$ only the upper or lower part. It means that we keep only $N/2$ variable from the vector $\boldsymbol{x}$. The other part is calculated by symmetry.

3.2 Balanced stacking

This rule requires to have the same number of the plies $+\theta$ or $-\theta$, where $\theta \in [0, 90]$. It eliminates the membrane coupling between shear and traction. This rule can be integrated by simply counting the number of the $+\theta$ and the $-\theta$ than we create one limitation associated for each angle such as

$$
\begin{aligned}
&\textbf{\textit{Begin}} \quad \{\theta = \theta_j \in [0, 90];\\
&\quad \textbf{\textit{Do}}\ (i = 1 \rightarrow i = NP)\\
&\quad \{if\ \theta_i = +\theta_j\ than\ N_{+\theta_j} + + ;\\
&\quad\ if\ \theta_i = -\theta_j\ than\ N_{-\theta_j} + +\\
&;\}\\
&\quad T_j = ||N_{+\theta_j} - N_{-\theta_j}||;\}\\
&\quad \textbf{\textit{If}}\ (count \geqslant 1)\ g_{\theta_i}(x) = -\sum T_j;\\
&\quad \textbf{\textit{Else}}\ g_{\theta_i}(x) = 1.0;
\end{aligned}
$$

NP is the number of plies or layers used in the laminated composite structure.

3.3 Maximum number of plies in one group

This rule consists in minimizing the number consecutive plies in a group of the same orientation. Depending on the ply thickness, the usual limit is fixed to 3 or 4 plies with the same orientation together. The Military Handbook recommends two different limits in terms of the total thickness depending on the orientations to the edge of the panels. The first limit is generally fixed to 0.8 mm when the orientation is parallel to the free edges. The second limit is fixed to 0.38 mm when the orientation is perpendicularly to the free edges. This rule aims at minimizing problems related to matrix cracking, such as the shear-out failure mode in bolted joints. This rule improves also the composite behaviour against impacts. It can be integrated as described in the following pseudo code.

$$
\begin{aligned}
&\textbf{\textit{Begin}} \quad \{test\ (NP \geqslant NPG);\\
&\quad \textbf{\textit{Do}}\ (j = 1 \rightarrow j = NP)\\
&\qquad \{\textbf{\textit{Do}}\ (i = j \rightarrow i = NPG)\\
&\qquad\quad \textbf{\textit{If}}\ (\sum ||(\theta_i - \theta_{i-1}|| \approx 0)\\
&\qquad\quad than\ count\ + +;\}\\
&\quad \}\ \textbf{\textit{If}}\ (count\ \geqslant 1)\ than\ g_{\theta i}(x) = -count;\\
&\quad \textbf{\textit{Else}}\ g_{\theta i}(x) = 1.0;\\
&\textbf{\textit{End}}
\end{aligned}
$$

NPG is the maximum number of plies in a group.

3.4 Maximum disorientation

The rule of disorientation imposes a maximal disorientation of 45 degrees between two successive plies. The aim of this conception rule is to minimize interlaminar shear effects and reduce delamination problems, especially around holes and free edges. In this perspective, the pertinence of this rule can be evaluated with the help of appropriated numerical tools to get the stress values around holes and the free edges. The pseudo code to describe this rule can be as follows:

$$
\begin{aligned}
&\textbf{\textit{Begin}} \quad \{\textbf{\textit{Do}}(i = 1 \rightarrow i = NP)\\
&\quad \{\textbf{\textit{If}}\ (||(\theta_i - \theta_{i-1})|| \geqslant 45)\ \textbf{\textit{than}}\ (count + +;\\
&\quad T_i = ||(\theta_i - \theta_{i-1})||/45;)\}\\
&\qquad \}\textbf{\textit{If}}\ (count \neq 0)\ \textbf{\textit{than}}\ g_\theta(x) =\\
&\qquad\qquad -\sum T_i/NP;\\
&\qquad \textbf{\textit{Else}}\ g_\theta(x) = 1.0;\\
&\textbf{\textit{End}}
\end{aligned}
$$

3.5 Homogeneous stacking sequences

This rule consists in distributing the different orientations along the thicknesses with respect to the other rules. This constraint is particularly hard to handle in the case of panel optimization for buckling. The principal objective of this rule is to reduce the twisting/bending coupling effects, particularly influents in the buckling of panels under shearing loads. The problem appears because engineers still do not master the conception in term of shearing and ignore most often the sign of this shearing.

3.6 Minimum percentage for each angle used in the stacking sequence

This rule imposes for instance a minimum of 10% of orientation for each 0°, ±45° and 90° orientations. The aim of this rule is to avoid obtaining a composite laminate whose behaviour is governed by the matrix in some directions (nonlinear case and creeps). In reality, it is mainly justified by the ignorance of the loadings and the necessity to insure a minimal stiffness and strength in all the directions. Therfore this rule can be reformulated in terms of minimum stiffness in each direction. Actually, the value of 8% is already employed for some aeronautic applications. The pseudo code to describe this rule can be as follows:

$$
\begin{aligned}
&\textbf{\textit{Begin}} \quad \{\textbf{\textit{Do}}(i = 1 \rightarrow i = NP)\\
&\quad g_{\theta i}(x) = (\theta_i \cdot 100/\ (Number\ of\ \theta_j))\\
&\quad \textbf{\textit{Than}}\ (count{+}{+}; T_i = ||(\theta_i - \theta_{i-1})||/45;)\}\\
&\quad g_{\theta i}(x) = (g_{\theta i}(x) - PAng)/PAng;\\
&\textbf{\textit{End}}
\end{aligned}
$$

PAng is the percent imposes for angles.

3.7 Use of integer orientation

The use of only specific orientations such as 0, ±45 and 90 seems to be no longer justified, as we are actually capable of manufacturing all kind of orientations. This rule

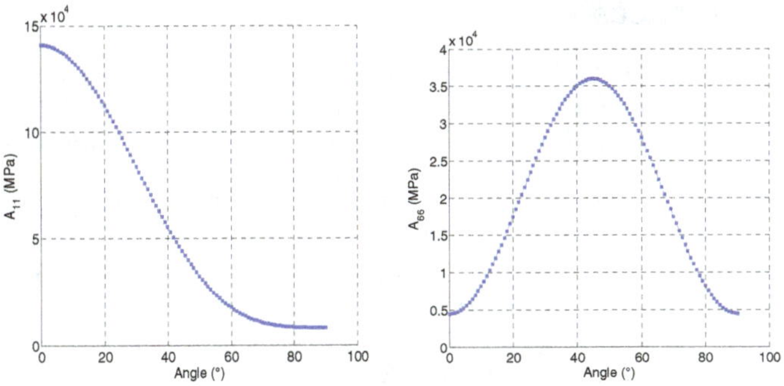

Fig. 4. A11 and A66 in function of the orientation angle.

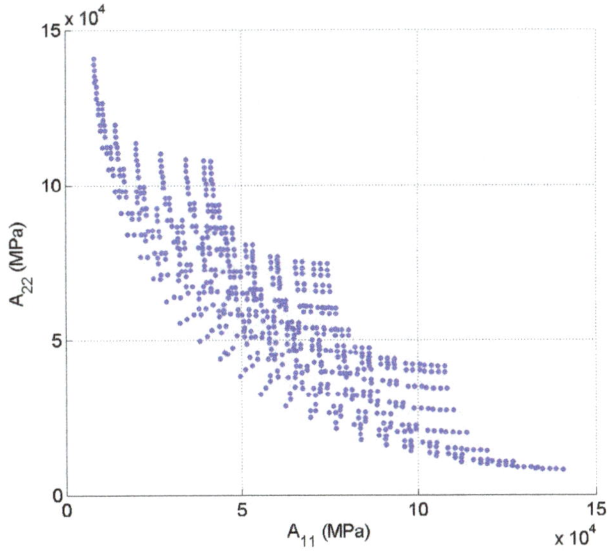

Fig. 5. Feasible domain in the case of a maximization of A11and A22 for 4 staking of 1 mm.

comes probably from the decomposition in elementary efforts of longitudinal and transverse tension/compression and shear that correspond to principal axes of the solicitations. In the case of a mixed variable the NSGA-II allows such combination (integer, real or mixed variables) [11, 18, 19].

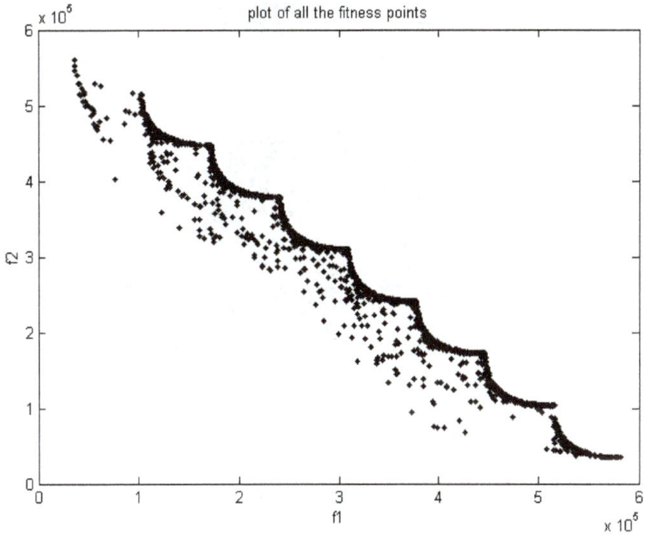

Fig. 6. Feasible domain for maximization problem of A11 (F_1) and A22 (F_2) for 8 staking.

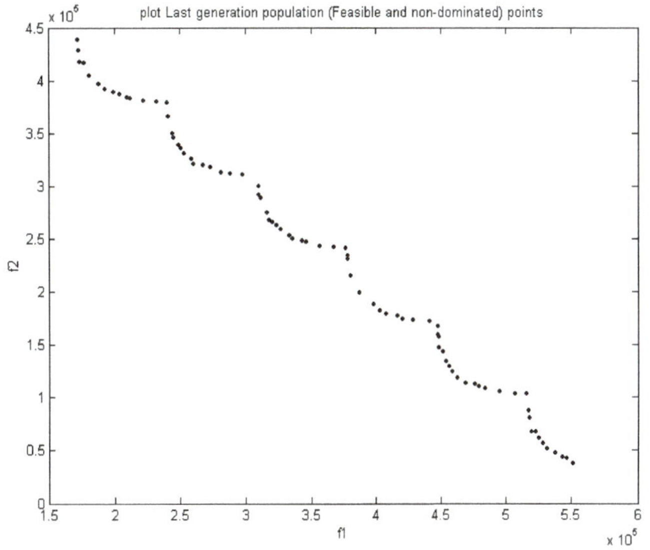

Fig. 7. Pareto Front at the convergence for maximization problem of A11 (F_1) and A22 (F_2) for 8 staking.

4 Numerical application to composite beam

To underline the complexity related to the multi-objective optimization in laminated composite structures we consider a basic composite beam with the classical laminate theory with constant thickness. In this case, we have the following $[ABD]$ matrix such as:

$$\begin{bmatrix} N \\ M \end{bmatrix} = \begin{bmatrix} A & B \\ B & D \end{bmatrix} \begin{bmatrix} \varepsilon \\ \rho \end{bmatrix} \tag{2}$$

where: $[A]$, $[D]$ and $[B]$ are respectively the membrane, bending and coupling terms of the stiffness matrix; (N, M) are the in-plane and out-plane internal forces; (ε, ρ) are respectively the in-plane and out-plane strains.

The objective functions that can be considered in the optimization process are the stiffness components A_{ij}, B_{ij} and D_{ij}. These functions are in general non-linear functions of the orientation angle (Fig. 4). In the following we consider a composed beam with the following material T700/M21 (E11 = 140GP, E22 = 8.2GP, $\upsilon 12 = 0.3$, G12 = 4.5GP).

If we consider for example the maximisation of (A11, A22) with $\theta_{(i=1-4)} \in [-\pi/2, \pi/2]$ the feasible domain is described in (Fig. 5).

In the case of the 8 staking (Fig. 6) we note clearly the increase of the non linearity, the discontinuity and the non convexity. At the convergence, the Pareto front is described in Figure 7. We can note that the final solutions are uniformly distributed along the Pareto front. This distribution is generally hard to obtain with common evolutionary strategies. If we integrate for example the following constraints (A11 must be greater than 350 GPa.mm

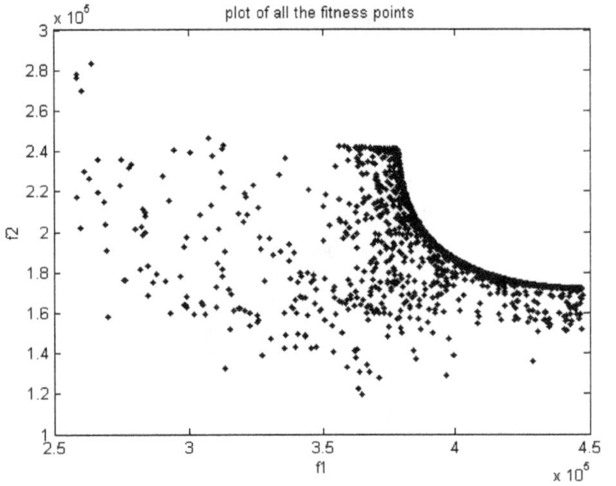

Fig. 8. Feasible domain with constraints on A11 (F_1) and A22 (F_2).

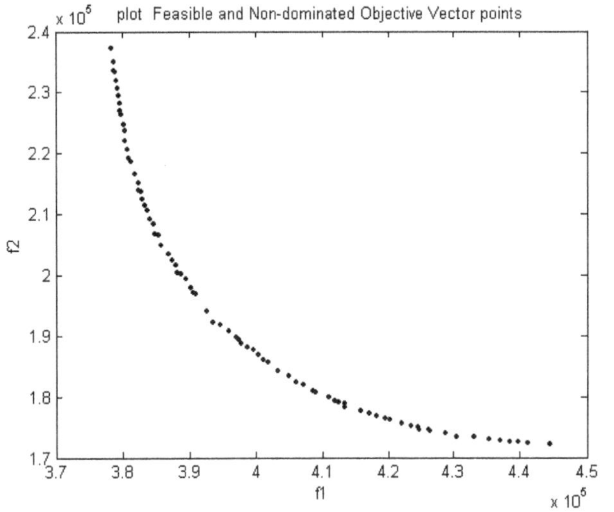

Fig. 9. Feasible domain with constraints on A11 and A22 with symetry stacking.

and A22 must be greater than 150 GPa.mm) we obtain the feasible domain in Figure 8. This domain is reduced in the case of symmetrical orientation Figure 9.

5 Conclusions

In this article, we have underlined the difficulties related to the multi-objective optimization of laminated composite under technological constraints. The handling method used in *NSGA-II* program has demonstrated its efficiency for a beam example to obtain a uniform spray of the Pareto solutions. However, further research is to be done concerning the integration of new technological constraints for composite materials.

This work was carried out under the AMERICO project (Multiscale Analysis: Innovating Research for CFRP) directed by ONERA (French Aeronautics and Space Research Center) and funded by the DGA/STTC (French Ministry of Defence) which is gratefully acknowledged.

References

1. J.M. Berthelot, *Matériaux composites: comportement mécanique et analyse des structures* (Masson, Paris, 1992)
2. S.W. Tsai, T. Hahn, *Introduction of composite materials* (Technomic Publ., Lancaster, 1980)
3. *Military Hand Book "MIL-HDBK-17-3E", Working Draft*, Design and Analysis, 23 Jan. 1997, Chap. 4
4. *Aircraft Crash Survival Design Guide*, USAAVSCOM TR 89-D-22A-E, Vol. I-IV, 1989
5. *Multicriteria Design Optimization*, edited by H. Eschenauer, J. Koski, A. Osyczka (Springer-Verlag, 1990)
6. S. Adali, M. Walker, V.E. Verijenko. Compos. Struct. **35**, 117 (1996)
7. M. Domaszewski, D.H. Bassir, W.H. Zhang, *Stress displacement and weight minimization by multicriteria optimization and game theory approach*, Computer Aided Optimum Design of Structures OPTI99 (Orlando, Florida, USA, WIT Press, 1999), pp. 171–181
8. R. Spallino, S. Rizzo. Mechanics Research Communications **29**, (2002)
9. P. Kere, M. Lyly, J. Koski. Compos. Struct. **62**, 329 (2003)
10. S. Adali, A. Richter, V.E. Verijenko. Microcomput. Civil Eng. **10**, 269 (1995)
11. K. Deb, A. Patrap, S. Agarwal, T. Meyarivan, *A fast and Elitist multi-objective Genetic algorithm: NSGA-II*, Technical Report No. 200001, 2005, Kanpur Genetic Algorithms Laboratory, India Institute of Technology
12. I. Das, Non linear multicriteria optimization and robust optimality, Ph.D. Thesis (Rice University, Houston, USA, 1997)
13. W.H. Zhang, M. Domaszwski, C. Fleury. Int. J. Numer. Meth. Eng. **52**, 889
14. J.H. Holland, *Adaptation in natural and artificial systems* (University of Michigan Press, Ann Arbor, 1975)
15. D.E. Goldberg, *Genetic algorithms in search, optimisation, and machine learning* (New York, Addison-Wesley, 1989)
16. Z. Michalewicz, *Genetic Algorithms + Data Structures = Evolution* (Springer-Verlag, Heidelberg, 1994)
17. C.M. Fonceca, P.J. Fleming, *Multiobjective optimization and multiple constraint handling with evolutionary algorithms*, part II: Application example, IEEE transactions on systems, man and Cybernetics: Part A: systems and humans, 1998, pp. 38–47
18. K. Deb, D.E. Goldberg, *An investigation of niche and species formation in genetic function optimization*, in *Proceedings of the Third International Conference on Genetic Algorithms*, edited by J.D. Schaffer (San Mateo, 1989), pp. 42–50
19. K. Deb, R.B. Agrawal, *Simulated binary crossover for continuous search space* (Complex Systems, 1995), pp. 115–148
20. F.X. Irisarri, D.H. Bassir, J.F. Maire, N. Carrere, Multiobjective stacking sequence optimisation strategy for laminated composite structures, *First International conference on multidisciplinary optimization and Applications*, April 17–20, 2007 (EDP Sciences), ISBN 978-2-7598-0023-0
21. K. Deb. International Journal of Simulation and Multidisciplinary Design Optimization **1**, 1 (2007)

25

Probabilistic approaches and reliability design of power modules

A. Micol[1,a], C. Martin[1], O. Dalverny[1], M. Mermet-Guyennet[2] and M. Karama[1]

[1] LGP-ENIT, 47 av. d'Azereix, BP 1629, 65016 Tarbes Cedex, France
[2] Power Electronics Associated Research Laboratory (PEARL), rue du Docteur Guiner, 65600 Semeac, France

Abstract – The weak point for the standard power IGBT modules in terms of reliability is thermal fatigue in solder joints due to the thermal stress induced by constitutive materials with different coefficients of thermal expansion (CTE). So far, many researches are aimed at defining accurate finite element simulation with constitutive equations of materials behaviour and fatigue failure relation connecting the inelastic strain and the number of cycles before failure. Even if these relations can be clearly identified, we can see that the validation of the finite element model is difficult due to the scatter of input data. In fact, the fatigue life of solder joints strongly depends on geometric shape, solders behaviour (due to the process) and applied load. The aim of this paper is to estimate the probability of failure of power module with the structural reliability methods. Thus the geometric, materials and loading variables are considered as random variables and the failure mode is modelled with the called limit state function. The two methods, response surface method and neural network method, are used here to evaluate the reliability of the lead-free solder. The sensitivities of the mean and the standard deviation for each random variable have been evaluated.

Key words: Reliability; FORM; response Surface; IGBT; Sn/Ag solder.

1 Introduction

Finite element modelling is more and more used to achieve the reliability of electronics packages. The power of the computer capacity increasing, it is now possible to integrate more accurate definitions in finite element model such as materials characterization, refined mesh and effects on the surrounding to evaluate this reliability [1]. This can also be said of power semiconductor modules where their reliability becomes a main focus on current research. Moreover, the use of lead-free solder, with an accurate behaviour which still remains to investigate, becomes a crucial matter in environmental concerns. To improve the accuracy of modelization, the natural scatter and uncertainties existing in the real power module such as geometric dimensions or material properties, can be integrated in fatigue analysis. According to a model of fatigue, the sensitivity analysis can determine the relevant parameters that influence the failure [2–4]. To refine this sensitivity analysis, we have to integrate the data with a complete probabilistic definition and no longer as values defined in a range.

The reliability of this module being conditioned by the fatigue of the solder joints, the main factors influencing the latter are the materials properties and the shape of the solder [5,6]. One of the weak points for these modules in terms of reliability was the wire-bonding connection, which is replaced here by a solution coming from the flip chip technology.

This paper presents the coupling between probabilistic and mechanical models to achieve fatigue life prediction for solders. First, we define and set up the probabilistic reliability methods and the mechanical-reliability coupling. Then, we present the mechanical problem by characterizing the mechanical and thermal behaviour of these connections with the constitutive law of the solder alloy. This behaviour is integrated as power law or hyperbolic sine law to calculate the stress-strain solution. The predictive fatigue life is estimated with the Coffin-Manson relation which uses equivalent inelastic strain in solder during one cycle. We analyse complete thermal and mechanical problems to achieve this reliability study, defining, on the one hand random variables as input of finite element model, and on the other hand a characteristic failure function.

2 Probabilistic design method

The probabilistic structural approach consists in determining, with a mathematical model, the probability of failure of a given system [7]. Indeed, when we build a traditional finite element model, all the model's input data are considered as fixed values and don't take into account

[a] Corresponding author: alexandre.micol@enit.fr

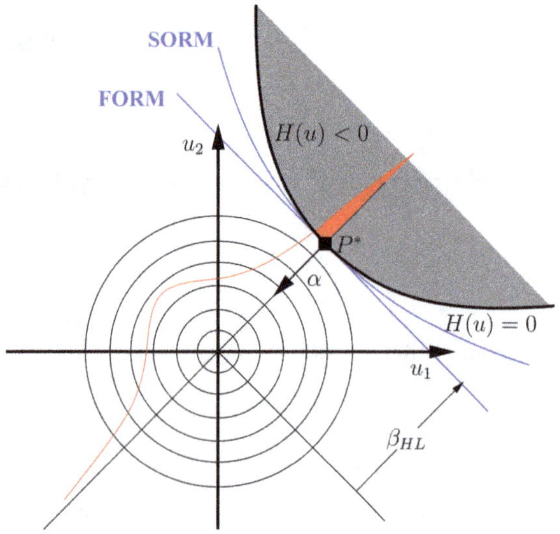

Fig. 1. Approximation methods FORM/SORM in standard space.

the natural scatter of the parameters. The solution to consider the natural variability and uncertainties of the input parameters on the model is to treat the input data as random variables defined by a law and its associated parameters. All relevant uncertainties influencing the probability of failure are then introduced in the vector $\boldsymbol{X}$ of basic random variables. In addition, the failure of the system is modelled by a functional relation $G(\boldsymbol{X})$, called limit state function, and defined to take a null or negative value in the failure domain:

$$G(\boldsymbol{X}) \leq 0, \boldsymbol{X} \in \text{failure domain}. \tag{1}$$

The components $x_k, 1 \leq k \leq n$ of the vector $\boldsymbol{X}$ are one realization of basic random variables. It's thus possible to define the probability of failure of the system as:

$$P_f = \int_{G(X)\leq 0} f_X(x)dx_1...dx_n \tag{2}$$

where $f_X(x)$ is the n-dimensional probability density of basic random variable vector $\boldsymbol{X}$. The problem to solve this integral comes from the limit state function which is not explicit because its evaluation is the result of finite element call. Approximation method can be established to compute the multi-dimensional integral of equation (2) by substituting the limit state function for a linear or second order hyper plane called respectively First Order and Second Order Reliability Method (FORM and SORM) (Fig. 1).

The calculation scheme of approximation method is as follow and the code to achieve this task was inspired by Ferum [8]:

- to obtain a probability of failure independent from the different ways of writing the same limit state function, Hasofer-Lind [9] suggest to map $x_i \rightarrow u_i$ the basic random variables from physical space to independent standardized Gaussian space (Gaussian variable with null mean and unity standard deviation) with the transformation T. The limit state function is also mapped in standardized space: $G(X) \rightarrow H(U)$;
- the most probable failure point P^*, i.e. the nearest point to the origin belonging to the limit state in standardized space, is computed by the improved Hasofer-Lind Rackwitz Fiessler (iHLRF) optimization algorithm [10, 11];
- an approximation of limit state function is built with a hyper plane or a quadratic surface to the *design* point P^* to compute the probability of failure with the relation:

$$P_f = \Phi(-\beta) \tag{3}$$

where β is called the reliability index and Φ is the standard cumulative distribution function.

The problem of the design point algorithm in non-linear finite element lies in the calculation of gradients. If the finite element code is not able to give the associated input derivatives, the latter have to be evaluated with the finite difference method. Indeed, the resolution of non-linear mechanical problems with finite element solver is achieved by integrating the non-linear materials laws. The use of an integration scheme involves an integration tolerance which defines the moment when the algorithm stops its iterations. For an accurate calculation of a stress increment, we define the error in the creep strain increment with:

$$\Delta\varepsilon^{cr}_{err} = (\dot{\varepsilon}^{cr}\mid_{t+\Delta t} - \dot{\varepsilon}^{cr}\mid_{t})\Delta t. \tag{4}$$

This integration scheme can lead to incorrect values of gradient if the step of finite difference approximation is not taken with judicious value [12]. The solution then consists in finding an explicit function or method which can approximate the finite element output: the response surface method and neural network. A numeric design of experiments (DOE) is made to construct this response surface or to train the neural network. The neural network, which has the advantage not to take a specific shape of the output function, can however have a long training. In our study, we define both methods and compare these in the following application.

2.1 Methodology

There are two ways to construct the DOE, each of these comes with its own advantages and disadvantages. A full factorial design can be constructed in physic space or in standard space. Constructing the numerical design of experiments in physic space ensures that points are physically realistic and don't have an impossible mechanical situation. However, the points of DOE aren't always equally spaced in standard space (an accurate solution with iHLRF algorithm isn't ensured in the whole domain). On the other hand, the DOE constructed in normal space ensures a good equalised fitting in the entire normal space (a good conditioning which certifies the accurate solution) but it is possible that some points in normal space diverge with the Hasofer-Lind transformation. For our study, we'll construct the DOE in standard space.

2.1.1 Building the response surface

The response surface is constructed with random sampling of basic random variables following the uniform distribution defined in [−3, 3] in standard space. These points are then mapped in physical space and sent to the finite element software.

The response surface method is very well adapted to parallel computing: the different output calculation of the DOE with several input parameters can easily be parallelised on cluster. Each node of the cluster has a finite element calculation to solve and to return the output to the master. As opposed to traditional FORM method, where one node computes the output solution and the other nodes compute the gradient, with this method, all nodes are used to compute the response surface. The output is stored in a library to build the response surface function.

Finally, the surface method consists in approximating the limit state function H with:

$$H_r = H_r(u_k) = \eta(u_k, p) + \varepsilon_r; \qquad \mathrm{E}(\varepsilon_r) = 0;\ k = 1, .., n;\ r = 1, .., N \tag{5}$$

where H_r is the rth evaluation of the exact limit state function at the u_k point, $\eta(u_k, p)$ the response surface with the p parameters at the u_k point and ε_r the error between the limit state function and the response surface.

For this study, we're going to construct different surface functions such as linear (polynomial) and non-linear regression models.

In the case of linear regression problems, the response surface for the rth sample can be written as follows:

$$\eta^{(r)}(u_k) = p_0 + \int_{j=1}^{m} p_j \Psi_j(u_k) \tag{6}$$

which, when using pure matrix notation for all sample, remains:

$$\eta = \Gamma\theta \tag{7}$$

where η contains the different responses $\eta^{(r)}(u_k)$, Γ represents the data matrix of DOE and θ the vector containing the parameters p.

The fitting method consists in finding $\bar{\theta}$, the vector of unknown coefficients p, by minimizing the sum of squares:

$$\bar{\theta} = \min_p \|H - \eta\|^2 \tag{8}$$

and is solved by:

$$\bar{\theta} = (\Gamma^T\Gamma)^{-1}\Gamma^T\eta. \tag{9}$$

In the case of nonlinear regression problems, the data matrix cannot be built and the Levenberg-Marquardt algorithm is used to solve equation (8).

The relevance of the prediction model can be evaluated with an analysis of variance. The multiple regression correlation coefficient $\overline{R}^2$ is a measure of the proportion of variability explained by, or due to the regression (linear relationship) in a sample of paired data:

$$\overline{R}^2 = 1 - \frac{\sum_i (H_r(u_k) - \eta^{(r)})^2/(N-m-1)}{\sum_i (E[H_r(u_k)] - \eta^{(r)})^2/(N-1)}. \tag{10}$$

This coefficient can lead error when the shape of response surface is unknow. We introduce for that the predictive quality of the regression can be evaluated by the Q^2 index:

$$Q^2 = 1 - \frac{\sum_{r=1}^{N} \epsilon_r^2}{\sum_r (E[H_r(u_k)] - \eta^{(r)})^2}$$

where:

$$\epsilon_r = \frac{H_r(u_k) - \eta^{(r)}}{1 - \mathbf{H}_{ii}}$$

and where $\mathbf{H}$ is the hat matrix defined by:

$$\mathbf{H} = \Gamma(\Gamma^T\Gamma)^{-1}\Gamma^T.$$

In the the case of non-linear regression problems, a linearization of the matrix of experiments is made around the supposed vector of parameter θ:

$$\Gamma = \left(\frac{\partial\eta(u,p)}{\partial p_1}, ..., \frac{\partial\eta(u,p)}{\partial p_n}\right).$$

This can correctly made with a prior knowledge of parameters p. The different iterations for finding the design point use then the previous parameters of the reponse surface for assess the current matrix of experiments. The fisrt iteration use as for it the defined the start θ vector.

We present here an optimization algorithm for establishing the design plan with continue and delimited input. The method presented here consists to solve the optimization problem:

$$\begin{aligned}\mathcal{L}(u,\lambda) = &- \min_{u^{(i)},u^{(j)}\forall i,j} (d(u^{(i)}, u^{(j)})) \\ &+ \lambda_0 \left[\mathbf{H}_{max} - Tr(\mathbf{H} - \mathbf{I}\frac{N}{m+1})\right] \\ &+ \sum_i \lambda_i \left[u_{max} - \|u^{(i)}\|\right]\end{aligned} \tag{11}$$

where:

- $\mathcal{L}(u,\lambda)$ represents the lagrangian with the λ multiplicators;
- $-\min_{u^{(i)},u^{(j)}} (d(u^{(i)}, u^{(j)}))$ represents the function which maximize the minimal distance between two points;
- $\mathbf{H}_{max} - Tr(\mathbf{H} - \mathbf{I}\frac{N}{m+1})$ represents the constrain to not obtain influent point;
- $u_{max} - \|u^{(i)}\|$ represents the constrain on the norm of the vector $u^{(i)}$ to prevent them from diverging beyond the cercle with a radius u_{max}.

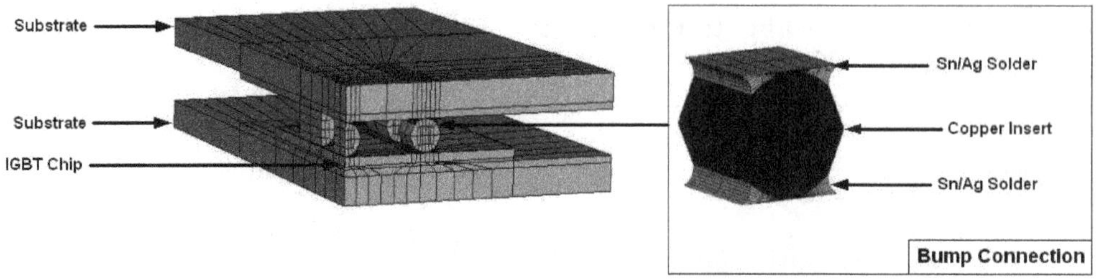

Fig. 2. The 3-D finite element model.

The algorithm starts to fit the points with a number of finite element call equal to the number of fitted parameters. These points are optimized in the domain by solving equation (11) with the COBYLA algorithm. At this point, the degree of freedom of the model (N-m-1) is equal to 0. The approximation cannot be relevant, the algorithm then optimizes and computes other points in the domain until it obtains Q^2 index higher than 0.9 to ensure a relevant model.

2.1.2 FORM analysis with explicit function

After having approximated the limit state function with response surface or neural network, the FORM method described above is used. To find an accuracy design point, the response surface method or neural network coupling with the design point optimisation is repeated with, at each step, a reduction of the DOE space. In fact, after finding the design point with the explicit function, a real finite element solution is computed at this point and, if this point doesn't respond to the convergence condition, another DOE is made around this point. The previous points stored in the library are used if their coordinates belong to the new design of experiments space.

2.1.3 Sensitivities

The main results available with approximation methods is then the probability of failure of the system (Eq. (3)) but other interesting results are the sensitivities of this probability regarding the different basic random variables [13]. In FORM formulation, the limit state function can be written $H(u) = au + \beta$ under normalized form. We can then define the sensitivity of the reliability index with respect to the standardized random variables as follows:

$$\alpha = \nabla_u \beta = \left[\frac{\partial \beta}{\partial u_1}, \frac{\partial \beta}{\partial u_2}, ..., \frac{\partial \beta}{\partial u_2}\right] = \left.\frac{\nabla_u H(u)}{\|\nabla_u H(u)\|}\right|_{u^*}. \tag{12}$$

We can study the sensitivity $\alpha_{p_{i_\gamma}}$ of the reliability index β, and then the probability of failure, with respect to the distribution parameters, such as mean or standard deviation, of the γth random variables to identify those which must have stricter quality control. The sensitivity $\alpha_{p_{i_\gamma}}$ is calculated with:

$$\nabla_p \beta = -J_{u^*,p}\alpha \tag{13}$$

where $J_{u^*,p}(i,j) = \frac{\partial T_i(x^*,p)}{\partial p_j}$ is the Jacobian matrix between the Hasofer-Lind transformation and the parameter p.

This sensitivity cannot be used directly for comparison, so the elasticity is defined by the normalisation of this sensitivity:

$$e_{p_i} = \frac{p_i}{\beta}\alpha_{i_\gamma} \text{ with } \alpha_{i_\gamma} = \left.\frac{\partial \beta}{\partial p_{i_\gamma}}\right|_{u^*}. \tag{14}$$

3 FE modelling

3.1 Geometry model

The device studied is a new water cooled IGBT power module. This module includes four IGBT chips and two diodes brazed on a substrate with a two-side cooling. The chip is cooled through a water blade on the top substrate and by a heat sink on the lower substrate. The power IGBT module FE model simulates one IGBT soldered on Aln substrate and attached with the top substrate with bump connection. A lead solder is used to solder the chip with the substrate and the eutectic Sn/Ag alloy is used to joint the two substrates via copper bumps cylinder-shape (1.45 mm diameter and 2 mm long) and two solders. Figure 2 shows a quarter of the assembly with bump technology. The bump connection is soldered to the IGBT with gold layer, and to the top substrate with 250 mm nickel metallization. Only a quarter of the entire multilayer switch is used because of the symmetric boundary condition and mesh optimization.

We define here the variability of the geometric dimensions. Only the shape of the solder is assumed to evolve during the process. This study focuses on the geometric dimension of the bump connection. The cross section of the bump connection in Figure 3 highlights the different geometric dimensions. Table 1 reports the characteristic geometric dimensions of the bump connection made on 20 specimens according to the dimension reported in Figure 3. Their mean value and standard deviation, calculated according to the geometric dimensions, follows a log-normal distribution. Only the thickness e between chip and copper insert seems to have a relevant influence on the failure. This dimension is then included in probabilistic analysis.

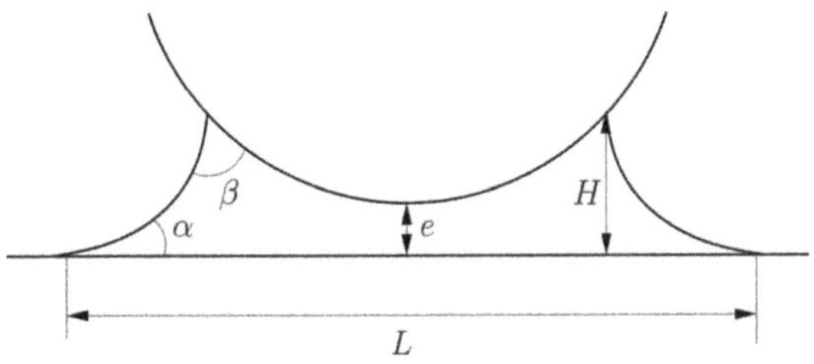

Fig. 3. Dimension characterising the solder bump.

Table 1. Geometric dimension results of cross section of bump connection according to Figure 3.

Geometric dimension	mean	Standard deviation
solder thickness (e)	12 mm	6 mm
solder height (h)	175 mm	16 mm
α	32	7
β	48	8
solder pad size (L)	1087 mm	39 mm

3.2 Materials characterisation

Lead free alloy is used as bump solder in the present module. The elastic modulus is dependent on the temperature, and can be modelled with the following equation:

$$E = E_0 - E_1 T(^\circ\mathrm{C}) \tag{15}$$

with E the temperature dependent Young modulus in MPa, the Young modulus at 0 °C, the decrease rate of the Young modulus according to the temperature and T the temperature in degree Celsius. For our solder, several nano-indentations are made in the solder to evaluate, which is taken as materials random variables in the probabilistic study.

The material non-linearity is modelled with the constitutive law characterizing the steady state creep rate. The power law can describe the mechanical behaviour of alloy in low or medium stress:

$$\dot{\varepsilon_{cr}} = A(B\sigma)^n \exp\left(\frac{-Q}{kT}\right) \tag{16}$$

with $\dot{\varepsilon_{cr}}$ the equivalent creep strain rate, σ the applied stress and T the temperature. Wiese identifies the law's different parameters for the eutectic Sn/Ag solder alloy in flip-chip configuration [14]. When high stress is reached the power law breaks down and can't follow the creep strain rate to increase more quickly with the applied stress. A second law was developed to find a closed form of creep behaviour. The sine hyperbolic law can describe the steady state creep for the low and high stress level:

$$\dot{\varepsilon}_{cr} = A\left(\sinh\left(B\sigma\right)\right)^n \exp\left(\frac{-Q}{RT}\right). \tag{17}$$

This sinh law, identified for Sn/Ag3.5 and Sn95.5/Ag3.8/Cu0.7 by Banerji and Darveaux [15], is presented with the different parameters. For our application and stress level, the best way should be to use the sine hyperbolic law but Wiese demonstrates the strong dependence of the law's parameters on the solder configuration during identification. To compute a best approach to the creep strain for flip-chip configuration, the identified flip-chip power law would be more suitable. The multiplier coefficient A and the power coefficient n are considered as random variables and are assumed to have a variation coefficient of 10%.

We consider the solder to have an isotropic linear strain hardening. The plasticity's parameters are incorporated in the study but are not considered as random variables.

3.3 Thermal load

The faces of the heat sink wings are subjected to thermal convection with a convective exchange coefficient calculated at the beginning, starting from a fluid software. The power IGBT module is subjected to a power cycling history. The active power cycling consists in computing a primary heat transfer analysis and the thermal temperature map output is sent to the mechanical analysis. The power injected is distributed in the volume of the IGBT. For this power cycling, temperature time history loading is only imposed to the element corresponding to the IGBT. Figure 4 shows the output temperature of the heat transfer analysis sent to the mechanical analysis to compute the difference of the output inelastic strain between the end and the beginning of the limit cycle.

3.4 Life prediction

The models proposed to evaluate predictive fatigue life of the solder joints can be divided in five major categories based on stress, plastic and/or creep strain, energy and damage accumulation [16]. For our study, we take the inelastic strain based model which considers plastic and creep phenomenon due to CTE mismatch. The following equation shows the Coffin-Manson [17,18] fatigue life model:

$$N_f = C\Delta\varepsilon_{in}^n \tag{18}$$

where is the accumulated equivalent inelastic strain during one cycle. For Sn/Ag3.5, $C = 23.78$ and $n = -1.08$ [19] and for Sn95.5/Ag3.8/Cu0.7, $C = 0.38$ and $n = -1.96$ [20]. The evaluation of the number of cycles before failure is investigated considering the whole volume of the solder. To calculate the accumulated equivalent inelastic strain, we use the following averaged equation with the volume weighted average (VWA) method:

$$\Delta\varepsilon_{in} = \frac{\sum_e \Delta\varepsilon_{in_e} V_e}{V_{tot}} \tag{19}$$

where V_{tot} is the total volume of the elements' set, V_e is the volume of one element and $\Delta\varepsilon_{in_e}$ the associated inelastic strain. This method ensures that the solution is less mesh dependent compared to using the maximum value of inelastic strain in the whole solder.

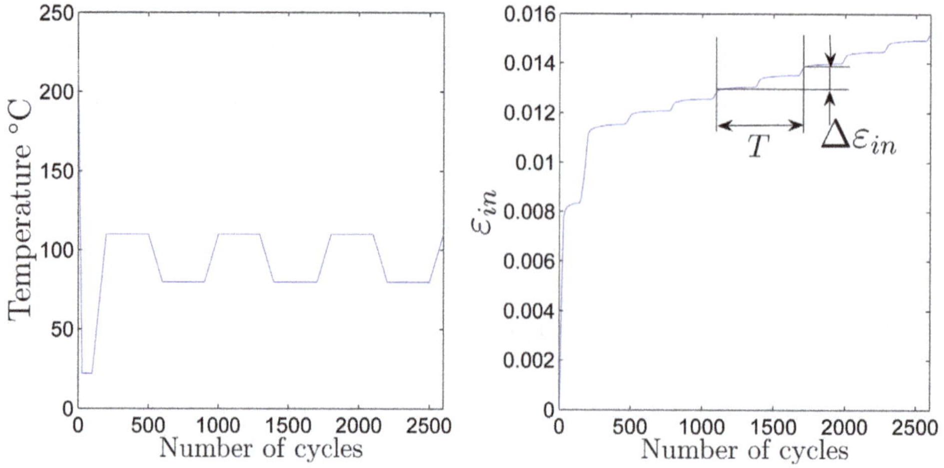

Fig. 4. Result temperature in chip and associated inelastic strain.

3.4.1 Random variable and limit state function

The limit state function, relating the thermo-mechanical fatigue of the solder, is written considering that the system falls into the failure field if the number of cycle doesn't reach a target value. The mechanical output in terms of inelastic strain is transformed by using the Coffin-Manson relation:

$$G(X) = C\Delta\varepsilon_{in}^{n} - N_{f_{target}} \tag{20}$$

where $C\Delta\varepsilon_{in}^{n}$ and $N_{f_{target}}$ are respectively the number of cycles before failure computed by the finite element code and the objective number of cycles that the system must reach.

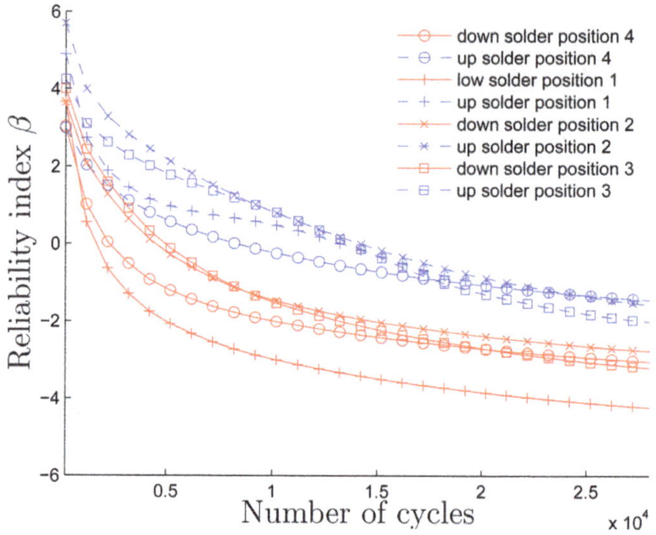

Fig. 5. Evolution of the reliability index β for the target number of cycles.

3.5 Results

The sensitivities study shows the evolution of the probability of failure and the importance of thermal solicitation. The main parameter influencing the failure remains the flux density imposed in the IGBT chip. The multiplier coefficients A of the constitutive law slightly act on the results of inelastic strain. The effort for an accurate identification must be made on the power coefficient n because of the influence of its mean value: the recommendation of the previous authors about a micro structure (varying with the temperature cooling and then the shape of solder due to this capacity of cooling) must be taken into account. Moreover, the sensitivity of this standard deviation indicates that a variability in the process or a variability on the quality of solder have a great impact in terms of reliability. The Sn/Ag CTE mean value and its standard deviation, which intervene in the failure, don't show a great dependence on the probability of failure.

The main design variability of this bump connection technology is identified on the thickness e because this is the only one of the geometric dimensions, identified with the experiments, which shows great correlation with the failure. It leads to a better definition of the shape and especially the problem of the location of the initial crack. Figure 5 shows the evolution of the β reliability index and Figure 6, the probability of failure of the module for different numbers of cycle and for the different solders of the module. It must keep in mind that the initial crack begins under the bump but in terms of electric functionality, the failure is not reached.

An investigation is now made to compare first mechanic formulations of the solder behaviour and then, different solders. The reliability is evaluated with the sinus hyperbolic and power behaviour formulation for Sn/Ag3.5 solder and with a sinus hyperbolic for Sn95.5/Ag3.8/Cu0.7. The number of cycles to reach to be in safe domain is taken here at $N_{f_{target}} = 2000$ cycles. The response surface is used here. For the same limit state function using the same coefficient of the Coffin-Mason relation, a difference is found according to the creep constitutive equation (Eqs. (16) and (17)). The main difference, highlighted by Figures 7 and 8, between the two

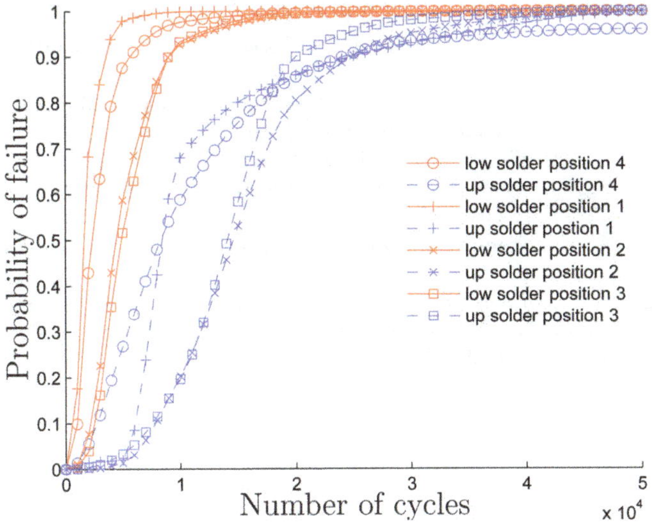

Fig. 6. Evolution of the probability of failure for the target number of cycles.

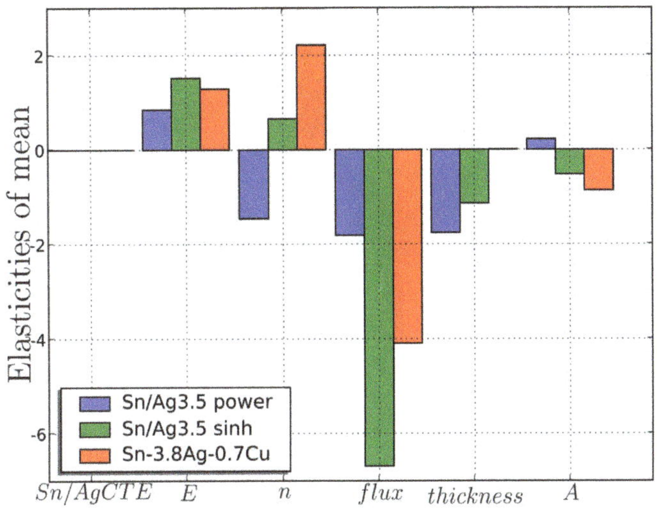

Fig. 7. Elasticities of the random variable's mean for different solder alloys.

formulations lies in the fact that the two laws are not identified on the same specimen. It is confirmed that the identification procedure of the law's parameters plays a major role. The Sn95.5/Ag3.8/Cu0.7 denotes a relatively better reliability in comparison to the Sn/Ag3.5 solder if we consider the same constitutive law that corresponds to Schubert's result. It is reminded that the Coffin-Masson parameters play a role here too.

4 Conclusion

We attempt in this article to estimate the reliability of power modules connections with structural reliability methods. The two approximation methods, response surface and neural network, are set up and used to define the probability of failure and the sensitivities of this latter ac-

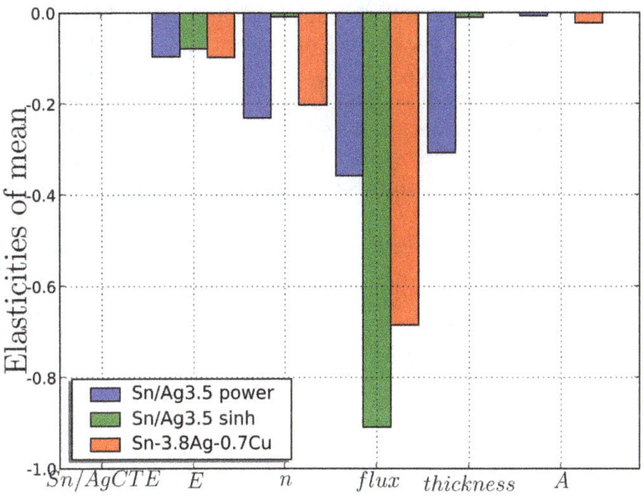

Fig. 8. Elasticities of the random variable's standard deviation for different solder alloys.

cording to the random variables. The response surface can model the behaviour of the inelastic strain if we can define the shape of the function. The advantage of neural network to not know the shape of the model is to solve this issue but it can be a problem because of the number of finite element call. The reliability analysis was made by taking as random variables the characteristics of the different materials. It shows the importance of the power law coefficient n but a variation of the loading action remains the major parameter changing the reliability. The future developments will have to include more sophisticated integration of load behaviour such as rainflow counting. We must keep in mind that this method evaluates the failure according to one failure mode: the fatigue of solders. The results of this reliability investigation remain sensitive to the constitutive law, its identification and to the fatigue failure relation.

References

1. P. Towashiraporn, G. Subbarayan, B. McIlvanie, B.C. Hunter, D. Love, B. Sullivan, in *Thermal and Thermomechanical Phenomena in Electronic Systems* (ITHERM, 2002), pp. 854–861
2. B. Vandevelde et al. Journal of Electronic Packaging **125**, 498 (2003)
3. D.G. Yang, J.S. Liang, Q.Y. Li, L.J. Ernst, G.Q. Zhang. Microelectronics Reliability **44**, 1947 (2004)
4. B. Wunderle, W. Nuchter, A. Schubert, B. Michel, H. Reichl. Microelectronics Reliability **44**, 1933 (2004)
5. L. Xu, T. Reinikainen, W. Ren, B.P. Wang, Z. Han, D. Agonafer. Microelectronics Reliability **44**, 1977 (2004)
6. L. Xingsheng, L. Guo-Quan, in *Components and Packaging Technologies, IEEE Transactions* (see also Components, Packaging and Manufacturing Technology)
7. R. Rackwitz. Structural Safety **23**, 365 (2001)
8. T. Haukass, *Finite element reliability using matlab*, http://www.ce.berkeley.edu/haukaas/FERUM/ferum.html

9. N.C. Lind, A.M. Hasofer. J. Eng. Mech. **100**, 111 (1974)
10. B. Fiessler, R. Rackwitz. Comput. Struct. **9**, 489 (1978)
11. A. Der Kuireghian, P.L. Liu. Struct. Safety **9**, 489 (1991)
12. A. Micol, M. Karama, O. Dalverny, C. Martin, M. Mermet-Guyennet, in *Reliability design of power modules using probabilistic approaches*, edited by B.H.V. Topping, G. Montero, R. Montenegro, *Proceedings of the Eighth International Conference on Computational Structures Technology* (Stirlingshire, United Kingdom, Civil-Comp Press, 2006)
13. M. Lemaire, *Fiabilité des structures* (Hermes, Collection Génie civil, 2005)
14. E. Meusel, S. Wiese, F. Feustel. Sensors and Actuators **99**, 188 (2002)
15. K. Banerji, R. Darveaux, Constitutive relations for tin-based solder joints, in *IEEE Proceeding* (1992), pp. 1013–1024
16. G.S. Selvaduray, W.W. Lee, L.T. Nguyen. Microelectronics Reliability **40**, 231 (2000)
17. L.F. Coffin. J. Mater. **6**, 388 (1971)
18. S.S. Manson, *Thermal Stress and Low-cycle Fatigue* (New York, Mc-Graw-Hill, 1966)
19. C. Kanchanomai. International Journal of Fatigue **24**, 57 (2002)
20. A. Schubert et al., in *Electronic Components and Technology Conference* (Proceedings 52nd, 2002), pp. 1246–1255

26

Analysis of a hybrid composite tank under pressure

A. Hocine[1,2,a], D. Chapelle[2], A. Benamar[3] and A. Bezazi[4]

[1] University Hassiba Benbouali of Chlef BP. 151, Chlef 02000, Algeria
[2] Institut FEMTO-ST, Dept. LMARC, 24, rue de l'épitaphe, 25000 Besançon, France
[3] ENSET, Department of mechanical engineering, BP 1523, Oran 31000, Algeria
[4] University 08 Mai 1945, BP. 401, Guelma 24 000, Algeria

Abstract – In this work, the analysis of a hybrid composite tank of storage of hydrogen is presented. This solution made of a carbon/epoxy envelope coated on a metal liner and an intermetallic material is proposed. Each layer of the laminate is assumed to be anisotropic, and liner and intermetallic were assumed to be isotropic. This work is concerned with the study of the elastic behaviour of this solution. Based on the three-dimensional (3-D) elasticity, an exact solution for the stresses, strain and displacement is presented. The effect of stacking sequences on the behaviour of hybride solution is presented.

1 Introduction

Hydrogen is considered as one of the more promising energy vectors of the future. It can be used as a fuel in many applications. However, this requires that several technological hurdles are cleared especially the one concerning its storage. Storage must offer a high degree of safety as well as allowing ease of use in terms of energy density and dynamics of fuel storage and controlled release [1].

As everyday use of such storage, we mention the cases of oxygen in hospitals and various gases in the university laboratories or industrial ones. The pressure of the compressed hydrogen is generally about 350 bars (35 MPa), which leads to reach a convincing mass density for the composite tanks. However, the density of storage remains low, and in order to make this technology competitive it is necessary to reach a pressure of 700 bars. At present, the thin-walled or thick tanks are largely used in several branches of engineering [3], such as the storage of compressed hydrogen, compressed and liquefied natural gas [4].

The hyperbare storage of hydrogen must take into account some important characteristics such as explosiveness, inflammability and the low value of hydrogen atom's radius. To improve the performances of this type of storage, research is mainly concerned with two aspects: powerful materials for both the structure and the internal coating of the tanks as well as economical manufacturing techniques [2].

Several storage models of hydrogen were proposed. Xia et al. [5] provide an analytical model for research and analysis of the mechanical properties of a multi-layer tube under mechanical loading. Chapelle and Perreux [6] present a paper which purpose is to study the cylindrical section of a Type 3 high-pressure hydrogen storage vessel, combining an aluminium liner which prevents gas diffusion and an overwrapped composite devoted to reinforce the structure. The laminate composite is assumed to be an elastic damageable material whereas the liner behaves as an elastic plastic material. Based on the classical laminate theory and on Hill's criterion to take into account the anisotropic plastic flow of the liner, the model provides an exact solution for stresses and strains on the cylindrical section of the vessel under thermomechanical static loading. Takeichia et al. [7] propose a novel hydrogen storage vessel with the lightness and smallness, a "hybrid hydrogen storage vessel", which is a combination of a lightweight high-pressure vessel and hydrogen storage alloy. This model is expected to solve the problem of hydrogen storage techniques in weight and volume other than the techniques using compressed hydrogen or hydrogen storage alloy, individually.

In this work, a hybrid prototype of storage is proposed. It consists of an carbon/epoxy envelope coated on a metal liner and hydride intermetallic material. The intermetallic aims to absorb hydrogen coming from microcracks, as those formed by hydrogen embrittlement of the aluminium liner. Interest of the site of the intermetallic directly placed at the contact of the aluminium liner, since the reinforced carbon fiber layer, the alloy expansion will mainly occurs towards the inside (e.g. the aluminium liner).

This work focus on the study of the elastic behaviour of a hybrid composite storage solution. The effect of the stacking sequence, on the structure stiffness may then be investigated. An analysis of the stresses, strains and the displacements through the wall's thickness is presented.

[a] Corresponding author: hocinea_dz@yahoo.fr

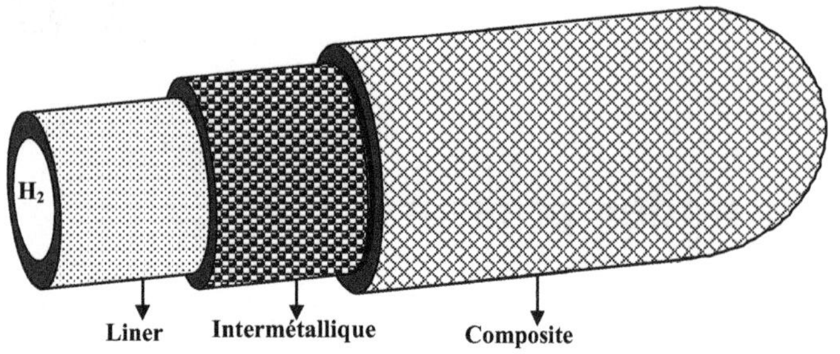

Fig. 1. Hybrid solution.

2 Analysis procedure

2.1 Stress and strain analysis

The stress and strain analysis of a cylindrical tube with internal and external radius r_0, r_a respectively subjected to an axisymmetric internal pressure is performed. The metal liner is reinforced with a composite material manufactured by the filament winding process, and an intermediate layer of intermetallic is added between the previous layers, as indicated in Figure 1.

The stress/strain relation of the kth layer of an anisotropic material can be written [8]:

$$\begin{Bmatrix} \sigma_z \\ \sigma_\theta \\ \sigma_r \\ \tau_{\theta r} \\ \tau_{z r} \\ \tau_{z\theta} \end{Bmatrix}^{(k)} = \begin{bmatrix} C_{11} & C_{12} & C_{13} & 0 & 0 & C_{16} \\ C_{12} & C_{22} & C_{23} & 0 & 0 & C_{26} \\ C_{13} & C_{23} & C_{33} & 0 & 0 & C_{36} \\ 0 & 0 & 0 & C_{44} & C_{45} & 0 \\ 0 & 0 & 0 & C_{45} & C_{55} & 0 \\ C_{16} & C_{26} & C_{36} & 0 & 0 & C_{66} \end{bmatrix}^{(k)} \begin{Bmatrix} \varepsilon_z \\ \varepsilon_\theta \\ \varepsilon_r \\ \gamma_{\theta r} \\ \gamma_{z r} \\ \gamma_{z\theta} \end{Bmatrix}^{(k)} \quad (1)$$

where σ_z, σ_θ, and σ_r are axial, circumferential, and radial stresses respectively; $\tau_{z\theta}$, τ_{zr}, and $\tau_{\theta r}$ are shear stresses in the planes z–θ, $z-r$, θ–r respectively; k is the number of the considered layer; $\boldsymbol{C}_{11}$-$\boldsymbol{C}_{66}$ are rigidity coefficients of the kth layer, ε_z, ε_θ, and ε_r are axial, circumferential, and radial strains respectively; $\gamma_{\theta r}$, γ_{zr}, and $\gamma_{z\theta}$ are the shear strains in the planes θ–r, $z-r$, and z–θ respectively.

The asymmetric loading of the hybrid tube allows reducing the equilibrium equations in the cylindrical coordinates; they are expressed as [5]:

$$\begin{cases} \frac{d\sigma_r^{(k)}}{dr} + \frac{\sigma_r^{(k)} - \sigma_\theta^{(k)}}{r} = 0 \\ \frac{d\tau_{\theta r}^{(k)}}{dr} + \frac{2}{r}\tau_{\theta r}^{(k)} = 0 \\ \frac{d\tau_{zr}^{(k)}}{dr} + \frac{\tau_{zr}^{(k)}}{r} = 0. \end{cases} \quad (2)$$

Strain/displacement relations are expressed as [5]:

$$\begin{cases} \varepsilon_r^{(k)} = \frac{dU_r^{(k)}}{dr}, \ \varepsilon_\theta^{(k)} = \frac{U_r^{(k)}}{r}, \ \varepsilon_z^{(k)} = \frac{dU_z^{(k)}}{dz} = \varepsilon_0 \\ \gamma_{zr}^{(k)} = 0, \ \gamma_{\theta r}^{(k)} = \frac{dU_\theta^{(k)}}{dr} - \frac{U_\theta^{(k)}}{r}, \ \gamma_{z\theta}^{(k)} = \frac{dU_\theta^{(k)}}{dz} = \gamma_0 r. \end{cases} \quad (3)$$

Substituting equation (1) into equation (2) and using equation (3), the following differential equation is obtained:

$$\frac{d^2U_r^{(k)}}{dr^2} + \frac{1}{r}\frac{dU_r^{(k)}}{dr} - \frac{N_1^{(k)}}{r^2}U_r^{(k)} = \alpha_1^{(k)}\frac{\varepsilon_0}{r} + \alpha_2^{(k)}\gamma_0 \quad (4)$$

with

$$\beta^{(k)} = \sqrt{N_1^{(k)}} = \sqrt{C_{22}^{(k)}/C_{33}^{(k)}}. \quad (5)$$

The solution of equation (4) takes the form:
If $\beta^{(k)} = 1$ then

$$U_r^{(k)} = D^{(k)}r + E^{(k)}/r \quad (6)$$

If $\beta^{(k)} \neq 1$ then

$$U_r^{(k)} = D^{(k)}r^{\beta^{(k)}} + E^{(k)}r^{-\beta^{(k)}} + \alpha_1^{(k)}\varepsilon_0 r + \alpha_2^{(k)}\gamma_0 r^2 \quad (7)$$

with

$$\begin{aligned} \alpha_1^{(k)} &= \frac{C_{12}^{(k)} - C_{13}^{(k)}}{C_{33}^{(k)} - C_{22}^{(k)}} \\ \alpha_2^{(k)} &= \frac{C_{26}^{(k)} - 2C_{36}^{(k)}}{4C_{33}^{(k)} - C_{22}^{(k)}}. \end{aligned} \quad (8)$$

2.2 Boundary conditions

The boundary conditions are imposed by the geometry conditions of the structure, due to continuity and volume conservation, and by the conditions of loading. It is assumed that there are no slips in the interfaces and that there is continuity of stresses and displacements. These boundary conditions allow to determine the integration constants $D^{(k)}$, $J^{(k)}$, γ_0 and ε_0.

The number of unknown factors, or integration constant of the system, to be solved is $2\,(z+1)$ for z layers of the hybrid solution; where D^k, E^k, γ_0 and ε_0 for $k \in [1,\ z]$.

The radius $r_{int}(k)$ and $r_{ext}(k)$ are introduced for each layer $k$and it is noted that

$$r_{int}^{(k=)} = r_0 \ et \ r_{ext}^{(k=z)} = r_a. \quad (9)$$

➢ The condition of continuity of radial displacements results in the relation is:

$$\forall k \in [1,\ z-1], \quad U^{(k)}(r_{ext}(k)) = U^{(k+1)}(r_{ext}(k)). \quad (10)$$

➢ The condition of continuity of radial stress results in the relation is:

$$\begin{aligned} &\forall k \in [1,\ z-1], \quad \sigma_r^{(k)}(r_{ext}(k)) = \sigma_r^{(k+1)}(r_{ext}(k)), \\ &\sigma_r^{(k=)}(r = r_0) = -p_0, \\ &\sigma_r^{(k=z)}(r_{ext}(z)) = 0. \end{aligned} \quad (11)$$

Table 1. Elastic parameters.

Properties	Carbon/Epoxy (T300 /934)	Liner Aluminium	Intermetallic (Zr_3Fe)
E_x (GPa)	141,6	69,5	122.5
E_y (GPa)	10,7	69,5	122.5
G_{xy} (GPa)	3,88	26,7	45.79
ν_{yx}	0,268	0,3	0.3375
ν_{zy}	0,495	–	–

➢ The equilibrium condition of axial force due internal pres-sure with end loading effect

$$2\pi \sum_{k=1}^{z} \int_{r_{k-1}}^{r_k} \sigma_z^{(k)}(r)\, r\, dr = \pi\, r_0^2\, p_0 + F \tag{12}$$

where F is the applied axial load.
➢ Torque balance is

$$2\pi \sum_{k=1}^{z} \int_{r_{k-1}}^{r_k} \tau_{z\theta}(r)\, r^2 dr = C \tag{13}$$

where C is the applied torque.

The hypothesis of this study neglects torque and axial loads, where $F = 0$ and $C = 0$. Thus one has 2 (z) equations to identify the whole integration constants. In the continuation, one endeavours to write the components of matrix A and of the vector B of the linear problem are equivalent such as:

$$A \times X = B \tag{14}$$

with: $$X = \begin{pmatrix} D^1 \\ D^2 \\ \cdot \\ \cdot \\ D^z \\ E^1 \\ E^2 \\ \cdot \\ \cdot \\ E^z \\ \varepsilon_0 \\ \gamma_0 \end{pmatrix}.$$

2.3 Geometry and resolution

The tube is characterised by an internal radius of 50 mm, a liner thickness of 0.5 mm, 0.1 mm for the intermetallic and 0.5 mm of each layer of the laminate. The properties of materials are presented in the table 1 below.

Table 2 presents the stacking sequence of the different studied laminates, where the order of the stacking of each laminate is taken from interior to exterior.

The internal wall of the hybrid tube is subjected to internal pressure of 10 MPa. All results are presented through wall's thickness of the solution.

Table 2. Different stacking sequences of the composite tube.

Composites Sequence types	Angle of wrap
Seq1	[+55/−55/+55/−55]
Seq2	[+55/−30/+30/−55]
Seq3	[+55/−55/+30/−30]

3 Results and discussions

3.1 Stress analysis

Figures 2–4 show the distribution of the axial, circumferential, radial and shear stresses σ_z, σ_θ, σ_r and $\tau_{z\theta}$ respectively through the wall's thickness for the three sequences. It can be also noticed that the change of the properties from one layer to another influences the overall mechanical behaviour.

3.1.1 Hoop and axial stress

As shown in Figure 2, it is obvious that the hoop and axial stresses of cylindrical tank have discontinuous variations at the interfaces with different materials. Figure 2 shows that the presence of intermetallic in this work induced maximal hoop and axial stress for ($52 \leqslant r \leqslant 52.1$) for the three sequences. The analysis shows for all sequence that the curves of the Hoop and axial Stress for the laminate Seq1 are below than the Seq2 and Seq3.

3.1.2 Shear stress

Figure 3 shows the shear stress $\tau_{z\theta}$ through the radius for the three sequences and has the same trend, but differs in values and occurs in five stages starting from the internal wall to the external. It is clear that $\tau_{z\theta}$ for all types show discontinuous variation, and the signs of the $\tau_{z\theta}$ are same variations with those of the winding angles. In addition the laminate Seq1 has the lowest values of the stress comparing with the Seq2 and Seq3.

3.1.3 Radial stress

Figure 4 shows the distribution of radial stress σr through the thickness of hybrid solution. The behaviour of radial stress through the thickness indicates the presence of maximal compression of –10 MPa in the internal wall and minimal (equal to zero) in the external wall for all the studied.

In addition, the distribution show nearly linear variation for the three sequences. The curves of radial stress σr for the laminates of the first and second sequence are below than the sequence of the third sequence. The distribution shows nearly linear variation for the three sequences. The curves of radial stress σr for the laminate of Seq1 are below than the laminates of the Seq2 and Seq3.

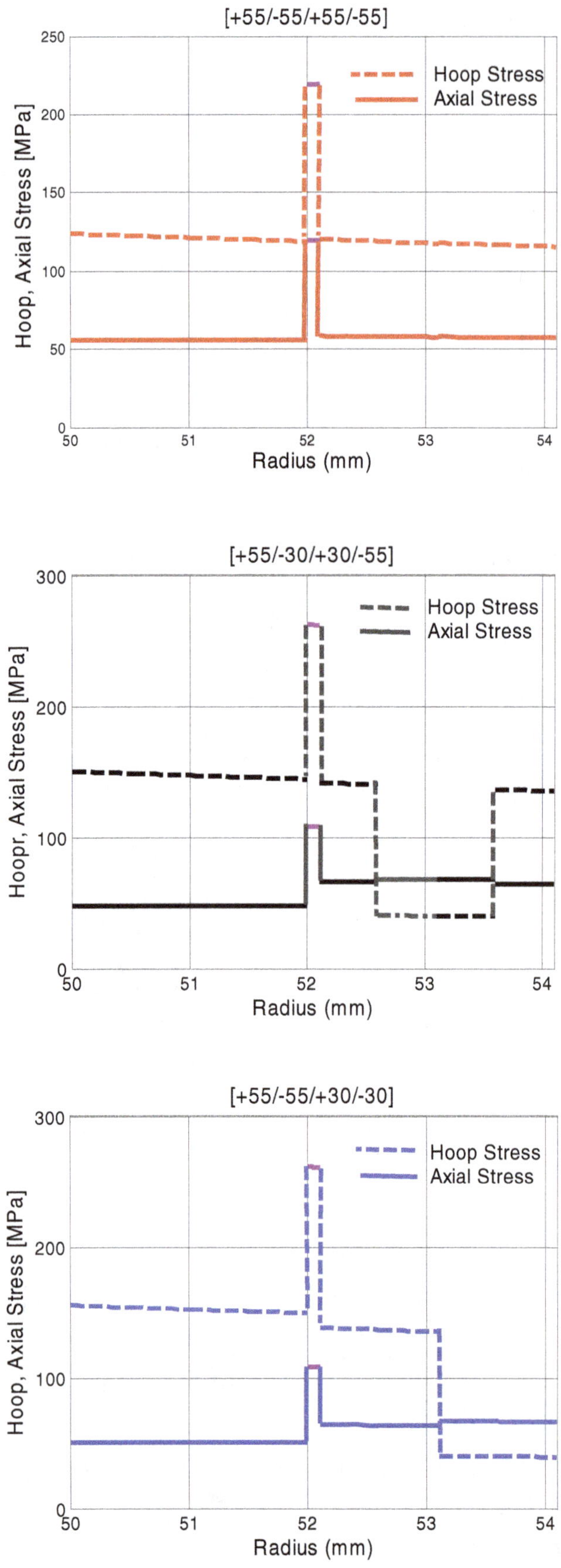

Fig. 2. The distribution of Hoop and axial stress through the thickness.

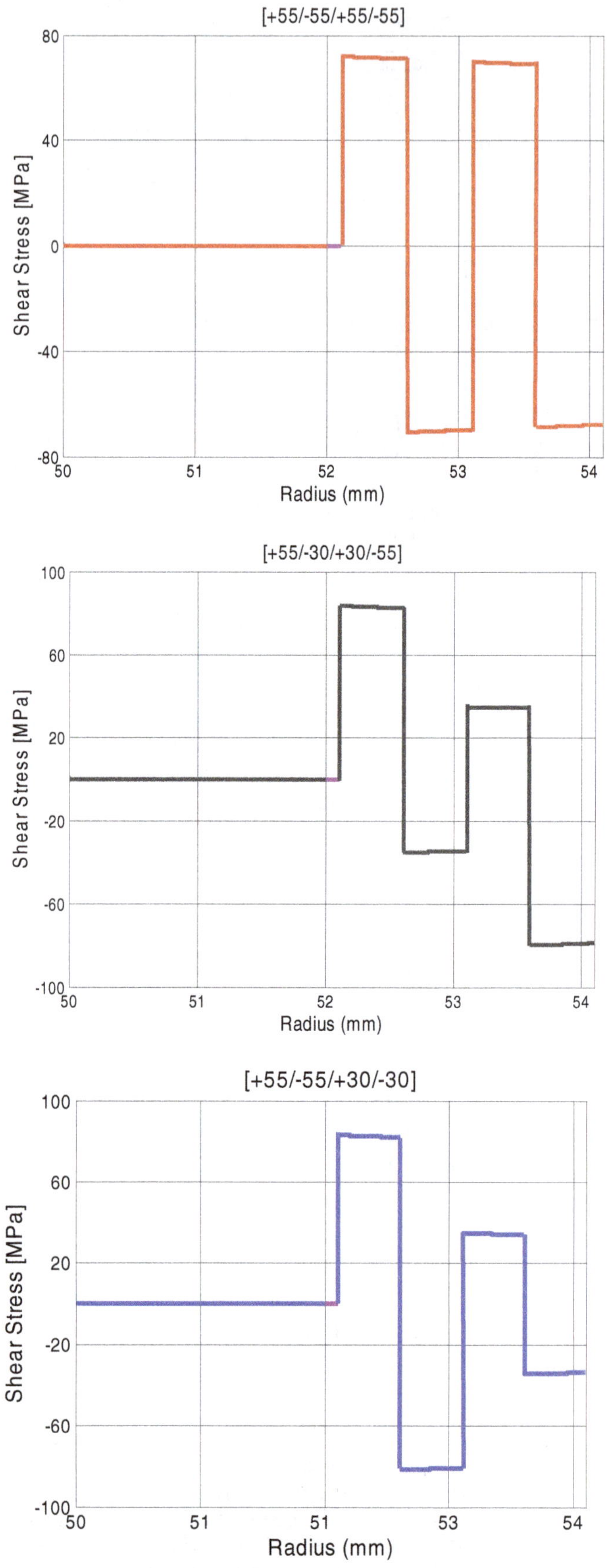

Fig. 3. The distribution of shear stress through the thickness.

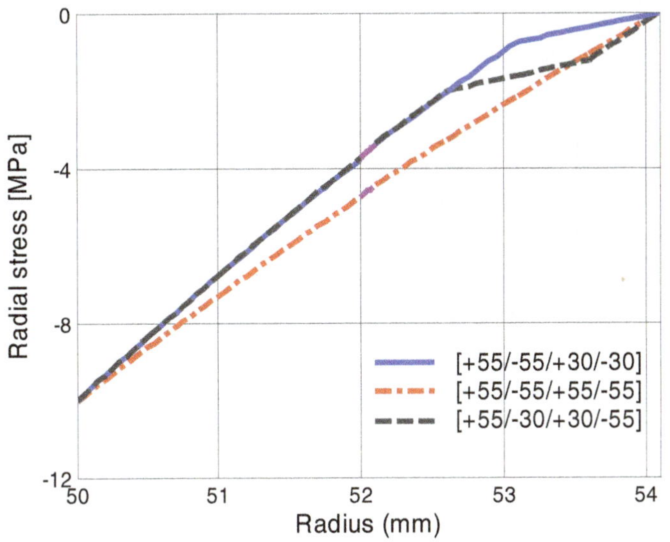

Fig. 4. The distribution of radial stress throught the thickness.

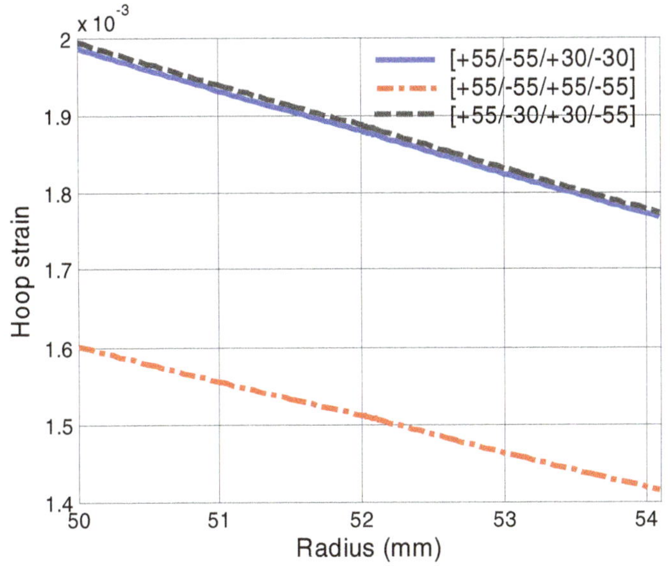

Fig. 5. The distribution of hoop strain through the thickness.

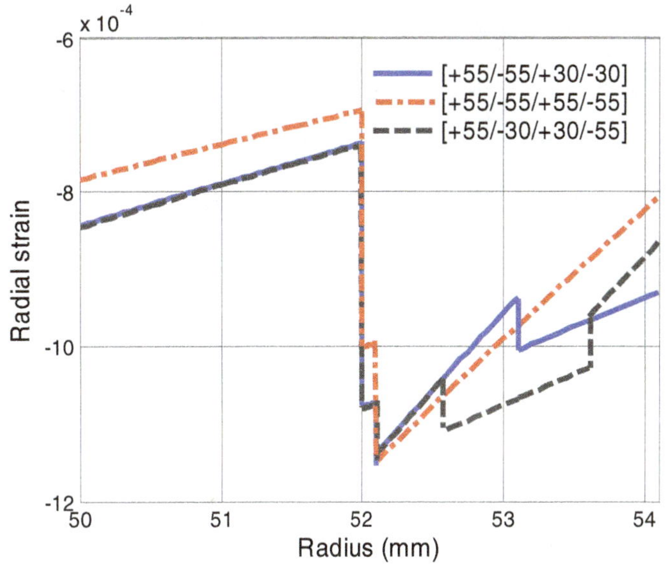

Fig. 6. The distribution of radial strain through the thickness.

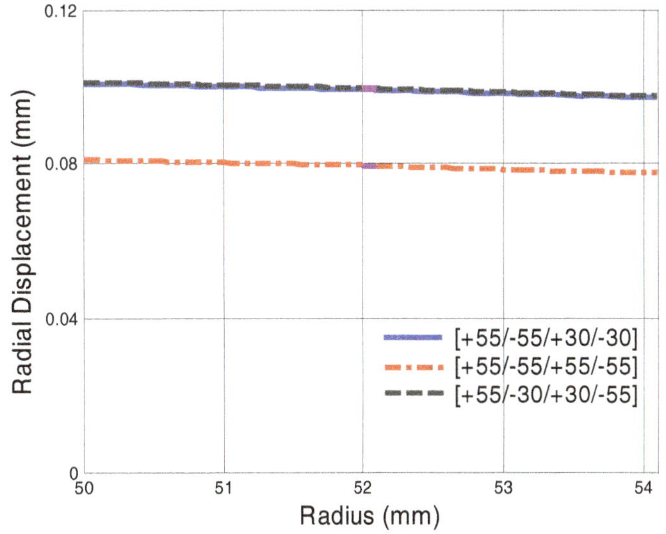

Fig. 7. The distribution of radial displacement through the thickness.

3.2 Strain analysis

➢ Hoop Strain:
Figure 5 shows the hoop strain through the radius of the cylindrical part of the hybrid solution. All hoop stain for the three sequences is characterised by continuous variation from the internal to the external wall. In addition the sequence Seq1 has the *lowest values* of the hoop strain compared with the Seq2 and Seq3.

➢ Radial strain:
Figure 6 shows the radial strain through the radius of the cylindrical part of the hybrid solution. All radial strain for the three sequences is characterised by discontinuous variation at the interfaces from the internal to the external wall. In the part of composite, the sequence Seq2 has the *lowest values* of the hoop strain compared with the Seq1 and Seq3.

3.2.1 Radial displacement

The variation of radial displacements Ur for the three stacking sequences is shown in Figure 7, where a similar trend for all sequence is observed and the maximum displacements are recorded at the internal wall. Starting from the internal wall, this variation decreases gradually to the minimum value at the external wall. The values of the radial displacement are larger for Seq2 and Seq3 compared to Seq1.

4 Conclusion

The study presents an elastic analytical modelling of a multilayer composite cylindrical tank coated on an aluminium liner and intermetallic for three stacking sequences: [+55/−55/+30/−30], [+55/−55/+55/−55] and

[+55/−30/+30/−55]. This analysis has shown the behaviour of hybrid solution depend strongly on the stacking sequences of the composite. The stacking sequence [+55/−55/+55/−55] presents the best results in term of stresses and displacements compared to other sequences.

The taking into account of the elasto-plastic behavior of the liner will be treated in second part, which is almost completed.

References

1. M. Latroche, *Le stockage de l'hydrogène: Du gaz au solide en passant par le liquide UPR 209* (LCMTR, ISCSA, CNRS, 2-8, rue Henri Dunant, 94320 Thiais, France)
2. M. Junker, L. Bocquet, M. Bendif, D. Karboviac. Ann. Chim. Sci. Mat. **26**, 117 (2001)
3. V.E. Verijenco, S. Adali, Y.P. Tabakov. Composite Structures **54**, 371 (2001)
4. V.V. Vasiliev, A.A. Krinakov, A.F. Razin. Composite Structure **62**, 449 (2003)
5. M. Xia, M. Takayanagi, K. Kemmochi. Composites structures **53**, 483 (2001)
6. D. Chapelle, D. Perreux. International Journal of Hydrogen Energy **31**, 627 (2006)
7. N. Takeichia, H. Senoha, T. Yokotab, H. Tsurutab, K. Hamadab, H.T. Takeshitac, H. Tanakaa, T. Kiyobayashia, T. Takanod, N. Kuriyamaa. International Journal of Hydrogen Energy **28**, 1121 (2003)
8. J.-M. Berthelot, *Composite Materials, Mechanical Behavior and Structural Analysis* (Springer, New York, 1999), p. 676

27

Current trends in evolutionary multi-objective optimization

Kalyanmoy Deb[a,b]
Department of Mechanical Engineering, Indian Institute of Technology Kanpur, PIN 208016, India

Abstract – In a short span of about 14 years, evolutionary multi-objective optimization (EMO) has established itself as a mature field of research and application with an extensive literature, many commercial softwares, numerous freely downloadable codes, a dedicated biannual conference running successfully four times so far since 2001, special sessions and workshops held at all major evolutionary computing conferences, and full-time researchers from universities and industries from all around the globe. In this paper, we make a brief outline of EMO principles, some EMO algorithms, and focus on current research and application potential of EMO. Besides, simply finding a set of Pareto-optimal solutions, EMO research has now diversified in hybridizing its search with multi-criterion decision-making tools to arrive at a single preferred solution, in utilizing EMO principle in solving different kinds of single-objective optimization problems efficiently, and in various interesting application domains which were not possible to be solved adequately due to the lack of a suitable solution technique.

1 Introduction

Evolutionary optimization (EO) methodologies are now popularly used in various problem solving tasks involving nonlinearities, large dimensionality, non-differentiable functions, non-convexity, multiple optima, multiple objectives, uncertainties in decision and problem parameters, large computational overheads, and various other complexities for which the classical optimization methodologies are known to be vulnerable. In a recent survey conducted before the World Congress on Computational Intelligence (WCCI) in Vancouver 2006, the evolutionary multi-objective optimization (EMO) field was judged as one of the three fastest growing field of research and application among all computational intelligence topics. For solving single-objective optimization problems or in other tasks focusing to find a single optimal solution, the use of a population of solutions in each iteration may at first seem like an overkill; in solving multi-objective optimization problems an EO procedure is a perfect fit [13]. The multi-objective optimization problems, by theory, give rise to a set of Pareto-optimal solutions which need a further processing to arrive at a single preferred solution. To achieve the first task, it becomes quite a natural proposition to use a modified EO, because the use of population in an iteration helps an EO to simultaneously find multiple Pareto-optimal solutions in a single simulation run.

In this paper, we briefly describe the principles of EMO and then discuss a few well-known computational procedures. Thereafter, we highlight the current interest of research and application of EMO. It is clear from the discussions that EMO is not only being found to be useful in solving multi-objective optimization problems, it is also helping to solve other kinds of optimization problems in a manner better than they are traditionally solved. As a by-product, EMO-based solutions are helping to reveal important hidden knowledge about a problem – a matter which is difficult to achieve otherwise. This paper should motivate interested readers to look into the extensive EMO literature indicated in the reference list and in appendix for more details.

[a] Corresponding author: deb@iitk.ac.in
[b] Currently visiting Helsinki School of Economics as a Finland Distinguished Professor (Email: Kalyanmoy.Deb@hse.fi).

2 Evolutionary Multi-objective Optimization (EMO)

As in the single-objective optimization problem, the multi-objective optimization problem usually has a number of constraints which any feasible solution (including the optimal solution) must satisfy:

$$\left.\begin{array}{lll} \text{Minimize/Maximize} & f_m(\mathbf{x}), & m = 1, 2, \ldots, M; \\ \text{subject to} & g_j(\mathbf{x}) \geq 0, & j = 1, 2, \ldots, J; \\ & h_k(\mathbf{x}) = 0, & k = 1, 2, \ldots, K; \\ & x_i^{(L)} \leq x_i \leq x_i^{(U)}, & i = 1, 2, \ldots, n. \end{array}\right\} \tag{1}$$

A solution $\mathbf{x}$ is a vector of n decision variables: $\mathbf{x} = (x_1, x_2, \ldots, x_n)^T$. The last set of constraints are called variable bounds, restricting each decision variable x_i to take a value within a lower $x_i^{(L)}$ and an upper $x_i^{(U)}$ bound.

Different solutions may produce trade-offs (conflicting scenarios) among different objectives. A solution that is extreme (in a better sense) with respect to one objective requires a compromise in other objectives. This prohibits one to choose a solution which is optimal with respect to only one objective. This clearly introduces two main goals of multi-objective optimization:

1. Find a set of solutions close to the optimal solutions, and
2. Find a set of solutions which are diverse enough to represent the true spread of optimal solutions.

Evolutionary multi-objective optimization (EMO) algorithms follow both the above principles more directly than their existing counterparts.

From a practical standpoint, a user needs only one solution, no matter whether the associated optimization problem is single-objective or multi-objective. In the case of multi-objective optimization, the user is now in a dilemma. Since a number of solutions are optimal, the obvious question arises: Which of these optimal solutions must one choose? This is not an easy question to answer. It involves many higher-level information which are often non-technical, qualitative and experience-driven. However, if a set of many trade-off solutions are already worked out or available, one can evaluate the pros and cons of each of these solutions based on all such non-technical and qualitative, yet still important, considerations and compare them to make a choice. Thus, in a multi-objective optimization, ideally the effort must be made in finding the set of trade-off optimal solutions by considering all objectives to be important. After a set of such trade-off solutions are found, a user can then use higher-level qualitative considerations to make a choice. In view of these discussions, the following principles are used in evolutionary multi-objective optimization (EMO) procedures:

Step 1 Find multiple trade-off optimal solutions with a wide range of values for objectives.

Step 2 Choose one of the obtained solutions using higher-level information.

3 State-of-the-art EMO methodologies

A number of non-elitist EMO methodologies [23, 26, 39] gave a good head-start to the research and application of EMO, but suffered from the fact that they did not use an important operator – an elite-preservation mechanism – in their procedures. An addition of elitism in an EO provides a monotonically non-degrading performance. The next-level EMO algorithms implemented an elite-preserving operator in different ways and gave birth to elitist EMO procedures, some of which we describe in the following subsections.

3.1 Elitist non-dominated sorting GA or NSGA-II

The NSGA-II procedure [14] for finding multiple Pareto-optimal solutions in a multi-objective optimization problem has the following three features: (i) it uses an elitist principle, (ii) it uses an explicit diversity preserving mechanism, and (iii) it emphasizes non-dominated solutions. In NSGA-II, the offspring population Q_t is first created by using the parent population P_t and the usual evolutionary operators. Thereafter, the two populations are combined together to form R_t of size $2N$. Then, a non-dominated sorting is used to classify the entire population R_t. Once the non-dominated sorting is over, the new population is filled by solutions of different non-dominated fronts, one at a time. The filling starts with the best non-dominated front and continues with solutions of the second non-dominated front, followed by the third non-dominated front, and so on. Since the overall population size of R_t is $2N$, not all fronts can be accommodated in N slots available in the new population. All fronts which could not be accommodated are simply deleted. When the last allowed front is being considered, there may exist more solutions in the last front than the remaining slots in the new population. This scenario is illustrated in Figure 1. Instead of arbitrarily discarding some members from the last front, the solutions which will make the diversity of the selected solutions the highest are chosen.

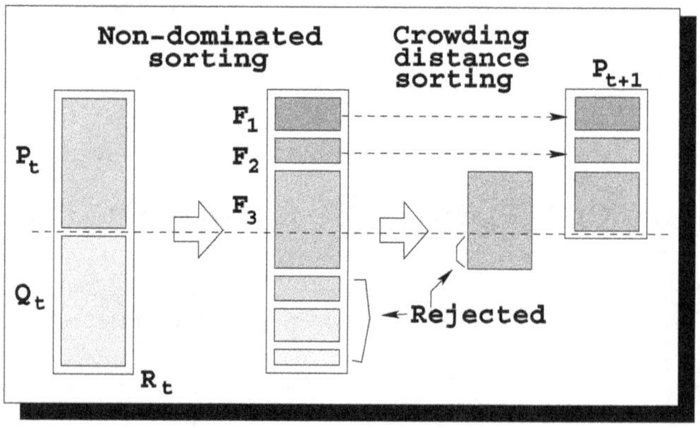

Fig. 1. Schematic of the NSGA-II procedure.

Due to the simplicity and efficient usage of its operators, NSGA-II procedure is probably the most popular EMO procedure today. A C code implementing the procedure is available from author's web site `http://www.iitk.ac.in/kangal/soft.htm`.

3.2 Strength Pareto EA (SPEA) and SPEA2

Zitzler and Thiele [44] suggested an elitist multi-criterion EA with the concept of non-domination in their strength Pareto EA (SPEA). They suggested maintaining an external population at every generation storing all non-dominated solutions discovered so far beginning from the initial population. This external population participates in evolutionary operations. At each generation, a combined population with the external and the current population is first constructed. All non-dominated solutions in the combined population are assigned a fitness based on the number of solutions they dominate. To maintain diversity and in the context of minimizing the fitness function, they assigned a higher fitness value to a non-dominated solution having more dominated solutions in the combined population. On the other hand, a higher fitness is also assigned to solutions dominated by more solutions in the combined population. Care is taken to assign no non-dominated solution a fitness worse than that of the best dominated solution. This assignment of fitness makes sure that the search is directed towards the non-dominated solutions and simultaneously diversity among dominated and non-dominated solution are maintained. On knapsack problems, they have reported better results than other methods used in that study.

In their subsequent improved version (SPEA2) [43], three changes have been made. First, the archive size is always kept fixed (thus if there are fewer non-dominated solutions in the archive than the predefined archive size, dominated solutions from the EA population are copied to the archive). Second, a fine-grained fitness assignment strategy is used in which fitness assignment to the dominated solutions are slightly different and a density information is used to resolve the tie between solutions having identical fitness values. Third, a modified clustering algorithm is used with k-th nearest neighbor distance measure and special attention is made to preserve the boundary elements.

3.3 Pareto Archived ES (PAES) and Pareto Envelope based Selection Algorithms (PESA and PESA2)

Knowles and Corne [28] suggested a simple possible EMO using evolution strategy (ES). In their Pareto-archived ES (PAES) with one parent and one child, the child is compared with respect to the parent. If the child dominates the parent, the child is accepted as the next parent and the iteration continues. On the other hand, if the parent dominates the child, the child is discarded and a new mutated solution (a new child) is found. However, if the child and the parent do not dominate each other, the choice of child or a parent considers the second task of keeping diversity among obtained solutions using a crowding procedure. To maintain diversity, an archive of non-dominated solutions found so far is maintained. The child is compared with the archive to check if it dominates any member of the archive. If yes, the child is accepted as the new parent and the dominated solution is eliminated from the archive. If the child does not dominate any member of the archive, both parent and child are checked for their *nearness* with the solutions of the archive. If the child resides in a least crowded region in the parameter space among the members of the archive, it is accepted as a parent and a copy of added to the archive. It is interesting to note that both features of (i) emphasizing non-dominated solutions, and (ii) maintaining diversity among non-dominated solutions are present in this simple algorithm. Later, they suggested a multi-parent PAES with similar principles as above.

In their subsequent version, they called Pareto Envelope based Selection Algorithm (PESA) [9], they combined good aspects of SPEA and PAES. Like SPEA, PESA carries two populations (an EA population and a comparatively larger archive population). Non-domination and the PAES's crowding concept is used to update the archive with the newly created child solutions.

In an extended version of PESA (or PESA2) [10], instead of applying the selection procedure on population members, hyperboxes in the objective space are selected based on the number of residing solutions in the hyperboxes. After hyperboxes are selected, a random solution from the chosen hyperboxes is kept. This region-based selection procedure has shown to perform better than individual-based selection procedure of PESA. In some sense, PESA2 selection scheme is similar in concept to the ϵ-dominance in which predefined ϵ values determine the hyperbox dimensions. Other ϵ-dominance based EMO procedures [18,30] have shown to be computationally faster and to better distribute solutions than NSGA-II or SPEA2.

There also exist other competent EMOs, such as multi-objective messy GA (MOMGA) [41], multi-objective micro-GA [6], neighborhood constraint GA [31], ARMOGA [36], and others. Besides, there exists other EA based methodologies, such as particle swarm EMO [7,35], ant-based EMO [24,33], and differential evolution based EMO [1].

4 Constraint handling in EMO

The constraint handling method modifies the domination principle. In the presence of constraints, each solution can be either feasible or infeasible. Thus, there may be at most three situations: (i) both solutions are feasible, (ii) one is feasible and other is not, and (iii) both are infeasible. We consider each case by simply redefining the domination principle as follows. A solution $\mathbf{x}^{(i)}$ is said to 'constrain-dominate' a solution $\mathbf{x}^{(j)}$, if any of the following conditions are true: (i) solution $\mathbf{x}^{(i)}$ is feasible and solution $\mathbf{x}^{(j)}$ is not, or (ii) solutions $\mathbf{x}^{(i)}$ and $\mathbf{x}^{(j)}$ are both infeasible, but solution $\mathbf{x}^{(i)}$ has a smaller constraint violation, or (iii) solutions $\mathbf{x}^{(i)}$ and $\mathbf{x}^{(j)}$ are feasible and solution $\mathbf{x}^{(i)}$ dominates solution $\mathbf{x}^{(j)}$ in the usual sense. The above change in the definition requires a minimal change in the NSGA-II or other EMO procedures described earlier. Figure 2 shows the non-dominated fronts on a six-membered population due to the introduction of two constraints. In the absence of the constraints, the non-dominated fronts (shown by dashed lines) would have been `((1,3,5), (2,6), (4))`, but in their presence, the new fronts are `((4,5), (6), (2), (1), (3))`. The first non-dominated front is constituted with the best infeasible solutions in the population and any feasible solution lies on a better non-dominated front than an feasible solution. This simple modification in domination principle allows to form an appropriate hierarchy among solutions in the population for them to move towards the true constrained Pareto-optimal front.

5 Current EMO research and practices

With the fast development of efficient EMO procedures, availability of free and commercial softwares (some of

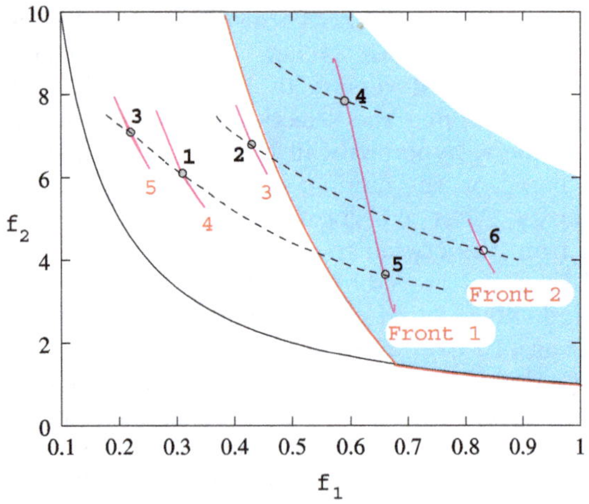

Fig. 2. Non-constrain-dominated fronts.

which are described in the Appendix), and applications to a wide variety of problems, EMO research and application is in its peak at the current time. It is difficult to cover every aspect of current research in a single paper. Here, we outline three main broad areas of research and application.

5.1 EMO and decision-making

It is important here to note that finding a set of Pareto-optimal solutions by using an EMO is only a part of multi-objective optimization, as choosing a particular solution for implementation is the remaining decision-making task which is also an equally important task. In the view of the author, the decision-making task can be considered from two main aspects:

1. **Generic consideration:** There are some aspects which most practical users would like to use in narrowing down their choice. For example, in the presence of uncertainties in decision variables and/or problem parameters, the users are usually interested in finding *robust* solutions which are relatively insensitive to the variation in decision variables or parameters. In the presence of such variations, no one is interested in Pareto-optimal but sensitive solutions. Practitioners do not hesitate sacrificing globally optimal solutions in order to achieve robust solutions which lie on relatively flat part of the objective function or away from the constraint boundaries. In such scenarios, instead of finding the globally Pareto-optimal front, the user may be interested in finding the robust frontier which may be partially or totally different from the true Pareto-optimal front. A robust EMO procedure [15] is shown to find only those Pareto-optimal solutions or solutions close to them which are robust and insensitive to parameter and variable perturbations.
   In addition, instead of finding the entire Pareto-optimal front, the users may be interested in finding some specific solutions on the Pareto-optimal front, such as *knee* points (requiring a large sacrifice in at least one objective to achieve a small gain in another), points having correlated relationship between objectives to decision variables (identifying the portion of the Pareto-optimal front which possesses simpler relationship between objective values and decision variable values), points having multiplicity (finding Pareto-optimal solutions corresponding to multiple (say at least two or more) decision variable vectors but each having identical objective values), points for which decision variable values are well within their allowable bounds and not near their lower or upper boundaries, points having some theoretical aspects such as all Lagrange multipliers having more or less identical magnitube (condition for having equal importance to each constraint), and others. These considerations are motivated from the fundamental and practical aspects of optimization and may be applied to most multi-objective problem solving tasks, without any consent of a decision-maker.
2. **Subjective consideration:** In this category, any problem-specific information can be used to narrow down the choices and the process may even lead to a single preferred solution at the end. Most decision-making procedures use some preference information (utility functions, reference point approaches [42], reference direction approaches [29], marginal rate of return and a host of other considerations [34]) to select a subset of Pareto-optimal solutions. A recent book is dedicated to the discussions of many such multiple criteria decision making (MCDM) tools and collaborative suggestions of using EMO with such MCDM tools [4]. Some hybrid EMO and MCDM algorithms are also suggested in the recent past [16,17,22,32,40].

5.2 Multi-objectivization

Interestingly, the act of finding multiple Pareto-optimal solutions using an EMO procedure has found its application outside the realm of solving multi-objective optimization problems per se. The concept of finding optimal trade-off solutions are applied to solve other kinds of optimization problems as well. For example, the EMO concept is used to solve constrained single-objective optimization problems by converting the task as a two-objective optimization task of additionally minimizing an aggregate constraint violation [8]. This eliminated the need of having any penalty parameter. If viewed this way, the usual penalty function approach used in the classical optimization studies is a special weighted-sum approach to the bi-objective optimization problem of minimizing the objective function and minimizing the constraint violation, for which the weight vector is a function of penalty parameter. The reduction of a well-known difficulty in the genetic programming studies, called the 'bloating', by minimizing the size of programs as an additional objective helped find high-performing solutions with a smaller size of the code [2]. Minimizing the intra-cluster distance and maximizing inter-cluster distance simultaneously in a bi-objective formulation of a clustering problem is found to yield better solutions than the usual single-objective minimization of the ratio of the intra-cluster distance to the inter-cluster distance [25]. A recent edited book [27] describes many such interesting applications in which EMO methodologies have help shown problems which are otherwise (or traditionally) not treated as multi-objective optimization problems.

5.3 Applications

EMO methodologies are being and must be applied to more interesting real-world problems to demonstrate the utility of finding multiple trade-off solutions. Although some recent studies are finding that EMO procedures are not computationally efficient to find multiple and widely distributed set of solutions on problems having a large number of objectives (more than five objectives or so) [11, 20], EMO procedures are still applicable in very large problems if the attention is changed to find a preferred region on the Pareto-optimal front, instead of the complete front. Some such preference based EMO studies [3, 16, 22] are applied to 10 or more objectives. In certain many-objective problems, the Pareto-optimal front can be low-dimensional mainly due to the presence of redundant objectives and EMO procedures can again be effective in solving such problems [5, 20, 37]. In addition, the use of reliability based EMO [12, 19] and robust EMO [15] procedures are ready to be applied to real-world multi-objective design optimization problems. Application studies are also of interest from the point of demonstrating how an EMO procedure and a subsequent MCDM approach can be combined in an iterative manner together to solve a multi-objective optimization problem. Such efforts may lead to development of GUI-based softwares and approaches for solving the task and will require addressing other important issues such as visualization of multi-dimensional data, parallel implementation of EMO and MCDM procedures, meta-modeling approaches, and others.

Besides solving real-world multi-objective optimization problems, EMO procedures are also found to be useful for a knowledge discovery task related to a better understanding of the problem. After a set of trade-off solutions are found by an EMO, these solutions can be compared against each other to unveil interesting principles which lie common to all these solutions. These common properties among high-performing solutions will provide useful insights about what makes a solution optimal in a particular problem. Also, investigating what properties are not common may provide information about what makes the solutions to differ from each other. Such useful information mined from the obtained EMO trade-off solutions were discovered in many real-world engineering design problems in the recent past [21] and the task of unveiling useful knowledge about a problem is termed as the task of 'innovization' in that study. The task is similar in concept to the product platform design task which uses multiple optimizations and statistical tools [38], but provides a systematic and direct mean of discovering common principles through optimization.

6 Conclusions

This paper has provided a brief introduction and current research focus to a fast-growing field of multi-objective optimization based on evolutionary algorithms. The EMO principle of solving multi-objective optimization has been to first find a set of Pareto-optimal solutions and then choose a preferred solution. Since an EO uses a population of solutions in each iteration, EO procedures are potentially viable techniques to find and capture a number of Pareto-optimal solutions in a single simulation run.

Besides their routine applications in solving multi-objective optimization problems, EMO has spread its wings in aiding other types of optimization problems, such as single-objective constrained optimization, clustering problems etc. EMO has been used to unveil important hidden knowledge about what makes a solution optimal. EMO techniques are increasingly being found to have tremendous potential to be used in conjunction with multiple criteria decision making (MCDM) tasks in not only finding a set of optimal solutions but also to aid in selecting a preferred solution at the end. In the reference and appendix sections, we provide some useful information about the EMO research, so that interested readers can get exposed to and involved with the vast literature of EMO.

References

1. B.V. Babu, M. Mathew Leenus Jehan, Differential Evolution for Multi-Objective Optimization. In *Proceedings of the 2003 Congress on Evolutionary Computation (CEC'2003)*, volume 4, pages 2696–2703, Canberra, Australia (December 2003). IEEE Press.
2. S. Bleuler, M. Brack, E. Zitzler, Multiobjective genetic programming: Reducing bloat using spea2. In *Proceedings of the 2001 Congress on Evolutionary Computation*, pages 536–543 (2001).
3. J. Branke, K. Deb, Integrating user preferences into evolutionary multi-objective optimization. In Y. Jin, editor, *Knowledge Incorporation in Evolutionary Computation*, pages 461–477. Hiedelberg, Germany: Springer (2004).
4. J. Branke, K. Deb, K. Mietinnen, R. Slowinski, *Multiobjective optimization: Interactive and evolutionary approaches.* Springer-Verlag (in press).
5. D. Brockhoff, E. Zitzler, Dimensionality Reduction in Multiobjective Optimization: The Minimum Objective Subset Problem. In K.H. Waldmann and U. M. Stocker, editors, *Operations Research Proceedings 2006*, pages 423–429. Springer (2007).
6. C.A.C. Coello, G. Toscano, *A micro-genetic algorithm for multi-objective optimization. Technical Report Lania-RI-2000-06*, Laboratoria Nacional de Informatica Avanzada, Xalapa, Veracruz, Mexico (2000).
7. C.A.C. Coello, M.S. Lechuga, MOPSO: *A Proposal for Multiple Objective Particle Swarm Optimization.* In *Congress on Evolutionary Computation (CEC'2002)*, volume 2, pages 1051–1056, Piscataway, New Jersey (May 2002). IEEE Service Center.

8. C.A.C. Coello, *Treating objectives as constraints for single objective optimization.* Engineering Optimization **32**, 275–308 (2000)
9. D. Corne, J. Knowles, M. Oates, *The Pareto envelope-based selection algorithm for multiobjective optimization,* In *Proceedings of the Sixth International Conference on Parallel Problem Solving from Nature VI (PPSN-VI),* pages 839–848 (2000).
10. D.W. Corne, N.R. Jerram, J.D. Knowles, M.J. Oates, *PESA-II: Region-based selection in evolutionary multiobjective optimization,* In *Proceedings of the Genetic and Evolutionary Computation Conference (GECCO-2001),* pages 283–290. San Mateo, CA: Morgan Kaufmann Publishers (2001).
11. D.W. Corne, J.D. Knowles, *Techniques for highly multiobjective optimisation: some nondominated points are better than others,* In *GECCO'07: Proceedings of the 9th annual conference on Genetic and evolutionary computation,* pages 773–780, New York, NY, USA (2007). ACM Press.
12. D. Daum, K. Deb, J. Branke, *Reliability-based optimization for multiple constraints with evolutionary algorithms,* In *Proceedings of the Congress on Evolutionary Computation (CEC-2007),* in press.
13. K. Deb, *Multi-objective optimization using evolutionary algorithms.* Chichester, UK: Wiley (2001).
14. K. Deb, S. Agrawal, A. Pratap, T. Meyarivan, *A fast and elitist multi-objective genetic algorithm: NSGA-II.* IEEE Transactions on Evolutionary Computation **6**, 182–197 (2002)
15. K. Deb, H. Gupta, Searching for robust Pareto-optimal solutions in multi-objective optimization, In *Proceedings of the Third Evolutionary Multi-Criteria Optimization (EMO-05) Conference (Also Lecture Notes on Computer Science 3410),* pages 150–164 (2005).
16. K. Deb, A. Kumar, *Interactive evolutionary multi-objective optimization and decision-making using reference direction method,* In *Proceedings of the Genetic and Evolutionary Computation Conference (GECCO-2007),* pages 781–788. New York: The Association of Computing Machinery (ACM), 2007.
17. K. Deb, A. Kumar, *Light beam search based multi-objective optimization using evolutionary algorithms,* Technical Report KanGAL Report No. 2007005, Indian Institute of Technology Kanpur, India, 2007.
18. K. Deb, M. Mohan, S. Mishra, *Towards a quick computation of well-spread pareto-optimal solutions.* In *Proceedings of the Second Evolutionary Multi-Criterion Optimization (EMO-03) Conference (LNCS 2632),* pages 222–236 (2003).
19. K. Deb, D. Padmanabhan, S. Gupta, A.K. Mall, Reliability-based multi-objective optimization using evolutionary algorithms. In *Proceedings of the Fourth International Conference on Evolutionary Multi-Criterion Optimization (EMO-2007) (LNCS, Springer),* pages 66–80 (2007).
20. K. Deb, D. Saxena, *Searching for pareto-optimal solutions through dimensionality reduction for certain large-dimensional multi-objective optimization problems.* In *Proceedings of the World Congress on Computational Intelligence (WCCI-2006),* pages 3352–3360 (2006).
21. K. Deb, A. Srinivasan, *Innovization: Innovating design principles through optimization.* In *Proceedings of the Genetic and Evolutionary Computation Conference (GECCO-2006),* pages 1629–1636, New York: The Association of Computing Machinery (ACM) (2006).
22. K. Deb, J. Sundar, N. Uday, S. Chaudhuri, *Reference point based multi-objective optimization using evolutionary algorithms.* International Journal of Computational Intelligence Research **2**, 273–286 (2006)
23. C.M. Fonseca, P.J. Fleming, *Genetic algorithms for multiobjective optimization: Formulation, discussion, and generalization,* In *Proceedings of the Fifth International Conference on Genetic Algorithms,* pages 416–423 (1993).
24. M. Gravel, W.L. Price, C. Gagné, *Scheduling continuous casting of aluminum using a multiple objective ant colony optimization metaheuristic.* Eur. J. Oper. Res. **143**, 218–229 (2002)
25. J. Handl, J. Knowles, *An evolutionary approach to multiobjective clustering.* IEEE T. Evolut. Comput. **11**, 56–76 (2007)
26. J. Horn, N. Nafploitis, D.E. Goldberg, *A niched Pareto genetic algorithm for multi-objective optimization,* In *Proceedings of the First IEEE Conference on Evolutionary Computation,* pages 82–87 (1994).
27. J. Knowles, D. Corne, K. Deb, *Multiobjective problem solving from nature,* Springer Natural Computing Series, Springer-Verlag (in press).
28. J.D. Knowles, D.W. Corne, *Approximating the nondominated front using the Pareto archived evolution strategy.* Evol. Comput. **8**, 149–172 (2000)
29. P. Korhonen, J. Laakso, A visual interactive method for solving the multiple criteria problem. Eur. J. Oper. Res. **24**, 277–287 (1986)
30. M. Laumanns, L. Thiele, K. Deb, E. Zitzler, *Combining convergence and diversity in evolutionary multi-objective optimization.* Evol. Comput. **10**, 263–282 (2002)
31. D.H. Loughlin, S. Ranjithan, *The neighborhood constraint method: A multiobjective optimization technique,* In *Proceedings of the Seventh International Conference on Genetic Algorithms,* pages 666–673 (1997).
32. M. Luque, K. Miettinen, P. Eskelinen, F. Ruiz, *Three different ways for incorporating preference information in interactive reference point based methods,* Technical Report W-410, Helsinki School of Economics, Helsinki, Finland (2006).
33. P.R. McMullen, *An ant colony optimization approach to addessing a JIT sequencing problem with multiple objectives.* Artificial Intelligence in Engineering **15**, 309–317 (2001)
34. K. Miettinen, *Nonlinear Multiobjective Optimization,* Kluwer, Boston (1999).
35. S. Mostaghim, J. Teich, *Strategies for Finding Good Local Guides in Multi-objective Particle Swarm Optimization (MOPSO),* In *2003 IEEE Swarm Intelligence Symposium Proceedings,* pages 26–33, Indianapolis, Indiana, USA (April 2003). IEEE Service Center.
36. D. Sasaki, M. Morikawa, S. Obayashi, K. Nakahashi, *Aerodynamic shape optimization of supersonic wings by adaptive range multiobjective genetic algorithms,* In *Proceedings of the First International Conference on Evolutionary Multi-Criterion Optimization (EMO 2001),* pages 639–652 (2001).
37. D. Saxena, K. Deb, *Trading on infeasibility by exploiting constraint's criticality through multi-objectivization: A system design perspective,* In *Proceedings of the Congress on Evolutionary Computation (CEC-2007),* in press.

38. C. Seepersad, F. Mistree, J.K. Allen, *A quantitative approach for designing multiple product platforms for an evolution portfolio of products*, In *Proceedings of ASME Design Engineering Technical Conferences*, pages 593–602 (2002).
39. N. Srinivas, K. Deb, *Multi-objective function optimization using non-dominated sorting genetic algorithms*. Evol. Comput. **2**, 221–248 (1994)
40. L. Thiele, K. Miettinen, P. Korhonen, J. Molina, *A preference-based interactive evolutionary algorithm for multiobjective optimization*, Technical Report Working Paper Number W-412, Helsingin School of Economics, Helsingin Kauppakorkeakoulu, Finland (2007).
41. D. Van Veldhuizen, G.B. Lamont, *Multiobjective evolutionary algorithms: Analyzing the state-of-the-art*. Evol. Comput. **8**, 125–148 (2000)
42. A.P. Wierzbicki, *The use of reference objectives in multiobjective optimization*, In G. Fandel and T. Gal, editors, *Multiple Criteria Decision Making Theory and Applications*, pages 468–486. Berlin: Springer-Verlag (1980).
43. E. Zitzler, M. Laumanns, L. Thiele, *SPEA2: Improving the strength pareto evolutionary algorithm for multi-objective optimization*, In K.C. Giannakoglou, D.T. Tsahalis, J. Périaux, K.D. Papailiou, and T. Fogarty, editors, *Evolutionary Methods for Design Optimization and Control with Applications to Industrial Problems*, pages 95–100, Athens, Greece, 2001. International Center for Numerical Methods in Engineering (Cmine).
44. E. Zitzler, L. Thiele, *Multiobjective optimization using evolutionary algorithms – A comparative case study*, In *Parallel Problem Solving from Nature V (PPSN-V)*, pages 292–301 (1998).

Appendix: EMO Repository

Here, we outline some dedicated literature in the area of multi-objective optimization. Further details can be found in the reference section.

Books in Print

- A. Abraham, L.C. Jain, R. Goldberg. *Evolutionary Multiobjective Optimization: Theoretical Advances and Applications*, London: Springer-Verlag, 2005.
  This is a collection of the latest state-of-the-art theoretical research, design challenges and applications in the field of EMO.
- C.A.C. Coello, D.A. VanVeldhuizen, G. Lamont. *Evolutionary Algorithms for Solving Multi-Objective Problems*. Boston, MA: Kluwer Academic Publishers, 2002.
  A good reference book with a good citation of most EMO studies. A revised version is in print.
- K. Deb. *Multi-objective optimization using evolutionary algorithms*. Chichester, UK: Wiley, 2001. (Third edition, with exercise problems)
  A comprehensive text-book introducing the EMO field and describing major EMO methodologies and salient research directions.
- A. Osyczka. *Evolutionary algorithms for single and multicriteria design optimisation*, Heidelberg: Physica-Verlag, 2002.
  A book describing single and multi-objective EAs with many engineering applications.
- Y. Collette and P. Siarry. *Multiobjective Optimization: Principles and Case Studies*, Berlin: Springer, 2004.
  This book describes multi-objective optimization methods including EMO and decision-making, and a number of engineering case studies.
- N. Nedjah and L. de Macedo Mourelle (Eds.). *Real-World Multi-Objective System Engineering*, New York: Nova Science Publishers, 2005.
  This edited book discusses recent developments and application of multi-objective optimization including EMO.
- M. Sakawa. *Genetic Algorithms and Fuzzy Multiobjective Optimization*, Norwell, MA: Kluwer, 2002.
  This book discusses EMO for 0-1 programming, integer programming, nonconvex programming, and job-shop scheduling problems under multiobjectiveness and fuzziness.
- K.C. Tan, E.F. Khor, T.H. Lee. *Multiobjective Evolutionary Algorithms and Applications*, London, UK: Springer-Verlag, 2005.
  A book on various methods of preference-based EMO and application case studies covering areas such as control and scheduling.

Some Review Papers

- C.A.C. Coello. Evolutionary Multi-Objective Optimization: A Critical Review. In R. Sarker, M. Mohammadian and X. Yao (Eds), *Evolutionary Optimization*, pp. 117–146, Kluwer Academic Publishers, New York, 2002.
- C. Dimopoulos. A Review of Evolutionary Multiobjective Optimization Applications in the Area of Production Research. *2004 Congress on Evolutionary Computation (CEC'2004)*, IEEE Service Center, Vol. 2, pp. 1487–1494, 2004.
- S. Huband, P. Hingston, L. Barone and L. While. A Review of Multiobjective Test Problems and a Scalable Test Problem Toolkit, *IEEE Transactions on Evolutionary Computation*, Vol. 10, No. 5, pp. 477–506, 2006.

Dedicated Conference Proceedings

- S. Obayashi, K. Deb, C. Poloni, and T. Hiroyasu (Eds.)., *Evolutionary Multi-Criterion Optimization (EMO-07) Conference Proceedings*, Also available as LNCS 4403. Berlin, Germany: Springer, 2007.
- C.A.C. Coello, A.H. Aguirre, E. Zitzler (Eds.), *Evolutionary Multi-Criterion Optimization (EMO-05) Conference Proceedings*, Also available as LNCS 3410. Berlin, Germany: Springer, 2005.
- Fonseca, C., Fleming, F., Zitzler, E., Deb, K., and Thiele, L. (Eds.), *Evolutionary Multi-Criterion Optimization (EMO-03) Conference Proceedings*. Also available as LNCS 2632. Heidelberg: Springer, 2003.

- Zitzler, E., Deb, K., Thiele, L., Coello, C. A. C. and Corne, D. (Eds.), *Evolutionary Multi-Criterion Optimization (EMO-01) Conference Proceedings*, Also available as LNCS 1993. Heidelberg: Springer, 2001.
- GECCO (LNCS series of Springer) and CEC (IEEE Press) annual conference proceedings feature numerous research papers on EMO theory, implementation, and applications.

Mailing lists

- emo-list@ualg.pt (EMO methodologies)
- MCRIT-L@LISTSERV.UGA.EDU (MCDM methodologies)

Public-domain source codes

- NIMBUS: `http://www.mit.jyu.fi/MCDM/soft.html`
- NSGA-II in C: `http://www.iitk.ac.in/kangal/soft.htm+`
- PISA: `http://www.tik.ee.ethz.ch/sop/pisa/`
- SPEA2 in C++: `http://www.tik.ee.ethz.ch/zitzler+`
- shark in C++: `http://shark-project.sourceforge.ne`
- Further information: `http://www.lania.mx/ccoello/EMOO/+`

Commercial codes implementing EMO

- iSIGHT and FIPER from Engineous (`http://www.engineous.com/`)
- GEATbx in Matlab (`http://www.geatbx.com/`)
- MAX from CENAERO (`http://www.cenaero.be/`)
- modeFRONTIER from Esteco (`http://www.esteco.com/`)

28

Reliability analysis based on gradient and heuristic optimization techniques of composite laminates using element-free Galerkin method

J.E. Rojas[1,2,a], A. El Hami and D.A. Rade[2]

[1] Institut National des Sciences Appliquées de Rouen, Laboratoire de Mécanique de Rouen, BP 08, Avenue de l'Université, 76801 Saint-Etienne du Rouvray, Cedex, France

[2] Federal University of Uberlândia, School of Mechanical Engineering, 2121, Av. Joao Naves de Avila, Campus Santa Mônica, CEP 38400-902 CP 593, Uberlândia, Brazil

Abstract – This work presents a reliability methodology that couples gradient and heuristic based reliability methods with element-free Galerkin method applied in composite laminates structures. First and second order reliability methods are the gradient based reliability methods. The heuristic based reliability method suggested by authors uses natured-inspired optimization method: genetic algorithms. Numerical examples on linear elasticity that involve random properties of laminated composite plates are presented in order to demonstrate the efficiency, capacity and robustness of the proposed methodology. The results show that the predicted reliability levels are accurate in comparison with similar approach that uses finite element analysis to evaluate implicit limit state functions.

Key words: Composite laminates; element-free Galerkin method; first order and second order reliability methods; heuristic based reliability method.

1 Introduction

Classical mesh-based methods, such as finite elements, depend clearly on their dependence on a mesh in the simulations of mechanical problems. In this sense, meshless methods have been proposed to eliminate the mesh dependent structure, of the classical mesh-based methods, by constructing the approximation entirely in terms of point, without element connectivity. From principles of ninety years, the meshless methods have experiment a fort computational development in solid mechanics. However, most of meshless methods have been focused on deterministic analysis [1]. In consequence, the research in meshless methods considering probabilistic analysis had not received much attention. The developed meshless method provides a rich, relatively unexplored area for research in computational structural reliability analysis.

Among meshless methods, element-free Galerkin (EFG) method is particularly appealing due to their simplicity and a formulation that corresponds to the well-established finite element method (FEM) [2]. The EFG method uses moving least squares approximation to construct shape functions based on a set of nodes scattered in the problem domain. The essential boundary conditions are imposed using their weak forms with Lagrange multipliers.

In this work the EFG method is used in structural reliability analysis for simulate laminated composite plates on linear elasticity. The reliability procedure proposed in this work, it is based on gradient and heuristic based reliability methods for estimate the reliability index. First and second order reliability methods (FORM and SORM, respectively) are the gradient based reliability methods which uses Newton-Raphson (N-R) procedure. The heuristic based reliability method (HBRM) suggested in [3] uses natured-inspired optimization methods: genetic algorithms (GA) [4,5], particle swarm optimization (PSO) [6,7] and ant colony optimization (ACO) [8,9]. In this work, GA it is used in numerical applications. When compared with FORM and SORM methods, HBRM it is different because not requires the initial guess and the computation of gradients of the limit state function because is based on multidirectional search.

The obtained results are compared with FEM solutions in order to evaluate the efficiency, capacity and robustness of the proposed methodology.

2 Stress-strain relations for plane stress in an orthotropic material

Consider the composite laminated plate shown in Figure 1 using a Cartesian coordinate system. The fibre orientation of a layer in indicated by θ as shown in Figure 1.

[a] Corresponding author: jhojan_enrique@yahoo.es

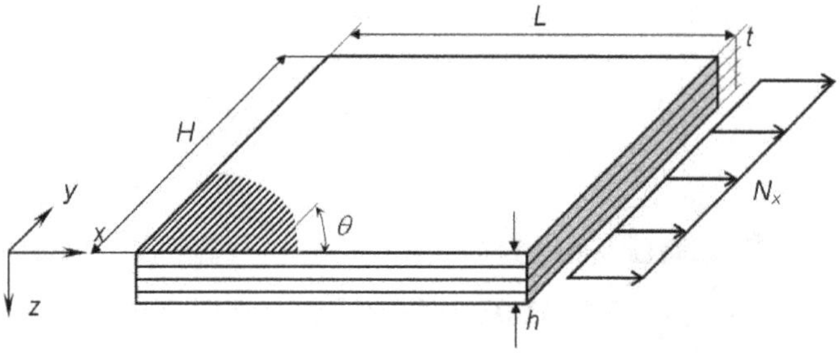

Fig. 1. Laminated composite plate and its fibre orientation.

The generalized Hooke's law for k layer relating stresses to strains can be expressed as follows [10]:

$$\{\sigma\}_k = \left[\bar{Q}\right]_k \{\varepsilon\}_k, \tag{1}$$

where $\{\sigma\}$ is the vector of stress components, $\left[\bar{Q}\right]$ is the stiffness matrix, and $\{\varepsilon\}$ is the strain components.

From classical lamination theory, based on Kirchhoff's assumptions, the kinematics of laminated composites are given by:

$$\left\{\begin{array}{c}\varepsilon_x\\ \varepsilon_y\\ \gamma_{xy}\end{array}\right\} = \left\{\begin{array}{c}\varepsilon_x^0\\ \varepsilon_y^0\\ \gamma_{xy}^0\end{array}\right\} + z\left\{\begin{array}{c}k_x\\ k_y\\ 2k_{xy}\end{array}\right\} = \left\{\begin{array}{c}\frac{\partial u_0}{\partial x}\\ \frac{\partial v_0}{\partial y}\\ \frac{\partial u_0}{\partial y}+\frac{\partial v_0}{\partial x}\end{array}\right\} + z\left\{\begin{array}{c}\frac{\partial^2 w_0}{\partial x^2}\\ \frac{\partial^2 w_0}{\partial y^2}\\ 2\frac{\partial^2 w_0}{\partial x \partial y}\end{array}\right\}, \tag{2}$$

where u_0, v_0 and w_0 are the displacements of the middle plane in the x, y and z directions, respectively. are denoted by u, v, w, respectively; k_{xy} is the total skew curvature.

The constitutive equations for k layer can be written as follows [10]:

$$\left\{\begin{array}{c}\sigma_x\\ \sigma_y\\ \tau_{xy}\end{array}\right\}_k = [T]_k^{-1}[Q]_k[T]_k^T = \left[\begin{array}{ccc}\bar{Q}_{11} & \bar{Q}_{12} & \bar{Q}_{16}\\ \bar{Q}_{12} & \bar{Q}_{22} & \bar{Q}_{26}\\ \bar{Q}_{16} & \bar{Q}_{26} & \bar{Q}_{66}\end{array}\right]_k \left\{\begin{array}{c}\varepsilon_x\\ \varepsilon_y\\ \gamma_{xy}\end{array}\right\}_k, \tag{3}$$

where:

$$\begin{aligned}
\bar{Q}_{11} &= Q_{11}\cos^4\theta + 2(Q_{12}+2Q_{66})\sin^2\theta\cos^2\theta + Q_{22}\sin^4\theta,\\
\bar{Q}_{12} &= (Q_{11}+Q_{22}-4Q_{66})\sin^2\theta\cos^2\theta + Q_{12}\left(\sin^4\theta+\cos^4\theta\right),\\
\bar{Q}_{22} &= Q_{11}\sin^4\theta + 2(Q_{12}+2Q_{66}) \times \sin^2\theta\cos^2\theta + Q_{22}\cos^4\theta,\\
\bar{Q}_{16} &= (Q_{11}-Q_{22}-2Q_{66})\sin\theta\cos^3\theta + (Q_{12}-Q_{22}+2Q_{66})\sin^3\theta\cos\theta,\\
\bar{Q}_{26} &= (Q_{11}-Q_{22}-2Q_{66})\sin^3\theta\cos\theta + (Q_{12}-Q_{22}+2Q_{66})\sin\theta\cos^3\theta,\\
\bar{Q}_{66} &= (Q_{11}+Q_{22}-2Q_{12}-2Q_{66})\sin^2\theta\cos^2\theta + Q_{66}\left(\sin^4\theta+\cos^4\theta\right).
\end{aligned} \tag{4}$$

In the last equations: $\nu_{12}E_2 = \nu_{21}E_1$, $Q_{66} = G_{12}$, $Q_{11} = E_1/(1-\nu_{12}\nu_{21})$, $Q_{22} = E_2/(1-\nu_{12}\nu_{21})$ and $Q_{12} = \nu_{12}E_2/(1-\nu_{12}\nu_{21}) = \nu_{21}E_1/(1-\nu_{12}\nu_{21})$ where G_{12} is the shear modulus, E_1 and E_2 are Young's modulus parallel and perpendicular to the fibres while ν_{12} and ν_{21} are the corresponding Poisson's ratios.

The coordinate transformation matrix is [11]:

$$[T] = \left[\begin{array}{ccc}\cos^2\theta & \sin^2\theta & 2\sin\theta\cos\theta\\ \sin^2\theta & \cos^2\theta & -2\sin\theta\cos\theta\\ -\sin\theta\cos\theta & -\sin\theta\cos\theta & \cos^2\theta-\sin^2\theta\end{array}\right]. \tag{5}$$

The relations between the resultants (forces N and moments M) and the strains (strains ε and curvatures $\mathbf{k}$) are given by:

$$\left\{\begin{array}{c}N_x\\ N_y\\ N_{xy}\end{array}\right\} = \left[\begin{array}{ccc}A_{11} & A_{12} & A_{16}\\ A_{12} & A_{22} & A_{26}\\ A_{16} & A_{26} & A_{66}\end{array}\right]\left\{\begin{array}{c}\varepsilon_x^0\\ \varepsilon_y^0\\ \gamma_{xy}^0\end{array}\right\} + \left[\begin{array}{ccc}B_{11} & B_{12} & B_{16}\\ B_{12} & B_{22} & B_{26}\\ B_{16} & B_{26} & B_{66}\end{array}\right]\left\{\begin{array}{c}k_x\\ k_y\\ 2k_{xy}\end{array}\right\}, \tag{6}$$

$$\left\{\begin{array}{c}M_x\\ M_y\\ M_{xy}\end{array}\right\} = \left[\begin{array}{ccc}B_{11} & B_{12} & B_{16}\\ B_{12} & B_{22} & B_{26}\\ B_{16} & B_{26} & B_{66}\end{array}\right]\left\{\begin{array}{c}\varepsilon_x^0\\ \varepsilon_y^0\\ \gamma_{xy}^0\end{array}\right\} + \left[\begin{array}{ccc}D_{11} & D_{12} & D_{16}\\ D_{12} & D_{22} & D_{26}\\ D_{16} & D_{26} & D_{66}\end{array}\right]\left\{\begin{array}{c}k_x\\ k_y\\ k_{xy}\end{array}\right\}. \tag{7}$$

Combine the above equations:

$$\left\{\begin{array}{c}\mathbf{N}\\ \mathbf{M}\end{array}\right\} = \left[\begin{array}{cc}\mathbf{A} & \mathbf{B}\\ \mathbf{B} & \mathbf{D}\end{array}\right]\left\{\begin{array}{c}\varepsilon\\ \mathbf{k}\end{array}\right\}, \tag{8}$$

where $\mathbf{A}$ is called the extensional stiffness matrix, $\mathbf{B}$ is called the coupling stiffness matrix and $\mathbf{D}$ is called the bending stiffness matrix of the laminate. The components of these three matrices are defined as follows:

$$\begin{aligned}
A_{ij} &= \sum_{k=1}^{n}\left(\bar{Q}_{ij}\right)_k (z_k - z_{k-1})\\
B_{ij} &= \frac{1}{2}\sum_{k=1}^{n}\left(\bar{Q}_{ij}\right)_k \left(z_k^2 - z_{k-1}^2\right)\\
D_{ij} &= \frac{1}{3}\sum_{k=1}^{n}\left(\bar{Q}_{ij}\right)_k \left(z_k^3 - z_{k-1}^3\right)
\end{aligned} \tag{9}$$

where z_k is the distance from the mid-plane to the centroid of the k layer (Fig. 1) [12].

3 Element-free Galerkin method

Before describing the EFG method, it is present briefly the governing equations for an incompressible, elastic and isotropic medium. Considering the two-dimensional problem of solid mechanics Ω bounded by $\Gamma : \nabla\sigma + b = 0$ in Ω, where σ is the stress tensor, which corresponds to the displacement field $\mathbf{u} = \{u\ v\}^T$, $\mathbf{b}$ the body force vector, and ∇ is the divergence operator. The boundary conditions are given by: $\sigma\mathbf{n} = \bar{t}$ on the natural boundary Γ_t, and $\mathbf{u} = \bar{\mathbf{u}}$ on the essential boundary Γ_u, where the superposed bar denotes the prescribed traction and displacement boundary values and $\mathbf{n}$ is the unit outward vector normal to the boundary Γ_t.

The variational or weak form of the equilibrium equation ($\nabla\sigma + b = 0$) is posed as follows. Consider trial functions $\mathbf{u} = (\mathbf{x}) \in H^1$ and Lagrange multipliers $\lambda \in H^0$ for all test functions $\delta\mathbf{v} = (\mathbf{x}) \in H^1$ and $\delta\lambda \in H^0$, as follows [13]:

$$\int_\Omega \delta\left(\nabla_S \mathbf{v}^T\right)\sigma d\Omega - \int_\Omega \delta\mathbf{v}^T \mathbf{b} d\Omega - \int_{\Gamma_t} \delta\mathbf{v}^T \bar{\mathbf{t}} d\Gamma - \int_{\Gamma_u} \delta^T.(\mathbf{u} - \bar{\mathbf{u}})\, d\Gamma - \int_{\Gamma_u} \delta\mathbf{v}^T \lambda d\Gamma = 0. \quad (10)$$

Then the equilibrium (Eq. (1)) and the boundary conditions (Eqs. (2) and (3)) are satisfied. Here $\nabla_S \mathbf{v}^T$ is the symmetric part of $\nabla\mathbf{v}^T$; H^1 and H^0 denote the Hilbert spaces of degree one and zero, respectively. It is observed that the trial functions do not satisfy the essential boundary conditions, so that they are imposed with Lagrange multipliers [2].

The EFG method employs recently developed approximation theories that permit the resultant shape functions to be constructed entirely in terms of arbitrarily placed nodes [13].

3.1 Moving least-square approximations

In this section a brief for moving least-square (MLS) [14] approximation is given. More details are referred to in references [2] and [13]. The moving least-squares interpolation is defined in a domain by:

$$\mathbf{u}^h(\mathbf{x}) = \sum_{j=1}^{n} p_j(\mathbf{x})\, a_j(\mathbf{x}) = \mathbf{p}^T(\mathbf{x})\mathbf{a}(\mathbf{x}) \quad (11)$$

where $\mathbf{p}^T(\mathbf{x}) = \{p_1(\mathbf{x}), p_2(\mathbf{x}), ..., p_m(\mathbf{x})\}$ is a vector for complete basis functions of order m and $\mathbf{a}^T(\mathbf{x}) = \{a_1(\mathbf{x}), a_2(\mathbf{x}), ..., a_m(\mathbf{x})\}$ is a vector of unknown parameters that depend on $\mathbf{x}$. For example, in two dimensions with x_1 and x_2 coordinates for $m = 3$ and $m = 6$, representing linear and quadratic basis functions, respectively:

$$\mathbf{p}^T(\mathbf{x}) = \left[1\, x_1\, x_2\right] \quad \text{and} \quad \mathbf{p}^T(\mathbf{x}) = \left[1\, x_1\, x_2\, x_1^2\, x_1 x_2\, x_2^2\right]. \quad (12)$$

In equation (11), the coefficient vector $\mathbf{a}(\mathbf{x})$ is obtained at point $\mathbf{x}$ by minimizing a weighted discrete least-squares norm as follows:

$$J = \sum_{I=1}^{n} w(\mathbf{x} - \mathbf{x}_I)\left[\mathbf{p}^T(\mathbf{x}_I)\mathbf{a}(\mathbf{x}) - u_I\right]^2 \quad (13)$$

where n is the number of points in the neighbourhood of $\mathbf{x}$ (called the domain of influence of $\mathbf{x}$) for which the weight function $w(\mathbf{x} - \mathbf{x}_I) \neq 0$, and u_I is the nodal value of $\mathbf{u}$ at $\mathbf{x} = \mathbf{x}_I$.

The stationarity of J with respect to $\mathbf{a}(\mathbf{x})$ leads to the following linear relation:

$$\mathbf{A}(\mathbf{x})\, \mathbf{a}(\mathbf{x}) = \mathbf{B}(\mathbf{x})\mathbf{u}, \text{ or } \mathbf{a}(\mathbf{x}) = \mathbf{A}^{-1}(\mathbf{x})\mathbf{B}(\mathbf{x})\mathbf{u}, \quad (14)$$

where $\mathbf{A}(\mathbf{x})$ and $\mathbf{B}(\mathbf{x})$ are the matrices defined by:

$$\mathbf{A}(\mathbf{x}) = \sum_{I=1}^{n} w(\mathbf{x} - \mathbf{x}_I)\mathbf{p}(\mathbf{x}_I)\,\mathbf{p}^T(\mathbf{x}_I), \quad (15)$$

$$\mathbf{B}(\mathbf{x}) = [w(\mathbf{x} - \mathbf{x}_1)\,\mathbf{p}(\mathbf{x}_1) w(\mathbf{x} - \mathbf{x}_2) \times \mathbf{p}(\mathbf{x}_2) w...(\mathbf{x} - \mathbf{x}_n)\,\mathbf{p}(\mathbf{x}_n)], \quad (16)$$

$$\mathbf{u}^T = \left[u_0\, u_1\, u_2\, ...\, u_n\right]. \quad (17)$$

Substituting (14) in (11), it is possible to write:

$$\mathbf{u}^h(\mathbf{x}) = \sum_{I=1}^{n} \Phi_I(\mathbf{x}) u_I = \Phi(\mathbf{x})\mathbf{u}, \quad (18)$$

where the MLS shape function $\Phi_I(\mathbf{x})$ is defined by:

$$\Phi_I(\mathbf{x}) = \sum_{j=0}^{m} p_j(\mathbf{x})(\mathbf{A}^{-1}(\mathbf{x})\,\mathbf{B}(\mathbf{x}))_{jI} = \mathbf{p}^T\mathbf{A}^{-1}\mathbf{B}_I. \quad (19)$$

The partial derivatives of $\Phi_I(\mathbf{x})$ can be obtained as follows:

$$\Phi_{I,i}(\mathbf{x}) = \sum_{j=0}^{m} \left\{p_{j,i}\left(\mathbf{A}^{-1}\mathbf{B}\right)_{jI} + p_j\left(\mathbf{A}^{-1}_{,i}\mathbf{B} + \mathbf{A}^{-1}\mathbf{B}_{,i}\right)_{jI}\right\}, \quad (20)$$

where: $\mathbf{A}^{-1}_{,i} = -\mathbf{A}^{-1}\mathbf{A}_{,i}\mathbf{A}^{-1}$ in which $()_{,i} = \partial()/\partial x_i$.

3.2 Shape functions

The EFG method modifies the weight functions to construct an approximation which is capable of reproducing the linear basis (Eq. (2)). This approximation is based upon MLS in curve and surface fitting [14].

The shape function in equation (19) is constructed as:

$$\Phi_I(\mathbf{x}) = \gamma^T(\mathbf{x})\,\mathbf{B}_I(\mathbf{x}) \quad (21)$$

in which:

$$\gamma(\mathbf{x}) = \mathbf{A}(\mathbf{x})^{-1}\mathbf{p}(\mathbf{x}). \quad (22)$$

The derivatives of $\gamma(\mathbf{x})$ are computed similarly:

$$\gamma_{,x}(\mathbf{x}) = \mathbf{A}^{-1}(\mathbf{p}_{,x} - \mathbf{A}_{,x}\gamma). \quad (23)$$

The vector $\gamma(\mathbf{x})$ is determined using an LU decomposition of $\mathbf{A}$ followed by back substitution, leading to a computationally efficient procedure for computing the derivates of $\mathbf{u}^h(\mathbf{x})$.

In this work, it is used the following cubic spline weighting function [13]:

$$w(r) = \begin{cases} \frac{2}{3} - 4r^2 + 4r^3 & \text{for } r \leqslant \frac{1}{2} \\ \frac{4}{3} - 4r + 4r^2 - \frac{4}{3}r^3 & \text{for } \frac{1}{2} < r \leqslant 1 \\ 0 & \text{for } r > 1 \end{cases} \tag{24}$$

in which r is the radius of the domain of influence. The tensor product of weighting functions at any point is defined as:

$$w(\mathbf{x} - \mathbf{x}_I) = w(r_x)w(r_y) \tag{25}$$

where $w(r_x)$ or $w(r_y)$ is given by equation (24) with r replaced by r_x or r_y respectively. These parameters are:

$$r_x = \frac{|\mathbf{x} - \mathbf{x}_I|}{d_{mx}}, \quad r_y = \frac{|\mathbf{y} - \mathbf{y}_I|}{d_{my}} \tag{26}$$

where $d_{mx} = d_{\max}c_{xI}$ and $d_{my} = d_{\max}c_{yI}$. The parameter $d_{\max}$ is a scaling parameter which is typically 2.0–4.0 for static analysis and c_{xI} and c_{yI} are determined at a particular node by searching for enough neighbor nodes such that $\mathbf{A}$ is non-singular everywhere in the domain, and thus invertible [15]. If the nodes are uniformly spaced, the values c_{xI} and c_{yI} correspond to the distance between the nodes in the x and y directions, respectively.

3.3 Discrete equations

In order to obtain the discrete equations form the weak form (Eq. (10)) the approximate solution $\mathbf{u}$ and test function $\delta\mathbf{v}$ are constructed according to equation (16). The Lagrange multiplier λ is expressed by [15]:

$$\lambda = \sum_K \psi_K(s)\lambda_K, \tag{27}$$

$$\delta\lambda = \sum_K \psi_K(s)\delta\lambda_K, \tag{28}$$

where $\psi_k(s)$ is a Lagrange interpolant and s is the arc-length along the boundary; the repeated indices designate summations. The final discrete equations can be obtained by substituting the trial functions, test functions and equations (27) and (28) into the weak form (Eq. (10)), yielding the following system of linear algebraic equations [15] for static analysis:

$$\begin{bmatrix} \mathbf{K} & \mathbf{G} \\ \mathbf{G}^T & 0 \end{bmatrix} \begin{Bmatrix} \mathbf{q} \\ \lambda \end{Bmatrix} = \begin{Bmatrix} \mathbf{g} \\ \mathbf{h} \end{Bmatrix} \tag{29}$$

and for modal analysis [11]:

$$\left(\begin{bmatrix} \mathbf{K} & \mathbf{G} \\ \mathbf{G}^T & 0 \end{bmatrix} - \begin{Bmatrix} \mathbf{f} \\ \lambda \end{Bmatrix} \begin{bmatrix} \mathbf{M} & \mathbf{G} \\ \mathbf{G}^T & 0 \end{bmatrix} \right) \begin{Bmatrix} \mathbf{\Delta} \\ \lambda \end{Bmatrix} = \{\mathbf{0}\} \tag{30}$$

where $\mathbf{K}$, $\mathbf{M}$ and $\mathbf{f}$ are the stiffness matrix, mass matrix and applied forces vector, respectively:

$$\mathbf{K}_{IJ} = \int_\Omega \mathbf{B}_I^T \mathbf{D} \mathbf{B}_J d\Omega \tag{31}$$

$$\mathbf{M}_{IJ} = \int_\Omega \rho \Phi_I \Phi_J d\Omega \tag{32}$$

$$\mathbf{G}_{IK} = -\int_{\Gamma_{ut}} \Phi_i \psi_K \mathbf{S} d\Gamma \tag{33}$$

$$\mathbf{g}_I = \int_{\Gamma_t} \Phi_I \bar{\mathbf{t}} d\Gamma + \int_\Omega \Phi_I \mathbf{b} d\Omega \tag{34}$$

$$\mathbf{h}_K = -\int_{\Gamma_{ut}} \psi_K \mathbf{S} \bar{\mathbf{u}} d\Gamma \tag{35}$$

where $\mathbf{S}$ is a diagonal matrix ($\mathbf{S}_{II} = 1$ if the displacement is imposed in $\mathbf{x}_I$ and 0 in other case [16]) and ρ is the density of material.

The differential operator matrix for two dimensions case is:

$$\mathbf{B}_I = \begin{bmatrix} \Phi_{I,x} & 0 \\ 0 & \Phi_{I,y} \\ \Phi_{I,y} & \Phi_{I,x} \end{bmatrix}. \tag{36}$$

For isotropic material [10]:

$$\mathbf{D} = \frac{E}{1-\nu^2} \begin{bmatrix} 1 & \nu & 0 \\ \nu & 1 & 0 \\ 0 & 0 & \frac{1-\nu}{2} \end{bmatrix}, \text{ for plane stress} \tag{37}$$

$$\mathbf{D} = \frac{E}{(1+\nu)(1-2\nu)} \begin{bmatrix} 1-\nu & \nu & 0 \\ \nu & 1-\nu & 0 \\ 0 & 0 & \frac{1-2\nu}{2} \end{bmatrix}, \text{ for plane strain.} \tag{38}$$

For orthotropic material $\mathbf{D}$ is given by equation (9) [11].

In order to perform the numerical integration a background mesh is needed, which can be independent of the arrangement of meshless nodes. Standard Gaussian quadratures were used to evaluate the integrals for assembling the stiffness, mass matrix and force vector.

4 Reliability index estimation as a general optimization problem

In traditional deterministic design optimization, the optimization problem is generally formulated in the physical space of the design variables and consists in minimizing or maximizing an objective function subject to geometrical, physical or functional constraints in the form:

$$\min f(\{y\}) \tag{39}$$

subjected to $g_k(\{y\}) \leqslant 0$, where $\{y\}$ designates the vector of deterministic design variables.

In reliability analysis, which involves random variables $\{x\}$, the deterministic optimal solution is not considered the exact solution of the optimum design but is one of the most probably design. In this case, the failure surface or limit state function is given by $G(\{x\},\{y\}) = 0$. This surface defines the limit between the safe region $G(\{x\},\{y\}) > 0$ and unsafe region of the design space. The failure occurs when $G(\{x\},\{y\}) < 0$, and the failure probability is calculated as $P_f = prob\,[G(\{x\},\{y\}) \leqslant 0]$.

The reliability index β is introduced as a measure of the reliability level of the system and is estimated in the so-called reduced coordinate system, where the random variables $\{u\}$ are statistically independent with zero mean and unit standard deviation. Thus a pseudo-probabilistic transformation $\{u\} = T[\{x\},\{y\}]$ must be defined for mapping the original space into the reduced coordinate system (see [17]). Considering that the probability density in the reduced space decays exponentially with the distance from the origin of this space, the point with maximum probability of failure (most probable point) on the limit state surface is the point of minimum distance from the origin. The reliability index is thus defined as the minimum distance between the origin of the reduced space and the hyper surface representing the limit state function $H(\{u\},\{y\})$. Hence, it is possible to find the most probable point or design point by solving a constrained optimization problem that is [18]:

$$\beta = \min \sqrt{\sum_{i=1}^{n} u_i^2} \tag{40}$$

subjected to safety constraints:

$$H(\{u\},\{y\}) = 0. \tag{41}$$

By formally introducing a cumulative density function (ϕ) of the normal probability distribution function, the first order approximation (tangent plane at the most probable point-MPP) to P_f can be written as [18]:

$$P_f = 1 - \phi(\beta). \tag{42}$$

In order to estimate the reliability index (Eq. (40)) it is possible make use of a variety of optimization methods.

FORM and SORM can be considered as gradient-based methods since they demand the evaluation of the partial derivatives of the limit state function with respect to the random variables at each iteration step.

FORM is based on linear (first order) approximation of the limit state surface tangent to the most probable point of the failure surface to the origin of a reduced coordinate system. Thus, the random variables are transformed to reduced variables in a reduced coordinate system. For estimating the reliability index based on FORM one can use the algorithm suggested in [19] in which the limit state function does not need to be solved because a N-R type recursive algorithm is introduced to find the design point [20]. This algorithm has been widely used in the literature [18] and used in this study.

SORM estimates the probability of failure by using a nonlinear approximation of the limit state function by a second order representation. The curvatures of the limit state function are approximated by the second-order derivatives with respect to the original variables. Thus, SORM improves FORM by including additional information about the curvature of the limit state function through a curvature parameter. SORM was explored in [21] using quadratic approximations. In that work the authors use a simple closed-form solution for the computation of failure probability using a second-order approach given by [22] based on the theory of asymptotic approximation. The SORM suggested by [22] is employ in this work. SORM uses as initial value the reliability index value estimated through FORM. Zhao and Ono [23] give more details of these classical techniques.

However, the solution of the optimization problem given by equation (40) by using classical gradient-based optimization methods, as a N-R, is not a simple task due to the existence of local minima in the design space and the necessity of computation of the gradients (partial derivatives). The existence of multiple MPPs is similar to multiple local minima in optimization. The solutions of many problems in structural optimization can be considered to be satisfactory once a local minimum is reached. However, this is an unacceptable procedure in reliability analysis since the local MPP may not represent the worst failure scenario and the actual failure may occur below the predicted level. Hence, only the global MPP represents the actual structural reliability [24].

Another difficulty that must be remembered is that traditional methods FORM and SORM require an initial guess of the solution (reliability index and random variables) and it is not always possible to assure global convergence. These aspects motivated the authors of this paper to explore an alternative approach for estimation of reliability index, which do not require the computation of gradients of the limit state function and are intrinsically based on multidirectional search. This approach was proposed initially in [3]. In this work the authors use the approach that uses finite element analysis and EFG method to evaluate implicit limit state functions and is based on a heuristic based reliability method (HBRM), which allows the use of optimization methods such as genetic algorithms [4,5], particle swarm optimization [6,7] and ant colony optimization [8,9]. It is believed that such approach can circumvent some of the difficulties mentioned above, and thus lead to improved results of reliability analysis.

Taking into account the performance of HBRM, it was observed that this methodology is able to handle multiple limit state functions based on numerical models and probabilistic variables related to geometrical, load and material properties parameters. Rojas et al. [3,25,26] give more details of HBRM. In this work was used GA algorithm in HBRM. The following section discusses the main ideas about this heuristic technique.

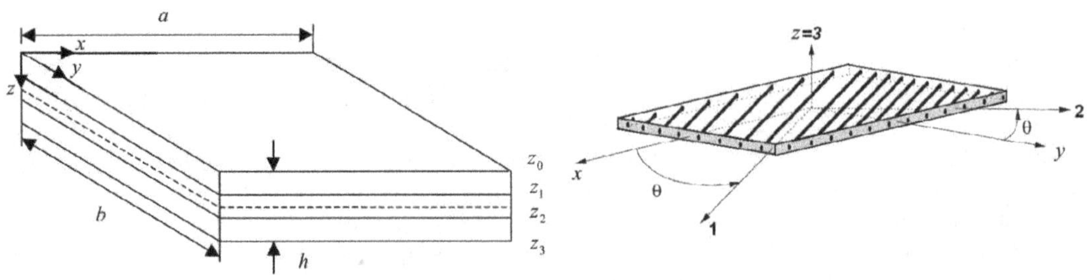

Fig. 2. Laminated composite plate.

5 Genetic algorithms

5.1 Back to nature

GA is an optimization algorithm used to find approximate solutions to difficult-to-solve problems through application of the principles of evolutionary biology to computer science. GA uses biologically-derived techniques such as inheritance, mutation, natural selection, and recombination (or crossover).

GA is based on Darwin's theory of survival and evolution of species, as explained in [4] and [5]. The algorithm starts from a population of random individuals, viewed as candidate solutions to the problem. During the evolutionary process, each individual of the population is evaluated, reflecting its adaptation capability to the environment. Some of the individuals of the population are preserved while others are discarded; this process mimics the natural selection in the Darwinism. The remaining group of individuals is paired in order to generate new individuals to replace the worst ones in the population, which are discarded in the selection process. Finally, some of them can be submitted to mutation, and as a consequence, the chromosomes of these individuals are altered. The entire process is repeated until a satisfactory solution is found.

5.2 Algorith description

The outline of a basic GA is as follows:

1. define the GA parameters (population size, selection method, crossover method, mutation rate, etc.);
2. create an initial population (it just allocates memory);
3. evaluate the objective function and take it as a fitness measure of each individual;
4. select the mates to the crossover;
5. reproduce and replace the worst individuals of the population;
6. mutate to avoid premature convergence;
7. go to step 3 and repeat until the stop criteria is achieved.

Although the initially proposed GA algorithm was dedicated to discrete variables only, nowadays improvements are available to deal with discrete and continuous variables. To obtain more information about GA see [4, 5].

Table 1. Geometrical properties of laminated composite plate.

L (m)	H (m)	h (m)	t (m)	θ	N_x (N/m)
5	5	0.4	0.1	−30 30 30 −30	1×10^6

Table 2. Mechanical properties of laminated composite plate.

E_1 (Pa)	E_2 (Pa)	G_{12} (Pa)	ν_{12}
4.5×10^{10}	1.2×10^{10}	4.5×10^9	0.3

Table 3. Displacements for EFG and FEM methods of the laminated composite plate.

	EFG	FEM	Error (%)
u_x (m)	5.05×10^{-4}	5.05×10^{-4}	0.006
u_y (m)	-3.43×10^{-4}	-3.43×10^{-4}	6.41×10^{-6}

Table 4. Statistical parameters of design variables of the laminated composite plate, case 1.

	Distribution	Mean	CV
h(m)	Normal	0.4	10%
θ (°)	Normal	30	10%
E_1 (N/m^2) $\times10^{10}$	Lognormal	4.5	10%
E_2 (N/m^2) $\times10^{10}$	Lognormal	1.2	10%
ν_{12}	Lognormal	0.30	10%
G_{12} (N/m^2) $\times10^9$	Lognormal	4.5	10%
N_x (N/m) $\times10^6$	Lognormal	1	10%

6 Numerical applications

Consider a laminated composite rectangular plate illustrated in Figure 2 and which geometric and mechanical properties are summarized in Tables 1 and 2, respectively.

The boundary conditions include: essential boundary conditions on the bottom ($u_y = 0$) and left edges ($u_x = 0$) and natural boundary conditions on the right edge on which traction N_x is applied.

In this case: $\nu_{21} = \nu_{12}E_2/E_1$.

For integration procedure a 4×4 Gauss quadrature is employed for a circular domain of influence. Concerning weight functions, the cubic spline function is used. The essential boundary condition is imposed using Lagrange interpolants.

Figure 3 illustrates Gauss points and deformed configuration of laminated composite plate.

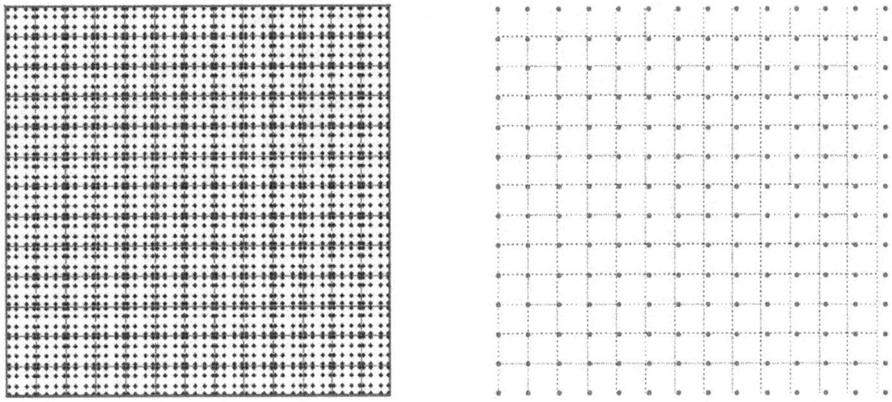

Fig. 3. Gauss points and deformed configuration of laminated composite plate.

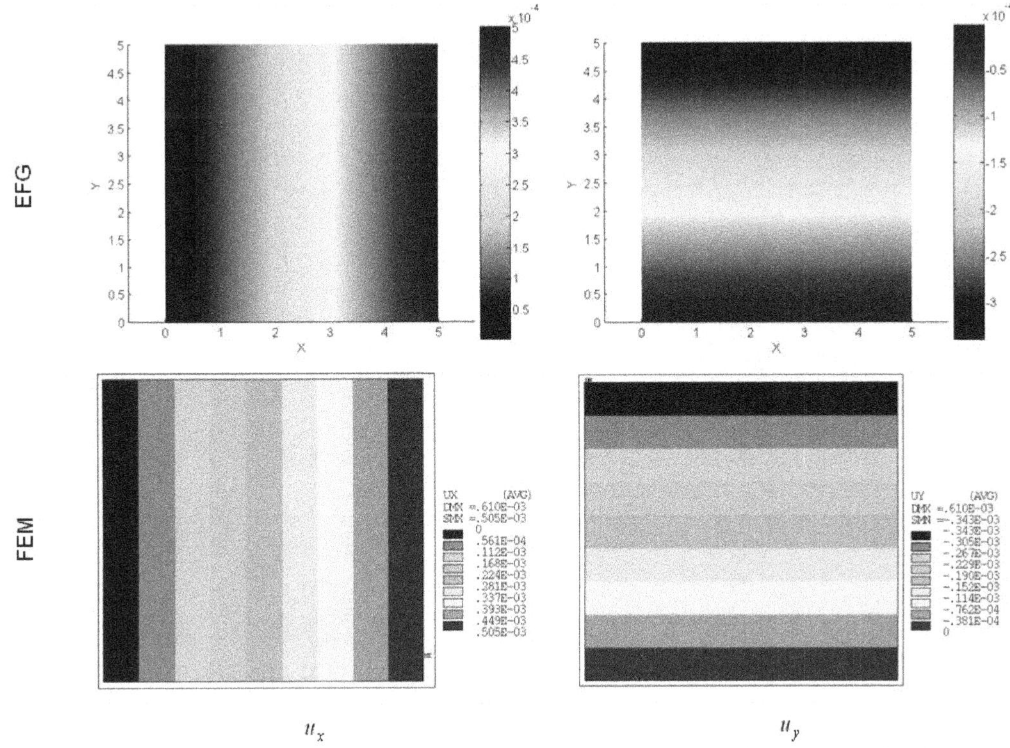

Fig. 4. M-H loop of the film.

The horizontal and vertical displacements were obtained using EFG and FEM analysis (Fig. 4). Numerical simulations through MEF were estimated using PLANE42 finite element of ANSYS®.

A comparison between maximal displacements obtained by EFG and FEM approaches are exposed in Table 6 as well as its differences. Taking to account these results it is possible to validate the EFG model.

Two cases are proposed for reliability analysis, the case 1 consider 7 design variables and case 2 consider 12 design variables in laminated composite plate. The statistical parameters (mean and coefficient of variation) of design variables for the first case are summarized in Table 4.

The limit state function is defined by the probability of failure $P_f[u_x(x_i) \leqslant u_x^{\lim}]$, where x_i are the design variables and $u_x^{\lim} = 1 \times 10^{-3}$ m.

Table 5 summarizes the results obtained by N-R in FORM and SORM approaches and GA in HBRM. The limit state function is evaluated by FEM and EFG methods. The number of evaluations of the limit state functions for FORM and SORM (until convergence) and for HBRM (until stop criterion) is represented by N_G.

The results calculated by FORM and SORM are better than HBRM counterparts for the same number of evaluations of limit state function using EFG and FEM approaches. For all reliability approaches, EFG and FEM analysis present close results. HBRM results can be taking as an initial guess to FORM and SORM procedures [27].

Table 5. Reliability analysis results, case 1.

	EFG			FEM		
	FORM	SORM	HBRM	FORM	SORM	HBRM
h(m)	0.32	0.32	0.39	0.32	0.32	0.42
θ (°)	35.85	35.85	29.03	35.91	35.91	29.50
E_1 (N/m^2) $\times 10^{10}$	4.20	4.20	4.86	4.21	4.21	4.60
E_2 (N/m^2) $\times 10^{10}$	1.19	1.19	1.30	1.19	1.19	1.07
ν_{12}	0.30	0.30	0.32	0.30	0.30	0.28
G_{12} (N/m^2) $\times 10^9$	4.09	4.09	4.53	4.08	4.08	4.36
N_x (N/m) $\times 10^6$	1.16	1.16	0.99	1.16	1.16	1.10
N_G	49	86	100	49	86	100
β	3.38	3.33	1.38	3.38	3.38	1.74
Pf (%)	0.04	0.04	8.45	0.04	0.04	4.07
Reliability (%)	99.96	99.96	91.55	99.96	99.96	95.93

Table 6. Statistical parameters of design variables of the laminated composite plate, case 2.

	Distribution	Mean	CV
t_1 (m)	Normal	0.1	10%
t_2 (m)	Normal	0.1	10%
t_3 (m)	Normal	0.1	10%
t_4 (m)	Normal	0.1	10%
θ_1 (°)	Normal	−30	10%
θ_2 (°)	Normal	30	10%
θ_3 (°)	Normal	30	10%
θ_4 (°)	Normal	−30	10%
E_1 (N/m^2)	Lognormal	4.5×10^{10}	10%
E_2 (N/m^2)	Lognormal	1.2×10^{10}	10%
G_{12} (N/m^2)	Lognormal	4.5×10^9	10%
N_x (N/m)	Lognormal	1×10^6	10%

Table 7. Reliability analysis results, case 2.

	EFG			FEM		
	FORM	SORM	HBRM	FORM	SORM	HBRM
t_1 (m)	0.09	0.09	0.09	0.09	0.09	0.10
t_2 (m)	0.09	0.09	0.11	0.09	0.09	0.10
t_3 (m)	0.09	0.09	0.10	0.09	0.09	0.10
t_4 (m)	0.09	0.09	0.10	0.09	0.09	0.10
θ_1 (°)	−32.64	−32.64	−31.46	−33.42	−33.42	−33.74
θ_2 (°)	33.24	33.24	30.92	32.62	32.62	30.89
θ_3 (°)	33.24	33.24	29.94	32.62	32.62	32.76
θ_4 (°)	−32.64	−32.64	−29.29	−33.42	−33.42	−32.21
E_1 (N/m^2) $\times 10^{10}$	3.80	3.80	4.47	3.82	3.82	4.18
E_2 (N/m^2) $\times 10^{10}$	1.19	1.19	1.28	1.20	1.20	1.31
G_{12} (N/m^2) $\times 10^9$	3.84	3.84	5.03	3.82	3.82	4.28
N_x (N/m) $\times 10^6$	1.37	1.37	0.98	1.36	1.36	0.99
N_G	73	165	100	73	165	100
β	4.74	4.72	1.99	4.72	4.72	2.23
Pf (%)	1.08×10^{-4}	1.18×10^{-4}	2.32	1.16×10^{-4}	1.16×10^{-4}	1.30
Reliability (%)	100	100	97.68	100	100	98.70

Table 6 summarizes the statistical parameters of the design variables considered in the second case (12 design variables). In this table, t_i and θ_i represent the thickness and fibre orientation of each layer of the laminated composite plate.

Reliability results obtained by FORM, SORM and HBRM using FEM and EFG methods for case 2 are summarized in Table 7.

Similar to case 1, FORM and SORM results are considered better than HBRM counterparts for the same number of evaluations of limit state function. In this case, the results obtained by EFG are better than FEM counterparts using FORM, very close using SORM but not using HBRM, where FEM have better performance in comparison with EFG. HBRM results can be used as initial guess to FORM and SORM algorithms. It is possible to obtain better results through HBRM for more evaluations of limit sate function.

7 Conclusions

A structural reliability methodology that couples FORM, SORM and HBRM with element-free Galerkin method to analyse laminated composite plates was suggested. Numerical examples on linear elasticity were used to illustrate the potentiality of the proposed procedure. Load, material and geometrical proprieties were considered as design parameters in numerical applications. Good agreement was observed between the results of the EFG method allied to RA. HBRM algorithm is not competing with traditional methods FORM and SORM, but it was tested because is able to solve global optimization problems efficiently. The obtained results demonstrate that the predicted reliability levels are accurate in comparison with similar approach that uses FEM to evaluate implicit limit state functions. Considering the results obtained it is possible to conclude that the methodology presented can be used in more complex applications, since mesh generation of these structures can be far more time-consuming and costly than the solution of a discrete set of equations. The EFG method provides an attractive alternative to FEM analyses in evaluating limit state functions in stochastic mechanics problems and encourage authors for applied this approach in problems where finite element analysis presents some difficulties.

Msc. Rojas are thankful to CAPES Foundation (Brazilian Ministry of Education) and Programme Alβan, the European Union Programme of High Level Scholarships for Latin America, for the support to their scholarship No. E06D103742BO. Dr. Rade gratefully acknowledges CNPq for the support to their research activities.

References

1. S. Rahman, B.N. Rao. International Journal of Solids and Structures **38**, 9313 (2001)
2. T. Belytschko, Y.Y. Lu, L. Gu. International Journal of Numerical Methods in Engineering **37**, 229 (1994)
3. J.E. Rojas, F.A.C. Viana, D.A. Rade, *7th Multidisciplinary International Conference Quality and Reliability* (Tangier, Morocco, March 20–22, 2007)
4. Z. Michalewicz, *Genetic Algorithms + Data Structures = Evolution Programs*, 2nd edn. (Springer-Verlag, New York, USA, 1994)
5. R.L. Haupt, S.E. Haupt, *Pratical Genetic Algorithms*, 2nd edn. (Wiley-Interscience Publication, New York, USA, 2004)
6. J. Kennedy, R.C. Eberhart, Particle Swarm Optimization, *Proceedings of the 1995 IEEE International Conference on Neural Networks* (Perth, Australia, 1995), pp. 1942–1948
7. G. Venter, J.S. Sobieski, Multidisciplinary Optimization of a Transport Aircraft Wing using Particle Swarm Optimization, *Proceedings of the 9th AIAA/ISSMO Symposium on Multidisciplinary Analysis and Optimization* (Atlanta, GA, September 4–6, 2002), Vol. AIAA. pp. 2002–5644
8. M. Dorigo, Optimization, *Learning and Natural Algorithms*, Ph.D. Dissertation, Politecnico di Milano, Italy, 1992
9. K. Socha, M. Dorigo, ACO for Continuous and Mixed-Variable Optimization. *Proceedings of the ANTS 2004 – Fourth International Workshop on Ant Colony Optimization and Swarm Intelligence*, Brussels, Belgium, 2004
10. R.M. Jones, *Mechanics of Composite Materials*, 2nd edn. (Taylor and Francis, Philadelphia, 1999)
11. X.L. Chen, G.R. Liu, S.P. Lim. Composite Structures **59**, 279 (2003)
12. J. Belinha, L.M.J.S. Dinis. Composite Structures **78**, 337 (2007)
13. J. Dolbow, T. Belytschko. International Journal for Numerical Methods in Engineering **46**, 925 (1999)
14. P. Lancaster, K. Salkauskas. Mathematics of Computation **37**, 147 (1981)
15. Y.Y. Lu, T. Belytschko, L. Gu. Computer Methods in Applied Mechanics and Engineering **113**, 397 (1994)
16. T. Belytschko, Y.Y. Lu, L. Gu. Engineering Fracture Mechanics **51**, 295 (1995)
17. A. Mohsine, *Contribution à l'optimisation Fiabiliste en Dynamique des Structures Mécaniques*, Thèse de Doctorat, Institut National des Sciences Appliquées de Rouen, France, 2006
18. A. Haldar, S. Mahadevan, *Reliability Assessment Using Stochastic Finite Element Analysis* (John Wiley & Sons, Inc., New York, USA, 2000)
19. R. Rackwitz, B. Fiessler. Computers and Structures **9**, 484 (1978)
20. B. Fiessler, H.J. Neumann, R. Rackwitz. Journal of Engineering Mechanics. ASCE **1095**, 661 (1979)
21. A. Der Kiureghian, A. De Stefano. Journal of the Engineering Mechanics Division, American Society of Civil Engineers **117** (1991)
22. K. Breitung. Journal of Engineering Mechanics ASCE **110**, 357 (1984)
23. Y.-G. Zhao, T.A. Ono. Structural Safety **21**, 95 (1999)
24. L. Wang, R.V. Grandhi. Comput. Struct. **59**, 1139 (1996)
25. J.E. Rojas, O. Bendaou, A. Elhami, D.A. Rade, *in Proceedings of the Eleventh International Conference on Civil, Structural and Environmental Engineering Computing*, edited by B.H.V. Topping (Civil-Comp Press, Stirlingshire, UK, 2007), paper 43
26. J.E. Rojas, F.A.C. Viana, A. Elhami, D.A. Rade, *in Proceedings of the Eleventh International Conference on Civil, Structural and Environmental Engineering Computing*, edited by B.H.V. Topping (Civil-Comp Press, Stirlingshire, United Kingdom, 2007), paper 41
27. J.E. Rojas, F.A.C. Viana, D.A. Rade DA, V.Jr. Steffen, A Procedure for Structural Reliability Analysis Based on Finite Element Modeling, *IV Congresso Nacional de Engenharia Mecânica*, 22 a 25 de Agosto 2006, Recife-PE, Brasil, 2006

29

Direct multi-objective optimization of parametric geometrical models stored in PLM systems to improve functional product design

J.-B. Bluntzer[1], S. Gomes[1], D.H. Bassir[2,a], A. Varret[1] and J.C. Sagot[1]

[1] Laboratoire SeT, Equipe ERCOS, University of Technology of Belfort-Montbéliard, 90010 Belfort Cedex, France
[2] Institut FEMTO-ST, UMR/CNRS 6174, 24, rue de l'Épitaphe, 25000 Besançon, France

Abstract – This paper presents a functional analysis and design method, including an optimization loop, which has been integrated into our self-developed, web-based Product Lifecycle Management (PLM) platform. Our design methodology, which includes advanced and parametric CAD modelling and direct multi-objective optimization is described here. Functional design and knowledge-based engineering features, such as functional parameters, expert rule definitions and design experience feedback are all developed in the product design process in order to reduce costs, lead time and also to improve product quality and value. To confirm our research hypotheses, an experimental case-study is chosen: the ground-link system of a racing car design and manufacturing project, including conceptual, embodiment, detailed design and manufacturing phases.

1 Introduction

Concurrent engineering methods and tools are increasingly used within industrial organizations in order to reduce costs, lead time and also to improve product quality and value. The design activity involves many contributors and experts throughout the product lifecycle, which starts with product task identification / functional definition / product modelling / manufacturing and ends with its destruction or recycling. These contributors and experts must collaborate using software, such as PLM (Product Lifecycle Management) systems, which can handle concurrent engineering, in order to help design team members to manage information and knowledge in their project tasks.

Our research activity is to develop a design methodology based on well-known design approach such as "Total design" [1], "Systematic design" [2] or the Ullman mechanical design process [3], but applying direct multi-objective optimization, linked to functional design and knowledge based engineering [4] features, such as expert rule definition or design experience feedback, in order to reduce costs, lead time and also improve product quality and value.

In our global market context, where management, designers, subcontractors and customers are geographically distant from each other, industrial companies have to innovate, often by defining complex products [5] whose design spans various engineering problems and disciplines. At the same time, companies have grown in complexity but have also reduced their competencies to be ultimately specialized in few disciplines and thus have to work more extensively with subcontractors. Besides the traditional financial considerations, more recent industrial requirements, such as robustness and performance of the design, but also marketing criteria have come into play and are fast becoming key characteristics of the design. In order to stay competitive, these too must be optimized.

Our design and optimization methodology helps the designer to progress parametric CAD models of an optimized product [6] through functional requirements, design rules and design objectives that can be, respectively, verified and reached using optimization loops. Nowadays, actual real-world engineering design problems involve simultaneous optimization of several objectives and compliance with limiting factors that have been determined by the design team. Engineering design of complex systems is a decision making process that must select from among a set of options leading to an irrevocable allocation of resources. It is inherently a multi-objective process. As products become increasingly complex, their design process, which is geographically distributed [7], is usually characterized by numbers of design variables, parameters, requirements, constraints and objective functions. Taking into consideration the product complexity and the concurrent engineering design context, our methodology is based on a collaborative design process [8] and a multi-objectives optimization loop, using a Product Lifecycle Management (PLM) environment [9].

[a] Corresponding author: hbassir@univ-fcomte.fr

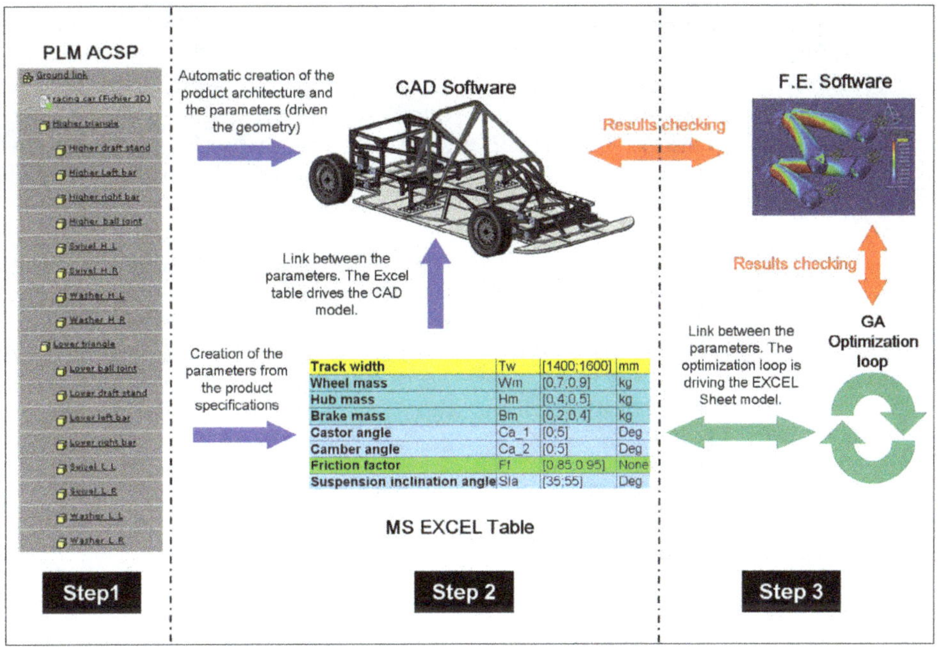

Fig. 1. Global methodology model for collaborative, functional and knowledge-based engineering using a PLM environment and integrating an optimization loop.

After this first presentation of the state of the art and of industrial requirements, we will present our functional and technical analysis and design approach, including an optimization loop, built into our self-developed web-based [10] PLM platform, called ACSP (in French: Atelier Coopératif de Suivi de Projets) [11]. Then, the main principles of multi-objective optimization using genetic algorithms are explained in the third section.

To confirm our research hypotheses, an experimental case study is chosen: the ground-link system of a racing car design and manufacturing project, including conceptual, embodiment and detailed design phases as well as manufacturing phases.

2 Proposed methodology and tools

In order to achieve our goals, our methodology includes three different steps as shown in Figure 2.

The first step is carried out by the designer. He has to input the customer specifications into the ACSP PLM system, in order to fill the functional parameter forms. At the same time, he has to establish the product architecture (also called product tree) [12]. In fact, he has to determine three different levels in order to characterize the product:

- Level 1: the global product, which is the assembly of all sub-products.
- Level 2: the sub-products, which are assemblies of parts which are kinematically fixed (no degrees of freedom).
- Level 3: the parts, which are the elementary components of the product.

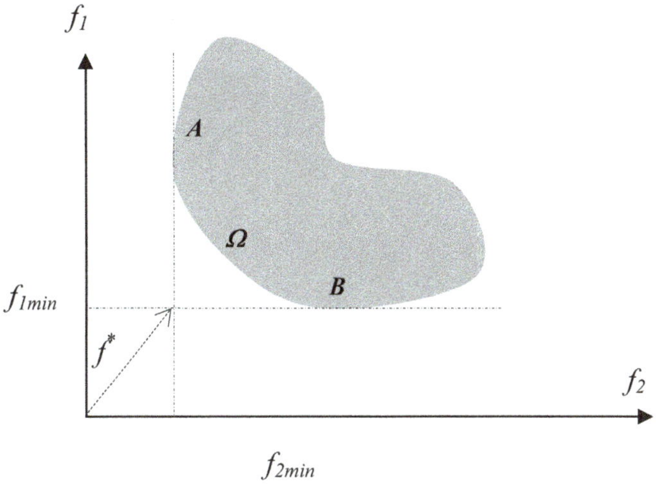

Fig. 2. Pareto curve (Min $(f_1$ *and* $f_2)$).

In the same step, the designer has to register all the design rules, the design terms, etc. in a shared knowledge database.

The second step which can be considered as the core of our methodology, is to carry out a functional/structural parameter (extracted from the first step) propagation taking into consideration the impact of each function (and also the corresponding value criteria, so called functional parameters), in each part of the product. Each parameter input into our CAD model is driven by an MS Excel file, also automatically generated from the PLM system. The main advantage of this kind of method is that we can modify the MS Excel files on other software in order to directly impact the CAD model.

In order to carry out the parameter propagation in the CAD model product tree, functional parameters of the product (global parameters) are copied into sub-products as local parameters. They are then linked to the previous global parameters through equality equations, in order to:

- generate a constraint/limitation/specification propagation from the top (global product) to the bottom (parts) of the CAD model product tree;
- at least, allow designers to carry out CAD modelling activities on a sample of the product Digital Mock-Up, after a check-out operation of this product sample (sub-product) using the PLM system. This approach preserves the impact of updated parameters on updated geometry and allows the other designers to work simultaneously on the other sub-products;
- interface our ACSP PLM system with commercial and parametric CAD software such as PTC's Pro/ENGINEER or Dassault Systems' CATIA v5. In this context, two requirements can be considered: data exchange between the PLM environment and the CAD tool, and also visualization of 3D models, using respectively, SQL and XML technology and VRML, 3DXML, X3D, U3D, DWF, open JT, etc. new CAD file formats, for the latter.

Finally, one last step can be included, as it is to optimize the parameters already defined in the MS Excel files. These parameters are directly sent into the FEM checking software, at each loop of the data computing process, in order to confirm the optimized results. The following figure describes graphically the previously described methodology.

In the next two sections, we will describe the optimization concepts using genetic algorithms and also how we can integrate this methodology in a real design case study that considers not only one but several objective functions at the same time.

3 Optimization concepts using genetic algorithms

In this paragraph, we will describe briefly the functioning of the genetic algorithms in the framework of multiobjective or multicriteria optimization.

3.1 Genetic algorithms

Genetic algorithm (GAs), is derived, as its name suggests, from natural selection. GAs provide an alternative to traditional optimization techniques for locating the optimal solutions in a complex search landscape. The first to introduce this paradigm is Holland in 1975 [13]. Since that time, many applications and publications had been described and especially after the last works of Goldberg in 1989 [14] which gave GAs this characteristic of providing global and efficient methods of overcoming complex optimization problems [14, 15].

The principle of GAs is to simulate the evolution of one population of individuals to which different production operators (*selection, crossover and mutation*) are applied. As the GAs start searching from different initial solutions, this gives them a global view of the problem. This global perspective prevents them from being trapped locally and allows them to explore the whole search landscape. Such algorithms identify the problem only through the value of the cost function and the values of the constraints. The behaviour of this algorithm is similar to a black box with several entries and one exit [20].

The functioning of GAs can be divided into three main parts [16, 17]: the coding of the parameters, the genetic operators and the choice of the objective functions. The general principles can be described as follows:

1. Assume an initial population of size N (containing N individuals or equivalent solutions).
2. Calculate the fitness of each individual in the population. The fitness is a given function related to the objective function.
3. Select randomly individuals from the current selection based on their fitness and form a new selection of size N. This is often called the reproduction operator.
4. Recombine the selected new population by identifying two individuals to represent the two parents, then choosing a crossing point and finally exchanging the values to the right of the crossing point in order to create two new sons. The crossing point is a location in the string representing the chromosome associated to each individual. Such step is represented by a crossover operator. It is carried out using a given probability independent of the choice of individuals.
5. The population obtained in the previous step is mutated with a given probability. Mutation is based on the string representation of the individuals. For instance, if the string representing the individual is coded as a sequence of 0 and 1, mutation is simply a switch of binary values to their opposite (0 to 1 and 1 to 0). Mutation introduces new individuals that are not in the initial set, which was defined in the first step.

The first problem to be faced during the use of the GAs is the representation of the individuals (coding of the parameters). It is the manner in which each variable of the optimization problem is coded. The coding can be binary coding or real coding [16, 17]. The binary coding is the common representation that is most often used in GAs. For continuum variables, many authors [18–20] prefer using real coding for its simplicity and efficiency of use in real problems. Real coding avoids the difficulties of achieving arbitrary precision in decision variables and the Hamming Cliff problem associated with binary string representation of real numbers. Whereas, in the case of discrete variables, the binary representation is more or less the unique solution to overcome certain problems. In this paper, real and binary coding were chosen to be implemented in our program.

A wise definition of the objective function is a very important task in the evolution process, because, GAs search the landscape for a solution often using only the discrete values of the objective function. If this function is not well defined, the GAs can not guarantee the location of the global solution.

Finally, the common operators used in GAs are selection, crossover and mutation. These operators are widely described in literature. However, in our program GAPS (Genetic Algorithm with Parallel Selection) [20], we have include the sharing concept that is inspired by nature. In a domain, two species can coexist and share the same resources. This idea [23] is introduced artificially in GAs through a sharing function. This function is to expel some individuals from a research-field that is reducing (*example of the niches or zones of local minima*) while the density of the population is increasing. The sharing function compares the relative distance between two individuals. Even if this function increases the calculation time of the GAs, it is very suitable in many real cases with many local optima because it improves the exploration capabilities of the GAs.

In the case of scalar optimization or mono-objective optimization, the selection operator is easy to implement. However, in the case of multi-objective optimization, the selection operator is based on a different concept [19,21]. In the following, we will make a brief introduction to multi-objective optimization and its characteristics, such as how to choose between the direct or post-priory approach and how to handle the limitations.

3.2 Multi-criteria/multi-objective optimization

A multi-criteria or multi-objective optimization problem is stated as follows: Find the vector of design variables $\mathbf{x} = [x_1, x_2, x_3, ., x_n]^T$ which minimizes the vector of objective functions $F(x)$.

$$\text{Min } \mathbf{F(x)} = \text{Min}[f_1(x), f_2(x). \dots . f_k(x)]. \quad (1)$$

Subject to linear or non-linear constraints: $\boldsymbol{g_j(x)} \geqslant 0$ $j = 1, 2, \dots, m$

The feasible domain (Fig. 3) defined by the constraints will be denoted by Ω. f_i: $\Omega \rightarrow R$, $i=1, 2,..,k$ are called criteria or objective functions and they represent the design objectives by which the performance is measured. A vector $\mathbf{x}^* \in \Omega$ is called Pareto-optimal solution, if there is no vector $\mathbf{x} \in \Omega$ which would decrease some criteria without causing a simultaneous increase in at least one criteria function. Usually several Pareto-optimal solutions exist.

The optimal solution from the Pareto domain can be reached by two main strategies: either by a direct approach or by a Posteriori approach (Fig. 4).

The Direct method is based on the transformation of the initial problem into a single optimization problem. This transformation can be made by using, for example, the weighting method. This reduces a vector-optimization problem to a scalar optimization problem, where for instance the scalar objective function $f(x)$ can be defined as the weighted sum of the individual objective functions. The Pareto-optimal solutions obtained by this method depend on the choice of the weighting factors α_i that will

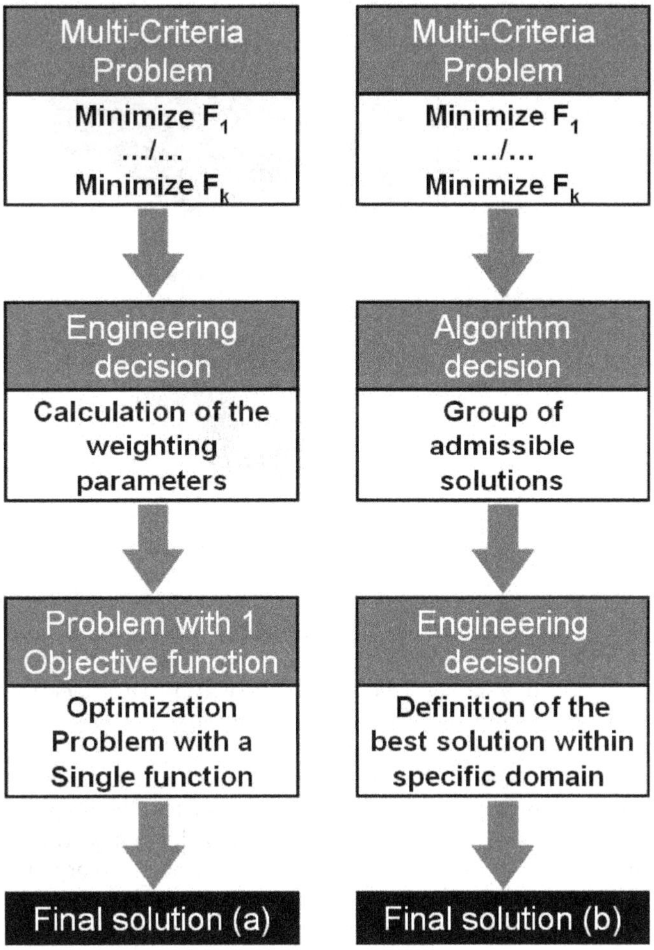

Fig. 3. Direct (a) and Posteriori (b) methods for multi-objective optimization.

generate a unique Pareto optimal solution of the original problem. This approach requires the definition of appropriate weighting factors that will guide the convergence of the optimal design process [22,23]. However, this definition depends on the decision-maker and the shape of the Pareto domain (continuity, convexity and the number of limitations to handle). If the weighting factors are not well chosen, this approach can converge into local zone of the Pareto domain, especially when the optimal solution is very sensitive to the weighting factors. This conclusion was described and demonstrated for some academic examples by various authors such as Das [24]. Other researchers have defined good strategies to calculate these weighting factors using sensitivity evaluation of the objective functions which is impossible in the case of the non differential functions. Even when the final solution is reached, there is no guarantee that it is the best one for the decision-maker, as he needs to run the process many times before making his decision.

As concerns the Posteriori method, it starts with a group of initial solutions that we spread uniformly to have a global idea about the Pareto front and to make the final decision easier to take for the decision-maker. In the case

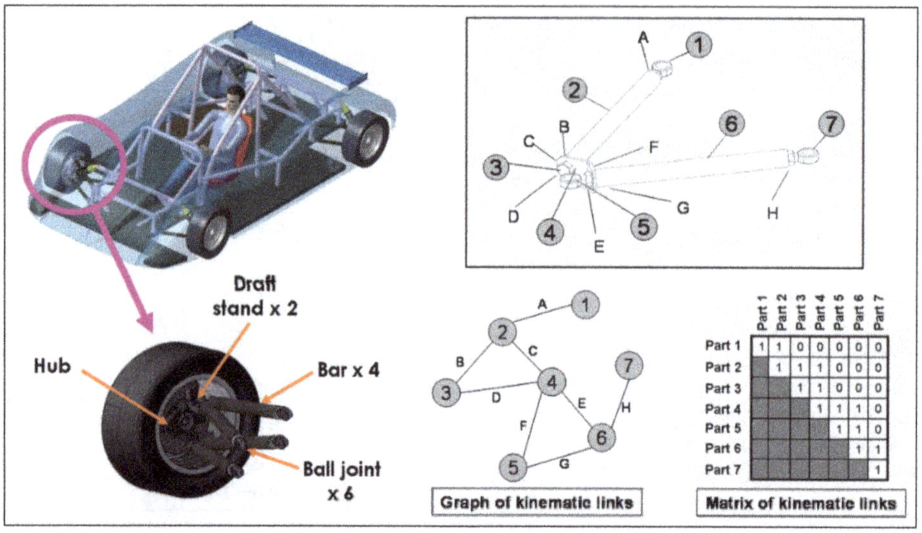

Fig. 4. Ground-link suspension system defined kinematically.

of a composite material, the difficulties in using a simple GA program are, firstly, that the comparison between two solutions can not be achieved easily in the selection process. Secondly, the constraints handling became difficult to represent as an inequality or equality equation. To overcome these difficulties, many researchers like Deb for instance, proposed approaches based on his previous investigations on the genetic algorithms [21]. He introduced the idea of no-dominated sorting to speed the convergence of his algorithm NSGA, to increase its performance and to obtain his actual algorithm NSGA-II. To choose between two solutions i and j in order to continue the iterative process of the GA, a non-dominate sorting approach is used in the selection process as follows.

- Between two admissible solutions i and j, we chose the one with the best objective function.
- If solution i is admissible and j non-admissible so we chose the solution i.
- If the solutions i and j are non-admissible, we chose the one with minimum violation of the limitations.

To handle the constraints in the GA program, we define the following composite fitness function for any solution x such as:

$$F(x) = f(x) \text{ if } x \text{ is feasible,} \qquad \text{otherwise } F(x) = f_{max} + CV(x). \quad (2)$$

f_{max} is the objective function value of the worst feasible solution in the population and $CV(x)$ is the overall normalized constraint violation of the solution x. Thus, there is no need to have any penalty parameter for handling the constraints as usually used in the common approaches. Constraints are normalized to avoid the scaling problems and are equal to one in the case of feasible solution.

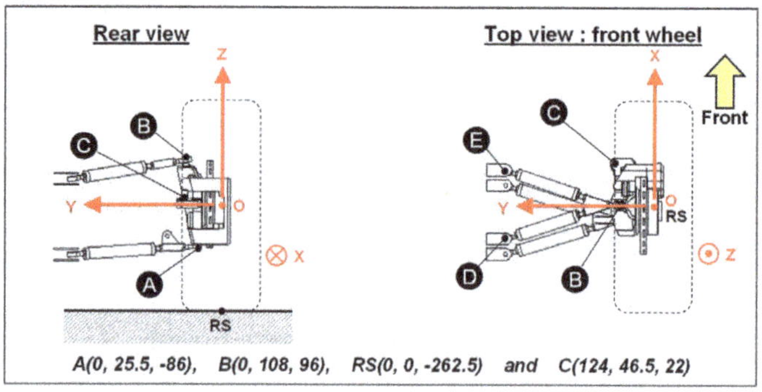

Fig. 5. Geometries of the wheel, hub and the brake.

4 Experimental design study

To demonstrate the efficiency of our approach, an experimental design case is chosen. Every year, our mechanical engineering and design department ERCOS has to develop and prototype an entire new racing vehicle. To simplify the demonstration, we choose to limit the experimental case study to a sub-product of the racing car: the ground-link suspension system (Fig. 4). This sub-product of the racing car includes many mechanical parts linking the wheel to the chassis. The design and optimization process will focus on the suspension triangles (wishbones) of a ground-link system. The main steps of our previously presented methodology are applied in this experimental case.

During the optimization process, the technical characteristics of external systems in interaction with the ground link suspension (chassis, hubs, brakes and wheels) have been considered as requirements (R) as follows.

The geometries of the wheel, hub and the brake are given below in Figure 5, with the associated coordinates.

The track width of the vehicle is fixed at 1600 mm, (Castor, camber angles, toe-out are equal to 0° (*simplified configuration of the sets*), the mass m of the wheel, hub and the brake is set at 15 kg, the vehicle suspended

Table 1. Table of the material variation.

Material		cost				
item	material	(€/T)	ρ (kg/m3)	E (Gpa)	Re (Mpa)	n
1	steel	700	7850	210	235	4
2	aluminium	4000	2900	75	180	4
3	stainless	4500	8700	203	185	4
4	carbon	5000	1530	50	555	1.5

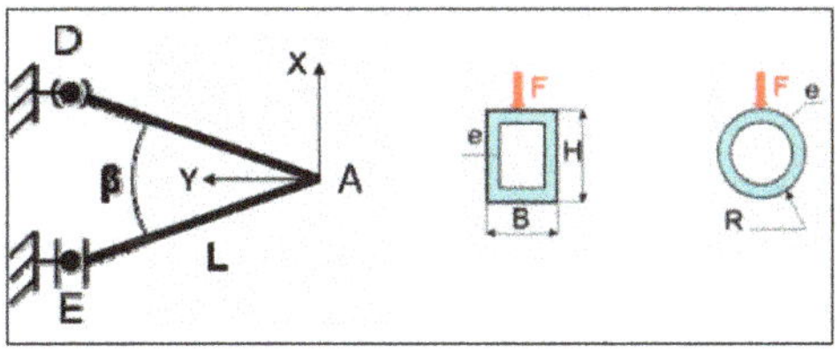

Fig. 6. Technical parameters scheme.

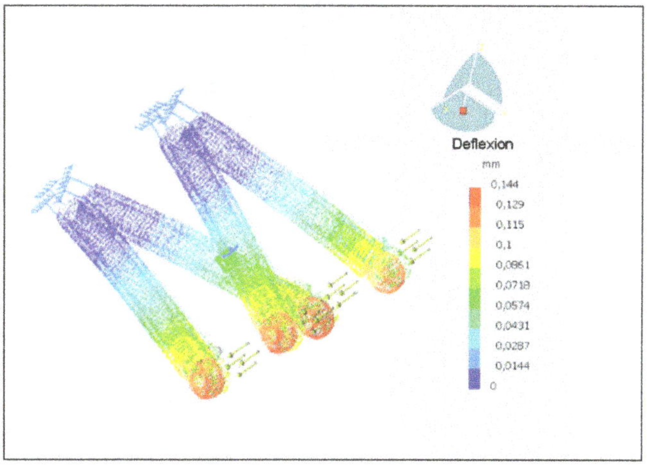

Fig. 7. Deflection representation using CATIA v5.

mass M is equal to 460 kg, the front/rear allocation of the suspended mass (M_v for the front and M_r for the rear are respectively equal to 160 kg and 300 kg), the vehicle center of gravity height above the ground h is fixed at 0.35 m, the vehicle wheel base is equal to 2.495 m, the suspension inclination angle in relation to the ground α_{am} is fixed at 45° and finally, the friction factor of the tyre on dry road is equal to 0.85. We should note also that the lower triangles are included in the horizontal plane. For each triangle, the bisectrix of the angle formed by the 2 tie-rods and the vehicle axis is perpendicular.

As concerns the requirements related to the mechanical characteristics of the ground link suspension, they are considered as a technical parameters (T) (Fig. 6). Based on these requirements, we obtain mixed variables for the optimization process (continuous and discrete variables).

The continuous variables are represented by the angle between the tie rods $\beta \in [20°;120°]$, the external diameter (or side) value H $\in$ [10 mm ; 40 mm], the thickness e $\in$ [1 mm;3 mm] and the tie-rod length $L = 300$ mm. The discrete variables are represented by the section type square shape (B=H or round shape: H = 2R) and the sets of material parameters (Tab. 1).

For the multi-objective optimization problem, we consider the three functions to be minimized that are described below in equations (3) to (5):

$$\text{Tie-rod mass} \quad m = \rho S L \tag{3}$$

$$\text{Global stresses} \quad \alpha_{max} = |\sigma_{tc}| + |\sigma_f| \tag{4}$$

$$\text{Global strains} \quad \Delta f = \sqrt{f^2 + dl^2}. \tag{5}$$

The optimization problem had been processed in two steps. First, we applied traditional mechanical analysis of undeformable solids, which enabled the maximum deceleration acceptable by the vehicle and loads generated at the various link points of the suspension triangles to be defined. Then, we defined a standard material strength calculation in order to identify the stresses and deformations corresponding to these loads.

The limitation or constraint G_j is represented by the maximum stress allowed in the normal section that should be lower than Re/n (Eq. (6)). Where Re and n are respectively the elastic limit and the safety margin

$$\alpha_{max} \leqslant \frac{Re}{n}. \tag{6}$$

The first step allows the highest load applied to the lower triangle at point A to be identified (the respective values of the load vectors in X and Y directions: – 1792 N; – 1482 N, in a constant deceleration phase, value $\gamma = 10$ m/s^2 corresponding to an emergency braking condition). However, this calculation identifies the front tie-rod of the lower triangle as the most highly stressed component. During the second step, we determine the normal and tangential load components in the front tie-rod and the corresponding stresses and strains.

As defined previously, during the project, requirements, technical parameters, objective functions, constraints and expert rules are integrated by the expert designers into the ACSP PLM environment and directly associated to the product part list. This data is then extracted and structured in order to create:

- knowledge archives in a knowledge management system, associated to ACSP;
- script files which automatically generate a CATIAv5 parametric CAD model and the Excel Files which describe the product parametric architecture without any solid features. These solid features will then be created by CAD designers, using the previous product architecture;
- the FEM model integrated into the CATIA v5 FEM Software. In this step, the new parameters connected with the design rules and the objective functions are

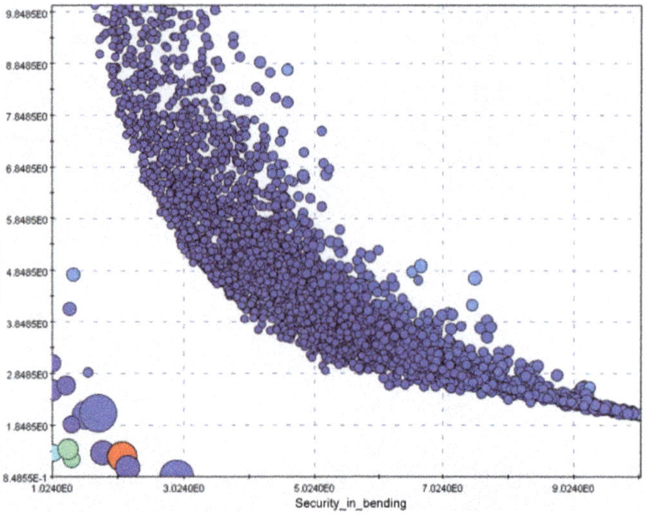

Fig. 8. Admissible solutions in the criteria space: deflection / safety in bending.

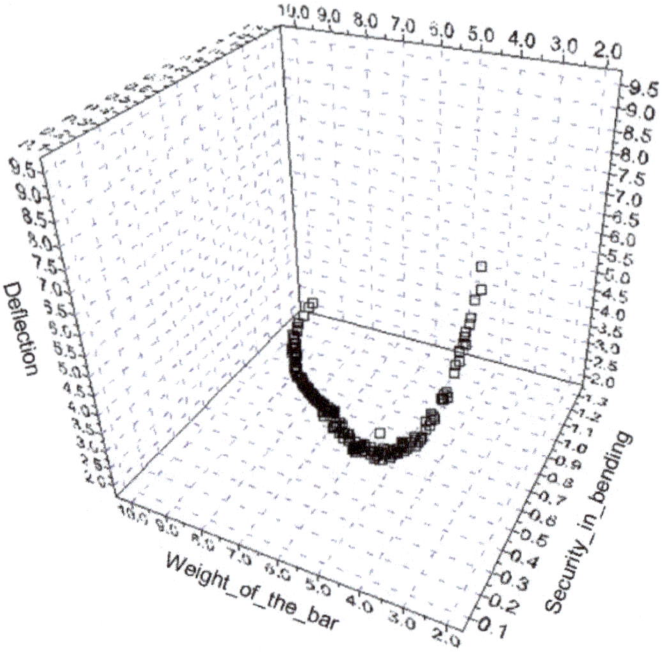

Fig. 9. Pareto domain in 3D with all functions: weight of the bar / safety in bending / Deflection.

checked in the software, for instance, we can check the deflection of the bar. The Figure 8 represents graphically the deflection using the FEM code integrated in CATIA v5.

With this method, we are now able to generate an output MS Excel file with the new calculated FEM constraints at each loop of the optimization algorithm, in order to give an input to the loop and perform the optimization computing process.

In Figures 8 and 9, we obtain the admissible solutions in the Bi-criteria space domain for deflection and safety in bending and then the Pareto curve at the end of the optimization process. Here, we observe the uniform distribution of the Pareto points in the criteria space with 3 functions: deflexion, safety in bending and the weight of the bar.

5 Conclusions

To conclude, parametric geometrical models stored in PLM systems and connected to knowledge-based engineering systems, for functional product design and automatic generation of design solutions, are becoming a reality in industry in order to improve productivity and quality in routine design engineering processes. Moreover these automatically generated solutions are not optimized taking into account the various, and sometimes contradictory design objectives and design constraints that must be considered in each project.

In this paper we have experimented with a direct, multi-objective optimization approach in order to help the designer involved in these routine design processes, to choose as quickly as possible the optimal design solutions, considering all the expert design rules, the objective functions (to be minimized or maximized) and also the design constraints.

We have also highlighted some difficulties related to multi-objective optimization and the handling method used in our genetic algorithm program. This approach has demonstrated its efficiency in the case of our model (the ground-link system of a racing car) in the obtaining of a uniform spray of the Pareto solutions.

The next step would be to improve the software design chain, in order to reduce time devoted to data exchange between the different software types involved in our partially automatic design methodology of optimal solutions (Product Lifecycle Management, Computer Aided Design, Knowledge Based Engineering, Finite Element modelling and simulation and Multi-objective Optimization).

The authors would like to thank Metropolitan Community of the region of Montbéliard, Franche-Comté Region Council, OSEO innovation, French Ministry of industry and the Automotive of the Future cluster, for their funding of this research activity.

References

1. S. Pugh, *Total Design - Integrated Methods for Successful Product Engineering*, 1st edn. (Addison-Wesley Publishing, 1990), p. 278
2. G. Pahl, W. Beitz, translated by K. Wallace, L. Blessing, F. Bauert, edited by K. Wallace. *Engineering Design, A Systematic Approach.* 2nd edn. (Springer-Verlag, New-York, 1995)
3. D. Ullman, *The mechanical design process* (MacGraw-Hill Higher Education, 2002), p. 409
4. D.E. Whitney, Q. Dong, J. Judson, C. Mascoli, *Introducing Knowledge-Based Engineering into an*

Interconnected Product development process. Procedings of the 1999 ASME Design Engineering Technical Conference, September, Las Vegas, Nevada. DET99-DTM8741, 1999
5. E. Rechting, Systems Architecting (Pentice Hall, New-York, NY, USA, 1991)
6. J. Eggers, D. Feillet, S. Kehl, M.O. Wagner, B. Yannou. European Journal of Operational Research 15 (2002)
7. D. Brissaud, O. Garro. Concurrent Engineering: Research and Applications **4**, 303 (1996)
8. T. Kvan. Automation in Construction **9**, 409 (2000)
9. W. Shen. Computers in Industry **52**, 1 (2003)
10. D.T. Liu, X.W. Xu. A Review of Web-based Product Data Management Systems. Computers in Industry **44**, 251 (2001)
11. S. Gomes, J.C. Sagot. *A concurrent engineering experience based on a cooperative and object oriented design methodology*, Integrated Design and Manufacturing in Mechanical Engineering (Kluwer Academic Publishers, Dordrecht, 2002)
12. S. Gomes, J.B. Bluntzer, J.-C. Sagot. Functional design through a PLM system for fastening routine definition of CAD models. PLEDM 2006, Playa Del Carmen, Mexico, 26 November – 1st December, 9p 2006
13. J.H. Holland, *Adaptation in natural and artificial systems* (University of Michigan Press, Ann Arbor, 1975)
14. D.E. Goldberg, *Genetic algorithms in search, optimisation, and machine learning* (Addison-Wesley, New York, 1989)
15. Z. Michalewicz, *Genetic Algorithms + Data Structures = Evolution* (Springer-Verlag, Heidelberg, 1994)
16. K. Deb, D.E. Goldberg, *An investigation of niche and species formation in genetic function optimization*, edited by J.D. Schaffer, Proceedings of the Third International Conference on Genetic Algorithms (Morgan Kauffman, San Mateo, 1989), pp. 42–50
17. K. Deb, R.B. Agrawal. Complex Systems, 115 (1995)
18. K. Deb, A. Patrap, S. Agarwal, T. Meyarivan. A fast and Elitist multi-objective Genetic algorithm : NSGA-II, Technical Report No. 200001 (Kanpur Genetic Algorithms Laboratory, India Institute of Technology, 2005)
19. F.X. Irisarri, D.H. Bassir, J.F. Maire, N. Carrere. *Multiobjective stacking sequence optimisation strategy for laminated composite structures*, First international conference on multidisciplinary optimization and applications (EDP Sciences), ISBN 978-2-7598-0023-0, 2007
20. D.H. Bassir, J.L. Zapico, M.P. González. International Journal of Simulation and Multidisciplinary Design Optimization **1**, 37 (2007)
21. K. Deb. International Journal of Simulation and Multidisciplinary Design Optimization **1**, 1 (2007)
22. M. Domaszewski, D.H. Bassir, W.H. Zhang, *Stress displacement and weight minimization by multicriteria optimization and game theory approach*, Computer Aided Optimum Design of Structures OPTI 99, Orlando, Florida, USA, WIT Press, 171 (1999)
23. W.H. Zhang, M. Domaszewski, C. Fleury. Int. J. Numer. Meth. Eng. **52**, 889 (2001)
24. I. Das, Non linear multicriteria optimization and robust optimality, Ph.D. Thesis, Rice University, Houston, USA, 1997

30

Bi-First: A simple and efficient algorithm to identify dissipated energy in impact problems

Z.-Q. Feng[1], B. Magnain[1], J.-M. Cros[1] and P. Joli[2]

[1] Université d'Évry-Val d'Essonne, Laboratoire de Mécanique d'Évry, 40 rue du Pelvoux, 91020 Évry, France
[2] Université d'Évry-Val d'Essonne, Laboratoire IBISC, 40 rue du Pelvoux, 91020 Évry, France

Abstract – The bi-potential method has been successfully applied for the modelling of frictional contact problems in static cases. This paper presents the extension of this method for dynamic analysis of impact problems with multiple deformable bodies. Instead of second order algorithms, a first order algorithm is applied for the numerical integration of the time-discretized equation of motion. The solution algorithm, named Bi-First, is simple and efficient. The principle of energy conservation for the given exemples is well preserved using the algorithm without any regularization. The numerical results also show clearly the physical energy dissipation introduced by frictional effects between the solids in contact.

Key words: Impact; energy dissipation; bi-potential method; time-integration

1 Introduction

Problems involving contact and friction are among the most difficult ones in mechanics and at the same time of crucial practical importance in many engineering branches. The main mathematical difficulty lies in the severe contact non-linearities because the natural first order constitutive laws of contact and friction phenomena are expressed by non-smooth multivalued force-displacement or force-velocity relations. In the last decade, substantial progress has been made in the analysis of contact problems using finite element procedures. A large number of algorithms for the numerical solution of the related finite element equations and inequalities have been presented in the literature. Review papers may be consulted for an extensive list of references [1–3]. See also the monographs by Kikuchi and Oden [4], Zhong [5] and Wriggers [6]. The popular penalty approximation and 'mixed' or 'trial-and-error' methods [7,8] appear, at first glance, suitable for many applications. But in this kind of method, the contact boundary conditions and friction laws are not satisfied accurately and it is tricky for the users to choose appropriate penalty factors. They may fail for stiff problems because of unpleasant numerical oscillations between contact statuses. The augmented Lagrangian method first appeared to deal with constrained minimization problems. Since friction problems are not minimization problems, the formulation needs to be extended. Alart and Curnier [9], Simo and Laursen [10] and De Saxcé and Feng [11] have obtained some extensions in mutually independent works. The first two works consist of applying Newton's method to the saddle-point equations of the augmented Lagrangian. De Saxcé and Feng proposed a theory called ISM (Implicit Standard Materials) and a bi-potential method, in which another augmented Lagrangian formulation was developed, which is essentially different from that of the first two works. In particular, in the bi-potential method, the frictional contact problem is treated in a reduced system by means of a reliable and efficient predictor-corrector solution algorithm. For the unilateral contact problems with friction, the classic approach is based on two minimum principles or two variational inequalities: the first for unilateral contact and the second for friction. The bi-potential method leads to a single displacement variational principle and a unique inequality. In consequence, the unilateral contact and the friction are coupled via a contact bi-potential. The application of the augmented Lagrangian method to the contact laws leads to an equation of projection onto Coulomb's cone, strictly equivalent to the original inequality [12]. For additional comments, see also the interesting discussion by Klarbring *et al.* [13,14].

For dynamic implicit analysis in structural mechanics, the most commonly used time integration algorithm is the second order algorithm such as Newmark, Wilson, HHT. Wriggers *et al.* [15] have developed a radial return mapping scheme to deal with impact-contact problems. Laursen *et al.* [16–18] have considered dynamic impact under the auspices of a conservative system and have proposed the means to address the dynamic contact conditions so that they preserve the global conservation properties. The integration scheme is based on the second order algorithm. Some first order algorithms have also been proposed by Zienkiewicz *et al.* [19] and Jean [20] for time stepping in structural dynamics.

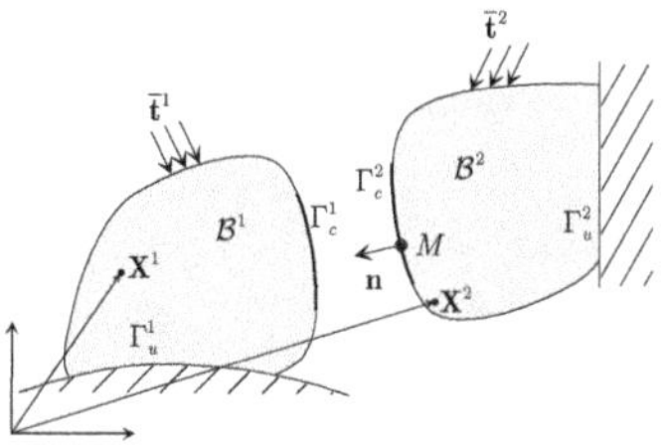

Fig. 1. Contact kinematics.

The aim of the present paper is to apply the bi-potential method for contact modeling in dynamic cases in the field of Non-Smooth Dynamics using the first order algorithm for integration of the equation of motion. The algorithm developed is named Bi-First and is implemented into the finite element code FER/Contact, using C++ with object oriented programming techniques. Two numerical examples are performed in this study to show the validity of the model developed. The first example concerns the oblique impact of an elastic plate onto a rigid surface with rebounding. The second example simulates the impact of two elastic cylinders in rigid walls. In order to show the physical energy dissipation by frictional effects and the behavior of the energy dissipation with respect to the friction coefficient, frictionless and frictional contact are considered for both examples.

2 Problem setting

2.1 Contact kinematics

In the following, basic definitions and notations used are described. Two deformable bodies $\mathcal{B}^\alpha$ (Fig. 1), $\alpha = 1, 2$, are considered. Each of them occupies the open, simply connected, bounded domain $\Omega^\alpha \subset \mathbb{R}^3$, whose generic point is denoted $\mathbf{X}^\alpha$. Furthermore, the solids are elastic and undergo large displacements. The boundary Γ^α of each body is assumed to be sufficiently smooth everywhere such that an outward unit normal vector, denoted by $\mathbf{n}^\alpha$, can be defined at any point M on Γ^α. At each time $t \in \mathbf{I}$, where $\mathbf{I} = [0, T]$ denotes the time interval corresponding to the loading process, the boundary Γ^α of the body $\mathcal{B}^\alpha$ can, in general, be split into three parts: Γ_u^α with prescribed displacements $\bar{\mathbf{u}}^\alpha$, Γ_t^α with prescribed boundary loads $\bar{\mathbf{t}}^\alpha$, and the potential contact surfaces Γ_c^α where the two bodies $\mathcal{B}^1$ and $\mathcal{B}^2$ may possibly come into contact at some time t:

$$\Gamma^\alpha = \Gamma_u^\alpha \cup \Gamma_t^\alpha \cup \Gamma_c^\alpha. \tag{1}$$

The successive deformed configurations of $\mathcal{B}^\alpha$ are described at each time t by the displacement fields $\mathbf{u}^\alpha$ defined on $\bar{\Omega}^\alpha$ (*i.e.* the closure of Ω^α). On the contact surface, a unique normal $\mathbf{n}$ directed towards $\mathcal{B}^1$ ($\mathbf{n} \equiv \mathbf{n}^2$) is defined and the tangential plane, orthogonal to $\mathbf{n}$ in $\mathbb{R}^3$, is denoted by $\mathbf{T}$. To construct an orthonormal local basis, two unit vectors $\mathbf{t}_x$ and $\mathbf{t}_y$ are defined within the plane $\mathbf{T}$. For describing the frictional contact interactions that may occur on Γ_c, we introduce the relative velocity with respect to $\mathcal{B}^2$

$$\dot{\mathbf{u}} = \dot{\mathbf{u}}^1 - \dot{\mathbf{u}}^2 \tag{2}$$

where $\dot{\mathbf{u}}^1$ and $\dot{\mathbf{u}}^2$ are the instantaneous velocities of $\mathcal{B}^1$ and $\mathcal{B}^2$, respectively. Let $\mathbf{r}$ be the contact force distribution exerted on $\mathcal{B}^1$ at M from $\mathcal{B}^2$. According to the action-reaction principle, $\mathcal{B}^2$ is subjected to the stress vector $-\mathbf{r}$. In the local coordinate system defined by the tangential plane $\mathbf{T}$ and the normal $\mathbf{n}$, any element $\dot{\mathbf{u}}$ and $\mathbf{r}$ may be uniquely decomposed as

$$\dot{\mathbf{u}} = \dot{\mathbf{u}}_t + \dot{u}_n\,\mathbf{n}, \qquad \dot{\mathbf{u}}_t \in \mathbf{T}, \qquad \dot{u}_n \in \mathbb{R}. \tag{3}$$

$$\mathbf{r} = \mathbf{r}_t + r_n\,\mathbf{n}, \qquad \mathbf{r}_t \in \mathbf{T}, \qquad r_n \in \mathbb{R}. \tag{4}$$

2.2 Contact law and friction rule

The unilateral contact law is characterized by a geometric condition of non-penetration, a static condition of no-adhesion and a mechanical complementary condition. These three conditions are known as the Signorini conditions. The non-penetration condition constraints the displacement fields $\mathbf{u}^\alpha$ and is given by

$$g(\mathbf{X}) = (\mathbf{X}^1 - \mathbf{X}^2) \cdot \mathbf{n} \geq 0 \tag{5}$$

where

$$\mathbf{X}^\alpha(t) = \mathbf{X}^\alpha(t = 0) + \mathbf{u}^\alpha. \tag{6}$$

The position vector $\mathbf{X}^2$ is found as the closest-point projection of the point $\mathbf{X}^1 \in \Gamma_c^1$ on the surface Γ_c^2. Denoting by h the initial gap obtained at the beginning of each time step.

$$h = (\mathbf{X}^1 - \mathbf{X}^2) \cdot \mathbf{n} \geq 0. \tag{7}$$

The impenetrability Signorini conditions are given by

$$u_n + h \geq 0, \quad r_n \geq 0, \quad (u_n + h)\, r_n = 0. \tag{8}$$

These conditions have to be satisfied at each time-instant $t \in \mathbf{I}$. Assume now that the bodies are initially in contact on a certain portion of Γ_c. On this part of Γ_c, the Signorini conditions turn into

$$u_n \geq 0, \quad r_n \geq 0, \quad u_n\, r_n = 0. \tag{9}$$

In general, at any time $t \in \mathbf{I}$, the potential contact surfaces Γ_c^α can be split into two disjoint parts: ${}^+\Gamma_c$ where the bodies are already in contact and ${}^-\Gamma_c^\alpha$ where the bodies are not in contact:

$$\Gamma_c^\alpha = {}^+\Gamma_c \cup {}^-\Gamma_c^\alpha. \tag{10}$$

In contrast to Γ_c^α, ${}^+\Gamma_c$ and ${}^-\Gamma_c^\alpha$ change in time t and can be empty at some $t \in \mathbf{I}$. We must stress that with the formulation (9) only a loss of contact is allowed and the extension of the contact area cannot be modelled with these relations. In the case of dynamic analysis such as impact problems, the Signorini conditions can be formulated, on ${}^+\Gamma_c$, in terms of relative velocity

$$\dot{u}_n \geq 0, \quad r_n \geq 0, \quad \dot{u}_n\, r_n = 0 \qquad \text{on } {}^+\Gamma_c. \tag{11}$$

$$\begin{array}{ll}\text{if} \quad r_n = 0 \text{ then } \dot{u}_n \geq 0 & \text{! separating} \\ \text{elseif } \mathbf{r} \in \text{int}\, K_\mu \text{then} \dot{u}_n = 0 \text{ and } -\dot{\mathbf{u}}_t = \mathbf{0} & \text{! sticking} \\ \text{else} \quad (\mathbf{r} \in \text{bd}\, K_\mu \text{ and } r_n > 0) & \\ \quad \left\{ \dot{u}_n \geq 0 \text{ and } \exists\, \dot{\lambda} > 0 \text{ such that } -\dot{\mathbf{u}}_t = \dot{\lambda} \frac{\mathbf{r}_t}{\|\mathbf{r}_t\|} \right\} & \text{! sliding} \\ \text{endif} & \end{array} \tag{13}$$

When $\dot{u}_n \geq 0$, the bodies are separating while they remain in contact for $\dot{u}_n = 0$. The previous formulation of the Signorini conditions (11) can be combined with the sliding rule to derive the complete frictional contact law applicable on the contacting part of Γ_c. This complete law specifies possible velocities of bodies that satisfy impenetrability, non-adhesion and the sliding rule. Obviously, for a strictly positive gap ($u_n \geq 0$), the normal relative velocity is arbitrary ($\dot{u}_n \in \mathbb{R}$) and the normal reaction force is equal to zero ($r_n = 0$). Motions of bodies that are not in contact are arbitrary until contact is made. This choice is motivated by the fact that the emphasis is put on the definition of admissible evolutions for contacting bodies where the time-integration has to be performed. In the rest of the paper, a "minus" sign will always precede the relative tangential velocity $-\dot{\mathbf{u}}_t$ to emphasize its opposite direction to the friction force.

Classically, a rate independent dry friction law is characterized by a kinematic slip rule. In this work, the classical Coulomb friction rule is used. The set of admissible forces, denoted by K_μ, is defined by

$$K_\mu = \left\{ \mathbf{r} \in \mathbb{R}^3 \quad \text{such that } \|\mathbf{r}_t\| - \mu r_n \leq 0 \right\}. \tag{12}$$

K_μ is the so-called Coulomb's cone and is convex.

2.3 Complete frictional contact law

We consider now the previous friction law embedding an impenetrability condition for completeness. On the contact surface Γ_c, the sliding rule can be combined with the rate form of the Signorini conditions to obtain the frictional contact law that specifies possible scenarios on the contact area (stick, slip, separation). The multivalued nature of this strongly non-linear law makes problems involving frictional contact among the most difficult ones in solid mechanics. Two overlapped "*if...then...else*" statements can be used to write it analytically:

see equation (13) above

where "int K_μ" and "bd K_μ" denote the interior and the boundary of K_μ, respectively. The multivalued character of the law lies in the first and the second part of the statement. If r_n is null then $\dot{\mathbf{u}}$ is arbitrary but its normal component $\dot{u}_n$ should be positive. In other words, one single element of $\mathbb{R}^3$ ($\mathbf{r} = \mathbf{0}$) is associated with an infinite number of velocity vectors $\dot{\mathbf{u}} \in \mathbb{R}^3$. The same arguments can be developed for the second part of the statement. The inverse law, *i.e.* the relationship $\mathbf{r}(-\dot{\mathbf{u}})$, can be written as:

$$\begin{array}{ll}\text{if} \quad \dot{u}_n > 0 \text{ then } r_n = 0 & \text{! separating} \\ \text{elseif } \dot{\mathbf{u}} = \mathbf{0} \text{ then } \mathbf{r} \in K_\mu & \text{! sticking} \\ \text{else} \quad (\dot{\mathbf{u}} \in \mathbf{T} - \{\mathbf{0}\}) & \\ \quad \left\{ \dot{u}_n \geq 0 \text{ and } \mathbf{r}_t = \mu\, r_n \frac{-\dot{\mathbf{u}}_t}{\|-\dot{\mathbf{u}}_t\|} \right\} & \text{! sliding} \\ \text{endif} & \end{array} \tag{14}$$

The complete form of the frictional contact law involves three possible states, which are separating, contact with sticking, and contact with sliding. Only the last state produces energy dissipation.

3 The bi-potential method

De Saxcé and Feng [12] have shown that the contact law (13) is equivalent to the following differential inclusion:

$$-\left(\dot{\mathbf{u}}_t + (\dot{u}_n + \mu\| - \dot{\mathbf{u}}_t\|)\, \mathbf{n}\right) \in \partial \bigcup_{K_\mu} \mathbf{r} \tag{15}$$

where $\bigcup_{K_\mu} \mathbf{r}$ denotes the so-called indicatory function of the closed convex set K_μ:

$$\bigcup_{K_\mu}(\mathbf{r}) = \begin{cases} 0 & \text{if } \mathbf{r} \in K_\mu \\ +\infty & \text{otherwise}. \end{cases} \tag{16}$$

The following contact bi-potential is obtained:

$$b_c(-\dot{\mathbf{u}}, \mathbf{r}) = \bigcup_{\mathbb{R}_-}(-\dot{u}_n) + \bigcup_{K_\mu}(\mathbf{r}) + \mu\, r_n \| - \dot{\mathbf{u}}_t\| \tag{17}$$

where $\mathbb{R}_- =]-\infty, 0]$ is the set of the negative and null real numbers.

In order to avoid nondifferentiable potentials that occur in nonlinear mechanics, such as in contact problems, it is convenient to use the Augmented Lagrangian Method [9–13]. For the contact bi-potential b_c, given by (17), provided that $\dot{u}_n \geq 0$ and $\mathbf{r} \in K_\mu$, we have:

$$\forall\, \mathbf{r}' \in K_\mu, \quad \varrho\mu(r_n' - r_n)\| - \dot{\mathbf{u}}_t\| + (\mathbf{r}' - (\mathbf{r} - \varrho\dot{\mathbf{u}})) \cdot (\mathbf{r}' - \mathbf{r}) \geq 0 \tag{18}$$

where ϱ is a solution parameter which is not user-defined. In order to ensure numerical convergence, ϱ can be chosen as the maximum value of the diagonal terms of the local contact stiffness matrix. Taking account of the decomposition (3,4), the following inequality has to be satisfied:

$$\mathbf{r}' \in K_\mu, \quad (\mathbf{r} - \boldsymbol{\tau}) \cdot (\mathbf{r}' - \mathbf{r}) \geq 0 \tag{19}$$

$$\begin{array}{lll} \text{if} & \mu|\boldsymbol{\tau_t}^{i+1}| < -\tau_n^{i+1} \text{ then } \mathbf{r}^{i+1} = 0 & \text{! separating} \\ \text{elseif} & |\boldsymbol{\tau_t}^{i+1}| < \mu\, \tau_n^{i+1} \text{ then } \mathbf{r}^{i+1} = \boldsymbol{\tau}^{i+1} & \text{! sticking} \\ \text{else} & \mathbf{r}^{i+1} = \boldsymbol{\tau}^{i+1} - \dfrac{(\|\boldsymbol{\tau}_t^{i+1}\| - \mu\, \tau_n^{i+1})}{(1+\mu^2)} \Big(\dfrac{\boldsymbol{\tau}_t^{i+1}}{\|\boldsymbol{\tau}_t^{i+1}\|} + \mu\, \mathbf{n} \Big) & \text{! sliding} \end{array} \tag{23}$$

where the modified augmented surface traction $\boldsymbol{\tau}$ is defined by

$$\boldsymbol{\tau} = \mathbf{r} - \varrho\big(\dot{\mathbf{u}}_t + (\dot{u}_n + \mu \| - \dot{\mathbf{u}}_t \|)\mathbf{n}\big). \tag{20}$$

The inequality (19) means that $\mathbf{r}$ is the projection of $\boldsymbol{\tau}$ onto the closed convex Coulomb's cone:

$$\mathbf{r} = \operatorname{proj}(\boldsymbol{\tau}, K_\mu). \tag{21}$$

For the numerical solution of the implicit equation (21), Uzawa's algorithm can be used, which leads to an iterative process involving one predictor-corrector step:

$$\begin{array}{l} \text{Predictor } \boldsymbol{\tau}^{i+1} = \mathbf{r}^i - \varrho^i\big(\dot{\mathbf{u}}_t^i + (\dot{u}_n^i + \mu \| - \dot{\mathbf{u}}_t^i \|)\,\mathbf{n}\big), \\ \text{Corrector } \mathbf{r}^{i+1} = \operatorname{proj}(\boldsymbol{\tau}^{i+1}, K_\mu). \end{array} \tag{22}$$

It is worth noting that, in this algorithm, the unilateral contact and the friction are coupled via the bi-potential. Another gist of the bi-potential method is that the corrector can be analytically found with respect to the three possible contact statuses: $\boldsymbol{\tau} \subset K_\mu$ (contact with sticking), $\boldsymbol{\tau} \subset K_\mu^*$ (no contact) and $\boldsymbol{\tau} \subset \mathbb{R}^3 - K_\mu \bigcup K_\mu^*$ (contact with sliding). K_μ^* is the polar cone of K_μ. This corrector step is explicitly given as follows:

see equation (23) above

It is important to emphasize the fact that this explicit formula is valid for both 2D and 3D contact problems with Coulomb's friction and allows us to obtain very stable and accurate results.

4 Finite element formulation of nonlinear structures

4.1 Total Lagrangian formulation

In the linear analysis, a linear relation is assumed between strains and displacements. However, if there are large displacements and rotations, such as in the case of dynamic multibody contact problems, the nonlinear relation between strains and displacements cannot be ignored. Also, the equilibrium equation of internal and external forces should be considered in the deformed configuration. See the monographs by Crisfield [21] and Simo and Hughes [22] for more details on computational aspects of nonlinear problems. The geometrically nonlinear analysis may be described by using the total or the updated Lagrangian formulations. The total Lagrangian formulation is derived with respect to the initial configuration. The updated Lagrangian formulation is derived with respect to the current configuration. In other words, the total Lagrangian formulation constructs the tangent stiffness matrix with respect to the initial configuration. On the other hand, the updated Lagrangian formulation constructs the tangent stiffness matrix with respect to the current configuration. The updated Lagrangian formulation is computationally effective because it does not include the initial displacement matrix. In the total Lagrangian formulation, the initial configuration remains constant. This simplifies the computation. Therefore, the total Lagrangian formulation was selected in this work for the finite element discretization. In order to describe the geometrical transformation problems, the deformation gradient tensor is defined by

$$\boldsymbol{\Phi} = \mathbf{Id} + \nabla \mathbf{u} \tag{24}$$

where $\mathbf{Id}$ is the unity tensor and $\nabla \mathbf{u}$ the displacement gradient tensor. Because of large displacements and rotations, Green-Lagrangian strain is adopted for the nonlinear relationships between strains and displacements. We note $\mathbf{C}$ the stretch tensor or the right Cauchy-Green deformation tensor ($\mathbf{C} = \boldsymbol{\Phi}^T\boldsymbol{\Phi}$). The Green-Lagrangian strain tensor $\mathbf{E}$ is defined by

$$\mathbf{E} = \frac{1}{2}(\mathbf{C} - \mathbf{I}). \tag{25}$$

In the context of the finite element method and from equations (24, 25), the Green-Lagrangian strain includes formally linear and nonlinear terms in function of nodal displacements:

$$\mathbf{E} = \Big(\mathbf{B}_L + \frac{1}{2}\mathbf{B}_{NL}(\mathbf{u})\Big)\mathbf{u} \tag{26}$$

where $\mathbf{B}_L$ is the matrix which relates the linear strain term to the nodal displacements, and $\mathbf{B}_{NL}(\mathbf{u})$, the matrix which relates the nonlinear strain term to the nodal displacements. From equation (26), the incremental form of the strain-displacement relationship is

$$\delta\mathbf{E} = \big(\mathbf{B}_L + \mathbf{B}_{NL}(\mathbf{u})\big)\delta\mathbf{u}. \tag{27}$$

In the case of elastic or hyperelastic laws, there exists an elastic potential function W (or strain energy density function) which is a scalar function of the strain tensors, whose derivative with respect to a strain component determines the corresponding stress component. This can be expressed by

$$\mathbf{S} = \frac{\partial W}{\partial \mathbf{E}} = 2\frac{\partial W}{\partial \mathbf{C}} \tag{28}$$

where $\mathbf{S}$ is the second Piola-Kirchhoff stress tensor. In the particular case of isotropic Saint-Venant-Kirchhoff material models, we have

$$W = \frac{1}{2}\mathbf{E} : \mathbf{DE}. \tag{29}$$

So $\mathbf{S}$ can be written by

$$\mathbf{S} = \mathbf{D}\mathbf{E} \tag{30}$$

where $\mathbf{D}$ denotes the usual material secant tangent. Using the principle of virtual displacement, the virtual work δU is given as

$$\delta U = \mathbf{M}\ddot{\mathbf{u}}\,\delta\mathbf{u} + \mathbf{A}\dot{\mathbf{u}}\,\delta\mathbf{u} + \int_{V_0} \mathbf{S}\,\delta\mathbf{E}\,dV - \mathbf{F}_{ext}\,\delta\mathbf{u} - \mathbf{R}\,\delta\mathbf{u} = 0 \tag{31}$$

where V_0 is the volume of the initial configuration, $\mathbf{F}_{ext}$ the vector of external loads, $\mathbf{R}$ the contact reaction vector, $\mathbf{M}$ the mass matrix, $\mathbf{A}$ the damping matrix, $\dot{\mathbf{u}}$ the velocity vector and $\ddot{\mathbf{u}}$ the acceleration vector. Substituting $\delta\mathbf{E}$ from equation (27) into equation (31) results in

$$\delta U = \mathbf{M}\ddot{\mathbf{u}}\,\delta\mathbf{u} + \mathbf{A}\dot{\mathbf{u}}\,\delta\mathbf{u} + \int_{V_0} \mathbf{S}(\mathbf{B}_L + \mathbf{B}_{NL}(\mathbf{u}))\delta\mathbf{u}\,dV - \mathbf{F}_{ext}\,\delta\mathbf{u} - \mathbf{R}\,\delta\mathbf{u} = 0. \tag{32}$$

The vector of internal forces is defined by

$$\mathbf{F}_{int} = \int_{V_0} \mathbf{S}(\mathbf{B}_L + \mathbf{B}_{NL}(\mathbf{u}))dV. \tag{33}$$

Since $\delta\mathbf{u}$ is arbitrary, a set of nonlinear equations can be obtained as

$$\mathbf{M}\ddot{\mathbf{u}} + \mathbf{A}\dot{\mathbf{u}} + \mathbf{F}_{int} - \mathbf{F}_{ext} - \mathbf{R} = 0. \tag{34}$$

It is noted that the stiffness effect is taken into account by the internal forces vector $\mathbf{F}_{int}$. Equation (34) can be transformed into

$$\mathbf{M}\,\ddot{\mathbf{u}} = \mathbf{F} + \mathbf{R}, \quad \text{where } \mathbf{F} = \mathbf{F}_{ext} - \mathbf{F}_{int} - \mathbf{A}\dot{\mathbf{u}} \tag{35}$$

with the initial conditions at $t = 0$

$$\dot{\mathbf{u}} = \dot{\mathbf{u}}_0 \text{ and } \mathbf{u} = \mathbf{u}_0. \tag{36}$$

Taking the derivative of $\mathbf{F}_{int}$ with respect to the nodal displacements $\mathbf{u}$ gives the tangent stiffness matrix as

$$\mathbf{K} = \frac{\partial \mathbf{F}_{int}}{\partial \mathbf{u}} = \int_{V_0} \Big(\frac{\partial \mathbf{S}}{\partial \mathbf{u}}(\mathbf{B}_L + \mathbf{B}_{NL}(\mathbf{u})) + \mathbf{S}\,\frac{\partial \mathbf{B}_{NL}(\mathbf{u})}{\partial \mathbf{u}}\Big)dV. \tag{37}$$

In addition, by using equations (27, 30), the tangent stiffness matrix is in fact the sum of the elastic stiffness matrix $\mathbf{K}_e$, the geometric stiffness (or initial stress stiffness) matrix $\mathbf{K}_\sigma$ and the initial displacement stiffness matrix $\mathbf{K}_u$:

$$\mathbf{K} = \mathbf{K}_e + \mathbf{K}_\sigma + \mathbf{K}_u \tag{38}$$

where

$$\mathbf{K}_e = \int_{V_0} \mathbf{B}_L^T \mathbf{D} \mathbf{B}_L\,dV, \tag{39}$$

$$\mathbf{K}_\sigma = \int_{V_0} \mathbf{S}\frac{\partial \mathbf{B}_{NL}}{\partial \mathbf{u}}\,dV, \tag{40}$$

$$\mathbf{K}_u = \int_{V_0} (\mathbf{B}_L^T\mathbf{D}\mathbf{B}_{NL} + \mathbf{B}_{NL}^T\mathbf{D}\mathbf{B}_L + \mathbf{B}_{NL}^T\mathbf{D}\mathbf{B}_{NL})dV. \tag{41}$$

4.2 First order integration algorithm

We can now integrate equation (35) between consecutive time configuration t and $t + \Delta t$. The most common method to do that is the Newmark method which is based on a second order algorithm. However, in impact problems, higher order approximation does not necessarily mean better accuracy, and may even be superfluous. At the moment of a sudden change of contact conditions (impact, release of contact), the velocity and acceleration are not continuous, and excessive regularity constraints may lead to serious errors. For this reason, Jean [20] has proposed a first order algorithm which is used in this work. This algorithm is based on the following approximations:

$$\int_t^{t+\Delta t} \mathbf{M}\,d\dot{\mathbf{u}} = \mathbf{M}\left(\dot{\mathbf{u}}^{t+\Delta t} - \dot{\mathbf{u}}^t\right) \tag{42}$$

$$\int_t^{t+\Delta t} \mathbf{F}\,dt = \Delta t\left((1-\xi)\,\mathbf{F}^t + \xi\,\mathbf{F}^{t+\Delta t}\right) \tag{43}$$

$$\int_t^{t+\Delta t} \mathbf{R}\,dt = \Delta t\,\mathbf{R}^{t+\Delta t} \tag{44}$$

$$\mathbf{u}^{t+\Delta t} - \mathbf{u}^t = \Delta t\left((1-\theta)\,\dot{\mathbf{u}}^t + \theta\,\dot{\mathbf{u}}^{t+\Delta t}\right) \tag{45}$$

where $0 \le \xi \le 1;\ 0 \le \theta \le 1$. In the iterative solution procedure, all the values at time $t+\Delta t$ are replaced by the values of the current iteration $i+1$; for example, $\mathbf{F}^{t+\Delta t} = \mathbf{F}^{i+1}$. A standard approximation of $\mathbf{F}^{i+1}$ gives

$$\begin{aligned}\mathbf{F}^{i+1} &= \mathbf{F}^i_{int} + \frac{\partial \mathbf{F}}{\partial \mathbf{u}}(\mathbf{u}^{i+1} - \mathbf{u}^i) + \frac{\partial \mathbf{F}}{\partial \dot{\mathbf{u}}}(\dot{\mathbf{u}}^{i+1} - \dot{\mathbf{u}}^i)\\ &= \mathbf{F}^i_{int} - \mathbf{K}^i\,\Delta\mathbf{u} - \mathbf{A}^i\,\Delta\dot{\mathbf{u}}.\end{aligned} \tag{46}$$

Finally, we obtain the recursive form of (19) in terms of displacements:

$$\begin{aligned}\bar{\mathbf{K}}^i\,\Delta\mathbf{u} &= \bar{\mathbf{F}}^i + \bar{\mathbf{F}}^i_{acc} + \mathbf{R}^{i+1},\\ \mathbf{u}^{i+1} &= \mathbf{u}^i + \Delta\mathbf{u}\end{aligned} \tag{47}$$

where the so-called effective terms are given by

$$\bar{\mathbf{K}}^i = \xi\,\mathbf{K}^i + \frac{\xi}{\theta\,\Delta t}\mathbf{A}^i + \frac{1}{\theta\,\Delta t^2}\mathbf{M}^i, \tag{48}$$

$$\bar{\mathbf{F}}^i_{acc} = -\frac{1}{\theta \Delta t^2}\,\mathbf{M}^i\left\{\mathbf{u}^i - \mathbf{u}^t - \Delta t\,\dot{\mathbf{u}}^t\right\}, \tag{49}$$

$$\bar{\mathbf{F}}^i = (1-\xi)\left(\mathbf{F}^t_{int} + \mathbf{F}^t_{ext}\right) + \xi\left(\mathbf{F}^i_{int} + \mathbf{F}^{t+\Delta t}_{ext}\right). \tag{50}$$

At the end of each time step, the velocity is updated by

$$\dot{\mathbf{u}}^{t+\Delta t} = \Big(1 - \frac{1}{\theta}\Big)\,\dot{\mathbf{u}}^t + \frac{1}{\theta\,\Delta t}(\mathbf{u}^{t+\Delta t} - \mathbf{u}^t). \tag{51}$$

By setting $\theta = \frac{1}{2}$, this scheme is then called the implicit trapezoidal rule and it is equivalent to the Tamma - Namburu method in which the acceleration need not be computed [23]. See [24] for the interesting comments on time stepping algorithms and on energy conservation.

It is noted that equation (47) is strongly non-linear, because of large rotations and large displacements of solid, for instance in multibody contact/impact problems. Besides, as mentioned above, the constitutive law of contact with friction is usually represented by inequalities and the contact potential is even nondifferentiable. Instead of solving this equation in consideration of all nonlinearities at the same time, Feng [25] has proposed a solution strategy which consists in separating the nonlinearities in order to overcome the complexity of calculation and to improve the numerical stability. As $\Delta \mathbf{u}$ and $\mathbf{R}$ are both unknown, equation (47) cannot be directly solved. First, the vector $\mathbf{R}$ is determined by the bi-potential method in a reduced system, which only concerns contact nodes. Then, the vector $\Delta \mathbf{u}$ can be computed in the whole structure, using contact reactions as external loading. It is very important to note that, as opposed to the penalty method or Lagrange multiplier method, the bi-potential method neither changes the global stiffness matrix, nor increases the degrees of freedom. One consequence of this interesting property is that it is easy to implement contact and friction problems in an existing general-purpose finite element code by this method. In addition, the solution procedure is more stable because of the separation of nonlinearities and improved numerical algorithms for calculation of contact reactions.

4.3 Energy computation

After determining the displacement and the velocity fields, we can calculate different energies. The total elastic strain energy of the contact bodies (discretized by n_{el} finite elements) is then written by

$$E_e = \sum_{e=1}^{n_{el}} \int_{\Omega_e} W_e \, d\Omega. \tag{52}$$

The total kinetic energy can be calculated at the global level by

$$E_k = \frac{1}{2} \dot{\mathbf{u}}^T \mathbf{M} \dot{\mathbf{u}}. \tag{53}$$

Finally, the total energy of the system of solids is

$$E_t = E_e + E_k. \tag{54}$$

The case of interest for the analysis presented below corresponds to the homogeneous Neumann problem, characterized by no imposed boundary displacements and no external loading. In addition, if frictionless contact is considered, the total energy should be conserved. For the given examples, this fundamental energy conservation property has been observed.

5 Numerical results

The algorithms presented above have been implemented and tested in the finite element code FER/Contact [26]. Many application examples, in static or quasi-static cases, have been carried out using the present method [25,27,28].

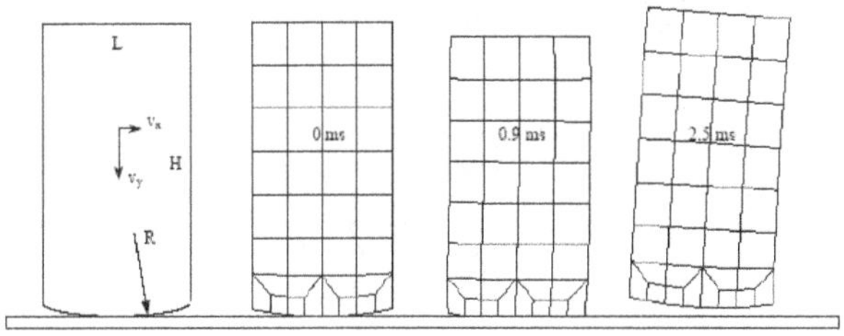

Fig. 2. Oblique impact of an elastic plate: geometry and deformed shapes vs time.

Table 1. Comparison of CPU time.

Method	Computer	CPU time (s)
Ko & Kwak [29]	CRAY 2S/4-128	19 000
Kim & Kwak [30]	HP 720	430
present	PC Pentium 4/2.8 GHz	7

To illustrate the behavior of a contact/impact simulation by the Bi-First algorithm described above, we consider two example applications. For both cases, we assume that no amortissement exists except for Coulomb friction between contact surfaces, *i.e.* $\mathbf{A} = 0$ in equations (34, 35, 48).

5.1 Oblique impact of an elastic plate with rebounding

The first example of dynamics analysis will be presented to show the validity and efficiency of the model developed. The problem concerns the oblique impact of an elastic plate onto a rigid surface with rebounding. This example has been proposed and studied by Kwak *et al.* [29, 30] using Linear Complementarity Problem (LCP) formulation. The geometric configuration and successive deformed meshes are displayed in Figure 2. For comparison purposes, we have used the same mesh as in [30]. The characteristics of this example are: Young's modulus $E = 10^7$ Pa, Poisson's ratio $\upsilon = 0.25$, mass density $\rho = 1000$ kg/m^3, Friction coefficient $\mu = 0.1$, Initial velocity: $v_x = 3$ m/s, $v_y = -5$ m/s. The geometric sizes are: $L = 0.04$ m, $H = 0.08$ m, radius $R = 0.101$ m, thickness $e = 0.01$ m. The total simulation time is 3.10^{-3} s and the solution parameters are: $\Delta t = 10^{-5}$ s, $\xi = \theta = 0.5$. The plate is modeled by 54 nodes and 37 linear quadrilateral plane stress elements (Fig. 2). The performance of the present approach in terms of CPU time, as compared to Kwak's solutions, is reported in Table 1, which shows the efficiency of the proposed method. It is noted from Figure 3 that the total energy is dissipated by frictional effects and the dissipated energy is calculated quantitatively. It is also interesting to examine another question: is the dissipated energy monotone to the friction coefficient? The answer is no according to numerical results. The proof is illustrated by Figure 4 which shows the evolution of the dissipated energy with respect to the friction coefficient. In fact, when the friction coefficient increases, the friction forces increase. However, the tangential slips will decrease. We know that the dissipated energy de-

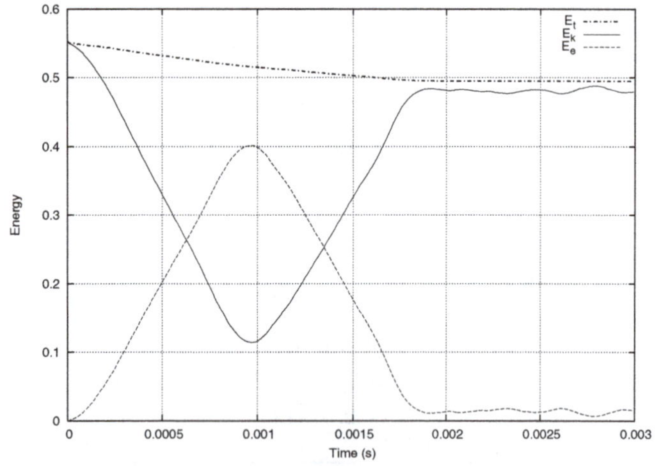

Fig. 3. Energy evolution vs time.

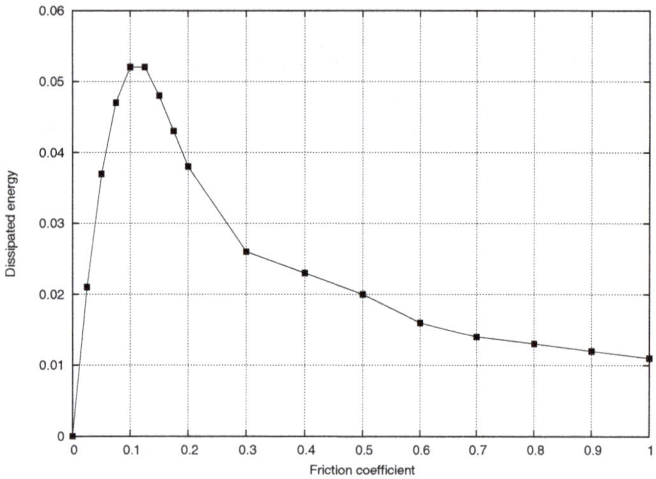

Fig. 4. Evolution of the dissipated energy with respect to the friction coefficient.

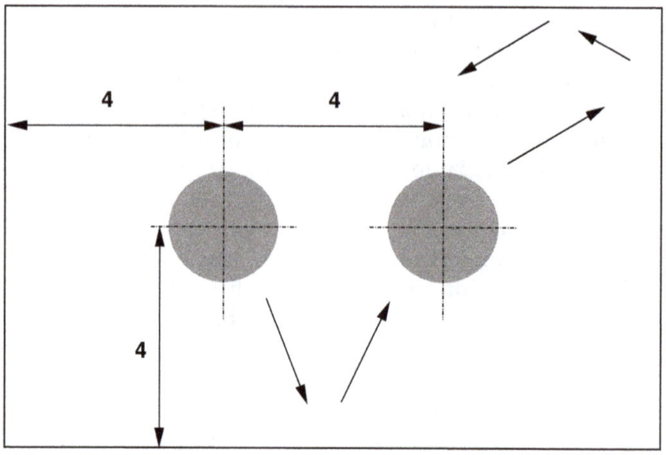

Fig. 5. Impact of two cylinders inside rigid walls.

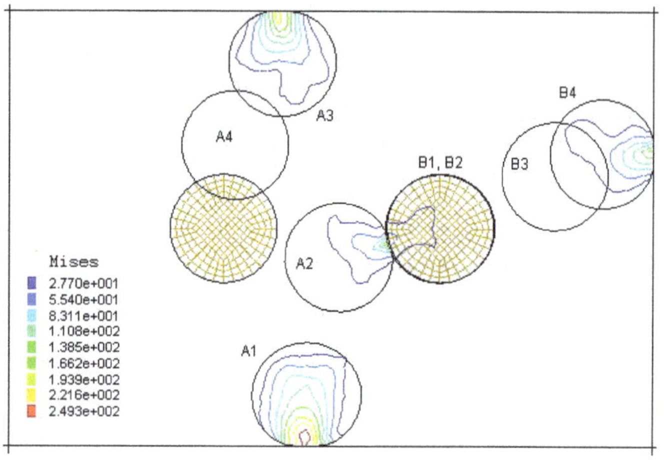

Fig. 6. Deformed configurations at different times (with friction).

pends not only on the friction forces but also on the tangential slips on the contact nodes, so it is understandable to have the behavior as shown in Figure 4.

5.2 Impact between two cylinders

The second example simulates the impact of two cylinders inside rigid walls. In doing so, we wish to further explore the performance of the present method in a general situation with complicated contact sequences. The problem is displayed in Figure 5. Dimensionless data are intentionally used. The cylinders have a diameter of 2. The Saint-Venant-Kirchhoff material model is assumed for both cylinders with material constants: $E = 2700, \upsilon = 0.33$, and mass density $\rho = 1$. The left cylinder is given an initial velocity of $v_x = 1, v_y = -2$, hitting the bottom rigid wall and afterwards the right cylinder as depicted in Figure 5. The total simulation time is 15 and the solution parameters are: $\Delta t = 10^{-3}$, $\xi = \theta = 0.5$. Figure 6 shows the deformed configurations of the two cylinders A and B at different times (1.558, 2.898, 5.278, 6.298) corresponding respectively to (A_1, B_1), (A_2, B_2), (A_3, B_3) and (A_4, B_4) as shown in the figure. The isolines represent the distribution of the Mises stress. Figure 7 shows the plots of the kinetic energy E_k, the elastic strain energy E_e and the total energy E_t. We can observe clearly that the total energy is perfectly conserved in the case of frictionless contact. However, in the case of frictional contact ($\mu = 0.2$), the total energy decreases at each shock (Fig. 8). So the energy is dissipated by frictional effects as expected.

It is interesting to note from Figure 7 that the left cylinder hits another one at $t = 2.45$. On the other hand, $t = 2.8$ in the case of frictional contact as indicated in Figure 8. This fact can be explained as follows: because of friction forces, the rebounding direction is changed such that the running distance of the left cylinder from the bottom wall to the right cylinder becomes longer. Thus, it takes more time to hit each other. Figure 9 shows the distribution of the shear stress of the left cylinder when it hits the bottom wall. Without friction, the distribution is symmetric, but this is not true with friction. Once again, the frictional effects are apparently demonstrated.

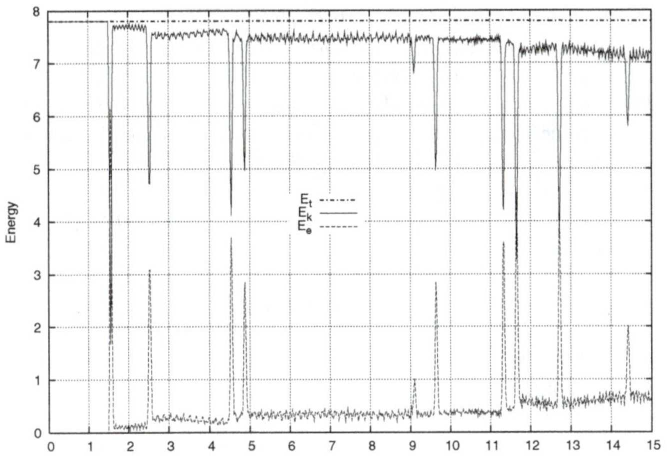

Fig. 7. Energy evolution without friction ($\mu = 0.0$).

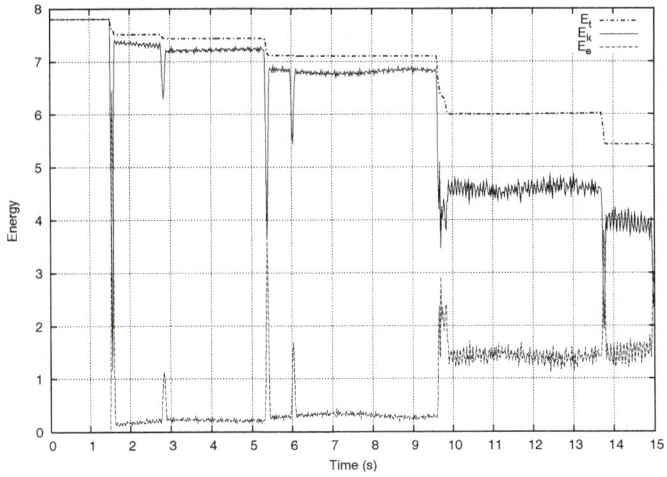

Fig. 8. Energy evolution with friction ($\mu = 0.2$).

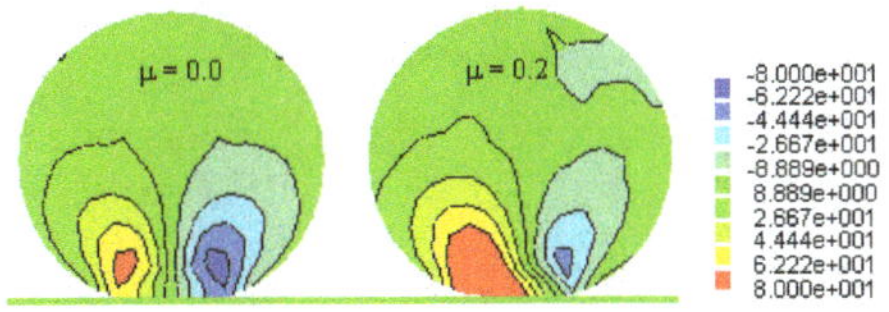

Fig. 9. Distribution of shear stress at t = 1.56.

6 Conclusion

In this paper, we have presented the recent development of the bi-potential method applied to dynamic analysis of two-dimensional contact problems with Coulomb friction. The Bi-First algorithm has been described and investigated numerically for two problems using different coefficients of friction. From numerical experiments, we have found that:

- The total energy is well conserved in the case of frictionless contact of solids.
- The Bi-First algorithm permits to determine quantitatively the physical energy dissipation by frictional effects.
- The dissipated energy is not monotone with respect to the friction coefficient.
- The Bi-First algorithm is simple and efficient:
  - no modification of the global stiffness matrix;
  - no regularization of contact and friction laws;
  - accurate calculation of contact forces in a reduced system;
  - first order time stepping instead of second or higher order integration.

We have felt that this approach could easily be extended to three-dimensional dynamic contact problems including nonlinear material constitutive laws and more complex frictional models [31]. This work is being undertaken.

References

1. Z.H. Zhong, J. Macherle, *Static contact problems - a review.* Eng. Computation **9**, 3–37 (1992)
2. A. Klarbring, Mathematical programming in contact problems, In MH. Aliabdali and CA. Brebbia, editors, *Computational methods in contact mechanics*, pages 233–263. Southampton: Computational Mechanics Publications (1993)
3. P. Wriggers, *Finite element algorithms for contact problems.* Arch. Comput. Method. E. **2**, 1–49 (1995)
4. N. Kikuchi, J.T. Oden, *Contact problems in elasticity: A study of variational inequalities and finite elements*, Philadelphia: SIAM (1988)
5. Z.H. Zhong, *Finite element procedures in contact-impact problems*, Oxford University Press (1993)
6. P. Wriggers, *Computational contact mechanics* (John Wiley & Sons, 2002)
7. A.B. Chaudhary, K.J. Bathe, *A solution method for static and dynamic analysis of three dimensional contact problems with friction.* Comput. Struct. **24**, 855–873 (1986)
8. H.A. Parisch, *Consistent tangent stiffness matrix for three-dimensional non-linear contact analysis.* Int. J. Numer. Meth. Eng. **28**, 1803–1812 (1989)
9. P. Alart, A. Curnier, *A mixed formulation for frictional contact problems prone to Newton like solution methods.* Comput. Method Appl. M. **92**, 353–375 (1991)
10. J.C. Simo, T.A. Laursen, *An augmented Lagrangian treatment of contact problems involving friction.* Comput. Struct. **42**, 97–116 (1992)
11. G. de Saxcé, Z.-Q. Feng, *New inequality and functional for contact with friction: The implicit standard material approach.* Mech. Struct. Mach. **19** 301–325 (1991)
12. G. de Saxcé, Z.-Q. Feng, *The bi-potential method: a constructive approach to design the complete contact law with friction and improved numerical algorithms.* Mathematical and Computer Modeling **28**, 225–245 (1998)
13. A. Klarbring, Mathematical programming and augmented Lagrangian methods for frictional contact problems. In A. Curnier, editor, *Contact Mechanics Int. Symp.* (1992) PPUR.
14. P.W. Chritensen, A. Klarbring, J.S. Pang, N.N. Strömberg, *Formulation and comparison of algorithms for frictional contact problems.* Int. J. Numer. Meth. Eng. **42**, 145–173 (1998)

15. P. Wriggers, T. Vu Van, E. Stein, *Finite element formulation of large deformation impact contact problems with friction.* Comput. Struct. **37**, 319–331 (1990)
16. T.A. Laursen, V. Chawla, *Design of energy conserving algorithms for frictionless dynamic contact problems.* Int. J. Numer. Meth. Eng. **40**, 863–886 (1997)
17. T.A. Laursen, G.R. Love, *Improved implicit integrators for transient impact problems geometric admissibility within the conserving framework.* Int. J. Numer. Meth. Eng. **53**, 245–274 (2002)
18. G.R. Love, T.A. Laursen, *Improved implicit integrators for transient impact problems: dynamic frictional dissipation within an admissible conserving framework.* Comput. Method. Appl. M. **192**, 2223–2248 (2003)
19. O.C. Zienkiewicz, W.L. Wood, L.W. Hine, R.L. Taylor, *A unified set of single step algorithms. part 1. general formulation and application.* Int. J. Numer. Meth. Eng. **20**, 1529–1552 (1984)
20. M. Jean, Dynamics with partially elastic shocks and dry friction: double scale method and numerical approach, In *4th Meeting on unilateral problems in structural analysis* (1989)
21. M.A. Crisfield, *Non-linear finite element analysis of solid and structures* (Wiley, 1991)
22. J.C. Simo, T.J.R. Hughes, *Computational inelasticity,* (Springer-Verlag, New York, 1998)
23. K.K. Tamma, R.R. Namburu, *A robust self - starting explicit computational methodology for structural dynamic applications: architecture and representations.* Int. J. Numer. Meth. Eng. **30**, 1441–1454 (1990)
24. J.C. Simo, K.K. Wong, *Unconditionally stable algorithms for rigid body dynamics that exactly preserve energy and momentum.* Int. J. Numer. Meth. Eng. **31**, 19–52 (1991)
25. Z.-Q. Feng, *2D or 3D frictional contact algorithms and applications in a large deformation context.* Commun. Numer. Meth. En. **11**, 409–416 (1995)
26. Z.-Q. Feng, `http://gmfe16.cemif.univ-evry.fr:8080/~feng/FerImpact.html`
27. Z.-Q. Feng, *Some test examples of 2D and 3D contact problems involving coulomb friction and large slip.* Mathematical and Computer Modeling **28**(4-8), 469–477 (1998) Special issue: Recent Advances in Contact Mechanics.
28. Z.-Q. Feng, F. Peyraut, N. Labed, *Solution of large deformation contact problems with friction between Blatz-Ko hyperelastic bodies.* Int. J. Eng. Sci. **41**, 2213–2225 (2003)
29. S.H. Ko, B. Kwak, *Frictional dynamic contact analysis using finite element nodal displacement description.* Comput. Struct. **42**, 797–807 (1992)
30. J.O. Kim, B. Kwak, *Dynamic analysis of two-dimensional frictional contact by linear complementarity problem formulation.* Int. J. Solids Struct. **33**, 4605–4624 (1996)
31. M. Hjiaj, Z.-Q. Feng, G. de Saxcé, Z. Mróz, *Theredimensional finite element computations for frictional contact problems with non-associated sliding rule.* Int. J. Numer. Meth. Eng. **60**, 2045–2076 (2004)

Permissions

All chapters in this book were first published in IJSMDO, by EDP Sciences; hereby published with permission under the Creative Commons Attribution License or equivalent. Every chapter published in this book has been scrutinized by our experts. Their significance has been extensively debated. The topics covered herein carry significant findings which will fuel the growth of the discipline. They may even be implemented as practical applications or may be referred to as a beginning point for another development.

The contributors of this book come from diverse backgrounds, making this book a truly international effort. This book will bring forth new frontiers with its revolutionizing research information and detailed analysis of the nascent developments around the world.

We would like to thank all the contributing authors for lending their expertise to make the book truly unique. They have played a crucial role in the development of this book. Without their invaluable contributions this book wouldn't have been possible. They have made vital efforts to compile up to date information on the varied aspects of this subject to make this book a valuable addition to the collection of many professionals and students.

This book was conceptualized with the vision of imparting up-to-date information and advanced data in this field. To ensure the same, a matchless editorial board was set up. Every individual on the board went through rigorous rounds of assessment to prove their worth. After which they invested a large part of their time researching and compiling the most relevant data for our readers.

The editorial board has been involved in producing this book since its inception. They have spent rigorous hours researching and exploring the diverse topics which have resulted in the successful publishing of this book. They have passed on their knowledge of decades through this book. To expedite this challenging task, the publisher supported the team at every step. A small team of assistant editors was also appointed to further simplify the editing procedure and attain best results for the readers.

Apart from the editorial board, the designing team has also invested a significant amount of their time in understanding the subject and creating the most relevant covers. They scrutinized every image to scout for the most suitable representation of the subject and create an appropriate cover for the book.

The publishing team has been an ardent support to the editorial, designing and production team. Their endless efforts to recruit the best for this project, has resulted in the accomplishment of this book. They are a veteran in the field of academics and their pool of knowledge is as vast as their experience in printing. Their expertise and guidance has proved useful at every step. Their uncompromising quality standards have made this book an exceptional effort. Their encouragement from time to time has been an inspiration for everyone.

The publisher and the editorial board hope that this book will prove to be a valuable piece of knowledge for researchers, students, practitioners and scholars across the globe.

List of Contributors

N. Pirc
CROMeP - École des Mines d'Albi-Carmaux, Campus Jarlard, 81013 Albi, Cedex 9, France
Université de Toulouse, LAAS-CNRS, and Institut de Mathématiques, UPS 31062 Toulouse Cedex 9, France

F. Schmidt
CROMeP - École des Mines d'Albi-Carmaux, Campus Jarlard, 81013 Albi, Cedex 9, France

M. Mongeau and F. Bugarin
Université de Toulouse, LAAS-CNRS, and Institut de Mathématiques, UPS 31062 Toulouse Cedex 9, France

D. Chamoret
Laboratoire M3M, UTBM, Site de Sévenans, 90010 Belfort Cedex, France

A. Rassineux
Laboratoire Roberval, FRE 2833 CNRS/UTC, BP 20.529, 60205 Compiégne Cedex, France

J.M. Bergheau
LTDS, UMR5513, CNRS/ECL/ENISE, 58 rue Jean Parot, 42023 Saint-Étienne, France

F. Peyraut
M3M, Belfort-Montbéliard University of Technology, Belfort, France

J.-L. Seichepine and C. Coddet
LERMPS, Belfort-Montbéliard University of Technology, Belfort, France

M. Hertter
MTU Aero Engines GmbH, Dachauer Straβe 665, 80995 München, Germany

A. Noriegaa, R. Vijande, E. Rodríguez, J.L. Cortizo and J.M. Sierra
Department of Mechanical Engineering, University of Oviedo, Gijón, Spain

H. Dehmous
Laboratory of Engineering Production, National Engineers School of Tarbes, 47 Avenue d'Azereix, BP 1629, 65016 Tarbes, France
Laboratory of Mechanics and Structures, University of Tizi-Ouzou, Algeria

Héléne Welemane and Moussa Karama
Laboratory of Engineering Production, National Engineers School of Tarbes, 47 Avenue d'Azereix, BP 1629, 65016 Tarbes, France

Kamel Aît Tahar
Laboratory of Mechanics and Structures, University of Tizi-Ouzou, Algeria

P. Fantini
Computational Aerodynamics, Aircraft Research Association Ltd., Bedford, UK

L.K. Balachandran and M.D. Guenov
Aerospace Engineering, Cranfield University, Bedford, UK

Dan Hea and Shutian Liu
State Key Laboratory of Structutal Analysis for Industrial Equipment, Dept. of Engineering Mechanics, Dalian Univerisity of Technology, 116023 Dalian, P.R. China

L. Guadagni
CNES, Rond-point de l'Espace, 91023 Evry, France

Yun-Kang Sui and Shan-Po Li
Centre of Numerical Simulation for Engineering, Beijing University of Technology, F100022 Beijing, P.R. China

Ying-Qiao Guo
Laboratory of Mechanics, Materials & Structures, University of Reims Champagne-Ardenne, 51687 Reims, France

Belarmino Adenso-Díaz, Pilar González-Torre and Verónica Ordóñez
Operations Management Department, University of Oviedo, Gijón, Spain

Juan Jośe del Coz
Civil Engineering Department, University of Oviedo, Gijón, Spain

W.L. Gambina and A. Zarzycki
Warsaw University of Technology, ul. Sw. A. Boboli 8, 02-525 Warszawa, Poland

David Hicham Bassir
Institute of FEMTO-ST, UMR / CNRS 6174, Dept. LMARC, 24, rue de l'Epitaphe, 25000 Besançon, France

José Luis Zapico, María Placeres González and Rodolfo Alonso
Universidad de Oviedo, Departamento de Construcción e Ingeniería de Fabricación, Campus de Gijón, 33203 Gijón, Spain

Shan-Suo Zheng, Lei Zeng, Jie Zheng, BinWang and and Lei Li
School of Civil Engineering, Xi'an University of Architecture and Technology, Xi'an, 710055, P.R. China

Wei-Hong Zhang
Northwestern Polytechnical University, Xi'an, P.R. China

J.H. Zhu
Sino-French Laboratory of Concurrent Engineering, Northwestern Polytechnical University 710072 Xi'an, Shaanxi, P.R. China
LTAS - Infographie, Universit´e de Li`ege, 4000 Liége, Belgium

W.H. Zhang
Sino-French Laboratory of Concurrent Engineering, Northwestern Polytechnical University 710072 Xi'an, Shaanxi, P.R. China

P. Beckers
LTAS - Infographie, Université de Liége, 4000 Liége, Belgium

G. Kharmanda
Faculty of Mechanical Engineering, Aleppo University, Syrian Arab Republic
INSA de Rouen, LMR, BP 08, Avenue de l'Universit´e 76801 St Etienne du Rouvray, France

A. Mohsine
INSA de Lyon, Lab. d'Automatique Industrielle, Bât. St Exupéry, 25, Avenue Jean Capelle, 69621 Villeurbanne, France
INSA de Rouen, LMR, BP 08, Avenue de l'Université 76801 St Etienne du Rouvray, France

A. Makloufi and A. El-Hami
INSA de Rouen, LMR, BP 08, Avenue de l'Université 76801 St Etienne du Rouvray, France

N.V. Banichuka, S.Yu. Ivanova and E.V. Makeev
Institute for Problems in Mechanics, Russian Academy of Scienses, Moscow, Russia

Nikos D. Lagarosa, Vagelis Plevris and Manolis Papadrakakis
Institute of Structural Analysis & Seismic Research, National Technical University of Athens, 9, Iroon Polytechniou Str., Zografou Campus, GR-15780 Athens, Greece

A. Zéanh
THALES Avionics Electrical Systems, 41 boulevard de la République, BP 53, 78401 Chatou, France
ENI de Tarbes - LGP, 47, avenue d'Azereix, BP 1629, 65016 Tarbes Cedex, France
Université Bordeaux - Laboratoire IMS, 351 cours de la Libération, 33405 Talence Cedex, France
PEARL, Alstom Transport Tarbes, rue du Docteur Guinier, BP4, 65600 Sém´eac, France

A. Bouzourene1, J. Casutt1
THALES Avionics Electrical Systems, 41 boulevard de la R´epublique, BP 53, 78401 Chatou, France

O. Dalverny and M. Karama
ENI de Tarbes - LGP, 47, avenue d'Azereix, BP 1629, 65016 Tarbes Cedex, France

E. Woirgard and S. Azzopardi
Université Bordeaux - Laboratoire IMS, 351 cours de la Libération, 33405 Talence Cedex, France

M. Mermet-Guyennet
PEARL, Alstom Transport Tarbes, rue du Docteur Guinier, BP4, 65600 Séméac, France

Shan-Suo Zheng, Lei Zeng, Jie Zheng, Lei Li and Bin Wang
School of Civil Engineering, Xi'an University of Architecture and Technology, Xi'an 710055, P.R. China

Wei-Hong Zhang
Northwestern Polytechnical University, Xi'an 710049, P.R. China

K.-E. Atcholi, E. Padayodi and K. Kadja
Laboratoire d'Études et de Recherches sur les Matériaux, les Procédés et les Surfaces (LERMPS) Équipe de Recherche sur les Agro-Matériaux et la Santé Environnementale (ERAMSE), Département de Génie Mécanique et Conception, Université de Technologie de Belfort-Montbéliard, 90010 Belfort Cedex, France

J. Vantomme
Civil Engineering Department, École Royale Militaire de Bruxelles, Belgique

D. Perreux
Laboratoire de Mécanique Appliquée, R. Chaléat (LMARC), Université de Franche-Comté, 25030 Besançon, France

Q.B. Nguyen
Université Paris-Est, Laboratoire de Modélisation et SimulationMulti-échelle, MSME FRE 3160 CNRS, 5 boulevard Descartes, 77454 Marne la Vallée, France

Institut National de l'Environnement Industriel et des Risques (INERIS), 10 boulevard Lahitolle, 18000 Bourges, France

A. Mebarki and R. Ami Saada
Université Paris-Est, Laboratoire de Modélisation et SimulationMulti-échelle, MSME FRE 3160 CNRS, 5 boulevard Descartes, 77454 Marne la Vallée, France

F. Mercier and M. Reimeringer
Institut National de l'Environnement Industriel et des Risques (INERIS), 10 boulevard Lahitolle, 18000 Bourges, France

B. Sid, M. Domaszewskia and F. Peyraut
M3M Laboratory, University of Technology of Belfort-Montbeliard, 90010 Belfort Cedex, France

E.H. Irhirane, J. Echaabi and M. Hattabi
Équipe de Recherche Appliquée sur les Polyméres, Département de Génie Mécanique, ENSEM, Université Hassan II Aïn Chok, BP 8118, Oasis, Casablanca, Morocco

M. Aboussaleh
Laboratoire de Mécanique, Département de Génie mécanique, ENSAM, Université Moulay Ismail, Présidence, Marjane 2, BP 298 Mekness, Morocco

A. Saouab
Laboratoire d'Ondes et Milieux Complexes, FRE 3102 CNRS, 53 rue Prony, BP 540, 76058 Le Havre Cedex, France

D.H. Bassir
Institut FEMTO-ST, UMR 6174, 24 rue de l'épitaphe, 25000 Besançon, France
Currently visiting Aerospace Structures, Faculty of Aerospace Engineering, TU Delft, The Netherlands

F.X. Irisarri, J.F. Maire and N. Carrere
ONERA, DMSC 29 avenue de la division Leclerc, 92322 Chatillon Cedex, France

A. Micol, C. Martin1, O. Dalverny and M. Karama
LGP-ENIT, 47 av. d'Azereix, BP 1629, 65016 Tarbes Cedex, France

M. Mermet-Guyennet
Power Electronics Associated Research Laboratory (PEARL), rue du Docteur Guiner, 65600 Semeac, France

A. Hocine
University Hassiba Benbouali of Chlef BP. 151, Chlef 02000, Algeria

Institut FEMTO-ST, Dept. LMARC, 24, rue de l'épitaphe, 25000 Besançon, France

D. Chapelle
Institut FEMTO-ST, Dept. LMARC, 24, rue de l'épitaphe, 25000 Besan¸con, France

A. Benamar
ENSET, Department of mechanical engineering, BP 1523, Oran 31000, Algeria

A. Bezazi
University 08 Mai 1945, BP. 401, Guelma 24 000, Algeria

Kalyanmoy Deb
Department of Mechanical Engineering, Indian Institute of Technology Kanpur, PIN 208016, India

J.E. Rojas
Institut National des Sciences Appliquées de Rouen, Laboratoire de Mécanique de Rouen, BP 08, Avenue de l'Université, 76801 Saint-Etienne du Rouvray, Cedex, France
Federal University of Uberlândia, School of Mechanical Engineering, 2121, Av. Joao Naves de Avila, Campus Santa Mônica, CEP 38400-902 CP 593, Uberlândia, Brazil

A. El Hami and D.A. Rade
Federal University of Uberlândia, School of Mechanical Engineering, 2121, Av. Joao Naves de Avila, Campus Santa Mônica, CEP 38400-902 CP 593, Uberlândia, Brazil

J.-B. Bluntzer, S. Gomes, A. Varret and J.C. Sagot
Laboratoire SeT, Equipe ERCOS, University of Technology of Belfort-Montbéliard, 90010 Belfort Cedex, France

D.H. Bassir
Institut FEMTO-ST, UMR/CNRS 6174, 24, rue de l'Épitaphe, 25000 Besançon, France

Z.-Q. Feng, B. Magnain and J.-M. Cros
Université d'Évry-Val d'Essonne, Laboratoire de Mécanique d'Évry, 40 rue du Pelvoux, 91020 Évry, France

P. Joli
Université d'Évry-Val d'Essonne, Laboratoire IBISC, 40 rue du Pelvoux, 91020 Évry, France

Index

O

P

R

S

T

U